Biomedical Application of Big Data and Artificial Intelligence

Biomedical Application of Big Data and Artificial Intelligence

Guest Editors

Yan Pei
Jijiang Yang

MDPI

Basel • Beijing • Wuhan • Barcelona • Belgrade • Novi Sad • Cluj • Manchester

Guest Editors
Yan Pei
Computer Science Division
University of Aizu
Aizuwakamatsu
Japan

Jijiang Yang
Department of Automation
Tsinghua University
Beijing
China

Editorial Office
MDPI AG
Grosspeteranlage 5
4052 Basel, Switzerland

This is a reprint of the Special Issue, published open access by the journal *Bioengineering* (ISSN 2306-5354), freely accessible at: https://www.mdpi.com/journal/bioengineering/special_issues/P71DI9G82A.

For citation purposes, cite each article independently as indicated on the article page online and as indicated below:

Lastname, A.A.; Lastname, B.B. Article Title. *Journal Name* **Year**, *Volume Number*, Page Range.

ISBN 978-3-7258-3569-0 (Hbk)
ISBN 978-3-7258-3570-6 (PDF)
https://doi.org/10.3390/books978-3-7258-3570-6

Contents

About the Editors

Yan Pei

Yan Pei obtained a Doctor of Engineering degree from Kyushu University, Fukuoka, Japan. He received his B.Eng. and M.Eng. degrees from Northeastern University, Shenyang, China. He is currently an Associate Professor at the University of Aizu. His research interests include computational intelligence, machine learning, interactive evolutionary computation, human–computer interaction, and humanized computing. He is a Senior Member of the IEEE Systems, Man, and Cybernetics Society (SMCS) and the IEEE Computational Intelligence Society (CIS), as well as a Member of the Association for Computing Machinery (ACM) and the Japanese Society for Evolutionary Computation. He serves as the Chair of the IEEE SMC Technical Committee on Soft Computing and the Technical Committee on Humanized Crowd Computing.

He has received several Best Paper Awards, including from ICGEC 2012, SOFT-Kyushu 2012, FC 2018, FC 2019, FC 2020, FC 2023, and BCD 2024-Summer. Additionally, he received the Most Active Technical Committee Award from the IEEE SMC Society, the Gold Award in the Multimedia Task at FC 2024, and the 2024 *Applied Soft Computing* Best Paper Award from *Applied Soft Computing*. He serves as an Editor for several prestigious journals, including *Applied Soft Computing* (Elsevier), *Engineering Applications of Artificial Intelligence* (Elsevier), *Scientific Reports* (Nature), *Current Pharmaceutical Biotechnology* (Bentham), and the *International Journal of Bio-Inspired Computation* (Inderscience). Additionally, he has served as a Guest Editor for numerous Special Issues.

Jijiang Yang

Professor Yang Ji-Jiang, PhD, is the director of the Center for Information Technology and Data Science Innovation in Drug Safety Surveillance at Tsinghua University and the director of the Digital Medical Health Engineering Research Center at the Department of Automation, Tsinghua University.

He is dedicated to enhancing medical service capabilities and quality through information technology. He has led or participated in over 10 national projects including 863 Science and Technology Support Programs, 973 Key Research and Development Programs, and the National Science and Technology Major Project. He won the Second Prize of Military Medical Science and Technology Achievements, the Third Prize of Chinese Medical Association Scientific and Technological Progress, the First Prize of Chinese Maternal and Child Health Care Association Scientific and Technological Achievements, and the First Prize of Beijing Archives Scientific and Technological Achievements. He has published over 190 papers. His key research fields include medical informatization and service industrialization, healthcare service and medical treatment models, big data analysis, artificial intelligence and its applications in medical care, key digital medicine technologies, etc.

He is a member of the Eighth and Ninth National Health Information Standardization Committee of the National Health Commission, the Deputy Chairperson of the Diabetes AI Special Committee of the Chinese Health Information/Healthcare Big Data Society, the Deputy Chairperson of the Health Engineering Branch of the Chinese Biotechnology Society, and an Advisor Member of the Heilongjiang Provincial Association for Elderly Education.

Preface

The rapid advancement of big data and artificial intelligence (AI) has profoundly transformed biomedical research and healthcare, enabling more precise diagnoses, personalized treatments, and innovative medical solutions. This Special Issue, Biomedical Applications of Big Data and Artificial Intelligence, aims to provide a comprehensive exploration of the intersection between AI, data science, and medicine. It presents cutting-edge research on topics such as disease diagnosis, medical data analysis, image processing, cognitive load assessment, survival prediction, and medical knowledge extraction, offering both theoretical insights and practical applications.

The motivation for this reprint stems from the growing need to bridge the gap between AI-driven analytics and biomedical challenges. While AI and big data techniques have shown immense potential in medical applications, challenges such as interpretability, data integration, and robustness remain critical. By curating state-of-the-art research, we aim to highlight the transformative power of AI and big data in healthcare, fostering deeper understanding and further advancements in the field.

This reprint is primarily intended for researchers, practitioners, and graduate students in biomedical engineering, AI, and healthcare technology. It serves as a valuable resource for those seeking to understand the latest developments in AI-driven biomedical applications and the potential of big data methodologies in improving patient outcomes.

We extend our heartfelt gratitude to all contributing authors for their insightful research and dedication. We express special thanks to the peer reviewers for their meticulous evaluations, ensuring the quality and integrity of the work herein. We also acknowledge the support from our academic institutions and research communities in fostering an environment conducive to interdisciplinary collaboration.

We hope that this reprint inspires future research and contributes to ongoing dialog at the intersection of biomedical science, AI, and big data.

Yan Pei and Jijiang Yang
Guest Editors

bioengineering

MDPI

Editorial

Biomedical Applications of Big Data and Artificial Intelligence

Yan Pei [1,*,†,‡] and Jijiang Yang [2,‡]

1 Computer Science Division, The University of Aizu, Aizuwakamatsu 965-8580, Japan
2 Department of Automation, Tsinghua University, Beijing 100084, China; yangjijiang@tsinghua.edu.cn
* Correspondence: peiyan@u-aizu.ac.jp; Tel.: +81-242-372765
† Current address: Tsuruga, Ikki-machi, Aizu-Wakamatsu 965-8580, Japan.
‡ These authors contributed equally to this work.

Abstract: This Special Issue of *Bioengineering* is dedicated to the profound impact of big data and artificial intelligence (AI) in the fields of biomedical research and healthcare. In an age defined by the rapid evolution of technology, this Issue explores the dynamic intersection of AI and data science with medicine. A total of 14 papers were accepted after a thorough review process, with their topics including disease diagnosis, medical data analysis, image processing, personalized medicine, pathological image segmentation, survival prediction, cognitive load assessment, and medical knowledge extraction. These studies aim to enhance medical image analysis, signal processing, data prediction, and interpretability to improve diagnostic accuracy, medical efficiency, and personalized treatment plans for patients. We hope the publication of this Special Issue can offer a comprehensive view of the transformative power of these innovative approaches and enrich research and investigations into the applications of big data and AI in biomedical research and healthcare.

Keywords: artificial intelligence; big data; image processing; data mining; deep learning; bioinformatics; bioengineering; healthcare

1. Introduction

The convergence of big data and artificial intelligence (AI) has catalyzed transformative advancements in the biomedical field, enabling innovative solutions for complex healthcare challenges [1]. This Special Issue, dedicated to exploring the "Biomedical Applications of Big Data and Artificial Intelligence", highlights significant advancements in research and development, such as disease diagnosis, medical data analysis, image processing, personalized medicine, pathological image segmentation, survival prediction, cognitive load assessment, and medical knowledge extraction, etc. From improving diagnostic accuracy to enhancing personalized treatment plans, these advancements underscore the potential of integrating AI and big data to revolutionize biomedical research and healthcare.

The 14 articles featured in this Special Issue, including 12 research papers and 2 review papers, encompass a wide range of topics such as healthcare big data analysis, diagnostic tools, imaging techniques, survival prediction, and explainable AI. Together, these contributions provide a comprehensive overview of how AI and big data are driving precision medicine, optimizing healthcare delivery, and advancing our understanding of complex biomedical phenomena. This editorial delves into the transformative power of these technologies and their profound implications for the future of biomedical research and healthcare.

Received: 8 February 2025
Accepted: 11 February 2025
Published: 19 February 2025

Citation: Pei, Y.; Yang, J. Biomedical Applications of Big Data and Artificial Intelligence. *Bioengineering* **2025**, *12*, 207. https://doi.org/10.3390/bioengineering12020207

2. An Overview of This Special Issue

The 14 papers accepted in this Special Issue can be categorized into four biomedical research directions: intelligent diagnosis and disease prediction, medical image processing and pathological analysis, medical signal processing and analysis, and survey and frontier research perspectives. We summarize these 14 papers from these four perspectives.

2.1. Intelligent Diagnosis and Disease Prediction

The critical issue of diagnostic errors in fuzzy processing, a leading cause of mortality, is addressed in the article "Trivial State Fuzzy Processing for Error Reduction in Healthcare Big Data Analysis towards Precision Diagnosis". This study introduces a Cooperative Trivial State Fuzzy Processing method that employs fuzzy logic to mitigate uncertainties and errors in healthcare data. By grouping diagnosis-relevant and irrelevant data, the proposed method reduces trivial state errors, enhancing the precision of diagnostic outcomes. This study demonstrates how optimization-based analysis can be effectively applied to structured healthcare data, marking a significant step forward in reducing diagnostic inaccuracies.

The study "Wearable 12-Lead ECG Acquisition Using a Novel Deep Learning Approach from Frank or EASI Leads with Clinical Validation" presents an innovative approach to portable cardiovascular diagnostics. By reconstructing 12-lead ECGs using a novel deep learning model, M2Eformer, this research overcomes the limitations of traditional portable ECG devices. The validation results, including a 96% diagnostic consensus among cardiologists, underscore the clinical utility of the reconstructed ECGs. This work exemplifies the potential of AI-driven solutions in enhancing accessibility and accuracy in cardiovascular diagnostics.

The study "Personalized Explanations for Early Diagnosis of Alzheimer's Disease Using Explainable Graph Neural Networks with Population Graphs" emphasizes the importance of explainability in AI-driven medical applications. By leveraging graph convolutional networks (GCNs) and population graphs, this research provides insights into amyloid-beta positivity prediction and the heterogeneity of Alzheimer's disease progression. The findings highlight the potential for more nuanced diagnoses and personalized therapeutic strategies.

"GNN-surv: Discrete-Time Survival Prediction Using Graph Neural Networks" explores the application of graph neural networks (GNNs) in survival prediction for cancer patients. By leveraging patient similarity networks built from genomic and clinical data, GNN-surv achieved significant improvements in survival prediction performance compared to traditional models. This study's adaptability across various datasets and its implications for personalized medicine underscore the transformative potential of GNNs in oncology.

The study "Prediction of Cognitive Load from Electroencephalography Signals Using Long Short-Term Memory Network" proposes an attention-based LSTM network for real-time cognitive load prediction using EEG signals. This method achieved the highest accuracy, outperforming traditional algorithms, and can enable adaptive systems to dynamically respond to users' mental states, enhancing applications like personalized learning and fatigue detection.

2.2. Medical Image Processing and Pathological Analysis

In "Elucidating Multimodal Imaging Patterns in Accelerated Brain Aging: Heterogeneity through a Discriminant Analysis Approach Using the UK Biobank Dataset", the authors explore the heterogeneity of accelerated brain aging (ABA). Utilizing multimodal imaging data and semi-supervised heterogeneity analysis (HYDRA), this study identified three distinct ABA subtypes, each associated with unique structural and functional characteristics. These findings highlight the potential for personalized neuroprotective treatments targeting age-related neurological and neuropsychiatric disorders.

The article "RGGC-UNet: Accurate Deep Learning Framework for Signet Ring Cell Semantic Segmentation in Pathological Images" addresses challenges in the diagnosis of signet ring cell carcinoma. The proposed RGGC-UNet framework incorporates residual ghost blocks and ghost coordinate attention to achieve high segmentation accuracy while minimizing computational overhead. By enriching existing datasets with annotated mask labels, this study provides a valuable resource for advancing pathological image analysis. The results demonstrate the efficacy of deep learning frameworks in enhancing diagnostic precision for complex pathological conditions.

The article "RGSB-UNet: Hybrid Deep Learning Framework for Tumor Segmentation in Digital Pathology Images" proposes a novel UNet-based architecture that combines residual ghost blocks and bottleneck transformers. This hybrid framework addresses the limitations of traditional CNN-based methods by capturing global features and achieving state-of-the-art segmentation performance. The integration of class-wise dice loss (CDL) further enhances the model's effectiveness, as demonstrated on multiple pathology datasets. Such advancements are pivotal in automating and improving tumor segmentation processes.

"Exploring the Possibility of Measuring Vertebrae Bone Structure Metrics Using MDCT Images: An Unpaired Image-to-Image Translation Method" tackles the challenge of evaluating bone structure metrics in vivo. By employing an unpaired image-to-image translation method, this study generated micro-CT-like images from MDCT scans, enabling accurate measurement of bone metrics. The proposed approach demonstrates significant improvements in both similarity and bone structure metrics, offering a practical solution for the early diagnosis of fragility fractures.

The paper "GSN-HVNET: A Lightweight, Multi-Task Deep Learning Framework for Nuclei Segmentation and Classification" presents GSN-HVNET, a compact deep learning framework for simultaneous nuclei segmentation and classification in pathology images. By integrating novel computational blocks, it improves efficiency and accuracy compared to state-of-the-art methods. The model demonstrates practical value in pathological diagnosis by reducing redundancy and computational costs.

2.3. Medical Signal Processing and Analysis

The paper "Physiological Noise Filtering in Functional Near-Infrared Spectroscopy Signals Using Wavelet Transform and Long-Short Term Memory Networks" proposes a method to filter physiological noise in fNIRS signals without using the desired hemodynamic response function (dHRF). It employs wavelet transform to extract low-frequency noise components from resting-state data, which are predicted and subtracted using an LSTM network during task sessions. The technique offers a reliable alternative when traditional methods are ineffective, particularly for passive brain–computer interfaces.

The study "Named Entity Recognition of Diabetes Online Health Community Data Using Multiple Machine Learning Models" introduces the RoBERTa-BiLSTM-CRF model to identify medical entities in diabetes online health community data. Using a dataset of 1889 texts, the model achieved 81.2% accuracy and outperformed traditional models. The results demonstrate its potential for constructing medical knowledge graphs to support personalized healthcare services.

2.4. Survey and Frontier Research Perspectives

The two review papers in this Special Issue provide a broader perspective on the integration of big data and AI into biomedical applications, from multimodal data and

AI technology [2] to machine learning and graph signal processing [3]. They discuss the current state of the art, challenges, and future directions, offering valuable insights for researchers and practitioners alike. These reviews serve as a foundation for further exploration and innovation in the field.

3. Conclusions

This Special Issue on the "Biomedical Applications of Big Data and Artificial Intelligence" showcases the profound impact of integrating AI and big data in healthcare. From reducing diagnostic errors and enhancing imaging techniques to enabling personalized treatments and explainable AI, the contributions highlighted here represent significant advancements in the field. These studies not only address critical challenges but also pave the way for future innovations in precision medicine and healthcare delivery.

These studies encompass areas such as intelligent diagnosis, medical image processing, signal analysis, and text mining. The key contributions of these works are as follows: First, numerous studies introduce novel models (e.g., UNet variants, GNN-surv, M2Eformer), expanding the application boundaries of AI in healthcare. Second, these studies primarily employ deep learning, graph neural networks, LSTM, and Transformer variants to enhance prediction and analysis capabilities in structured and unstructured medical data. Finally, some research concepts, such as a focus on precision medicine, personalized diagnosis, explainable AI, and automated medical image analysis, which address challenges in accuracy, efficiency, and explainability that traditional methods struggle to overcome, are implemented in these works. These studies demonstrate that AI is increasingly becoming a core driving force in biomedical computing and healthcare. Future challenges will include improving explainability, reducing computational costs, enhancing model generalization, and promoting the practical implementation of AI in clinical applications.

As we move forward into this new era of healthcare, our focus must remain on responsible innovation, patient well-being, and equitable access. The intersection of compassion and cutting-edge technology will truly unlock the potential of big data and AI in biomedical applications, making a healthier world not just a dream, but an attainable reality [4]. The diverse methodologies and applications presented in this Issue underscore the potential of interdisciplinary research in advancing biomedical science. We hope this Special Issue inspires further research and innovation in this dynamic and impactful field.

Conflicts of Interest: The authors declare no conflicts of interest.

References

1. Wang, H.; Fu, T.; Du, Y.; Gao, W.; Huang, K.; Liu, Z.; Chandak, P.; Liu, S.; Katwyk, P.V.; Deac, A.; et al. Scientific discovery in the age of artificial intelligence. *Nature* **2023**, *620*, 47–60. [CrossRef]
2. Xu, X.; Li, J.; Zhu, Z.; Zhao, L.; Wang, H.; Song, C.; Chen, Y.; Zhao, Q.; Yang, J.; Pei, Y. A Comprehensive Review on Synergy of Multi-Modal Data and AI Technologies in Medical Diagnosis. *Bioengineering* **2024**, *11*, 219. [CrossRef]
3. Calazans, M.A.A.; Ferreira, F.A.B.S.; Santos, F.A.N.; Madeiro, F.; Lima, J.B. Machine Learning and Graph Signal Processing Applied to Healthcare: A Review. *Bioengineering* **2024**, *11*, 671. [CrossRef]
4. Topol, E.J. As artificial intelligence goes multimodal, medical applications multiply. *Science* **2023**, *381*, eadk6139. [CrossRef]

bioengineering MDPI

Article

Trivial State Fuzzy Processing for Error Reduction in Healthcare Big Data Analysis towards Precision Diagnosis

Mohd Anjum [1], Hong Min [2,*] and Zubair Ahmed [3]

1 Department of Computer Engineering, Aligarh Muslim University, Aligarh 202002, India; mohdanjum@zhcet.ac.in
2 School of Computing, Gachon University, Seongnam 13120, Republic of Korea
3 Department of Zoology, College of Science, King Saud University, Riyadh 11451, Saudi Arabia
* Correspondence: hmin@gachon.ac.kr

Abstract: There is a significant public health concern regarding medical diagnosis errors, which are a major cause of mortality. Identifying the root cause of these errors is challenging, and even if one is identified, implementing an effective treatment to prevent their recurrence is difficult. Optimization-based analysis in healthcare data management is a reliable method for improving diagnostic precision. Analyzing healthcare data requires pre-classification and the identification of precise information for precision-oriented outcomes. This article introduces a Cooperative-Trivial State Fuzzy Processing method for significant data analysis with possible derivatives. Trivial State Fuzzy Processing operates on the principle of fuzzy logic-based processing applied to structured healthcare data, focusing on mitigating errors and uncertainties inherent in the data. The derivatives are aided by identifying and grouping diagnosis-related and irrelevant data. The proposed method mitigates invertible derivative analysis issues in similar data grouping and irrelevance estimation. In the grouping and detection process, recent knowledge of the diagnosis progression is exploited to identify the functional data for analysis. Such analysis improves the impact of trivial diagnosis data compared to a voluminous diagnosis history. The cooperative derivative states under different data irrelevance factors reduce trivial state errors in healthcare big data analysis.

Keywords: big data; data grouping; fuzzy process; healthcare

Citation: Anjum, M.; Min, H.; Ahmed, Z. Trivial State Fuzzy Processing for Error Reduction in Healthcare Big Data Analysis towards Precision Diagnosis. *Bioengineering* **2024**, *11*, 539. https://doi.org/10.3390/bioengineering11060539

Academic Editors: Yan Pei and Jijiang Yang

Received: 11 April 2024
Revised: 16 May 2024
Accepted: 20 May 2024
Published: 24 May 2024

1. Introduction

Big data analysis is a process that helps organize a massive amount of data. It also correlates raw data to processed data, minimizing the complexity of the decision-making process [1]. Healthcare applications contain various datasets that require proper analysis processes to enhance the performance range of the systems [2]. Big data analysis in healthcare improves the overall development and feasibility level of healthcare applications. Medical health records maintained in healthcare contain necessary patient details [3], including personal data, health conditions, types of diseases, medications, and the process of diagnosing diseases for patients. An intelligent-enabled big data analysis technique is commonly used in healthcare applications [4]. This technique analyzes structured and unstructured healthcare data in the management system. The analyzed information produces relevant data for further disease detection and diagnosis processes. The analysis technique also increases the accuracy of disease prediction, reducing latency in providing services to patients [5].

Error reduction is a crucial task to perform in the big data analytics process. Big data analytics eliminates unwanted raw data from the database [6]. Error reduction in data analytics is mainly used to improve the data quality required for further data processing. Errors such as noisy data, negative data, and inconsistent data are present in healthcare management systems [7]. Big data analytics produces optimal information for healthcare applications. Natural language processing (NLP)-based big data analysis is used in healthcare to reduce errors in computational processes [8]. The NLP data processing tool analyzes the critical clinical data necessary for diagnosis [9]. The NLP identifies the negative data that causes errors during data processing and analysis. The identified errors are eliminated immediately to reduce unwanted challenges or issues in big data analytics systems. The NLP-based technique improves the overall functional capability level of the analytics process in healthcare applications [10,11].

Fuzzy methods are widely used in various fields to solve the problems presented in systems. They are also employed in big data analytics to enhance the effectiveness of the systems [12]. A fuzzy-optimized data management (FDM) approach is utilized in big data analytics. This approach employs an extraction technique to extract useful information from datasets [13]. The extracted information provides accumulated data to perform tasks for healthcare systems. The FDM approach analyzes the exact relationship between data and produces feasible data for further analysis [14], thereby improving the feasibility and significance of the analytics process. Additionally, a novel big data analytic technique using a fuzzy similarity measure model is employed in healthcare applications [15]. This analytic technique analyzes the potential data required to perform specific tasks in healthcare centers [16] and manages critical information collected from various divisions. The fuzzy-based technique reduces the data analysis inaccuracy ratio, enhancing the significance of healthcare applications [17].

Now, the key objectives and highlights of the research are stated as:

- To design a Cooperative-Trivial State Fuzzy Processing (CTSFP) method for significant data analysis with possible derivatives.
- To apply fuzzy optimization techniques for grouping data based on functional and irrelevant factors.
- To enhance diagnostic progression by employing various fuzzy derivatives to minimize analysis errors.
- Conduct data and metrics analysis to assess the effectiveness and validate the proposed method.

Hypothesis 1: *Big data analytics positively impacts the innovation system of medical diagnosis.*

Hypothesis 2: *There will be a positive correlation between learning objectives and healthcare data analytics abilities and between learning objectives and performance results in data analytics.*

2. Related Works

This literature review highlights significant strides in healthcare data analytics and machine learning applications. In [18], the authors developed a specialized data analytics suite for the management of type 2 diabetes. This suite comprised multi-tier classifiers and advanced analytical methods such as exploratory, predictive, and visual analyses to elucidate the complex interplay between patients' biological markers, enabling more accurate disease classification and streamlined decision-making processes. A sensor-based data analytics (SBDA) model for real-time patient monitoring in connected healthcare systems addresses the growing need for timely emergency detection and improved efficiency in healthcare applications [19]. Big data analytics gathers various biomedical sensor data for disease detection and prediction. The proposed model is commonly used for real-time patient monitoring, providing feasible emergency detection datasets. These advancements underscore the potential of data-driven approaches to revolutionize disease management and enhance patient care outcomes. Similarly, another study [20] crafted a specialized big data analytics technique for facilitating decision-making within healthcare centers. This technique was designed to gather data from structured and unstructured sources, streamlining the complexity of detection processes. Big data management ensures the generation of optimal datasets, which are essential for effective decision-making. Consequently, the developed technique has been shown to improve the accuracy rate in clinical decision-making, thereby enhancing the feasibility of diagnostic services.

In addressing the pressing concern of data privacy in healthcare, Elayan et al. presented a novel privacy-preserving framework, namely deep federated learning [21]. This framework ensures the safety and confidentiality of patient data while maximizing performance and reducing operational costs by leveraging Internet of Things-enabled devices. Furthermore, a hybrid deep learning technique for healthcare data analytics focuses mainly on disease diagnosis [22]. This technique demonstrates promising results in improving diagnostic accuracy and enhancing the efficiency of diagnostic services, highlighting the transformative impact of machine learning in healthcare. In recent years, wearable sensors have emerged as vital technology applications for monitoring users' physiological signs, offering valuable insights into health trends. This capability to gather and analyze physiological data has significant implications for enhancing healthcare solutions. An Un-Synchronized Sensor Data Analytics (USDA) model has been developed [23], addressing the need to effectively manage wearable device data, particularly in time-sensitive healthcare scenarios. By classifying data based on timing and occurrence frequency, coupled with a diagnosis module, the USDA model identifies defects and addresses missing sensor data crucial for accurate analyses. Utilizing sophisticated machine learning methods enhances diagnostic accuracy and enables timely healthcare solutions, ultimately improving system efficiency and reducing complications in healthcare performance assessment.

In recent literature, researchers have unveiled a groundbreaking healthcare facility management approach by integrating Building Information Modeling (BIM) with big data analytics. This innovative method, rooted in BIM, is designed to harness the information available within building models, thereby generating optimal data for improved detection and diagnosis within healthcare facilities [24]. By leveraging this BIM-based approach, healthcare systems can be transformed, offering enhanced efficiency and effectiveness in patient care delivery. Moreover, a sophisticated smart health monitoring system, grounded in big data principles, has been developed to advance patient care [25]. The proposed model applies Hybrid Dingo Coyote Optimization (HDCO) for optimal feature selection and utilizes a Deep Ensemble Learning algorithm (DEL). This model (HDCO-DEL) accurately classifies various types and classes of medical data, ensuring precise analysis. Integration with Internet of Medical Things devices enables seamless data collection from wireless sensors, thereby minimizing latency in the classification process. Through these innovations, the proposed model significantly elevates the performance standards of healthcare monitoring systems, promising enhanced efficiency and effectiveness in the delivery of patient care. Similarly, Feng et al. introduced a pioneering approach, the confidential information coverage hole prediction, tailored for collecting healthcare big data [26]. Primarily applied within large-scale hybrid wireless sensor networks, this model monitors essential data for detection. Its core objective is leveraging sensor nodes to forecast prior information, mitigating energy consumption in subsequent processes.

Nowadays, healthcare policies worldwide are increasingly emphasizing the importance of leveraging information instruments and digital technologies to enhance public health and quality of life. Therefore, health policies have evolved to incorporate big data analytics as a key driver of digital social innovation in healthcare. In the study [27], the authors introduced digital social innovation-based big data analytics for healthcare applications. Critical analysis is used here to analyze the necessary features and data from medical information. Digital social innovation is mainly used to enhance society's overall well-being, increasing the efficiency of digital data in smart cities. The introduced method increases the development range of healthcare centers.

Similarly, Khan et al. [28] introduced a systematic analysis framework for healthcare big data analytics to enhance the accuracy and efficiency of disease diagnosis. The model integrates a feature extraction technique to extract pertinent medical data crucial for disease prediction and detection. Notably, the proposed model demonstrates an expanded effectiveness scope for healthcare centers compared to existing approaches.

The prevalence of diabetes mellitus, a chronic metabolic disorder, remains a significant global health challenge, with a concerning proportion of cases going undiagnosed. Early detection and effective management are pivotal in preventing complications and improving patient outcomes. Many studies have focused on the early detection of diabetes by leveraging various machine learning and deep learning models, including stacking algorithms. Therefore, a diabetes patient classification model utilizing the stacking ensemble method is introduced for local healthcare centers [29]. Employing cross-validation techniques enhances the precision of patient classification. By leveraging medical data sourced from local healthcare centers, the model bolsters the robustness of the detection process, ultimately elevating the accuracy of disease detection. In [30], the authors have presented a novel approach to predicting obstructive sleep apnea visit costs in healthcare settings. The method generates viable data inputs for prediction by leveraging electronic healthcare records and Transformer models. These inputs are derived from short visit histories maintained within healthcare centers. Introducing this method maximizes the precision ratio in obstructive sleep apnea prediction, significantly enhancing the robustness of the predictive systems.

Furthermore, Razzak et al. introduced multimedia big data analytics tailored for healthcare centers to improve outcomes in healthcare applications [31]. This automated data processing framework analyzes patients' health condition data, commonly employed for decision-making and prediction tasks to mitigate computational costs. The experimental findings indicate that the proposed model enhances the quality of patient care, marking a significant advancement in healthcare analytics.

Data storage and transmission formats used by healthcare information technology systems based on CTSFP might be inconsistent and based on diverse standards. Obstacles to interoperability must be overcome so that different systems can communicate data without difficulties and integrate these systems. Unstructured text data, such as patient reports, social media conversations, clinical notes, and medical literature, may be analyzed, and insights may be extracted using NLP algorithms. NLP techniques make it possible to glean useful information from unstructured text, making it easier to accomplish things like sentiment analysis, entity identification, and document summarization.

3. Data Collection

The data used in this article is acquired from the "health score" electronic health record for assessment (https://www.kaggle.com/datasets/hansaniuma/patient-health-scores-for-ehr-data, accessed on 15 March 2024). The temperature, pulse, respiratory, blood pressure, dialysis, and imagery information are stored under 79,540 entries. This data is used to classify patient health as severe (or) normal using individual score values. Figure 1 illustrates the acquisition and utilization of healthcare data.

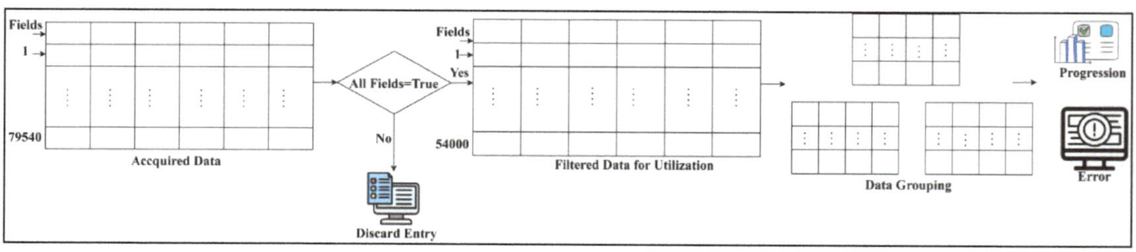

Figure 1. Illustration of healthcare data acquisition and utilization.

This analysis filters the complete data based on its continuity and field availability. This filtered data is utilized for data grouping and invertible assessments. Of the acquired 79,540 entries, 54,000 are used for this analysis as they possess full values (refer to Figure 1). The progression is analyzed if the detected "severity" is similar to the filtered data entry.

The mis-detected (severity as normal (or) vice versa) is regarded as an error in analysis. The CTSFP approach is implemented in real-time scenarios to preprocess the healthcare big data and reduce errors. This process involves noise reduction, outlier detection, data normalization, or fuzzy logic-based processing to enhance data quality and accuracy.

4. Cooperative-Trivial State Fuzzy Processing Method

In healthcare management systems, the trivial state reduces big data analytics errors. It defines some significance or uncertainty in the medical data, which decreases the precision of diagnosis. The data points and states are deployed to this trivial state, reducing the importance of the healthcare data. It involves data grouping in this approach where the trivial state data are detected in the healthcare management system and provide efficient results. This concept indicates the big data input and forwards to the detection phase, where the state and data points are recognized. The key idea of this work is to reduce errors and improve the precision of diagnosis. This analysis (CTSFP) is proposed along with this fuzzy model to address the error rate and enhance the diagnosis. In Figure 2, the functional parts of the proposed method are described. Computer resources for the proposed trivial state fuzzy processing include processing power with multi-core processors and adequate memory space for storing healthcare data.

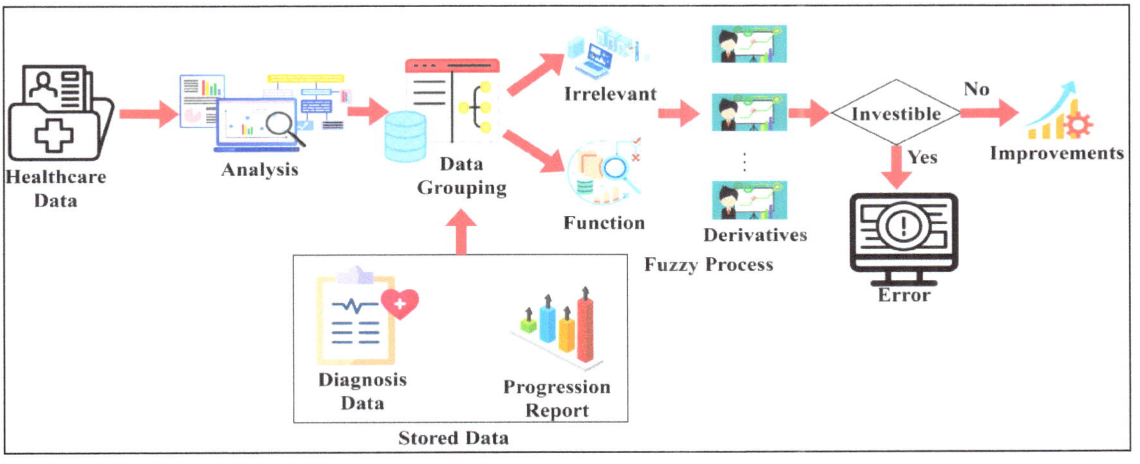

Figure 2. Functional parts of the proposed CTSFP method.

This detection category illustrates the relevant and irrelevant data that are matched with the history of medical data. This analysis is carried out from the stored data, including the diagnosis, which holds the patient's previous history and the progression report, which illustrates the patient's update of the prior observation. Based on this stored data, the grouping is performed to handle trivial states better. In this process, a fuzzy processing model is developed to find the n-number of derivations; based on this, invertible and non-veritable are differentiated and the result is provided. The preliminary step in this paper is to handle the trivial state of the healthcare data formulated below.

$$h_n = (m_g + o_t + d_p) * H_a + \left({}^{m_g + o_t + d_p/g_i}/\sum H_a\right) * \left[(H_a + g_i) * (m_g + o_t)\right] - d_p + \left({}^{H_a}/\sum_{g_i}(m_g + o_t + d_p)\right) \quad (1)$$

The trivial state of healthcare data processing, including precision diagnosis and handling, is represented as h_n, the healthcare data is H_a, which includes the missing value, outlier, and data points, and they are symbolized as m_g, o_t, and d_p. The diagnosis is described as g_i. Based on this approach, missing and inconsistent data are detected. This is used to identify the healthcare data better, allowing for better data quality and reliability in this proposed work. This processing step involves the data points where the outliers are detected in this methodology, and from this processing, reduction is observed. In this computation step, trivial data is handled to detect better precision among inconsistent and missing values.

The missing value, outlier, and data points are used to deploy the healthcare data and provide the efficient handling of constraints, which is formulated as $({}^{m_g+o_t+d_p}/{}_{g_i}/\sum H_a)$. In this category, the analysis examines the trivial state and provides a better progression report for diagnosing the healthcare data. Medical history is analyzed for precision diagnosis, and based on this, the fuzzy model is developed for the n-number of derivations. This trivial state illustrates the derivation from the fuzzy process where the analysis is carried out appropriately. In this method, handling is based on the trivial state of the healthcare data and illustrates the missing value and outlier in the data. From the trivial state handling, the analysis is performed for the varying healthcare data in the stored format, which is equated below.

$$\alpha = \left({}^{h_n+H_a}/{}_{o_t}+d_p+{}^{m_g}/\textstyle\prod_{\omega}(T_s+g_i)\right)+\left\{\left[(T_s+H_a)*\omega+{}^{c_t}/\textstyle\sum_{m_g}H_a\right]\right\}*[(g_i+H_a)+h_n]-(c_t+m_g)+\omega(H_a+g_i) \quad (2)$$

The missing and inconsistent data are reasonable in the handling phase, and from this, the analysis is performed, and it is equated as α. A trivial state is represented as T_s, detection is ω, and inconsistent is described as c_t. Here, it states that better processing of the diagnosis of healthcare data is needed and provides reliable computation. In this case, the analysis is carried out to improve the data processing. By examining this analysis, the healthcare data are fetched from the database, and from that, matching is achieved with the previous history, and the data is detected similarly to this analysis procedure.

The handling of the trivial state is observed in this approach where the error rate is included, and from this, diagnosis is carried out and is represented as $\left({}^{h_n+H_a}/{}_{o_t}+d_p+{}^{m_g}/\prod_{\omega}(T_s+g_i)\right)$ In this category, the pragmatic data is analyzed for the better output from this trivial state of healthcare data. The existing process detects missing values, and the upcoming methodology is addressed here. In this concept, the trivial state is handled to deploy the quality and reliability of the medical data. Thus, the analysis is carried out for the varying data computation methods in this field, and from this $\omega(H_a+g_i)$ is performed. Then comes the data grouping process in this work, illustrated in the Equation below.

$$\beta = {}^{1}/{}_{\omega(p\prime+g_i)}+\begin{cases}(d_v+a_q)*\|T_s+\alpha\|+\left(H_a+{}^{a_q}/\prod_{h_n}T_s\right),\ \forall\ I_r\\ \sum_{\alpha}^{d_v}(a_q+d_n)*\langle m_g+T_s\rangle+c_t-\alpha,\ \forall\ F_u\end{cases} \quad (3)$$

The data grouping classification is derived from irrelevant data and functional data, which are symbolized as I_r and F_u. The data grouping classification is labelled as β, acquiring is a_q, the progression report is represented as $p\prime$, and the n-number of derivations is d_n. The first stage regards the irrelevant data acquired from the progression report and diagnosis. This history of data and the update of the current scenarios are stored in the database and used for the classification process. The first case indicates the irrelevant process that deploys the acquisition of the data from the dataset and performs a better analysis rate. Handling trivial data involves healthcare data, and it is represented as $(H_a + {}^{a_q}/_{\prod_{h_n} T_s})$. The pseudocode for data grouping is presented in Algorithm 1.

Algorithm 1: Pseudocode for β

The inputs are: H_a, d_p
Initialized $\alpha\ \forall H_a$ *towards* T_s
Start of loop: 1 *to* $d_p \in D_a$: *Perform*
Verification $(g_i + H_a) == T_s$
Achieving β *as* $(m_g + T_s) = (m_g + g_i + H_a)$
New H_a *and verify* $m_g || O_t$
Extract $d_p \forall H_a$ *and move to new* w
Verification fails $(g_i + H_a) \neq T_s$
Computing β *as* $(d_v + a_q) = (O_t + H_a)$
Declare β *as* c_t *under* I_r
End Loop

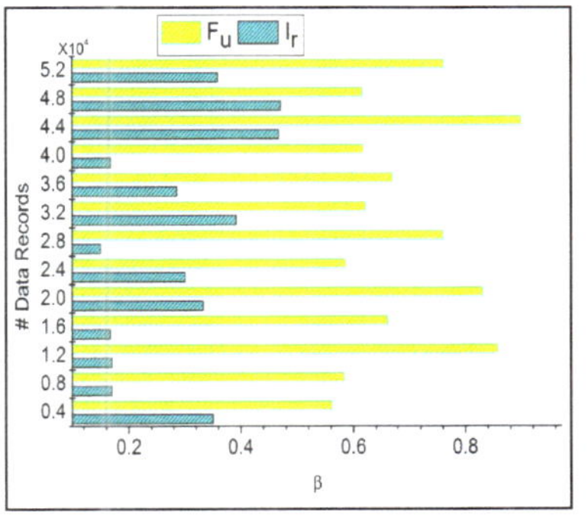

This trivial state indicates healthcare data that includes inconsistent and missing values. The acquisition of the desired data is used to provide trivial data, and the analysis is followed up for the missing value, in which the inconsistency is examined. The second condition is derived from the stored data in which the trivial data are used to better detect the diagnosis among the patients. This approach uses detection to classify the irrelevant and function values. Both classifications are grouped and fetch the data from the storage space. Thus, the data grouping is classified in this concept, and from this, the stored data is observed in the Equation below.

$$\rho = \left\{ \left(\sum_{\beta}^{H_a/g_i} (d_n + \alpha) \right) * \left[(T_s + h_n) * (g_i + p') \right] + a_q \right\} * \left(\sum (H_a + a_q)/T_s \right) \quad (4)$$

This is used to acquire specific data from storage derived from the data grouping. The n-number of derivations is used to perform the better role in this healthcare data processing step, which is described as ρ. This approach includes the classification model for reliable computation of trivial data. Acquiring this data indicates the patient's previous history, where the matching is processed with the current scenario. In this case, the computation rate is improved by deploying the progression report and diagnosis. This progression report analyses the regular updates for the upcoming process. In Table 1, the input data state is classified and represented.

Table 1. Input data state classification.

	F1	F2	F3	F4	F5	F6	State
	Low/Normal/ High	Normal/ Abnormal	Low/High	Low/Normal/ High	Normal/ Abnormal	Available/ Unavailable	
d_p	●●●	●●	●●	●●●	●●	●	β
I_r	●●●	●●	●●	●●●	●●	●	α
F_u	●●●	●●	●●	●●●	●●	●	β
c_t	●●●	●●	●●	●●●	●●	●	α

Table 1 represents the β or α state analysis of the healthcare data accumulated. The $F1$ to $F6$ indicates the fields used in the filtered data; their range values are regarded in red (or) green or yellow. Green color denotes the health state low, normal and available ranges from [0,1,2]. Red color indicates the health state high, abnormal and unavailable ranges from [3,4,5]. Considerably, the $F6$ determines the overall output of the state as α or β. If the abnormal case is high, then α is yet to be completed, for which further derivations are required. In the alternating case, if c_t is the maximum possibility, then the data is T_s to be denied. Therefore, the state of the data is functional/irrelevant for grouping. Here, the diagnosis and the progressive report indicate a better computation factor and provide efficient derivation matching with the processing history. The n-number of derivations is associated with the analysis where the diagnosis and progressive report are included for further processing, and it is formulated as $\left(\sum_{\beta}^{H_a/g_i}(d_n+\alpha)\right)$. Thus, stored data are used in this trivial processing state where the error is reduced. Since the referred stored data are used in this case, and from this step, the irrelevant and function are differentiated and mapped using the fuzzy optimization model, which is deliberated in the section below.

5. Fuzzy Optimization for F_u and I_r Derivatives

The fuzzy optimization method finds reliable results in the healthcare system where the irrelevant and function are differentiated. This fuzzy logic states whether it is 0 or 1; this paper describes whether it is invertible or non-invertible. This process uses the decision-making method, leading to error identification or improvement. Based on this decision-making approach, the required data are acquired from the stored data, providing a better result. Here, it deploys the trivial state of computation where it indicates the healthcare data for the data grouping methodology. From this, the fuzzification is performed in Equation (5):

$$u_c = \left[\left(H_a + a_q\right) * w_d\right] + d_n * \sum_{\alpha}(\rho + \beta) \tag{5}$$

The fuzzification is performed where the input is fetched from the previous step and forwards to the n-number of derivations. The fuzzification is described as u_c, the forwarding is w_d. This category uses the analysis to forward the necessary process and provide a reliable classification. This classification states the progressive report and the diagnosis of the healthcare data. Based on this section, the n-number of derivations provides efficient processing in trivial states. For the n-number of derivations, the fuzzification is pragmatic and from which the forwarding is examined in five stages, e.g., $H_a + a_q$ is large positive, medium positive, small, medium negative, and significant negative.

All these are split in this fuzzification methodology, where healthcare data are acquired for reliable computation from the existing step, including the data grouping. The data grouping indicates the diagnosis and the progressive report, which deploys invertible and non-invertible data. Based on this detection, the performance is measured to process reliable results among the derivation values. The derivation is used to provide better processing, which uses the fuzzy model in this Equation. This fuzzification output equates to the defuzzification in Equation (6).

$$d_z = (u_c - H_a) + d_n * \left(\sum_{a_q}(g_i + p')/{h_n/c_t}\right) + (T_s + \omega) \tag{6}$$

The defuzzification model is the reverse of the process from the fuzzification where the crisp value is obtained. Equation (6) relies on the healthcare data acquisition and derives the stored data as a diagnosis and progressive report, and the defuzzification is labelled as d_z. The fuzzification and defuzzification for the grouping are illustrated in Figure 3.

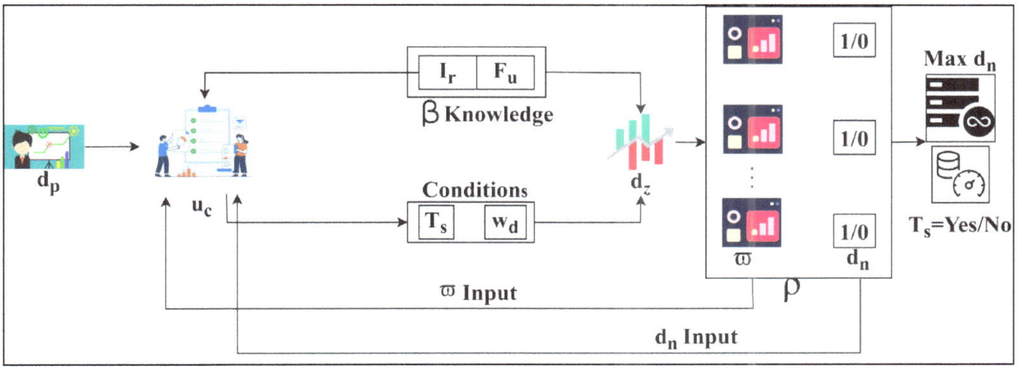

Figure 3. Fuzzification and defuzzification for β.

The u_c process is initiated with d_p input for $F1$ *to* $F6$ entries. This process is later defuzzified through (T_s, w_d) conditions: $(H_a + a_q) == T_s$ (or) $(g_i + H_a) \neq T_s$. The satisfying conditions generate ω and d_n which are the defuzzified derivatives used to analyze data states. From these ρ sequences, maxd_n and $T_s = yes/no$ is obtained. The ρ outputs are further used for sequence classification from u_c to verify β knowledge satisfaction (Figure 3). The inconsistent data are acquired from this category and forward the derivation to the next level that indicates the handling phase, and it is formulated as $\left(\sum_{a_q}(g_i + p')/_{h_n/c_t}\right)$. Detecting this trivial state is associated with fuzzification, which computes the error reduction. The membership function is accomplished by formulating Equation (7) from above fuzzification and defuzzification.

$$\mu = \{(H_a, \mu(m)(H_a))|H_a \in \beta\} \tag{7}$$

The membership function in fuzzy logic is used to find the 0 or 1 here, and it is defined as either an error or not. If it is not an error, then the improvement is performed. The membership function is m, and the value range from [0, 1] is invertible data and is represented as μ. Universal information is healthcare data; the set of ordered pairs refers to the grouped classifications from the stored data. This observation is performed in the membership function to find the ordered pairs in the trivial state. Here, the invertible data are acquired from the derivation, where the analysis is carried out appropriately. Posted to this membership function in the fuzzy model, the separation is examined from the derivation, whether invertible, non-invertible, or formulated in Equation (8).

$$\vartheta = \left.\begin{matrix} (g_i + p') * \left(\alpha + d_n/(H_a)F_u\right) + T_s = 0 \\ \sum_{T_s}(I_r + d_n) * (g_i + p') + \beta \neq 0 \end{matrix}\right\} \tag{8}$$

The separation of trivial states depends on the data that is invertible and non-invertible. Based on this process, the stored data are acquired and perform better data grouping for the irrelevant and function data. The separation is described as ϑ, where the value equal to 0 states the invertible, whereas the value not equal to 0 defines non-invertible. The separation process pseudocode is presented in Algorithm 2.

Algorithm 2: Pseudocode for ϑ

$The\ inputs\ are\ (\rho, d_n)$

$Initialize\ \beta\ for\ all\ \alpha\ data$

$Start\ of\ Loop: 1\ to\ \rho: Perform$

$Estimation: d_z = T_s + \omega; (u_c - H_a) > 0$

$Achieving: u = [H_a, u(m) | H_a \in \beta]$

$Output\ achievable: v = (g_i + p'), T_s = 0$

$Estimation\ fails\ (u_c - H_a) = 0$

$Achieving\ (I_r + d_n)\ \forall\ \omega\ in\ \rho$

$Output\ achievable: v = (I_r + d_n) + \beta; v \neq 0$

$End\ of\ Loop$

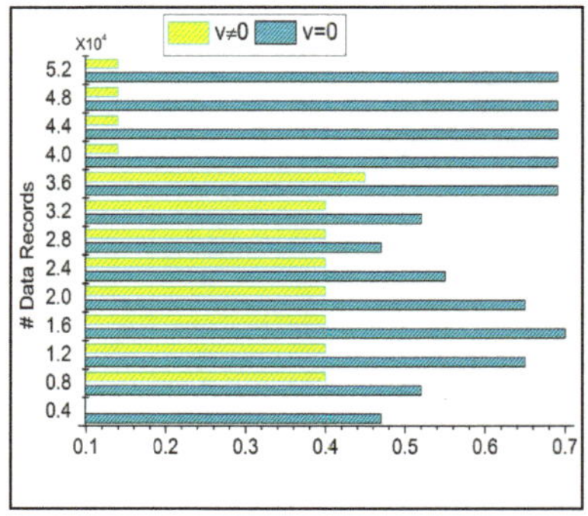

Based on this classification, trivial data are obtained. In this category, detection is observed on the irrelevant function from which the stored data are acquired. Here, data grouping classification is performed for the varying trivial state that deploys the big data as the input for this processing. This derivation is pragmatic for separating invertible and non-invertible data computation. This separation process evaluates the decision-making to find whether it is invertible, equated below.

$$\delta = \begin{cases} 1,\ if \prod_{\beta} (\alpha + H_a) * (\mu - \mu\prime/g_i + T_s) + \vartheta * m \\ 0,\ otherwise \end{cases} \tag{9}$$

The separation of invertible and non-invertible data follows up the decision-making and provides the changes that occur during the trivial state dispensation, where non-invertible is represented as $\mu\prime$. The decision-making is δ, defined in *if* and *otherwise* conditions where it deploys the trivial state for the pragmatic healthcare data in this methodology. These functional data are associated with the analysis where the derivations are used to provide better separation among the errors that occur due to the uncertainty due to the missing value. Based on the separation, the fuzzy outputs for T_s are analyzed for F1 to F6 in Figure 4.

In Figure 4 above, the classifications under β for different data input fields are validated. The Y-axis denotes the trivial state T_s and the X-axis indicates the data grouping process β. Based on the available d_z and it is corresponding u_c, the H_a the analysis is presented. If the u_c and d_z processes are tallied, then T_s is high, otherwise it is low. This demands data acquisition for further ω and g_i processes (Figure 4). These derivations are used to examine the invertible, where it is associated with it, and the condition of the *otherwise* function. This condition is followed through the decision-making process, in which fuzzy optimization plays a significant role in this derivation. Here, the invertible condition is satisfied after this identification of functional data runs through the analysis and is equated below.

$$\varphi = \langle [(\vartheta + \mu) * (I_r - F_u)] + \delta \rangle + u'(H_a) + d_n \tag{10}$$

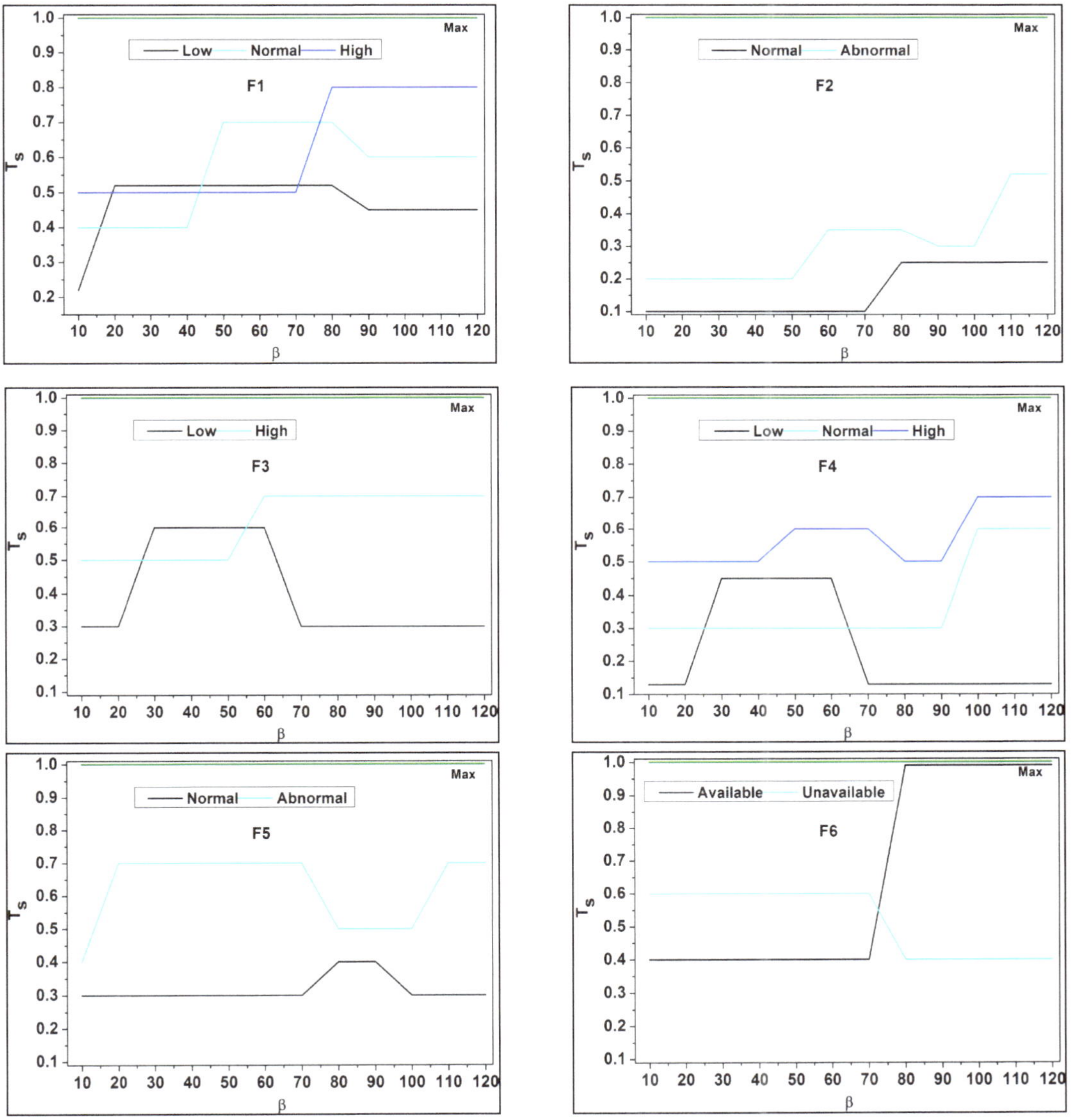

Figure 4. T_s Analysis for F1 to F6.

The identification of functional data is derived and is formulated as φ; in this stage, the decision-making is carried out for the reliable computation based on the invertible data, and the uncertainty is labelled as $u\prime$. The decisions are made from the n-number of derivations and followed up. The healthcare data are integrated with the fuzzy processing in which the membership function is introduced for precision diagnosis. The derivations are associated with irrelevant and functional data processing in this stage. Here, extensive data analysis is performed for the trivial state computation and continues towards the data grouping. Thus, the identification of functional data is examined with the decision-making approach. Then the diagnosis and progression from the stored data are identified and updated based on the current user details. This diagnosis and progression are expressed in Equation (11).

$$\varphi(g_i, p\prime) = 1/d_n * \left[\left(u' + H_a\right) + T_s + \delta/\mu * w_d\right] \quad (11)$$

Here, the identification runs through the diagnosis and progression report that includes the status of the patient information. These processes are built into the stored data function. The stored data reflects the changes if any update or changes occur during a short time interval. These changes are identified based on the uncertainty occurrences in the healthcare data for the varying derivations. The relationship between φ and $\varphi(g_i, p\prime)$ for invertible analysis is tabulated in Table 2.

Table 2. Invertible analysis using φ and $\varphi(g_i, p\prime)$.

φ	F1	F2	F3	F4	F5	F6	$\varphi(g_i,p')$
	Low/Normal/ High	Normal/ Abnormal	Low/ High	Low/Normal/ High	Normal/ Abnormal	Available/ Unavailable	
F1	0.65	0.52	0.58	0.65	0.71	0.82	Low
	1	0.61	0.65	0.74	0.75	0.85	Normal
	0.18	0.68	0.82	0.87	0.65	0.93	High
F2	0.25	1	0.91	0.91	0.74	0.87	Normal
	0.36	0.15	0.86	0.85	0.72	0.87	Abnormal
F3	0.25	0.17	1	0.96	0.81	0.95	Low
	0.41	0.21	0.25	0.87	0.92	0.97	High
F4	0.39	0.35	0.31	0.92	0.98	0.97	Low
	0.42	0.36	0.42	1	0.97	0.95	Normal
	0.39	0.41	0.52	0.41	0.84	0.97	High
F5	0.42	0.5	0.42	0.48	1	0.99	Normal
	0.45	0.36	0.11	0.32	0.5	0.87	Abnormal
F6	0.48	0.29	0.28	0.36	0.48	1	Available
	0.39	0.31	0.32	0.24	0.42	0.49	Unavailable

In Table 2, the relationship between ψ and $\psi(g_i, P')$ based on valued d_z is presented. The field-to-field with normal/available conditions represents the highest relationship (i.e., $\psi = \psi(g_i, P')$). The rest of the cases are validated based on d_n across $v = 0$ and $v \neq 0$ separations. These two cases are derived from multiple u_c and d_z derivatives such that the $u\prime$ is mitigated. Thus, the separation for $\psi = \psi(g_i, p')$ incurring instances (above) are high compared to $u\prime$ incurring cases (below). This process is repeated until the least possible derivatives $u\prime$ are extracted/identified. The n derivations are associated with the trivial state in which the decision-making concept is used for the state forwarding. Based on the update, the changes in the derivation are observed, and the trivial state is used to define the processing step for the different states of the approach. Thus, the identification is carried out for the diagnosis and the progressive report, and the invertible data is detected from the decision-making approach and is equated in the Equation below.

$$\mu(\alpha) = (d_n + w_d) * \prod[(o_a + u\prime) + (\beta * \delta)] \quad (12)$$

The invertible data analysis is pragmatic based on the diagnosis and progressive report update. In this approach, functional data is identified, and it is labelled as o_a; in this classification, it is followed up for the decision-making concept from the fuzzy model. This analysis is executed for the data grouping model and the decision-making of data, whether it is invertible or not. This Equation is used to acquire the input from the big data where the function changes to irrelevant, and then it is said to be invertible; on the other hand, if irrelevant changes to function are observed, it is non-invertible. Thus, the analysis is processed for the invertible data in a trivial state, and the error is detected from this data computation which is equated below.

$$\omega = \left[(T_s + F_u) * (h_n + o_a) / \textstyle\sum_{\vartheta}^{\alpha} (\mu + H_a) + (\varphi * m_v) \right] - e_r \tag{13}$$

The trivial state error is detected from Equation (13), where the functional data are handled for functional usage. The detection is formulated as ω, the error is symbolized as e_r, and improvement is described as m_v, where the diagnosis is executed from the derivation from the fuzzy process. In this, the membership function is used to provide the reduction phase for the derivation and order of the pairs. Thus, the error rate is detected in the first derivative and addressed to the upcoming derivations. Based on this error detection in the trivial state, precision diagnosis is improved using this method. With the v for O_a and e_r using ρ under data grouping towards progression, an error is analyzed, as shown in Figure 5.

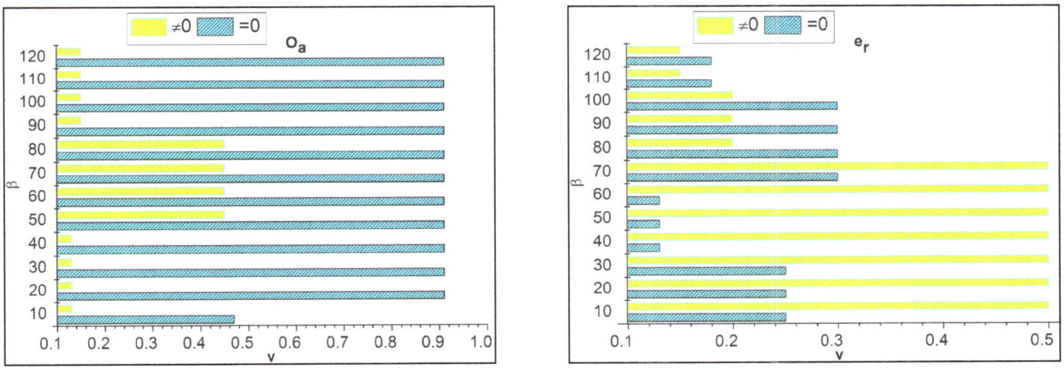

Figure 5. v analysis for O_a and e_r.

In the above assessment for O_a and e_r under $v = 0$ and $v \neq 0$, the β impacts the performance using ρ. The derivatives are abrupt through the max d_n and $T_s = yes$ conditions for ω improvements. Under different β knowledge conditions, the v separation is performed such that $u_c = d_z \; \forall \rho$ is satisfied. Hence, the case of β for $v = 0$ is high (O_a) and *low* (e_r) under different data inputs. The filtered data thus is utilized for its intense field-to-field matching for different β (refer to Figure 5). The possible biases caused by fuzzy processing in healthcare data include sampling, measurement, patient selection, and algorithmic biases.

6. Performance Assessment

The performance assessment uses analysis rate, data grouping, irrelevance estimation, error, and analysis time. The number of data records considered is 52K, filtered from the dataset inputs, and a maximum of 120 groups are formed. In this assessment, the existing USDA [23], SBDA [19], and HDCO-DEL [25] methods are added along with the proposed CTSFP method for efficacy verification. The data used in this article were acquired from the "health score" electronic health record for assessment (https://www.kaggle.com/datasets/hansaniuma/patient-health-scores-for-ehr-data, accessed on 15 March 2024). The temperature, pulse, respiratory, blood pressure, dialysis, and imagery information are stored under 79,540 entries. This data is used to classify patient health as severe or normal using individual score values.

In healthcare data analysis, the CTSFP method outperforms existing methods like USDA [23], SBDA [19], and HDCO-DEL [25] in terms of error reduction capabilities Improved diagnosis accuracy and reliability are the results of the proposed method's use of fuzzy logic-assisted processing to reduce the impact of uncertainties and mistakes in the structure of healthcare data.

By surpassing previous methods in detecting and classifying data important to diagnosis and unrelated data, the technique demonstrates notable advancements in data grouping and irrelevant estimation. The suggested approach improves data classification accuracy and relevance identification by relying on a recent understanding of diagnosis progression and using fuzzy logic in decision-making processes. CTSFP outperforms state-of-the-art approaches regarding concerns about managing healthcare data with trivial states. Compared to more traditional methods, this method can identify and fix data issues, including missing values, inaccurate information, and errors, making it a better foundation for accurate diagnosis and evaluation.

6.1. Analysis Rate

The analysis rate for the proposed work increases for the varying trivial state that deploys the fuzzy processing to detect the healthcare data. This approach is based on functional data, which provides the n-number of derivations. Here, the invertible concept is used to find the improvement in the detection process. The big data input is fetched from the healthcare system, and the diagnosis is processed precisely. In this stage, the analysis rate is enhanced by detecting the invertible in the data, which is proposed by the fuzzy optimization method. In this work, data storage is developed for the diagnosis that includes the history of patient data and the progression reports that hold the update of the patient's health. The trivial state handling of healthcare data addresses the error or the outlier in the input data. This trivial state includes missing or inconsistent datapoints, and it is discussed in Equation (1) and is represented as $(H_a/\sum_{g_i}(m_g+o_t+d_p))$. The execution of the analysis rate in this work is improved by processing this fuzzy processing under healthcare data (Figure 6).

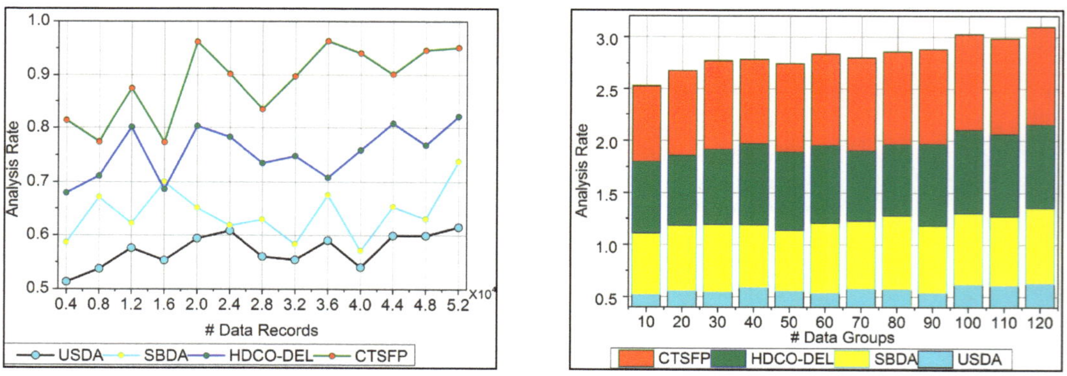

Figure 6. Analysis rate.

6.2. Data Grouping

In Figure 7, data grouping is improved by the classification phase equated in Equation (3). Here, this computation relates to the n-derivation in the fuzzy process and determines the invertible under the input data. In this case, data forwarding to the end process relies on the precision of diagnosis. This concept is proposed to provide irrelevant and functional data in the healthcare system. The evaluation step includes the invertible and non-invertible data used to find the n-number of derivations and provides a better precision diagnosis. The detection process is used to improve the recommendation for the data grouping for healthcare data. The invertible is input to the improvement if the fuzzy process runs accurately for the big data analysis, and it is formulated as $(\alpha + H_a) * (\mu - \mu\prime/g_i + T_s)$. Here, the invertible and non-invertible data are examined to find the trivial state. This data grouping is used to represent the diagnosis and the progression report from the invertible data processing. This data grouping is extracted from the stored data, which shows better improvement.

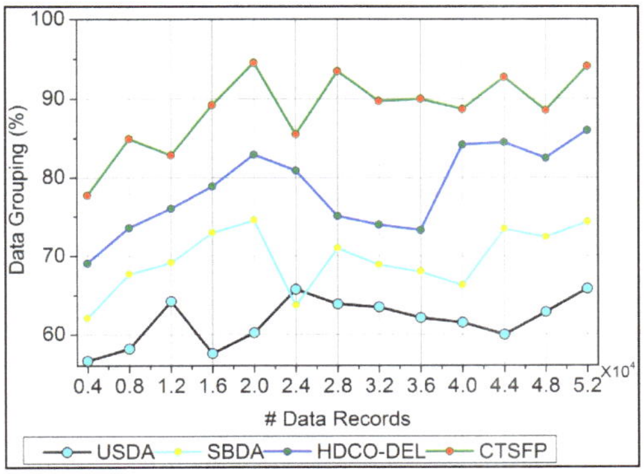

Figure 7. Data grouping.

6.3. Irrelevance Estimation

The irrelevant estimation shows better improvement based on the trivial state data computation. Here, the decision-making approach, followed by fuzzy logic, is deployed. This fuzzy logic illustrates the membership function that defines the invertible and the data grouping. The irrelevant estimation is identified in this category based on these two structures. This is one approach; the precision diagnosis is used to define the better-stored data and find the trivial state by examining the missing value. The inconsistent and missing values are detected from the trivial state, providing better identification. The diagnosis and progressive report are based on the trivial state handling under healthcare data management, and it is represented as $\langle[(\vartheta + \mu) * (I_r - F_u)] + \delta\rangle$. The classification phase is used in this work for the functional data and irrelevant identification from the data grouping. From this case, the irrelevant estimation shows the higher value range extracted from the data grouping concept. This estimation phase relies on detecting trivial states for the healthcare data (Figure 8).

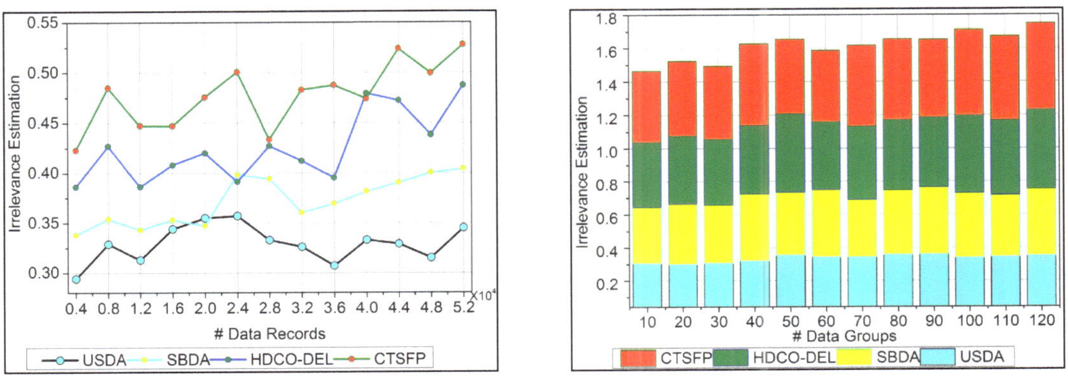

Figure 8. Irrelevance estimation.

6.4. Error

In Figure 9, the error is detected, and a lesser range is shown for the trivial state by deploying recommendations under the healthcare data. This analysis defines the irrelevant data extraction from the healthcare data and provides the invertible process. This approach is observed for the n-derivation, where the trivial state is examined to find the missing and the inconsistent data from the trivial state handling. This error detection is used to illustrate the data point and the classification under the reduction of constraint for the healthcare data. The detection is followed up for uncertainty and improves the invertible data computation in this approach. The n-derivation is associated with invertible data processing, where the trivial state is used to deploy the fuzzy model. Fuzzification and defuzzification are better used to reference the fuzzy model's trivial state. Equation (13) is used to find the trivial state of detection and reduce the error factor, and it is equated as $(T_s + F_u) * (h_n + o_a) / \sum_{\theta}^{\alpha}(\mu + H_a)$. Here, the trivial state and functional data are acquired, and the invertible is found, reducing the error.

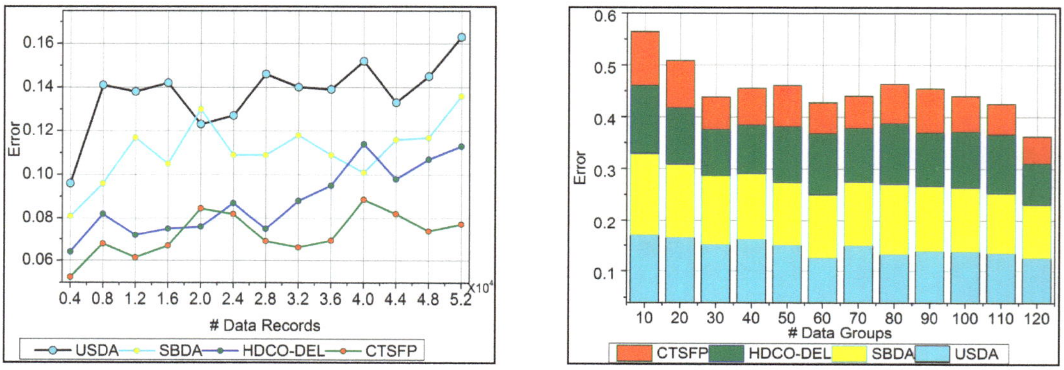

Figure 9. Error.

6.5. Analysis Time

In Figure 10, the analysis time for the proposed work is reduced based on the data grouping concept. This approach indicates the n-derivation forwarding and provides reliable fuzzy processing. The diagnosis precision is improved in this work by reducing the error in this healthcare data. The uncertainty means the healthcare big data analysis where the identification provides functional data under fuzzy processing. This fuzzy processing is used to provide the invertible under the functional data. This stored data is used to define the diagnosis and the progression report to acquire the data grouping. The derivation is computed by classifying irrelevant or functional data in this optimization process. This identification is used to develop a precision diagnosis with reduced errors. From this stage, the analysis is used to propose the trivial state handling under the irrelevant computation. The analysis time is calculated for the healthcare data processing, and it is reduced and represented as $[(g_i + H_a) + h_n] - (c_t + m_g)$.

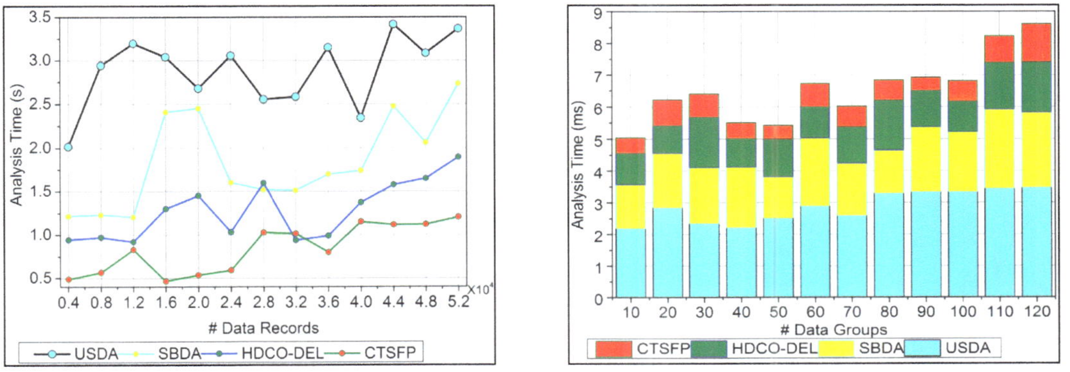

Figure 10. Analysis time.

This study utilizes *p*-values to determine whether the sample estimate differs considerably from a hypothesized value. If there was no actual impact, the *p*-value indicates the probability that the observed effect within the research occurred by chance. Statistical significance is traditionally conferred upon data with a *p*-value of <0.05 or <0.01. Within a specified confidence level (e.g., 95%), a confidence interval gives a range of values, one of which is the precise value of the statistical constraint within the specified population. A confidence interval is a range that includes the most likely lower and upper bounds of a connection or difference for a given population. Confidence intervals, as opposed to p-values, provide greater evidence about the accuracy of an estimate; for example, a 95% confidence interval would mean that the range would include the real value in 95% of cases.

In this performance evaluation, the recommended CTSFP is selected for comparison assessment with the existing state-of-the-art methods like USDA, SBDA, and HDCO-DEL methodologies because of their significant advantages to healthcare big data analytics. Important vital aspects covered by each approach include managing data from wearable devices, monitoring patients in real-time, and using an advanced selection of features for accurate evaluation. Essential insights for healthcare management decision-making can be derived from analyzing their performance concerning accuracy, analysis rate, data grouping efficiency, irrelevance estimation, error detection and appropriateness of analysis time for healthcare applications.

In this comparison evaluation, the results across the considered metrics reveal the advantages and disadvantages of each strategy, giving helpful information about the best method to put them into practice in analyzing healthcare big data. For efficient and precise processing of missing information from sensors, the USDA model performs exceptionally well in the real-time monitoring of patients. Disease detection and forecasting are two areas where the SBDA model shines, demonstrating its strength in times of crisis. In contrast, the HDCO-DEL model is scalable and displays remarkable accuracy in medical data classification, making healthcare surveillance techniques more dependable.

Despite the methods' strengths, they all have drawbacks, such as computational inefficiency and problems with data analysis. In contrast, the suggested CTSFP approach is used to preprocess healthcare big data to increase accuracy, data analysis time, and resilience to overcome these restrictions. The CTSFP method is an innovative new direction for healthcare management practices since it uses data normalization, noise reduction, and outlier detection to enhance healthcare analytics. Additional empirical testing is required to determine the method's practical usefulness. Still, it offers a fresh perspective on the challenges associated with big data healthcare analysis and could lead to more reliable decision-making in the healthcare domain.

7. Conclusions

This article introduces the CTSFP method to improve the efficacy of healthcare data analytics. The proposed method separates healthcare data analysis toward invertible improvement and error reduction. This method utilizes fuzzy optimization to identify irrelevant and functional data based on real-time measures. The fuzzy derivatives satisfying invertible conditions are utilized under error reduction and diagnosis-oriented improvements. Based on the separation, data grouping for irrelevant and functional input is validated under fuzzification and defuzzification processes that extract data deviations separately. Therefore, the functional data for diagnosis improvements are augmented for further irrelevant data reduction. Thus, the proposed method is introduced for significant data analysis with the possibility of derivatives detected using the fuzzy optimization method. Since the trivial state error rate is reduced, healthcare big data analysis is enhanced. Data storage and transmission formats used by healthcare information technology systems might be inconsistent and based on diverse standards. The proposed scheme has overcome obstacles to interoperability so that different systems can communicate data without any problems, integrating these systems.

Author Contributions: Conceptualization, M.A. and H.M.; methodology, M.A. and H.M.; software, M.A. and H.M.; validation, Z.A. and H.M.; formal analysis, M.A. and Z.A.; resources, Z.A. and H.M.; data curation, Z.A.; writing—original draft preparation, M.A. and H.M.; writing—review and editing, M.A. and Z.A.; visualization, Z.A. and H.M.; funding acquisition, Z.A. and H.M. All authors have read and agreed to the published version of the manuscript.

Funding: This research was supported by Basic Science Research Program through the National Research Foundation of Korea (NRF) funded by the Ministry of Education (No. 2021R1F1A1055408).

Institutional Review Board Statement: Not applicable.

Informed Consent Statement: Not applicable.

Data Availability Statement: The data used in this study is available from Kaggle at the following link: https://www.kaggle.com/datasets/hansaniuma/patient-health-scores-for-ehr-data.

Acknowledgments: The authors express their sincere appreciation to the Researcher Supporting Project Number (RSPD2024R1113) King Saud University, Riyadh, Saudi Arabia.

Conflicts of Interest: The authors declare no conflicts of interest.

References

1. Yan, L.; Huang, W.; Wang, L.; Feng, S.; Peng, Y.; Peng, J. Data-enabled digestive medicine: A new big data analytics platform. *IEEE/ACM Trans. Comput. Biol. Bioinform.* **2019**, *18*, 922–931. [CrossRef] [PubMed]
2. Biswas, R. Outlining big data analytics in health sector with special reference to COVID-19. *Wirel. Pers. Commun.* **2022**, *124*, 2097–2108. [CrossRef] [PubMed]
3. Batko, K.; Ślęzak, A. The use of Big Data Analytics in healthcare. *J. Big Data* **2022**, *9*, 1–24. [CrossRef] [PubMed]
4. Hassan, S.; Dhali, M.; Zaman, F.; Tanveer, M. Big data and predictive analytics in healthcare in Bangladesh: Regulatory challenges. *Heliyon* **2021**, *7*, e07179. [CrossRef] [PubMed]
5. Chao, K.; Sarker, M.N.I.; Ali, I.; Firdaus, R.R.; Azman, A.; Shaed, M.M. Big data-driven public health policy making: Potential for the healthcare industry. *Heliyon* **2023**, *9*, e19681. [CrossRef] [PubMed]
6. Koeck, J.A.; Young, N.J.; Kontny, U.; Orlikowsky, T.; Bassler, D.; Eisert, A. Interventions to reduce pediatric prescribing errors in professional healthcare settings: A systematic review of the last decade. *Pediatr. Drugs* **2021**, *23*, 223–240. [CrossRef] [PubMed]
7. Ahsani-Estahbanati, E.; Sergeevich Gordeev, V.; Doshmangir, L. Interventions to reduce the incidence of medical error and its financial burden in health care systems: A systematic review of systematic reviews. *Front. Med.* **2022**, *9*, 875426. [CrossRef] [PubMed]
8. Gates, P.J.; Hardie, R.A.; Raban, M.Z.; Li, L.; Westbrook, J.I. How effective are electronic medication systems in reducing medication error rates and associated harm among hospital inpatients? A systematic review and meta-analysis. *J. Am. Med. Inform. Assoc.* **2021**, *28*, 167–176. [CrossRef] [PubMed]
9. Schiavone, F.; Sabetta, A.; Leone, D.; Chiao, B. Industrial convergence and industrial crisis: A situational analysis about precision medicine during the Covid-19 pandemic. *IEEE Trans. Eng. Manag.* **2021**, *70*, 1456–1467. [CrossRef]

10. Fioretos, T.; Wirta, V.; Cavelier, L.; Berglund, E.; Friedman, M.; Akhras, M.; Botling, J.; Ehrencrona, H.; Engstrand, L.; Helenius, G.; et al. Implementing precision medicine in a regionally organized healthcare system in Sweden. *Nat. Med.* **2022**, *28*, 1980–1982. [CrossRef]
11. Geissler, J.; Makaroff, L.E.; Söhlke, B.; Bokemeyer, C. Precision oncology medicines and the need for real world evidence acceptance in health technology assessment: Importance of patient involvement in sustainable healthcare. *Eur. J. Cancer* **2023**, *193*, 113323. [CrossRef] [PubMed]
12. Yan, F.; Huang, H.; Pedrycz, W.; Hirota, K. A disease diagnosis system for smart healthcare based on fuzzy clustering and battle royale optimization. *Appl. Soft Comput.* **2024**, *151*, 111123. [CrossRef]
13. Ghorbani, A.; Davoodi, F.; Zamanifar, K. Using type-2 fuzzy ontology to improve semantic interoperability for healthcare and diagnosis of depression. *Artif. Intell. Med.* **2023**, *135*, 102452. [CrossRef] [PubMed]
14. Ahmed, A.; Xi, R.; Hou, M.; Shah, S.A.; Hameed, S. Harnessing big data analytics for healthcare: A comprehensive review of frameworks, implications, applications, and impacts. *IEEE Access* **2023**, *11*, 112891–112928. [CrossRef]
15. Rehman, A.; Naz, S.; Razzak, I. Leveraging big data analytics in healthcare enhancement: Trends, challenges and opportunities. *Multimed. Syst.* **2022**, *28*, 1339–1371. [CrossRef]
16. Taipalus, T.; Isomöttönen, V.; Erkkilä, H.; Äyrämö, S. Data Analytics in Healthcare: A Tertiary Study. *SN Comput. Sci.* **2022**, *4*, 1–13. [CrossRef] [PubMed]
17. Badawy, M.; Ramadan, N.; Hefny, H.A. Healthcare predictive analytics using machine learning and deep learning techniques: A survey. *J. Electr. Syst. Inf. Technol.* **2023**, *10*, 40. [CrossRef]
18. Philip, N.Y.; Razaak, M.; Chang, J.; O'Kane, M.; Pierscionek, B.K. A data analytics suite for exploratory predictive, and visual analysis of type 2 diabetes. *IEEE Accesss* **2022**, *10*, 13460–13471. [CrossRef]
19. Harb, H.; Mansour, A.; Nasser, A.; Cruz, E.M.; de la Torre Diez, I. A sensor-based data analytics for patient monitoring in connected healthcare applications. *IEEE Sens. J.* **2020**, *21*, 974–984. [CrossRef]
20. Hussain, F.; Nauman, M.; Alghuried, A.; Alhudhaif, A.; Akhtar, N. Leveraging Big Data Analytics for Enhanced Clinical Decision-Making in Healthcare. *IEEE Access* **2023**, *11*, 127817–127836. [CrossRef]
21. Elayan, H.; Aloqaily, M.; Guizani, M. Sustainability of healthcare data analysis IoT-based systems using deep federated learning. *IEEE Internet Things J.* **2021**, *9*, 7338–7346. [CrossRef]
22. Shafqat, S.; Fayyaz, M.; Khattak, H.A.; Bilal, M.; Khan, S.; Ishtiaq, O.; Abbasi, A.; Shafqat, F.; Alnumay, W.S.; Chatterjee, P. Leveraging deep learning for designing healthcare analytics heuristic for diagnostics. *Neural Process. Lett.* **2021**, *55*, 1–27. [CrossRef] [PubMed]
23. Alfarraj, O.; Tolba, A. Unsynchronized wearable sensor data analytics model for improving the performance of smart healthcare systems. *J. Ambient. Intell. Humaniz. Comput.* **2021**, *12*, 3411–3422. [CrossRef]
24. Demirdöğen, G.; Işık, Z.; Arayici, Y. BIM-based big data analytic system for healthcare facility management. *J. Build. Eng.* **2023**, *64*, 105713. [CrossRef]
25. Abidi, M.H.; Umer, U.; Mian, S.H.; Al-Ahmari, A. Big Data-based Smart Health Monitoring System: Using Deep Ensemble Learning. *IEEE Access* **2023**, *11*, 114880–114903. [CrossRef]
26. Feng, J.; Chen, H.; Deng, X.; Yang, L.T.; Tan, F. Confident information coverage hole prediction and repairing for healthcare big data collection in large-scale hybrid wireless sensor networks. *IEEE Internet Things J.* **2020**, *8*, 16801–16813. [CrossRef]
27. Batko, K. Digital social innovation based on Big Data Analytics for health and well-being of society. *J. Big Data* **2023**, *10*, 171. [CrossRef]
28. Khan, S.; Khan, H.U.; Nazir, S. Systematic analysis of healthcare big data analytics for efficient care and disease diagnosing. *Sci. Rep.* **2022**, *12*, 22377. [CrossRef] [PubMed]
29. Reza, M.S.; Amin, R.; Yasmin, R.; Kulsum, W.; Ruhi, S. Improving diabetes disease patients' classification using stacking ensemble method with PIMA and local healthcare data. *Heliyon* **2024**, *10*, e24536. [CrossRef]
30. Chen, Z.; Siltala-Li, L.; Lassila, M.; Malo, P.; Vilkkumaa, E.; Saaresranta, T.; Virkki, A.V. Predicting Visit Cost of Obstructive Sleep Apnea using Electronic Healthcare Records with Transformer. *IEEE J. Transl. Eng. Health Med.* **2023**, *11*, 306–317. [CrossRef]
31. Razzak, I.; Eklund, P.; Xu, G. Improving healthcare outcomes using multimedia big data analytics. *Neural Comput. Appl.* **2022**, *34*, 15095–15097. [CrossRef] [PubMed]

bioengineering

Article

Wearable 12-Lead ECG Acquisition Using a Novel Deep Learning Approach from Frank or EASI Leads with Clinical Validation

Fan Fu [1,†], Dacheng Zhong [1,†], Jiamin Liu [1], Tianxiang Xu [1], Qin Shen [2], Wei Wang [3], Songsheng Zhu [1,*] and Jianqing Li [4,5,*]

1 School of Biomedical Engineering and Informatics, Nanjing Medical University, Nanjing 211166, China
2 The First Affiliated Hospital of Nanjing Medical University, Nanjing 210029, China; echo_yaya77@163.com
3 The Jiangsu Engineering Research Center of Province Intelligent Wearable Monitoring and Rehabilitation Device, Nanjing Medical University, Nanjing 211166, China
4 The Engineering Research Center of Intelligent Theranostics Technology and Instruments, Ministry of Education, Nanjing Medical University, Nanjing 211166, China
5 The State Key Laboratory of Bioelectronics, School of Instrument Science and Engineering, Southeast University, Nanjing 210096, China
* Correspondence: zhu@njmu.edu.cn (S.Z.); jqli@njmu.edu.cn (J.L.)
† These authors contributed equally to this work.

Abstract: The 12-lead electrocardiogram (ECG) is crucial in assessing patient decisions. However, portable ECG devices capable of acquiring a complete 12-lead ECG are scarce. For the first time, a deep learning-based method is proposed to reconstruct the 12-lead ECG from Frank leads (V_X, V_Y, and V_Z) or EASI leads (V_{ES}, V_{AS}, and V_{AI}). The innovative ECG reconstruction network called M2Eformer is composed of a 2D-ECGblock and a ProbDecoder module. The 2D-ECGblock module adaptively segments EASI leads into multi-periods based on frequency energy, transforming the 1D time series into a 2D tensor representing within-cycle and between-cycle variations. The ProbDecoder module aims to extract Probsparse self-attention and achieve one-step output for the target leads. Experimental results from comparing recorded and reconstructed 12-lead ECG using Frank leads indicate that M2Eformer outperforms traditional ECG reconstruction methods on a public database. In this study, a self-constructed database (10 healthy individuals + 15 patients) was utilized for the clinical diagnostic validation of ECG reconstructed from EASI leads. Subsequently, both the ECG reconstructed using EASI and the recorded 12-lead ECG were subjected to a double-blind diagnostic experiment conducted by three cardiologists. The overall diagnostic consensus among three cardiology experts, reaching a rate of 96%, indicates the significant utility of EASI-reconstructed 12-lead ECG in facilitating the diagnosis of cardiac conditions.

Keywords: deep neural network; EASI lead system; electrocardiogram; 12-lead ECG reconstruction

Citation: Fu, F.; Zhong, D.; Liu, J.; Xu, T.; Shen, Q.; Wang, W.; Zhu, S.; Li, J. Wearable 12-Lead ECG Acquisition Using a Novel Deep Learning Approach from Frank or EASI Leads with Clinical Validation. *Bioengineering* **2024**, *11*, 293. https://doi.org/10.3390/bioengineering11030293

Academic Editors: Andrea Cataldo, Yan Pei and Jijiang Yang

Received: 27 February 2024
Revised: 11 March 2024
Accepted: 18 March 2024
Published: 21 March 2024

1. Introduction

Heart disease is the leading cause of mortality worldwide [1]. Electrocardiogram (ECG) monitoring serves as an effective means for the early detection of cardiovascular disease [2]. In clinical practice, the 12-lead ECG plays a pivotal role in assessing and guiding patient management decisions [3]. In order to record prolonged cardiac activity, ambulatory ECG was introduced in 1961 [4]. However, due to its influence on daily life, which stems from the number and placement of recording points and its relatively short recording duration (most 20 to 48 h [5]), there is an urgent need for new measurement methods to capture long-term cardiac activity.

The wearable ECG, while meeting long-term monitoring and comfort demands [6], falls short of meeting clinical requirements as the standard 12-lead ECG. Existing wearable ECG devices predominantly capture single leads (two electrodes [7] or optical sensors [8]) or three-lead ECG (four electrodes [2] or five electrodes [9]). Compared with standard 12-lead

ECG, wearable ECG offers limited intuitive cardiac information (as shown in Figure 1) and is currently primarily used to diagnose arrhythmias [10–12]. To our knowledge, no specific diagnostic standards have yet been established for wearable ECG in clinical practice. The reconstruction of a standard 12-lead ECG from wearable ECG data can enhance the clinical utility of wearable ECG. As a result, the reconstruction of 12-lead ECG from a reduced number of leads has become a research hotspot.

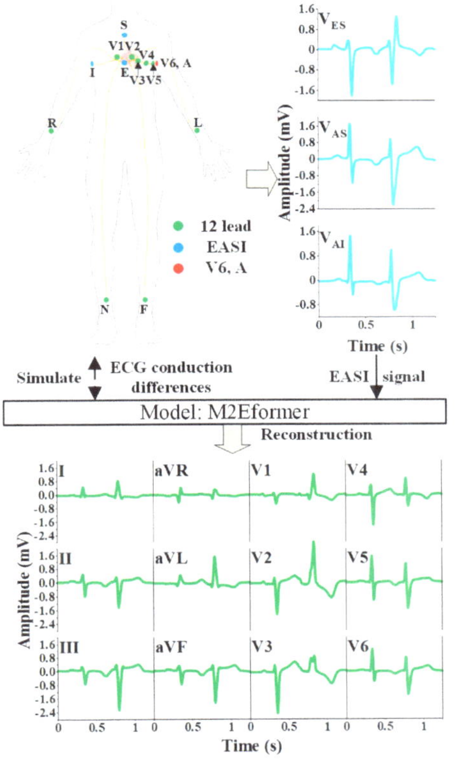

Figure 1. Schematic illustration of the 12-lead ECG reconstruction.

The theoretical foundation for the reconstruction of ECG was established by Frank [13–15] and Dower et al. [9,16–18]. Frank et al. introduced the Frank-XYZ orthogonal spatial vector ECG (V_X, V_Y, and V_Z) [13], but it is not suitable for dynamic cardiac monitoring. Based on Frank et al.'s theory, Dower et al. proposed the EASI lead system (V_{ES}, V_{AS}, and V_{AI}), which is suitable for dynamic acquisition, and theoretically demonstrated the feasibility of reconstructing the 12-lead ECG using EASI leads [16]. As shown in Figure 1, the EASI system consists primarily of four electrodes (E-A-S-I) that can capture three bipolar leads (V_{ES}, V_{AS}, and V_{AI}), each containing information from both the transverse and coronal planes. Notably, not all three-lead systems can reconstruct a 12-lead ECG. For example, limb leads only contain information from the coronal plane and do not provide the necessary information to derive chest leads, theoretically lacking the feasibility to derive precordial leads [17,19,20]. Dower et al. introduced the "Dower universal transform" method, which achieves a linear transformation of EASI data to derive 12-lead ECG using a biased matrix [16]. Field et al. then enhanced the "Dower universal transform" coefficients originally proposed by Dower et al. [18], and Nelwan et al. observed significant differences between 12-lead ECGs reconstructed using improved EASI coefficients and the recorded ones [21]. Similarly, Schreck et al. employed a straightforward nonlinear approach to construct a universal matrix to reconstruct missing leads [22,23]. The foundational leads used to reconstruct the remaining 12 ECG leads initially included I, aVF, and V2 [22], with subsequent work incorporating I, II, and V2 [23]. This method represents an ideal "one-size-fits-all" solution but may not adapt well to interferences from factors such as equipment, biological variations, and environmental conditions [24].

The least squares regression (LSR) method was used by Trobec et al. to estimate the transformation to generate a 12-lead ECG from three differential leads (DLs) [25]. This method yielded the best results in generating 12 leads from the three DLs proposed by the authors, with an average correlation coefficient of 0.954. However, this method exhibited a lower correlation coefficient of 0.71 in lead aVL, and the root mean square error reached 115.3 μV in lead V5. Their study aligns with the approach of Dower et al., resulting in limited generalization capabilities. Mulyadi et al. proposed reconstructing the 12-lead ECG using a segment-based approach (divided into P, QRS, and T segments) through LSR [26]. Unfortunately, abnormal ECG can exhibit phenomena such as P wave disappearance, QRS-wave distortion, and low-amplitude T wave, which can cause reconstruction failure. Despite attempts to use neural networks to synthesize ECG [27], including the application of focused time-delay neural networks used for speech recognition to ECG reconstruction [28], as reported in their results, the generalizability of ECG reconstruction has improved but still requires further enhancement.

The EASI-lead ECG represents a simplified expression of cardiac status, while the 12-lead ECG provides a richer and more clinically informative representation. This result is analogous to the task of machine translation, where understanding the semantics of one language and translating it into another is required. The Transformer model and its variants are currently among the state-of-the-art models in the field of machine translation. Furthermore, they have also shown good performance in time-series forecasting [29–32]. With the assistance of attention mechanisms, they can uncover hidden pairwise temporal dependencies between time points. Zhou et al. introduced the application of the Transformer model to the prediction of long sequence time series, using its attention mechanism to capture long-term dependencies within the sequence [29]. However, it is challenging for attention mechanisms to directly identify reliable dependencies from scattered time points [31].

In this study, we analyze ECG signals from a multicycle perspective. Sinus ECG exhibits quasiperiodic behavior. However, the conduction of abnormal cardiac electrical activity is influenced by the current cardiac cycle and the increased excitability of ectopic rhythm points or the reentrant excitement from the last cycle, presenting a multicycle pattern. Consequently, the detected abnormal ECG signal results from the superimposition of sinus rhythms and ectopic rhythms, exhibiting multicycle characteristics. However, raw ECG sequences have a one-dimensional structure that captures changes only between adjacent time points, making it challenging to explicitly extract both types of variation simultaneously. We employ Fourier Transformation to dissect 1D time series into several segments based on the ECG frequency composition, stacking them into a 2D structure. At this juncture, the rhythms within each segment predominantly represent within-cycle variations, whereas the variances in the ECG at identical positions across segments are shaped by between-cycle variations. This enables us to represent within-cycle and between-cycle variations concurrently in a 2D space, resulting in temporal 2D variations.

Motivated by the abovementioned considerations, we propose a multichannel-based 2D-variation ECG reconstruction network (M2Eformer). This network comprises two primary modules as follows: the 2D-ECGblock and the ProbDecoder. With the support of the 2D-ECGblock, M2Eformer can identify the multicyclic nature of ECG sequences and fuse information into the attention-based ProbDecoder to achieve target leads. We evaluated the algorithm's performance in publicly available databases [33] using quantitative metrics such as the Pearson coefficient r (*Pr*) and mean absolute error (*MAE*), as well as macro-level evaluations provided by cardiologist annotations. Furthermore, regarding practical application value, we collected synchronous EASI and 12-lead ECG from cardiac patients who required 12-lead ECG monitoring. We analyzed the consistency in the diagnoses made by cardiac experts for the reconstructed and recorded 12-lead ECG.

As illustrated in Figure 1, this document establishes a mapping relationship between EASI and the 12-lead ECG using the deep learning model M2Eformer.

- For the first time, a deep learning-based ECG reconstruction network is presented, which deeply extracts latent cardiac information from EASI leads and reconstructs a standard 12-lead ECG consistent with the diagnostic practices of cardiac experts. This provides a feasible approach to the application of wearable ECG for clinical diagnosis.
- We propose a 2D-ECGblock module for the reconstruction network that transforms time-domain signals into multiperiod 2D tensors based on spectral energy. This module simultaneously extracts dependent information from both within-cycle and between-cycle components in the ECG. Additionally, we designed the ProbDecoder module, which employs a sparse attention mechanism to achieve ECG reconstruction in a residual-like manner.
- We conducted a clinical diagnostic validation study of 25 cases using a 12-lead ECG reconstructed from EASI leads. Next, focusing on four cardiac conditions, namely, atrial fibrillation, atrial flutter, coronary artery disease, and myocardial infarction, which require 12-lead ECG monitoring, three experts were invited to participate in a double-blind diagnostic experiment comparing the reconstructed 12-lead ECG with standard recorded ones. The overall consistency coefficient reached 96%.

The remaining parts of the paper are structured as follows: Section 2 outlines the framework of this paper, encompassing the composition of the dataset, the network architecture, and the evaluation methodologies employed. Section 3 presents the results. Then, Section 4 provides the discussion. Finally, Section 5 summarizes the conclusion.

2. Materials and Methods

The general framework of this study, as depicted in Figure 2, comprises three modules as follows: data preparation, model construction, and results analysis. The aim is to reconstruct a standard 12-lead ECG using EASI leads (V_{ES}, V_{AS}, and V_{AI}). The following two databases were used in this research: a publicly available database (Frank-XYZ + 12-lead ECG) [33] and a self-constructed database (EASI leads + 12-lead ECG), each serving different experimental purposes including the algorithm comparison experiment (Task 1) and the EASI practicality analysis experiment (Task 2).

To construct the M2Eformer model, we initially calculated the correlation coefficient distribution between input signals (V_X, V_Y, and V_Z, or V_{ES}, V_{AS}, and V_{AI}) and target signals on the training set. The lead with the highest correlation was selected as the input for the corresponding ProbDecoder model. Subsequently, we used M2Eformer to reconstruct the 12-lead ECG. Finally, a results analysis was conducted. The details of each component are further elaborated below.

2.1. Databases

In this study, we used the publicly available PhysioBank Physikalisch-Technische Bundensanstalt Diagnostic (PTB-DN) ECG database [33] to compare the performance of the Task 1 algorithm. The main reasons for this choice are as follows: 1. the PTB-DN database includes synchronous Frank-XYZ leads and standard 12-lead ECG, with Frank-XYZ leads forming the theoretical basis for EASI; 2. the PTB-DN database is the largest publicly available database known to contain both synchronous Frank-XYZ leads and standard 12-lead ECG, comprising 549 records from 290 subjects; and 3. many previous studies on ECG reconstruction have also utilized this database [16,22,23,27,28], which facilitates our algorithm comparison experiments.

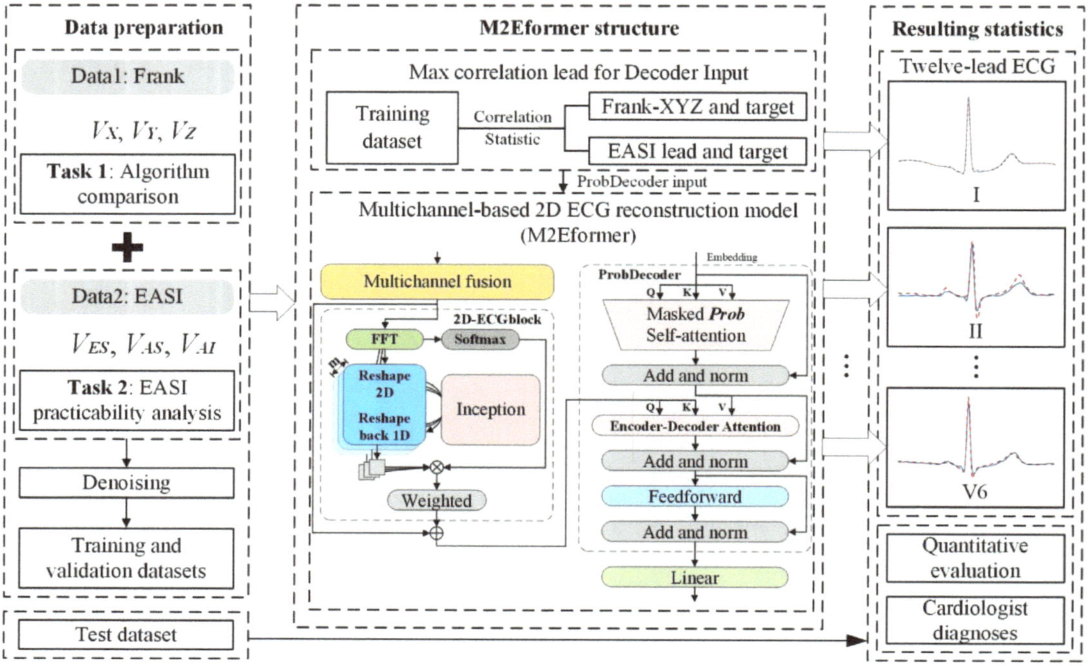

Figure 2. The research framework of this paper. Data1 (Frank-XYZ and 12-lead ECG) was employed to optimize the M2Eformer model and to conduct a comparative performance analysis with prior algorithms. Subsequently, M2Eformer was validated on Data2 to ascertain the reliability of the 12-lead ECG reconstructed from EASI leads for clinical diagnosis.

PTB-DN data were sampled at a rate of 1000 Hz with a 16-bit resolution, and the least significant bit represented 0.5 μV. Before use, all ECG records were preprocessed in 8 s windows, involving a 50 Hz notch filter and 20th-order polynomial filtering to eliminate powerline noise and baseline drift. In order to eliminate the influence of high-frequency noise, local regression smoothing filtering was applied with a smoothing window of 10 sample points. Furthermore, despite preprocessing, some records still contained significant artifacts (ECG drowned by noise or existing severe wandering baseline) or missing information (missing leads or diagnostic information) and were excluded from this study. The data composition used for the algorithm evaluation is detailed in Table 1.

Table 1. Details of the PTB-DN database. The ratio of training to validation to test sets: 3:1:1.

	Training	Validation	Test	Total
Healthy controls	35	11	11	57
Myocardial infarction	127	43	43	213
Bundle branch block	7	2	2	11
Myocardial hypertrophy	6	2	2	10
Valvular heart disease	2	1	1	4
Cardiomyopathy	4	1	1	6
Total	183	61	61	305

Among these, each category of ECG records was roughly divided into training, validation, and test sets in a ratio of approximately 3:1:1 [29,34]. It should be noted that the data for the training and test sets were strictly derived from different individuals.

In order to validate the reliability of the 12-lead ECG reconstructed by EASI for monitoring purposes, Task 2 involved the collection of synchronized EASI leads (V_{ES}, V_{AS}, and V_{AI}) and standard 12-lead ECG. As depicted in Figure 1, thirteen electrodes were attached to the patient's body, where ten electrodes were used to capture the 12-lead ECG, and four electrodes, with one overlapping electrode A and V6, were utilized for capturing EASI leads ($V_{ES} = V_E - V_S$, $V_{AS} = V_A - V_S$, and $V_{AI} = V_A - V_I$). Heart disease patients were arranged for ECG collection at the First Affiliated Hospital of Nanjing Medical University. As shown in Table 2, the types of heart diseases among the patients included atrial flutter (1 case), atrial tachycardia (2 cases), myocardial infarction (3 cases), and coronary heart disease (9 cases). There were also ten healthy participants from Nanjing Medical University. Data were collected using medical equipment (NaLong RAGE-18P) with a sampling rate of 1000 Hz. Healthy individuals were monitored for 10 min, while patients were monitored for 5 min. It is important to emphasize that the selected types of heart disease required joint assessment using a 12-lead ECG. Ethics approval was obtained from the Nanjing Medical University Ethics Committee.

Table 2. Details of the EASI database. The ratio of the training to test sets is 4:1.

	Age ± Std	Training	Test	Total
Healthy controls	26.2 ± 7.2	8	2	10
Atrial flutter	73	12	3	15
Atrial tachycardia	76.5 ± 7.5			
Myocardial infarction	66.2 ± 7.9			
Coronary heart disease	59.3 ± 17.6			
Total	50.5 ± 22.2	20	5	25

2.2. Multichannel 2D-Variation ECG Reconstruction Network (M2Eformer)

Figure 3 illustrates the network architecture of the proposed 12-lead ECG reconstruction model. M2Eformer consists of two modules, namely, the 2D-ECGblock and the ProbDecoder module. In the 2D-ECGblock module, the ECG data are adaptively transformed into a 2D representation based on frequency domain energy, thus enabling simultaneous extraction of within-cycle and between-cycle variations. In the ProbDecoder module, initial sparse-attention calculations are performed on the input signal (Max Correlation Lead) to extract relevant information from the ECG. Subsequently, in the Encoder–Decoder Attention layer, the extracted data are fused, providing the foundational knowledge for the reconstruction of target leads.

As shown in Figure 3, the input to M2Eformer consists of a three-lead ECG represented by V_X, V_Y, and V_Z. For a cardiac sequence of length L, the original 1D structure is denoted as $X_{1D} \in R^{L \times 3}$. The collected ECG vectors represent the projection of the vectorcardiography at that moment onto the coordinate axes of the electrodes and the cardiac dipole. Therefore, to extract cardiac information at time t, we designed the multichannel fusion layer, and the computational method is as follows:

$$X_{1D}^{dmodel} = \mathrm{Conv1d}_{3\times3}\left(X_{1D}^{\mathrm{Ein}}\right) \quad (1)$$

By mapping the original three-channel ECG into a high-dimensional vector X_{1D}^{dmodel} and simulating the distribution of vectorcardiograms at time t, we enhanced the model's generalization capability.

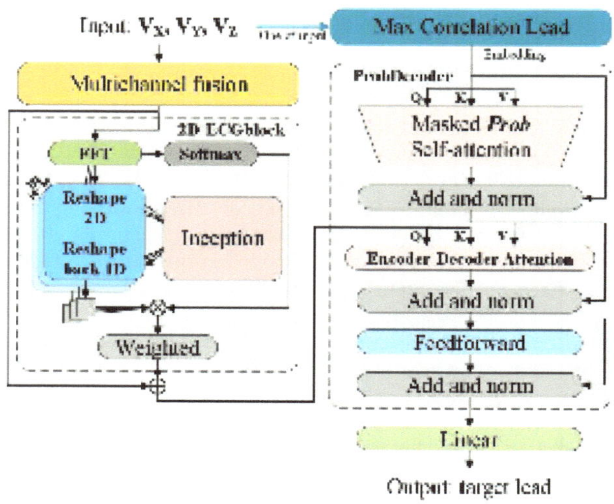

Figure 3. Schematic illustration of the 12-lead ECG reconstruction. Max correlation lead refers to the lead exhibiting the highest correlation with the target lead among the three input leads, as statistically determined based on the training set.

In order to capture between-cycle variations in the ECG sequences, it is essential to first identify their periods. Inspired by the work of Hu et al. [34], we designed the adaptive 2D unfolding module, referred to as the 2D-ECGblock. This method utilizes the Fast Fourier Transform to identify the highest m frequency bands with the highest energy in the ECG sequence, as shown below:

$$A_{f_*} = \mathrm{Amp}\left(\mathrm{FFT}\left(X_{1\mathrm{D}}^{dmodel}\right)\right),\ f_* \in \{1, \cdots, L/2\} \tag{2}$$

$$\{f_1, \cdots, f_m\} = \mathrm{Top}\left(\mathrm{Avg}\left(A_{f_*}\right)\right) \tag{3}$$

$$p_i = \frac{T}{f_i},\ i \in \{1, \cdots, m\} \tag{4}$$

In the above context, FFT(·) represents the Fast Fourier Transform, and Amp(·) is used for the calculation of the amplitude. A_{f_*} denotes the amplitudes calculated for each frequency band, and their mean across the *dmodel* dimensions is obtained through the Avg(·) function. Given the sparsity in the frequency domain, we sought to avoid the noise impact of irrelevant high frequencies; thus, we selected only the top m amplitudes, obtaining the most significant frequency bands $\{f_1, \ldots, f_m\}$ along with their corresponding amplitudes $\left\{A_{f_1}, \cdots, A_{f_m}\right\}$. These selected frequency bands correspond to the durations of k period lengths $\{p_1, \ldots, p_m\}$. Due to the conjugate symmetry in the frequency domain, we only use frequencies within the $\{1, \ldots, L/2\}$ range. Based on the selected period lengths $\{p_1, \ldots, p_m\}$ and frequencies $\{f_1, \ldots, f_m\}$, we can reconstruct the 1D time sequence $X_{1\mathrm{D}} \in \mathbb{R}^{T \times dmodel}$ into a 2D tensor using the following formula:

$$X_{2\mathrm{D}}^{i} = \mathrm{Reshape}_{2\mathrm{D},\ p_i,\ f_i}\left(\mathrm{Padding}\left(X_{1\mathrm{D}}^{dmodel}\right)\right),\ i \in \{1, \cdots, m\} \tag{5}$$

In the above formula, Padding(·) extends the time sequence by padding zeros along the time dimension to evenly divide $X_{1\mathrm{D}}^{dmodel}$ into f_i segments along the time dimension. Next, p_i and f_i represent the number of rows and columns in the resulting 2D tensor after transformation, where each row represents between-cycle variation and each column represents within-cycle variation. $X_{2\mathrm{D}}^{i} \in \mathbb{R}^{p_i \times f_i \times dmodel}$ is the i-th 2D tensor obtained based on frequency f_i. After transformation, an efficient Inception block was applied [35] to process the 2D tensor, denoted as Inception(·). In our implementation of Inception(·), we include 2D convolution kernels of three scales including 1, 3, and 5. The calculation formula is as follows:

$$\widehat{X^i_{2D}} = \text{Inception}\left(X^i_{2D}\right),\ i \in \{1, \cdots, m\} \tag{6}$$

The Inception(·) module here is shared among m layers of X^i_{2D} tensors to improve parameter efficiency.

Finally, we need to transform the $\left\{\widehat{X^1_{2D}}, \cdots, \widehat{X^m_{2D}}\right\}$ back into 1D representations for the next layer and perform information fusion. Inspired by Wu et al. [31], the amplitude of each frequency band reflects its relative importance. Here, we base the fusion on the transformed m 1D tensors after amplitude-based fusion. The formula is as follows:

$$\widehat{X^i_{1D}} = \text{Reshape}_{1D,\, p_i,\, f_i}\left(\widehat{X^i_{2D}}\right),\ i \in \{1, \cdots, m\} \tag{7}$$

$$X^{\text{Eout}}_{1D} = \sum_{i=1}^{m} \widehat{A_{f_i}} \times \widehat{X^i_{2D}},\ \widehat{A_{f_*}} = \text{Softmax}\left(A_{f_1}, \cdots, A_{f_m}\right) \tag{8}$$

Due to the within-cycle and between-cycle dependency information encapsulated in the m highly structured 2D tensors, the 2D-ECGblock can extract multiscale temporal 2D variations through the Inception module. Compared with the original Transformer, which obtains interelement dependencies through attention mechanisms, the 2D-ECGblock enables more efficient representation learning.

The ProbDecoder has two input components. The first part of the input consists of one of the Frank-XYZ leads (V_X, V_Y, and V_Z). In the training dataset, we computed the Pr between the Frank-XYZ leads and the target lead, as shown in Figure 4. When training the corresponding model, the lead from V_X, V_Y, or V_Z with the highest correlation coefficient to the target lead is selected as the input for the ProbDecoder. According to the statistical results in Figure 4, the final correspondence for the ProbDecoder input is as follows: I-X, II-Y, III-Y, aVR-X, aVL-X, aVF-Y, V1-Z, V2-Z, V3-Z, V4-Z, V5-X, and V6-X. From the graph, it can be observed that the leads with the highest correlation are negatively correlated with the standard 12 leads, specifically aVR, V1, V2, V3, and V4. This is because aVR-X, V1-Z, V2-Z, V3-Z, and V4-Z represent vectors located on the opposite side of the heart with opposite polarities.

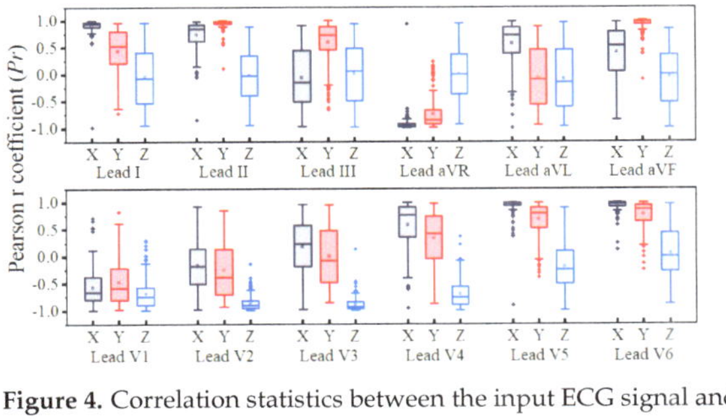

Figure 4. Correlation statistics between the input ECG signal and the target lead.

First, we encode the input X^{Din}_{1D} for the ProbDecoder:

$$Q, K, V = \text{Linear}\left(\text{Conv1d}\left(X^{\text{Din}}_{1D}\right)\right) \tag{9}$$

where Q, K, and V, respectively, represent the query, key, and value matrices in the Transformer, with K being the same size as Q ($L_K = L_Q = L$). Since the input ECG for the ProbDecoder itself is sparse, with a small portion of physiologically significant cardiac signals and a larger portion of baseline signals, we were inspired by Zhou et al. [29] to propose a Probsparse self-attention calculation for the encoded cardiac data. Q only needs to perform dot products with $\ln(L_K)$ key matrices randomly, and the remaining L_K–$\ln(L_K)$ pairs are filled with zeros. The calculation process is as follows:

$$\begin{bmatrix} x_1^1 0 & \cdots & 0 x_1^L \\ \vdots \vdots & \ddots & \vdots \vdots \\ x_L^1 0 & \cdots & 0 x_L^L \end{bmatrix} = \text{Padding}\left(Q \times K_{\ln L_K}\right) \tag{10}$$

$$Q = \begin{bmatrix} q_1^1 & \cdots & q_1^{dmodel} \\ \vdots & \ddots & \vdots \\ q_{L_Q}^1 & \cdots & q_{L_Q}^{dmodel} \end{bmatrix}, K = \begin{bmatrix} k_1^1 & \cdots & k_1^{dmodel} \\ \vdots & \ddots & \vdots \\ k_{\ln L_K}^1 & \cdots & k_{\ln L_K}^{dmodel} \end{bmatrix} \tag{11}$$

In the computed $L \times L$ matrix, only $\ln(L)$ columns have numerical values. Therefore, in the ProbDecoder, self-attention only needs to calculate $O(L \times \ln(L))$ dot products. Max-mean measurements are performed on the computed $L \times L$ matrices:

$$M_h = \max\left\{x_h^1, \cdots, x_h^L\right\} - \frac{1}{L}\sum_{j=1}^{L} x_h^j,\ h \in \{1, \cdots, L\} \tag{12}$$

Next, based on the sorting of $\{M_1, \cdots, M_L\}$, we select the top-u $\left\{x_u^1, \cdots, x_u^{dmodel}\right\}$ vectors of Q to form $\overline{Q}$, where $u = C \times \ln L$.

Here, C is a hyperparameter, and it was chosen as $C = 5$ based on results from [29]. The self-attention matrix computed for the sparse matrix $\overline{Q}$, K, and V, is also sparse, with the remaining rows of the V matrix filled with the mean of that row. This approach helps to emphasize the importance of the positions, where the ECG waveforms are located while reducing the model's focus on baseline waveforms. The final Probsparse self-attention matrix still has a size of $L \times L$, which is calculated as follows:

$$Attention = \left\{\text{Softmax}\left(\frac{\overline{Q}K^T}{\sqrt{dmodel}}\right)V, \text{Mean}(V)\right\} \tag{13}$$

As mentioned above, periodic variations are extracted from the V_X, V_Y, and V_Z three-lead ECG signals through the 2D-ECGblock. Based on this information, we perform attention mechanism calculations in the Encoder–Decoder layer and correct central ECG waveforms in the value matrix, achieving reconstruction of the target lead electrocardiogram in a residual-like manner. Therefore, based on this attention calculation, a new value matrix $\overline{V}$ is computed as follows:

$$\widehat{V} = \text{Norm}\left(Attention + X_{1D}^{\text{Din}}\right) \tag{14}$$

$$\widehat{Q}, \widehat{K} = \text{Linear}\left(\text{Conv1d}\left(X_{1D}^{\text{Eout}}\right)\right) \tag{15}$$

The second part of ProbDecoder's input is the output X_{1D}^{Eout} from the 2D-ECGblock. After a linear transformation, new query $\overline{Q}$ and key $\overline{K}$ matrices are obtained. The ECG waveform correction is performed in the Encoder–Decoder Attention layer, and after passing through a feedforward layer and a linear layer, the target lead ECG is obtained as follows:

$$\text{Target} = \text{Linear}\left(\text{Feed}\left(\text{Softmax}\left(\frac{\widehat{Q}\widehat{K}^T}{\sqrt{dmodel}}\right)\widehat{V}\right)\right) \tag{16}$$

In order to ensure that the training process of each lead ECG reconstruction network does not interfere with each other, we trained 12 separate M2Eformer models, each dedicated to reconstructing the corresponding lead ECG signal.

2.3. Evaluation

Based on previous research, we used the Pearson coefficient r (Pr) and the mean value of absolute error value (MAE) to quantify the differences between the predicted ECG leads and the recorded leads. Pr measures the degree of linear correlation between two sets of data, variables X and Y, and is calculated as follows:

$$Pr = \frac{\sum_{i=1}^{n}(X_i - \overline{X})(Y_i - \overline{Y})}{\sqrt{\sum_{i=1}^{n}(X_i - \overline{X})^2}\sqrt{\sum_{i=1}^{n}(Y_i - \overline{Y})^2}} \tag{17}$$

where X represents the reconstructed ECG leads, Y represents the recorded ECG leads, and n denotes the duration of each record.

MAE provides a better reflection of the actual amplitude error in the reconstructed ECG, with smaller values indicating greater reconstruction precision. It is calculated as follows:

$$MAE = \frac{1}{n}\sum_{i=1}^{n}|X_i - Y_i| \tag{18}$$

3. Results

3.1. A Comparison of Training Results

The proposed M2Eformer model employs an attention-based Transformer architecture. It utilizes a single layer of the 2D-ECGblock module as the Encoder and a single layer of the ProbDecoder module as the Decoder. The embedding dimension was set to 512. We initially conducted grid search experiments on the PTB-DN validation set with epochs of 100 and 200, and learning rates of 0.001, 0.0001, and 0.00001. An epoch of 100 and a learning rate of 0.0001 were selected, taking into account both training speed and reconstruction performance. Subsequent experiments were conducted on the validation set to evaluate various configurations, including the number of layers in the 2D-ECGblock module (0, 1, and 2 layers) and the ProbDecoder module (1, 2, and 4 layers), as well as different embedding dimensions (64, 256, and 512). Based on these tests, we ultimately selected a setup with one layer for the 2D-ECGblock module, one layer for the ProbDecoder module, and an embedding dimension of 512. The batch size for training was set to 200, determined by the GPU's memory capacity of 24 G. To prevent model overfitting and enhance the generalization capability of the training model, we employed the Dropout function as the regularization method, with the dropout rate set to 0.1. The training process utilized the Adam optimizer and the MSE (Mean Squared Error) Loss function [30–32,34].

The loss curves on the validation set are depicted in Figure 5, where gray represents the original Transformer, blue represents the T-Transformer, which embeds the 2D-ECGblock into the Transformer while keeping the Decoder unchanged, and red represents the proposed M2Eformer.

From Figure 5, we can observe that in leads aVR, aVL, V1, and V4, the proposed M2Eformer achieves a lower validation loss in the validation set, significantly outperforming both the Transformer and T-Transformer. With other leads, the convergence results are relatively close. The minimum validation loss values and their corresponding best epochs for each of the three models are listed in Table 3. As indicated in Table 3, the proposed M2Eformer has a slightly higher validation loss on lead V5 compared with the Transformer (0.0001) but achieves better or consistent results on the remaining leads. Moreover, Figure 5 demonstrates that M2Eformer does not exhibit a noticeable overfitting phenomenon in all leads despite its slower convergence compared with the other two frameworks.

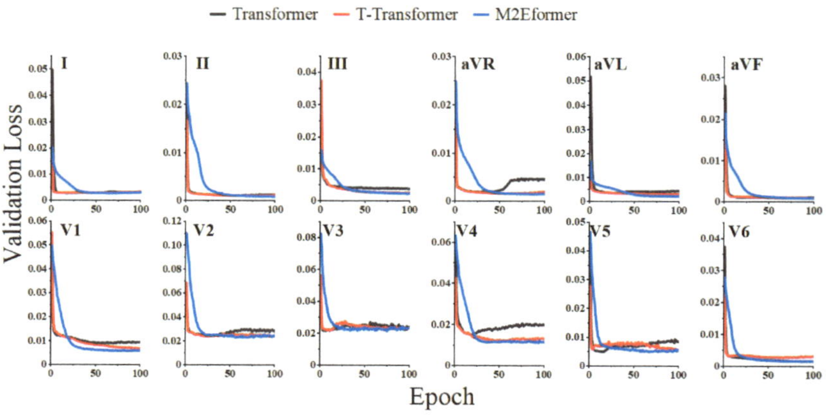

Figure 5. Validation loss of M2Eformer (blue), T-Transformer (red), and Transformer (gray).

Table 3. Best epoch and minimal validation loss for M2Eformer, T-Transformer, and Transformer.

Model	Best Epoch/Min Loss							
	I	II	III	aVF	V2	V3	V5	V6
M2Eformer	64/0.0027	98/0.0009	97/0.0022	98/0.0008	55/0.0229	74/0.0215	66/0.0050	85/0.0016
T-Transformer	12/0.0028	70/0.0010	91/0.0024	95/0.0009	93/0.0236	10/0.0219	94/0.0054	66/0.0027
Transformer	18/0.0027	51/0.0010	97/0.0038	70/0.0009	28/0.0243	10/0.0215	15/0.0049	89/0.0016

In Figure 5, when comparing the validation loss between the Transformer and the T-Transformer, we notice that the Transformer exhibits a more pronounced overfitting issue (especially in leads aVR, V2, and V4). The cause of overfitting may be attributed to the attention mechanism in the encoder failing to capture reliable temporal dependencies within the signal [31]. Our parameter analysis revealed that the total number of parameters in the Transformer (4.2 million) is smaller than that in the T-Transformer (13.4 million). This result suggests that the phenomenon of overfitting is not caused by excessively large model parameters, further confirming the effectiveness of the 2D-ECGblock in extracting hidden cardiac information.

3.2. ECG Reconstruction Effect Comparison

To validate the performance of M2Eformer, we compared it with various algorithms using two key metrics including *Pr* and *MAE*. The algorithms compared included Transformer, T-Transformer, as well as algorithms mentioned in previous studies, such as Linear transformation [16,22,23] and least squares regression (LSR) [25]. We also included the commonly used Long Short-Term Memory (LSTM) network for comparison in time series tasks [36–38].

The results demonstrate that several methods used in the experiments can reconstruct the 12-lead ECG, with superior overall performance achieved by deep learning-based approaches. Tables 4 and 5 present the *Pr* and *MAE* between the reconstructed ECG and the recorded ECG obtained using these six algorithms in the test dataset, where the ratio of training, validation, and testing was set at 3:1:1.

Table 4 reveals that the proposed M2Eformer exhibits the best overall reconstruction performance for the 12-lead ECG (total *Pr* = 0.8785), followed by the T-Transformer (total *Pr* = 0.8579). M2Eformer surpasses the Transformer in performance across leads II-V1 and V3-V6 for each lead, underscoring the efficacy of the 2D-ECGblock.

Table 4. Pearson's correlation r (*Pr*) in test dataset.

Model	I	II	III	aVR	aVL	aVF	V1	V2	V3	V4	V5	V6	Total
M2Eformer (ours)	0.8465	0.9588	0.7817	0.8921	0.8447	0.8321	0.9105	0.8930	0.9420	0.8641	0.8554	0.9215	0.8785
M2Eformer (ours)	0.8441	0.9207	0.7496	0.8534	0.7875	0.8647	0.8932	0.9052	0.9220	0.8409	0.8326	0.8814	0.8579
Transformer	0.8568	0.9063	0.6705	0.7983	0.7036	0.8748	0.8865	0.9133	0.9217	0.7883	0.7895	0.9004	0.8342
LSTM	0.6573	0.6357	0.4625	0.6181	0.6273	0.6366	0.7009	0.5212	0.5600	0.5341	0.5612	0.6926	0.6006
LSR	0.8507	0.8266	0.5891	0.9224	0.6456	0.6741	0.9015	0.8835	0.9516	0.8552	0.8594	0.9621	0.8268
Linear	0.8087	0.9485	0.6381	0.8642	0.6155	0.8534	0.9068	0.8795	0.9440	0.7759	0.6735	0.9220	0.8192

Table 5. Mean absolute error (*MAE*) in test dataset.

Model	I	II	III	aVR	aVL	aVF	V1	V2	V3	V4	V5	V6	Total
M2Eformer (ours)	0.0399	0.0215	0.0395	0.0266	0.0370	0.0241	0.0619	0.1044	0.0815	0.0730	0.0629	0.0406	0.0511
M2Eformer (ours)	0.0401	0.0260	0.0368	0.0316	0.0465	0.0252	0.0632	0.0972	0.0879	0.0827	0.0688	0.0465	0.0544
Transformer	0.0380	0.0265	0.0430	0.0343	0.0474	0.0241	0.0638	0.1003	0.0933	0.1045	0.0635	0.0457	0.0570
LSTM	0.0479	0.0576	0.0507	0.0488	0.0373	0.0464	0.0789	0.1410	0.1403	0.1147	0.0853	0.0620	0.0759
LSR	0.0352	0.0422	0.0491	0.0234	0.0401	0.0487	0.0517	0.0980	0.0655	0.0756	0.0570	0.0311	0.0515
Linear	0.0516	0.0212	0.0470	0.0322	0.0478	0.0257	0.0432	0.0753	0.0582	0.0954	0.0997	0.0358	0.0528

Although LSR achieves the highest *Pr* in a few leads (aVR, V3, V5, V6), the average correlation coefficient in leads III, aVL, and aVF is less than 0.7. This finding indicates that the ECG reconstructed by LSR in leads III, aVL, and aVF deviates significantly from the recorded ECG (as shown in Figure 6), especially in lead III, where the amplitude difference in the S wave reaches 1mV. This divergence could potentially lead to a misdiagnosis (e.g., patients with reduced ECG amplitudes suggestive of myocardial injury in cases of coronary artery disease).

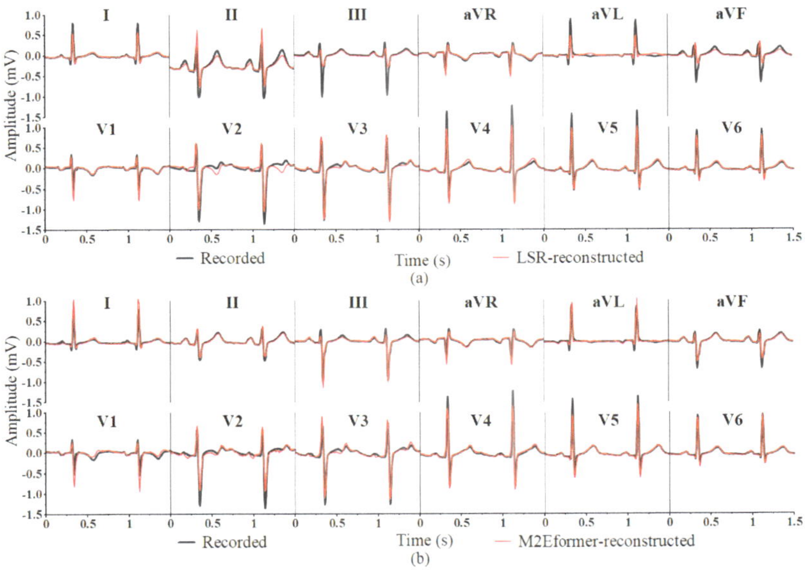

Figure 6. Comparison of the 12-lead ECG reconstructed by (**a**) LSR (**above**) and (**b**) M2Eformer (**below**) in test dataset.

The *MAE* can characterize the actual errors in the predicted values, with smaller values indicating a smaller amplitude difference between the reconstructed ECG and the recorded ECG. In Table 5, M2Eformer only exhibits the lowest *MAE* in a few leads (aVL, aVF, V4). However, its overall *MAE* to reconstruct the 12-lead ECG is the lowest (total *MAE* = 0.0511). Although the Linear method achieves the lowest *MAE* in leads II, V1-V3, its performance in terms of *Pr* in Table 4 is not outstanding. This is because the Linear method better fits the waveforms with larger amplitudes (Q and S waves) in these four leads. As depicted in Figure 7, the amplitudes of the R and S waves reconstructed in leads III and aVL for Linear assessment differed significantly. This result can also result in a misdiagnosis by cardiologists (e.g., diagnosing coronary artery disease as myocardial injury).

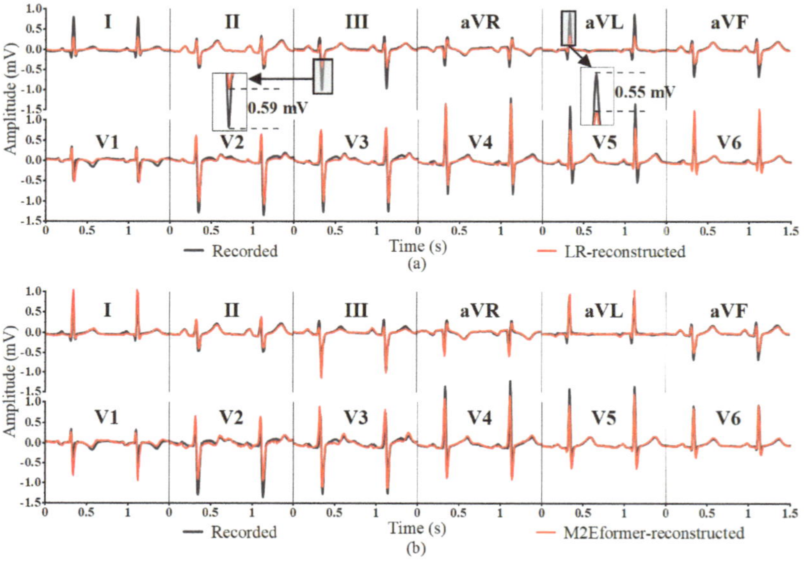

Figure 7. Comparison of 12-lead ECG reconstructed by (**a**) Linear (**above**) and (**b**) M2Eformer (**below**) in the test dataset. (The same segment of the ECG is shown in Figure 6.)

In general, considering the results in Tables 4 and 5, M2Eformer achieves the best overall performance in reconstructing the 12-lead ECG, and the T-Transformer shows improvement compared with the original Transformer. This result demonstrates that the 2D-ECGblock meets our expectations for effectively extracting ECG information.

Figure 8 presents box plots of *Pr* in the test set for the Transformer, the T-Transformer, and M2Eformer. The mean *Pr* for each lead in Table 4 is also represented in the figure as squares (□). In Figure 8, M2Eformer shows a more concentrated *Pr* distribution in most leads (II, III, aVR, aVL, and V3-V6), with higher mean and median values. This result indicates that the ECG reconstructed by M2Eformer shows more consistent waveform changes (synchronously rising and falling) with the recorded ECG.

The Transformer performs better in lead I, but compared with the other two algorithms, it does not show statistically significant differences at a confidence level of $p = 0.05$. The Transformer only shows statistically significant superiority (higher mean) in leads aVF and V2. By comparing the Transformer and M2Eformer training processes (Figure 5 and Table 3), we observe that M2Eformer achieves a lower loss in the validation set and does not exhibit overfitting. We believe that this might be due to the limited size of the validation dataset, which may not fully reflect the real training process.

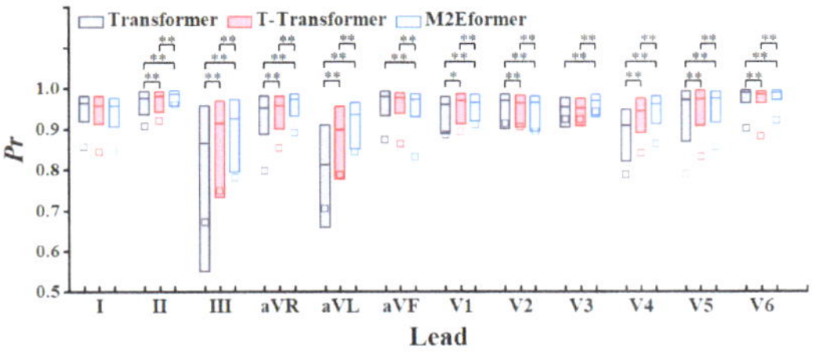

Figure 8. Performance of M2Eformer compared to the Transformer and T-Transformer as correlation coefficient. *: $p < 0.05$, **: $p < 0.01$. The box's upper edge is the upper quartile; the box's lower edge is the lower quartile; the box's middle line is the median value; and the square (□) is the mean value.

In summary, Figure 8 further demonstrates the superior performance of M2Eformer.

The performance of M2Eformer in reconstructing the ECG is shown in Figure 9. This segment (1.25 s) of the ECG data was collected from a patient with a myocardial infarction. The red line represents the reconstructed ECG, while the black line represents the recorded ECG. We presented a 10 s segment containing this ECG snippet to a cardiac specialist for diagnosis. The diagnosis based on the reconstructed ECG (red line) indicates "old anterior myocardial infarction (V1-3 leads exhibit QS morphology)" and "lateral myocardial ischemia (ST-segment depression in leads I, V5-6)". The diagnosis based on the recorded ECG (black line) is "anterior myocardial injury (poor R-wave progression in V1-3 leads)" and "lateral myocardial ischemia (ST-segment depression in leads I, V5-6)". Among them, "old anterior myocardial infarction" and "anterior myocardial injury" correspond to the same cardiac injury, but the expression is different. In the reconstructed ECG, there is a noticeable discrepancy in the 0.5–1 s region of leads V2 and V3 compared with the recorded ECG. However, these differences are mainly in terms of amplitude, with their waveforms being nearly synchronous, indicating that M2Eformer captures the periodic variations in the ECG signal and reflects them in the output. However, there is room for improvement in M2Eformer in terms of the extraction and representation of waveform amplitude information.

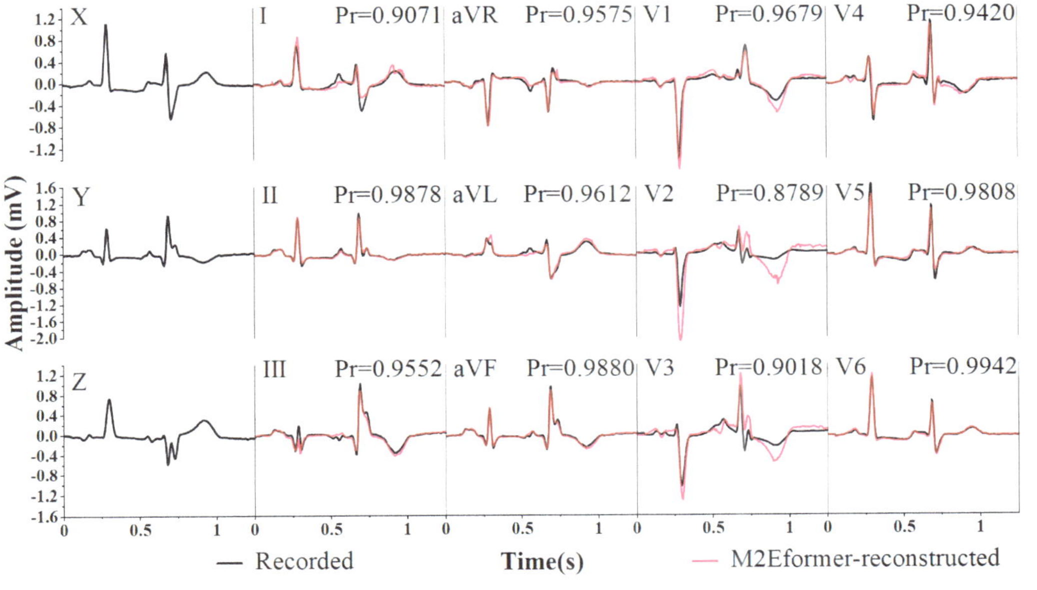

Figure 9. Example of reconstructed 12-lead ECG with Frank-XYZ. Reconstructed (red) and recorded (black) ECG. The test data were collected from MI individuals.

In the test dataset, we performed an analysis of the consistency between the diagnostic results of the reconstructed ECG and the recorded ECG, as shown in Table 6. We selected 10 s ECG segments that demonstrated the highest average Pr between the reconstructed and recorded 12-lead ECG on that record, resulting in 61 segments of reconstructed ECG and 61 segments of recorded ECG. Diagnostics were performed using a double-blind method. When the diagnostic results of the recorded ECG and the reconstructed ECG for the same segment were consistent with the cardiologists, we considered the reconstructed ECG to have no impact on clinical diagnosis. Overall agreement (OvA) is a method used to assess the consistency in diagnoses among three experts. For example, in the case of the reconstructed and recorded ECG segments, if two experts arrive at the same diagnostic conclusion, the segment is considered to have consistent OvA, even if the third expert's diagnosis diverges.

Table 6. Diagnostic results from the cardiologists in the test dataset.

Data	Cardiologist 1		Cardiologist 2		Cardiologist 3		Overall Agreement (OvA)	
	CS/AS	PoC	CS/AS	PoC	CS/AS	PoC	CS/AS	PoC
Healthy controls	9/11	81.8%	9/11	81.8%	10/11	90.9%	9/11	81.8%
Myocardial infarction	41/43	95.3%	41/43	95.3%	39/43	90.7%	43/43	100%
Dysrhythmia	2/2	100%	2/2	100%	2/2	100%	2/2	100%
Bundle branch block	2/2	100%	2/2	100%	2/2	100%	2/2	100%
Myocardial hypertrophy	1/1	100%	1/1	100%	1/1	100%	1/1	100%
Valvular heart disease	1/1	100%	1/1	100%	0/1	0%	1/1	100%
Cardiomyopathy	1/1	100%	1/1	100%	1/1	100%	1/1	100%
Total	57/61	93.4%	57/61	93.4%	55/61	90.2%	59/61	96.7%

AS: all sample; CS: consistent sample; PoC, percentage of consistency.

As shown in Table 6, the percentage of consistency for cardiologist 1 was 93.4%, for cardiologist 2 was 93.4%, and for cardiologist 3 was 90.2%. We calculated the overall consistency among the three experts, which reached 96.7%, with only two cases of inconsistency among the diagnoses of healthy individuals; the reasons for these inconsistencies are examined in the Section 4.

3.3. EASI Leads to 12-Lead ECG

In order to further validate the reliability of the 12-lead ECG reconstruction through the EASI lead configuration for monitoring purposes, we conducted simultaneous data collection of EASI leads and standard 12-lead ECG. Furthermore, to comply with clinical requirements, we selected patients with various cardiac conditions that require a combined 12-lead diagnosis. Ultimately, we obtained effective ECG data from 10 healthy individuals and 15 patients, including those with atrial fibrillation, atrial flutter, coronary artery disease, and myocardial infarction. Moreover, due to the limited sample size, we employed 5-fold cross-validation for our analysis. The hyperparameters of the M2Eformer model (epoch = 100, learning rate = 0.00001, batch size = 200) remained unchanged.

The experimental results are presented in Figure 10, where (a) shows the histogram distribution of Pr between the reconstructed ECG and recorded ECG, (b) displays the boxplot distribution of Pr and MAE between the reconstructed ECG and recorded ECG, and (c) illustrates the consistency results of the annotations by cardiac experts for the reconstructed ECG and recorded ECG.

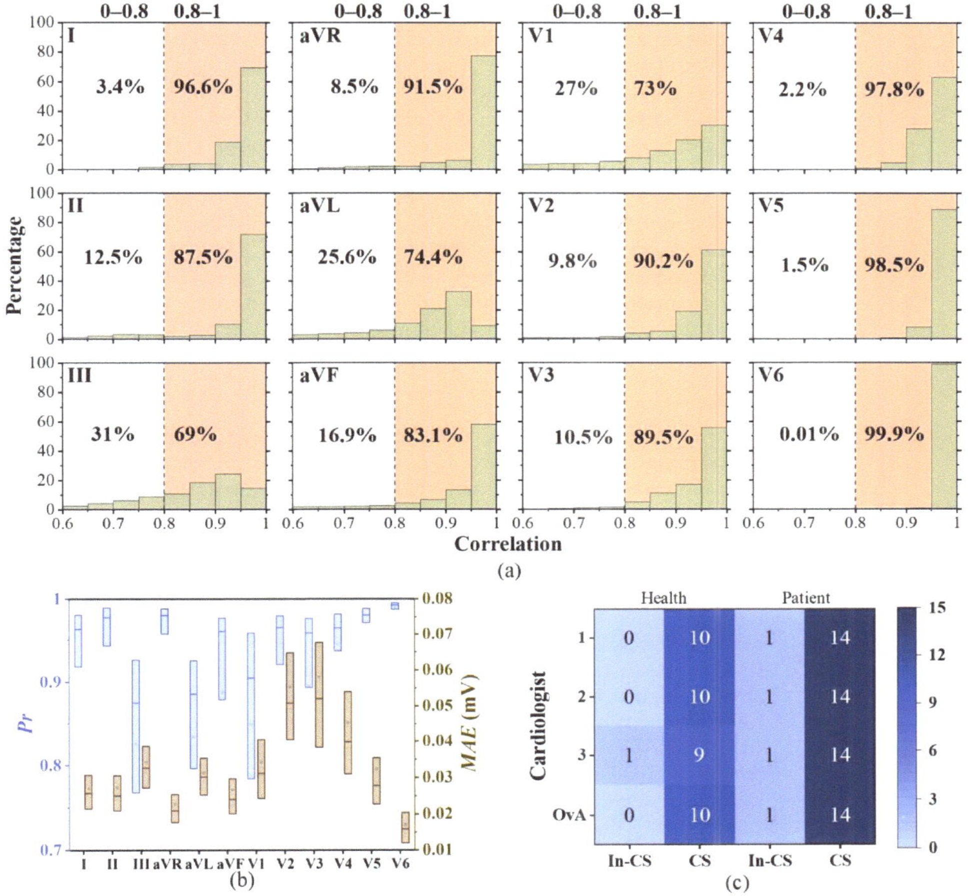

Figure 10. Comparison of the 12-lead reconstructed ECG with EASI vs. the recorded signal. (**a**) The histogram distribution of Pr; (**b**) median (interquartile range) of Pr and MAE; and (**c**) expert labeling results. (In-CS: inconsistent, CS: consistent, OvA: overall agreement). OvA is a method used to analyze the consistency in diagnostic outcomes among three experts as a majority voting mechanism.

Figure 10a reveals that in more than half of the leads (lead I, aVR, V2, V4, V5, and V6), the proportion of *Pr* greater than 0.8 exceeds 90%. Among the remaining leads, in leads II, aVF, and V3, more than 80% of the *Pr* values are greater than 0.8, while in leads III, aVL, and V1, the proportion of *Pr* values exceeding 0.8 is around 70%. Combining Figure 10a,b, we observe that in more than half of the leads (lead I, II, aVR, aVF, and V2–V6), the median Pr exceeds 0.9, with even lead V1 having a median *Pr* of 0.9044. Although Figure 10b shows that the median *Pr* values for leads III and aVL are below 0.9, their median *MAE* values are 0.0326 and 0.0302 mV, indicating small differences in amplitude.

We engaged three cardiologists to annotate the reconstructed ECG and the recorded ECG for a macro evaluation. We selected 10 s ECG segments for each record, comprising 25 segments of reconstructed ECG and 25 segments of recorded ECG. The diagnostic results are shown in Figure 10c, with individual diagnosis consistency rates of 96%, 96%, and 92% for the three experts. Notably, among them, inconsistent samples from cardiologist 2 and cardiologist 3 are interlaced. Importantly, the samples identified as In-CS by cardiologists 1 and 2 are identical, indicating that the OvA classification for this particular sample is marked as In-CS. Conversely, the samples identified as In-CS by cardiologist 3 do not overlap with those deemed In-CS by cardiologists 1 and 2, thereby not impacting the final OvA analysis. Therefore, the OvA among the diagnostic results from three cardiac experts achieved 96% (24/25).

In Figure 10c, for the only sample with In-CS OvA outcomes, the reconstructed ECG interpretations varied as follows: healthy individuals with variants or old anterior interwall myocardial infarction (cardiologist 1), coronary heart disease (cardiologist 2), and healthy (cardiologist 3). Conversely, the recorded ECG was unanimously classified as healthy by all three cardiologists. This ECG was obtained from a patient who had returned to sinus rhythm following ablation for atrial flutter. Therefore, in this study, there is a certain discrepancy between the reconstructed and recorded 12-lead ECG. Nonetheless, the high consistency observed in the one-versus-all (OvA) outcomes (96%) underscores the substantial adjunctive value of EASI-reconstructed 12-lead ECGs in the clinical diagnosis of atrial fibrillation, atrial flutter, and coronary artery disease.

4. Discussion

This study is the first of its kind to propose a deep learning-based ECG reconstruction network that reconstructs 12-lead ECG from EASI leads, enabling EASI leads to help diagnose a wider range of cardiac diseases. In this study, the designed novel ECG reconstruction network involves the following key components: 1. the 2D-ECGblock, which simultaneously extracts within-cycle and between-cycle dependencies from input ECG, and 2. the ProbDecoder, which is a carefully designed generation component using Probsparse self-attention mechanisms to achieve residual-like ECG reconstruction. Furthermore, we conducted clinical diagnostic validation of the reconstructed 12-lead ECG on our self-established database. The diagnostic results of the cardiologists indicate that the EASI-reconstructed 12-lead ECG has the potential to assist in the diagnosis of atrial flutter, atrial fibrillation, coronary artery disease, and myocardial infarction. Conversely, the use of EASI leads in isolation offers minimal assistance in the diagnosis of these four conditions.

Linear regression (Linear) [15–18] and least square regression (LSR) [15,25,26] are commonly used methods for the reconstruction of 12-lead ECG. Attention-based deep learning networks have achieved promising results in time-series prediction tasks [29–32]. Consequently, this study presents M2Eformer, a novel attention-based model for 12-lead ECG reconstruction, and conducts a comprehensive performance comparison with traditional methods, including Linear and LSR, widely utilized in prior research. The results in Tables 4 and 5 and Figures 6 and 7 demonstrate that the proposed M2Eformer outperforms LSR and linear methods in the overall performance of Frank-XYZ reconstruction of 12 leads on the PTB-DN database. In Table 4, M2Eformer performs best in the reconstruction of the ECG for most leads, although its performance is slightly lower than that of LSR in the reconstruction of the aVR, V3, V5, and V6 leads. In future studies, the complexity of the parameters required for each lead's reconstruction can be explored. Additionally, the dataset can be further expanded to optimize the M2Eformer model and enhance the reconstruction performance of each lead model.

In Table 6, there were differing opinions among the three cardiologists regarding the annotations for two healthy individuals. For the first inconsistent sample, the cardiologists provided different annotations for the reconstructed ECG including bundle branch block (cardiologist 1) and healthy (cardiologists 2 and 3). However, their annotations for the recorded ECG were myocardial infarction (cardiologist 1), incomplete right bundle branch block (cardiologist 2), and healthy (cardiologist 3). According to the PTB-DN database records, this ECG was collected from a healthy individual. The inconsistency in the diagnostic results for this sample is primarily attributed to cardiologist 2's interpretation of the recorded ECG as RBBB, due to a significant error in the diagnosis by cardiologist 1 for this sample. In the second inconsistent sample, the cardiologists provided different annotations for the reconstructed ECG including possible high lateral myocardial ischemic injury (cardiologist 1), myocardial ischemia (cardiologist 2), and possible myocardial injury (cardiologist 3). However, their annotations for the recorded ECG were possible myocardial ischemic injury (cardiologist 1) and healthy (cardiologists 2 and 3). According to the database records of the PTB-DN, this ECG was also obtained from a healthy individual. Based on the comprehensive annotations of the three experts, the reconstructed ECG was annotated as "myocardial injury" (indicating myocardial infarction), while the recorded ECG was annotated as "healthy". The diagnostic inconsistencies observed in this sample may stem from the data imbalance within the PTB-DN database, characterized by a discrepancy between myocardial infarction cases (213) and healthy individuals (57). This imbalance, favoring myocardial infarction instances, might have led to the inadvertent integration of myocardial infarction-related features into the reconstructed ECG, culminating in erroneous annotations.

The primary limitation of the proposed model is that the loss function employs a generic calculation method and does not adjust specifically for abnormal ECG waveforms, such as incorporating the error between the R waves of the reconstructed and recorded ECG into the training loss for backpropagation. Another limitation of this study is the relatively small sample size of our self-constructed database, and the issue of data balance needs further resolution. In our subsequent work, we will address this issue by collecting a more diverse range of clinical data.

5. Conclusions

This paper explores the clinical diagnostic value of using 12-lead reconstructed ECG through EASI leads for wearable ECG monitoring. A novel network architecture designed for ECG reconstruction, called M2Eformer, is proposed. This model utilizes the 2D-ECGblock to synchronously extract information regarding within-cycle and between-cycle dependencies. Information fusion is achieved through a specially designed ProbDecoder, enabling the reconstruction of a 12-lead ECG. The experimental results demonstrate that M2Eformer achieves the best overall reconstruction performance for 12 leads ($Pr = 0.8785$ and $MAE = 0.0511$), with Pr values higher than traditional methods such as the LSR and Linear methods (0.0517 and 0.0593, respectively). Expert annotations obtained from the recorded data (overall consistency of 96%) suggest the potential value of the reconstructed 12-lead ECG in aiding the clinical diagnosis of conditions such as atrial flutter, atrial fibrillation, coronary artery disease, and myocardial infarction.

Author Contributions: Conceptualization, F.F. and D.Z.; methodology, F.F., D.Z. and S.Z.; software, F.F.; validation, W.W. and S.Z; resources, W.W., S.Z. and J.L. (Jiamin Liu); data curation, F.F. and Q.S.; writing—original draft preparation, F.F. and D.Z.; writing—review and editing, S.Z., W.W. and J.L. (Jianqing Li); visualization, T.X.; supervision, J.L. (Jiamin Liu); project administration, J.L. (Jianqing Li) and S.Z.; funding acquisition, J.L. (Jianqing Li). All authors have read and agreed to the published version of the manuscript.

Funding: This research was funded by the National Key Research and Development Program of China (2022YFC2405600), NSFC (62071241, 81871444, 62075098, 62001240), the Leading-edge Technology and Basic Research Program of Jiangsu (BK20192004D), the Key Research and Development Program of Jiangsu (BE2022160), and the Postgraduate Research and Practice Innovation Program of Jiangsu Province (KYCX22_1779).

Institutional Review Board Statement: This study was conducted in accordance with the Declaration of Helsinki and approved by the Nanjing Medical University Ethics Committee (No.112, 2022).

Informed Consent Statement: Informed consent was obtained from all subjects involved in this study.

Data Availability Statement: The raw data are available on request. For any requests, please contact Songsheng Zhu.

Acknowledgments: We would like to thank Lei Guo for her help in the data collection process.

Conflicts of Interest: The authors declare no conflicts of interest.

References

1. Benjamin, E.J.; Muntner, P.; Alonso, A.; Bittencourt, M.S.; Callaway, C.W.; Carson, A.P.; Chamberlain, A.M.; Chang, A.R.; Cheng, S.; Das, S.R. Heart disease and stroke statistics—2019 update: A report from the American Heart Association. *Circulation* **2019**, *139*, e56–e528. [CrossRef] [PubMed]
2. Liu, C.; Yang, M.; Di, J.; Xing, Y.; Li, Y.; Li, J. Wearable ECG: History, key technologies and future challenges. *Chin. J. Biomed. Eng.* **2019**, *38*, 641–652.
3. Zhang, X.; Jiang, M.; Polat, K.; Alhudhaif, A.; Hemanth, J.; Wu, W. Detection of Atrial Fibrillation from Variable-Duration ECG Signal Based on Time-Adaptive Densely Network and Feature Enhancement Strategy. *IEEE J. Biomed. Health Inform.* **2022**, *27*, 944–955. [CrossRef] [PubMed]
4. Holter, N.J. New method for heart studies: Continuous electrocardiography of active subjects over long periods is now practical. *Science* **1961**, *134*, 1214–1220. [CrossRef]
5. Rajanna, R.R.; Sriraam, N.; Prakash, V.; Vittala, P.R.; Arun, U.; Sahoo, S. External cardiac loop recorders: Functionalities, diagnostic efficacy, challenges and opportunities. *IEEE Rev. Biomed. Eng.* **2021**, *15*, 273–292. [CrossRef]
6. Liu, H.; Gao, T.; Liu, Z.; Shu, M. FGSQA-Net: A Weakly Supervised Approach to Fine-grained Electrocardiogram Signal Quality Assessment. *IEEE J. Biomed. Health Inform.* **2023**, *27*, 3844–3855. [CrossRef]
7. Gladstone, D.J.; Wachter, R.; Schmalstieg-Bahr, K.; Quinn, F.R.; Hummers, E.; Ivers, N.; Marsden, T.; Thornton, A.; Djuric, A.; Suerbaum, J. Screening for atrial fibrillation in the older population: A randomized clinical trial. *JAMA Cardiol.* **2021**, *6*, 558–567. [CrossRef]
8. Perez, M.V.; Mahaffey, K.W.; Hedlin, H.; Rumsfeld, J.S.; Garcia, A.; Ferris, T.; Balasubramanian, V.; Russo, A.M.; Rajmane, A.; Cheung, L. Large-scale assessment of a smartwatch to identify atrial fibrillation. *N. Engl. J. Med.* **2019**, *381*, 1909–1917. [CrossRef] [PubMed]
9. Teferra, M.N.; Hobbs, D.A.; Clark, R.A.; Reynolds, K.J. Electronic-textile 12-lead equivalent diagnostic electrocardiogram based on the EASI lead placement. *IEEE Sens.* **2022**, *22*, 5994–6001. [CrossRef]
10. Fang, B.; Chen, J.; Liu, Y.; Wang, W.; Wang, K.; Singh, A.K.; Lv, Z. Dual-channel neural network for atrial fibrillation detection from a single lead ECG wave. *IEEE J. Biomed. Health Inform.* **2021**, *27*, 2296–2305. [CrossRef] [PubMed]
11. Rajakariar, K.; Koshy, A.N.; Sajeev, J.K.; Nair, S.; Roberts, L.; Teh, A.W. Accuracy of a smartwatch based single-lead electrocardiogram device in detection of atrial fibrillation. *Heart* **2020**, *106*, 665–670. [CrossRef]
12. Himmelreich, J.C.; Karregat, E.P.; Lucassen, W.A.; van Weert, H.C.; de Groot, J.R.; Handoko, M.L.; Nijveldt, R.; Harskamp, R.E. Diagnostic accuracy of a smartphone-operated, single-lead electrocardiography device for detection of rhythm and conduction abnormalities in primary care. *Ann. Fam. Med.* **2019**, *17*, 403–411. [CrossRef] [PubMed]
13. Frank, E. An accurate, clinically practical system for spatial vectorcardiography. *Circulation* **1956**, *13*, 737–749. [CrossRef] [PubMed]
14. Frank, E. General theory of heart-vector projection. *Circ. Res.* **1954**, *2*, 258–270. [CrossRef]
15. Levkov, C.L. Orthogonal electrocardiogram derived from the limb and chest electrodes of the conventional 12-lead system. *Med. Biol. Eng. Comput.* **1987**, *25*, 155–164. [CrossRef] [PubMed]
16. Dower, G.E.; Yakush, A.; Nazzal, S.B.; Jutzy, R.V.; Ruiz, C.E. Deriving the 12-lead electrocardiogram from four (EASI) electrodes. *J. Electrocardiol.* **1988**, *21*, S182–S187. [CrossRef] [PubMed]
17. Nelwan, S.P.; Kors, J.A.; Meij, S.H.; van Bemmel, J.H.; Simoons, M.L. Reconstruction of the 12-lead electrocardiogram from reduced lead sets. *J. Electrocardiol.* **2004**, *37*, 11–18. [CrossRef] [PubMed]
18. Feild, D.Q.; Feldman, C.L.; Hor, B.M. Improved EASI coefficients: Their derivation, values, and performance. *J. Electrocardiol.* **2002**, *35*, 23–33. [CrossRef]
19. Li, S.F.; Jing, Y.; Guo, S.; Yu, A.X.; Zhu, T.; Li, Z.J. A comparative analysis of Cabrera and Wilson lead electrocardiogram in normal subjects. *Clin. Meta-Anal.* **2010**, *25*, 1149–1150.
20. Drew, B.J.; Pelter, M.M.; Wung, S.-F.; Adams, M.G.; Taylor, C.; Evans, G.T., Jr.; Foster, E. Accuracy of the EASI 12-lead electrocardiogram compared to the standard 12-lead electrocardiogram for diagnosing multiple cardiac abnormalities. *J. Electrocardiol.* **1999**, *32*, 38–47. [CrossRef]
21. Nelwan, S.P.; Kors, J.A.; Crater, S.W.; Meij, S.H.; van Dam, T.B.; Simoons, M.L.; Krucoff, M.W. Simultaneous comparison of 3 derived 12-lead electrocardiograms with standard electrocardiogram at rest and during percutaneous coronary occlusion. *J. Electrocardiol.* **2008**, *41*, 230–237. [CrossRef]
22. Schreck, D.; Brotea, C.; Shah, S. Derivation of the 12-lead electrocardiogram using abstract factor analysis and simplex optimization. *Int. J. Bioelectromagn.* **2002**, *4*, 337–338.
23. Schreck, D.M.; Fishberg, R.D. Derivation of the 12-lead electrocardiogram and 3-lead vectorcardiogram. *Am. J. Emerg. Med.* **2013**, *31*, 1183–1190. [CrossRef] [PubMed]
24. Feild, D.Q.; Zhou, S.H.; Helfenbein, E.D.; Gregg, R.E.; Lindauer, J.M. Technical challenges and future directions in lead reconstruction for reduced-lead systems. *J. Electrocardiol.* **2008**, *41*, 466–473. [CrossRef]
25. Trobec, R.; Tomašić, I. Synthesis of the 12-lead electrocardiogram from differential leads. *IEEE Trans. Inf. Technol. Biomed.* **2011**, *15*, 615–621. [CrossRef] [PubMed]
26. Mulyadi, I.H.; Supriyanto, E. Segment-specific EASI coefficients for improving accuracy of derived 12-lead electrocardiography. In Proceedings of the 2018 International Conference on Applied Engineering, Batam, Indonesia, 3–4 October 2018; pp. 1–4.

27. Atoui, H.; Fayn, J.; Rubel, P. A neural network approach for patient-specific 12-lead ECG synthesis in patient monitoring environments. In Proceedings of the Computers in Cardiology, Chicago, IL, USA, 19–22 September 2004; pp. 161–164.
28. Smith, G.H.; Van den Heever, D.J.; Swart, W. The reconstruction of a 12-lead electrocardiogram from a reduced lead set using a focus time-delay neural network. *Acta Cardiol. Sin.* **2021**, *37*, 47.
29. Zhou, H.; Zhang, S.; Peng, J.; Zhang, S.; Li, J.; Xiong, H.; Zhang, W. Informer: Beyond efficient transformer for long sequence time-series forecasting. In Proceedings of the AAAI Conference on Artificial Intelligence, Vancouver, BC, Canada, 2–9 February 2021; pp. 11106–11115.
30. Liu, S.; Yu, H.; Liao, C.; Li, J.; Lin, W.; Liu, A.X.; Dustdar, S. Pyraformer: Low-complexity pyramidal attention for long-range time series modeling and forecasting. In Proceedings of the International Conference on Learning Representations, Virtual, 3–7 May 2021.
31. Wu, H.; Xu, J.; Wang, J.; Long, M. Autoformer: Decomposition transformers with auto-correlation for long-term series forecasting. *Adv. Neural Inf. Process. Syst.* **2021**, *34*, 22419–22430.
32. Zhou, T.; Ma, Z.; Wen, Q.; Wang, X.; Sun, L.; Jin, R. Fedformer: Frequency enhanced decomposed transformer for long-term series forecasting. In Proceedings of the International Conference on Machine Learning, Baltimore MD, USA, 17 July 2022; pp. 27268–27286.
33. Bousseljot, R.; Kreiseler, D.; Schnabel, A. Nutzung der EKG-Signaldatenbank CARDIODAT der PTB über das Internet. *Biomed. Eng./Biomed. Tech.* **1995**, *40*, 317–318. [CrossRef]
34. Wu, H.; Hu, T.; Liu, Y.; Zhou, H.; Wang, J.; Long, M. Timesnet: Temporal 2d-variation modeling for general time series analysis. *arXiv* **2022**, arXiv:2210.02186.
35. Szegedy, C.; Liu, W.; Jia, Y.; Sermanet, P.; Reed, S.; Anguelov, D.; Erhan, D.; Vanhoucke, V.; Rabinovich, A. Going deeper with convolutions. In Proceedings of the IEEE Conference on Computer Vision and Pattern Recognition, Boston, MA, USA, 7–12 June 2015; pp. 1–9.
36. Wang, J. An intelligent computer-aided approach for atrial fibrillation and atrial flutter signals classification using modified bidirectional LSTM network. *Inf. Sci.* **2021**, *574*, 320–332. [CrossRef]
37. Almutairi, H.; Hassan, G.M.; Datta, A. Classification of Obstructive Sleep Apnoea from single-lead ECG signals using convolutional neural and Long Short Term Memory networks. *Biomed. Signal Process. Control* **2021**, *69*, 102906. [CrossRef]
38. Andersen, R.S.; Peimankar, A.; Puthusserypady, S. A deep learning approach for real-time detection of atrial fibrillation. *Expert Syst. Appl.* **2019**, *115*, 465–473. [CrossRef]

bioengineering

Article

Personalized Explanations for Early Diagnosis of Alzheimer's Disease Using Explainable Graph Neural Networks with Population Graphs

So Yeon Kim [1,2]

[1] Department of Artificial Intelligence, Ajou University, Suwon 16499, Republic of Korea; jebi1771@ajou.ac.kr
[2] Department of Software and Computer Engineering, Ajou University, Suwon 16499, Republic of Korea

Abstract: Leveraging recent advances in graph neural networks, our study introduces an application of graph convolutional networks (GCNs) within a correlation-based population graph, aiming to enhance Alzheimer's disease (AD) prognosis and illuminate the intricacies of AD progression. This methodological approach leverages the inherent structure and correlations in demographic and neuroimaging data to predict amyloid-beta (Aβ) positivity. To validate our approach, we conducted extensive performance comparisons with conventional machine learning models and a GCN model with randomly assigned edges. The results consistently highlighted the superior performance of the correlation-based GCN model across different sample groups in the Alzheimer's Disease Neuroimaging Initiative (ADNI) dataset, suggesting the importance of accurately reflecting the correlation structure in population graphs for effective pattern recognition and accurate prediction. Furthermore, our exploration of the model's decision-making process using GNNExplainer identified unique sets of biomarkers indicative of Aβ positivity in different groups, shedding light on the heterogeneity of AD progression. This study underscores the potential of our proposed approach for more nuanced AD prognoses, potentially informing more personalized and precise therapeutic strategies. Future research can extend these findings by integrating diverse data sources, employing longitudinal data, and refining the interpretability of the model, which potentially has broad applicability to other complex diseases.

Keywords: graph neural networks; alzheimer's disease; amyloid-beta positivity; population graph; explainable graph neural networks

Citation: Kim, S.Y. Personalized Explanations for Early Diagnosis of Alzheimer's Disease Using Explainable Graph Neural Networks with Population Graphs. *Bioengineering* **2023**, *10*, 701. https://doi.org/10.3390/bioengineering10060701

Academic Editor: Larbi Boubchir

Received: 22 May 2023
Revised: 5 June 2023
Accepted: 6 June 2023
Published: 8 June 2023

1. Introduction

Alzheimer's disease (AD) is a progressive neurodegenerative disorder characterized by a series of changes in the brain that occur years or even decades before the first symptoms of cognitive decline become evident [1–4]. Amyloid-beta (Aβ) is a protein that is implicated in AD, one of the most common forms of dementia. Aβ deposition, which can be observed via amyloid positron emission tomography (PET) imaging, is one of the earliest detectable pathological changes and a pathological hallmark of AD. It precedes other biomarkers, such as tau pathology, neuronal injury or neurodegeneration, and cognitive symptoms [5]. Once amyloid plaques start to build up, there is a cascade of events, including the accumulation of tau tangles inside neurons and eventual cell death, leading to brain atrophy, which can be observed through magnetic resonance imaging (MRI). Detecting the presence or predicting the onset of Aβ positivity can, therefore, be instrumental in the early diagnosis and prevention of this debilitating disease.

Numerous studies have employed machine learning and deep learning methodologies for predicting amyloid pathology and other AD phenotypes [1,6–20], typically classifying individuals into the categories of cognitively normal (CN), mild cognitive impairment (MCI), and Alzheimer's disease (AD) [11–14]. Furthermore, several studies have proposed machine learning frameworks designed to predict the conversion from MCI to AD [15–19], highlighting the potential of these methodologies in forecasting AD progression. Despite these advances, a crucial area of focus remains the early detection of AD pathology, especially in cognitively unimpaired individuals. Aβ deposition, distinguishable before any cognitive impairment becomes apparent, serves as a pivotal biomarker for identifying individuals predisposed to AD. It is noteworthy that cognitively unimpaired individuals may not yet demonstrate significant Aβ deposition and are often categorized as Aβ negatives. However, AD typically harbors an extended preclinical phase where, despite an absence of overt cognitive symptoms, individuals may already possess considerable Aβ deposition [21,22]. Thus, predicting Aβ positivity could significantly expedite AD diagnosis, optimize participant selection for clinical trials of disease-modifying therapies, and facilitate proactive monitoring and potential early intervention. Many researchers are increasingly targeting the early stages and the preclinical phase of AD in an attempt to curtail disease progression [1,8,20,23].

Various biomarkers, including demographic traits, genetic factors, and MRI imaging features, are key tools in predicting Aβ positivity [24,25]. Age, a significant demographic factor, is intrinsically linked with the risk of developing AD. As age advances, the probability of both AD and other dementia forms, as well as Aβ plaque accumulation, increases. Sex is another demographic characteristic of note, with women statistically more likely to develop AD compared to men, a fact which is an active area of investigation. Educational attainment, gauged as years of formal education completed, is associated with AD risk. Those with higher education levels often exhibit a lower risk of AD, potentially attributed to a bolstered cognitive reserve, which allows for increased tolerance of brain damage before dementia symptoms manifest.

Genetic factors, particularly the apolipoprotein E (APOE) $\epsilon 4$ allele, have a significant role in AD risk prediction [26–28]. Carriers of this allele are at elevated risk of developing AD and often experience earlier onset of symptoms. This allele is believed to influence AD by modifying Aβ processing or clearance in the brain. MRI features, including brain atrophy and other structural changes linked to AD, offer powerful predictive tools for Aβ positivity [29,30]. Changes such as specific regional brain shrinkage, ventricular expansion, and alterations in white matter integrity can be detected by MRI. The combined utilization of these biomarkers offers a comprehensive approach to predicting Aβ positivity. The multifaceted strategy facilitates earlier, more precise diagnosis; improved prognostic predictions; and the potential for personalized treatment plans. It can also guide the design of clinical trials and the development of new therapeutic interventions, underlining the enhanced predictive model offered by their combined use.

Effective capture of the collective power of these biomarkers for the efficient diagnosis and prognosis of AD can be achieved through graph-based machine learning, particularly graph neural networks (GNNs). GNNs have made significant strides in the healthcare sector, modeling interactions between biological entities, predicting potential disease-associated genes, constructing patient similarity networks, and even playing a crucial role in drug discovery [31–41]. GNNs have been utilized to model intricate correlations between multiple biomarkers, such as genetic, clinical, and neuroimaging features, offering valuable insights into the underlying mechanisms of AD progression [42–46]. The strength of GNNs lies in their ability to interpret the interconnections between brain regions and the impacts of changes in these regions on AD progression. These capabilities can enhance understanding of the complex disease trajectory, allowing for more precise prediction of AD prognosis.

One such tool that has proven to be powerful in graph-based deep learning is the subset of GNNs known as graph convolutional networks (GCNs) [47]. They extend the concept of convolutional operations from regular, grid-like structures—typical in images—to irregular graph structures. Notably, GCNs have been shown to be useful in disease prediction, particularly for autism spectrum disorder and AD [38]. They operate on a population graph, where nodes represent individuals and edges symbolize similarity in certain characteristics, thereby facilitating the deciphering of population-level patterns and individual variations in brain images.

Despite these advances, their inherent black-box nature poses a challenge due to limited transparency. This opacity can hinder understanding of the models' internal decision-making processes—a significant concern in the medical field, where model interpretability is crucial. To combat this issue, researchers are turning to explainable artificial intelligence (XAI) to foster more comprehensible and transparent GNNs. Various techniques, such as GradCAM-based explanation [48,49], PGExplainer [50], PGMExplainer [51], XGNN [52], and GNNExplainer [53], have been developed to enhance the interpretability of GNNs. However, the use of explainable GNNs remains largely restricted to medical image analysis [54–56] and drug discovery [57–60], suggesting a need for broader application and integration. Moreover, the comprehensive prioritization of personalized biomarkers, crucial for personalized medicine in AD diagnosis and treatment, is a largely unexplored area.

This study aims to bridge this gap by leveraging GCNs [47] to offer accurate predictions alongside interpretable results, thus contributing to a more holistic understanding of individual AD prognoses. This study is motivated by the hypothesis that cohorts with analogous clinical or neuroimaging characteristics may show a correlation that extends beyond the influence of prevalent biomarkers, such as the APOE genotype. Thus, we can build a population graph where nodes symbolize individuals at risk and edges depict similarities in demographic, genetic, and neuroimaging attributes. Our study highlights the utility of GCNs in predicting Aβ positivity, a crucial early indicator of AD, by demonstrating our proposed correlation-based population graph of cognitively unimpaired individuals. Furthermore, we utilize GNNExplainer [53], an explainable GNN model, which optimizes a subgraph within an individual's neighborhood and pinpoints a set of crucial features integral for the prediction. For each individual, we prioritize personalized AD risk factors, allocating risk scores derived from the average importance values garnered from their neighbors. This elucidation process further unveils a significant variation in the biomarkers identified for AD prognosis across different sample groups. The overview of the proposed model is illustrated in Figure 1. The main contributions of this study include:

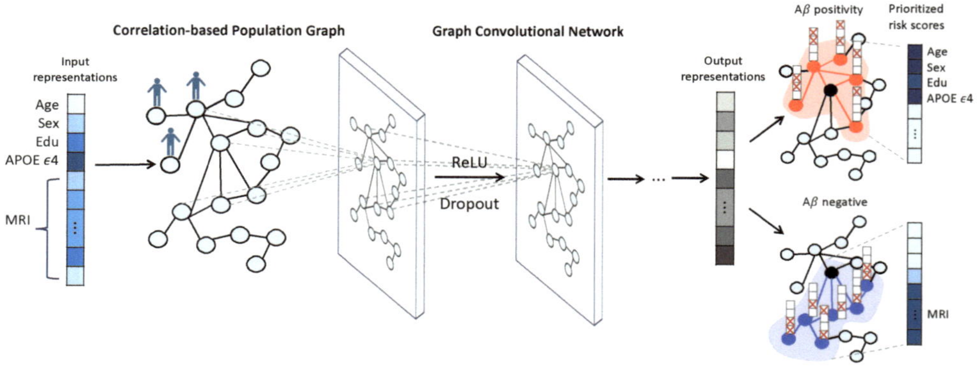

Figure 1. Overview of the proposed model. First, we construct a population graph in which the vertices represent individuals and are characterized by demographic features (age, sex, years of education), genetic information (APOE ϵ4 status), and MRI imaging features (average cortical thickness values for 69 brain regions of interest (ROIs). Edges are assigned when there is a high correlation between a pair of individuals. Next, we employ graph convolutional networks (GCNs) to analyze the population graph and predict the Aβ positivity for each individual. Finally, GNNExplainer provides an explanation for each prediction, optimizing a subgraph of the individual's neighborhood and identifying a set of crucial features for the prediction. For each individual, we prioritize the top 10 personalized biomarkers by assigning risk scores based on the average importance values obtained from their neighbors.

- We demonstrate the effectiveness of using graph neural networks on population graphs for early AD diagnosis in cognitively unimpaired individuals;
- We provide explanations of graph neural network predictions, offering sample-level interpretations using demographic and neuroimaging features;
- We prioritized personalized risk factors for AD by explaining graph neural network predictions, thereby characterizing groups of individuals based on their risk factors in AD prognosis.

2. Materials and Methods

2.1. Dataset

In this study, we leveraged the Alzheimer's Disease Neuroimaging Initiative (ADNI) GO/2 dataset (adni.loni.usc.edu, accessed on 21 May 2023). The ADNI was launched in 2003 as a public–private partnership led by principal investigator Michael W. Weiner, MD. The primary goal of ADNI is to test whether serial MRI, PET, other biological markers, and clinical and neuropsychological assessment can be combined to measure the progression of mild cognitive impairment (MCI) and early Alzheimer's disease (AD). For up-to-date information, see www.adni-info.org (accessed on 21 May 2023). Our selection targeted individuals identified as either cognitively normal (CN) or with mild cognitive impairment (MCI) status. We utilized a total of 506 samples from the ADNI cohort encompassing 214 CN and 292 MCI samples, with each sample characterized by 73 features.

These features included 3 demographic aspects (age, sex, and years of education), APOE $\epsilon 4$ status, and 69 neuroimaging features derived from MRI scans. The APOE $\epsilon 4$ status, denoted by the count of $\epsilon 4$ alleles (0, 1, or 2), functions as a critical genetic biomarker associated with elevated AD risk, with two $\epsilon 4$ alleles conferring the highest susceptibility. The neuroimaging features derived from MRI provide a thorough representation of neuroanatomical alterations, encompassing metrics such as cortical thickness. In this study, we utilized quantitative ADNI MRI data. The quantitative MRI data specifically represent the cortical thickness from T1-weighted MRI images obtained from the University of California, San Francisco, and archived at the LONI Image and Data Archive (IDA). We used average cortical thickness values for each of the 69 brain regions of interest (ROIs). A comprehensive description of the MRI image data acquisition process can be found here: https://adni.loni.usc.edu/methods/mri-tool/mri-analysis/ (accessed on 21 May 2023). As a result, there were 69 distinct numerical MRI features available for each individual.

Our study primarily focuses on predicting Aβ positivity in cognitively unimpaired individuals. The levels of Aβ were quantified from ^{18}F-florbetapir PET scans, which specifically bind to Aβ plaques present in the brain. The measurements of Aβ levels were obtained from the LONI Image and Data Archive (IDA) at the University of California, Berkeley. The burden of Aβ deposits was evaluated using the averaged value of the standardized uptake value ratio (SUVR). A detailed description of the PET image analysis method can be found here: https://adni.loni.usc.edu/methods/pet-analysis-method/ (accessed on 21 May 2023). If the averaged value exceeded a cutoff of 1.11, the individual was classified as Aβ-positive, indicative of AD pathology [61,62]. The ADNI cohort contained 291 Aβ-negatives and 215 Aβ-positives. Summary statistics of the clinical features for the CN and MCI samples are presented in Table 1.

Table 1. Demographic and neuroimaging characteristics of cognitively normal (CN) and mild cognitive impairment (MCI) groups from the Alzheimer's Disease Neuroimaging Initiative (ADNI) dataset.

	CN (N = 214)	MCI (N = 292)
Age	74.18 ± 6.14	73.14 ± 7.48
Sex	96 F/118 M	173 F/119 M
Education (years)	16.7 ± 2.54	16.13 ± 2.64
APOE $\epsilon 4$	144 neg/70 pos	173 neg/119 pos
Aβ positivity	139 neg/75 pos	152 neg/140 pos

N: Number of samples; F: Female; M: Male; pos: Positive; neg: Negative

2.2. Population Graph Construction

In our study, we constructed a graph denoted by $\mathcal{G} = (\mathcal{V}, \mathcal{E})$, where the vertex set $\mathcal{V} = (v_1, \ldots, v_n)$ signifies the collection of individuals. Each vertex v_i is associated with the i-th individual and is characterized by a feature vector x_{v_i}, which is composed of demographic and neuroimaging features. Specifically, for each individual, we concatenate features such as age, sex, years of education, APOE ϵ4 status (0, 1, or 2), and quantitative MRI imaging features (average cortical thickness values for 69 brain ROIs), resulting in a 73-feature vector. The set $\mathcal{E}$ denotes the collection of undirected edges linking the vertices in $\mathcal{V}$.

Pearson's correlation coefficient and associated p-values are computed for each vertex pair's feature vectors x_v and x_w. These metrics elucidate the magnitude, direction, and statistical significance of the linear relationship between the data linked to each vertex pair. Following this calculation, we rank all vertex pairs using two criteria: the absolute correlation coefficient value (in descending order) and the p-value (in ascending order). We identify the top M pairs, those exhibiting the highest absolute correlation and lowest p-values, and assign them edges, $e_{vw} = (v, w) \in \mathcal{E}$. These assigned edges signify the most statistically significant and the strongest correlations within our population graph.

We introduced the proposed model, which employs GCNs and was termed GCN-corr, to this correlation-based population graph. Its performance was contrasted with an equivalently sized graph but with edges assigned randomly termed GCN-random. This comparison elucidated the advantages of utilizing a correlation-based graph for predicting Aβ positivity in cognitively unimpaired individuals as opposed to a graph with randomly assigned edges. In the GCN-corr model, edges mirror the correlations between individual nodes, thus enabling more precise prediction congruent with the population structure.

Through an ablation study, we ascertained the optimal number of edges (M) by manipulating the network density. This density, calculated as $\frac{2M}{N(N-1)}$, represents the proportion of actual edges compared to the maximum possible in a fully connected network of N nodes. We varied the density from a sparse 1% connectivity network to a maximally interconnected network (100% connectivity), with each increment representing a 10% increase. This methodology aids in striking an equilibrium between network complexity and predictive accuracy, thereby refining the graph's structure for enhanced prediction efficacy using the GCN model. Additional discussions on the ablation study are expounded upon in the Supplementary Materials.

2.3. Graph Convolutional Networks

In this study, we leveraged graph convolutional networks (GCNs), which were proposed by [47], with a population graph $\mathcal{G}$ to discern the connections between the demographic and neuroimaging features of individuals. This study addresses the problem of predicting Aβ positivity as a node (individual)-level prediction task within the GCN model. An adjacency matrix $A \in [0,1]^{N \times N}$ representing pairwise correlations between nodes in a population graph $\mathcal{G}$, a feature matrix $X \in \mathbb{R}^{N \times p}$, and labels $y \in [0,1]^N$ (Aβ positivity) are used as input to train the model.

GCNs are designed to learn robust node representations by aggregating information from the local neighborhoods within the graph [47]. The core operation in GCNs, graph convolution, operates as a message-passing mechanism that facilitates information exchange between adjacent nodes. Within each GCN layer, nodes gather, process messages from their neighbors, and subsequently transform this information using a learnable weight matrix. Diverging from the convolution concept used in convolutional neural networks (CNNs), which apply a filter to localized input data segments, GCNs redefine convolution as the process of aggregation of neighboring node information. This notion retains the integral principle of incorporating local information, a concept fundamental to CNNs, thus justifying its appellation as graph convolution. Within a single GCN layer, the operation can be denoted as:

$$H^{(l+1)} = \sigma\left(\tilde{D}^{-\frac{1}{2}} \tilde{A} \tilde{D}^{-\frac{1}{2}} H^{(l)} W^{(l)}\right) \quad (1)$$

In this equation, $H^{(l)}$ and $H^{(l+1)}$ represent the feature matrices at the l-th and $(l+1)$-th layers, respectively, each encapsulating node representations. The matrix $\tilde{A}$, derived by adding the identity matrix I (representing self-loops, which are edges from nodes to themselves) to the original adjacency matrix A, is referred to as a self-loop inclusive adjacency matrix. The degree matrix of $\tilde{A}$, $\tilde{D}$, is a diagonal matrix representing the degree (number of connections) of each node. The learnable weight matrix at the l-th layer, $W^{(l)}$, serves to transform the aggregated neighbor node information. σ signifies the activation function; in this study, a rectified linear unit (ReLU).

The model is then trained to minimize the loss $\mathcal{L}$ using the ADAM optimizer and learn the optimal parameters for the prediction task. The loss $\mathcal{L}$ represents a cross-entropy function, which is computed as follows:

$$\mathcal{L} = -(w_{y_0} y \log(h) + w_{y_1}(1-y)\log(1-h)) \tag{2}$$

In this equation, h denotes the model's prediction output following softmax activation, and y represents the ground-truth label. Given the higher number of Aβ-negative samples in the cognitively unimpaired individuals dataset, the model could be impacted by class imbalance. To address this issue, we assigned rescaling weights during model training, which are inversely proportional to class frequencies as follows:

$$w_{y_c} = \frac{N}{2N_{y_c}} \tag{3}$$

Here, N represents the total number of samples, while N_{y_c} denotes the number of samples belonging to class c.

2.4. Interpretation with GNNExplainer

GNNExplainer [53] is a model-agnostic method specially designed to elucidate the predictions made by graph neural networks (GNNs), including the GCN model. The primary goal of GNNExplainer is to offer insights into the model's decision-making process, highlight significant features and relationships within the graph, and foster trust in the model's predictions.

The GNNExplainer method works by learning to extract a concise subgraph from the original input graph. This subgraph is optimized to best explain the GNN model's prediction for a specific target node or graph. This is accomplished by formulating an optimization problem where the objective is to minimize the difference between the original GNN model's prediction and the prediction made using the extracted subgraph. This optimization process involves the use of a binary mask for nodes, edges, and features. This mask determines whether to include or exclude graph elements, contingent on their contribution to the prediction.

In this research, we employed GNNExplainer to provide interpretable explanations for the predictions generated by our GCN model. This tool allowed us to explore individual biomarkers that significantly contribute to the prediction of Aβ positivity. Additionally, we identified distinct groups of individuals who share common biomarkers yet contain unique prioritized features that differentiate them from other groups. We also detailed the characteristics of each group based on their significant biomarkers.

2.5. Performance Evaluation

In this study, we formulated a supervised node classification problem, where our objective was to predict the Aβ positivity for each individual within the test sets. We assessed the classification performance by employing stratified five-fold cross-validation repeated 10 times. This method ensured that the ratio of Aβ-positive to Aβ-negative samples was maintained across all sets. During each iteration, we set aside one fold as the test set, while the remaining four folds were randomly divided into an 80% training set and a 20% validation set. We carried out this stratified partitioning to optimize and validate our model. Consequently, the data were split into a 64% training set, a 16% validation set, and a 20% test set. We report the final classification performance as the mean area under the curve (AUC) values over a total of 50 iterations.

3. Results

3.1. Experimental Setting

In the experiments, we assessed the performance of our proposed correlation-based population graph model (GCN-corr) and compared it with an equivalently sized graph but with edges assigned randomly (GCN-random). Furthermore, to provide a broader perspective on the efficacy of GCN models applied to population graphs, our comparison was not limited to GCN-random and GCN-corr. We expanded our analysis to include a comparison with traditional machine learning algorithms. These encompassed the support vector machine (SVM) using a radial basis function (RBF) kernel, the random forest (RF) classifier, logistic regression (LR) with the L2 penalty (also known as ridge regularization), and the multi-layer perceptron (MLP).

To achieve optimal performance in our analyses, we set certain hyperparameters empirically for both the GCN and the MLP models. The GCN model was structured as a two-layer network, which included a hidden layer comprising 32 units. The model underwent training for a maximum of 550 epochs, with a learning rate set at 0.005. The L2 loss was set to 5×10^{-10}, and a dropout rate of 0.5 was employed to prevent overfitting and enhance generalization. Simultaneously, the MLP model was set up with the same hyperparameters as the GCN model for a fair comparison.

Due to the predominance of Aβ-negative samples in our dataset, we utilized class weights as outlined in Equation (3) to balance the classes. For fairness in comparison, the same rescaling weights were applied to the conventional machine learning models previously mentioned. This adjustment in weights ensured a balanced evaluation of each model's performance, despite the unequal sample sizes between the Aβ-positive and Aβ-negative groups.

It is worth noting that, although we employed stratified cross-validation, a technique which guarantees a proportionate class distribution across all validation folds, it did not directly influence the model's learning process. Conversely, class weights were incorporated during model training, conferring more emphasis to the less represented class and thereby ameliorating the effects of imbalanced data. Therefore, these two methodologies, while serving distinct roles, complement each other: stratified cross-validation augments the accuracy of performance estimation, while class weighting refines the model's capacity to learn from imbalanced datasets.

3.2. Performance of Prediction of Aβ Positivity

We evaluated the performance in predicting Aβ positivity with comparative analyses conducted across three distinct groups: cognitively normal (CN) individuals, individuals with mild cognitive impairment (MCI), and a combined group of both CN and MCI individuals. The performances for each model were evaluated across these sample groups and are comprehensively displayed in Table 2 and Figure 2.

Upon analysis, we observed that the GCN models, when applied to the population graph constructed in this study, consistently outperformed the conventional machine learning models across all sample groups. Moreover, the correlation-based population graph model (GCN-corr) consistently demonstrated superior performance compared to the GCN model with randomly assigned edges (GCN-random). The GCN-corr model outperformed all other models across all groups, achieving the highest mean AUCs of 0.8851, 0.8741, and 0.8632 for the CN, MCI, and CN + MCI groups, respectively. This was particularly evident in the combined CN + MCI group, where the performance of the GCN-random model showed a significant drop (0.7160 ± 0.0135), not just in comparison to the GCN-corr model but also against most of the conventional machine learning models. This suggests the potential of GCN models in significantly enhancing the predictive accuracy for

Aβ positivity when used with well-crafted population graphs, affirming their proficiency in handling complex biological data.

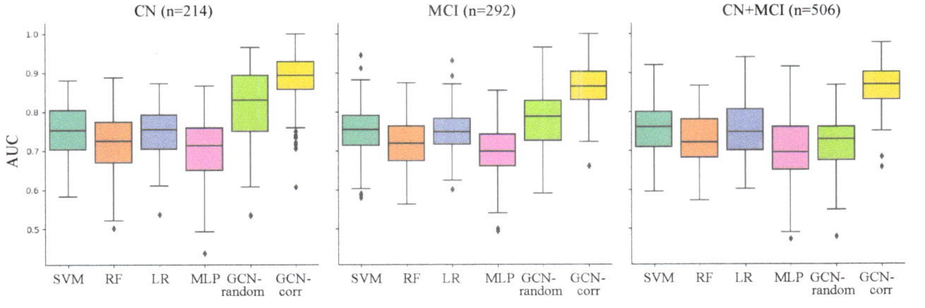

Figure 2. Comparison of the performance of each model (represented in x-axis and different colors) in predicting Aβ positivity across three sample groups: cognitively normal (CN) individuals, those with mild cognitive impairment (MCI), and a combined group (CN + MCI). The performance was measured with the area under the ROC curve (AUC) derived from 10 repetitions of five-fold cross-validation. The compared models are support vector machine using a radial basis function kernel (SVM), random forest (RF) classifier, logistic regression with ridge regularization (LR), multi-layer perceptron (MLP), GCN model on a graph with randomly assigned edges (GCN-random), and GCN model on a correlation-based population graph (GCN-corr).

Table 2. Summary of the performance of each model in predicting Aβ positivity across three sample groups: cognitively normal (CN) individuals, those with mild cognitive impairment (MCI), and a combined group (CN + MCI). The performance metrics were computed as the mean area under the ROC curve (AUC), along with a 95% confidence interval, derived from 10 repetitions of five-fold cross-validation. The highest performing result for each sample group is highlighted in bold text.

	AUC (Mean ± 95% CI)		
Model	**CN**	**MCI**	**CN + MCI**
SVM (RBF)	0.7515 ± 0.0131	0.7531 ± 0.0137	0.7537 ± 0.0129
RF	0.7205 ± 0.0143	0.7226 ± 0.0140	0.7238 ± 0.0134
LR (ridge)	0.7490 ± 0.0129	0.7480 ± 0.0112	0.7500 ± 0.0144
MLP	0.7009 ± 0.0158	0.7013 ± 0.0137	0.7009 ± 0.0159
GCN-random	0.8110 ± 0.0185	0.7768 ± 0.0153	0.7160 ± 0.0135
GCN-corr	**0.8851 ± 0.0154**	**0.8741 ± 0.0114**	**0.8632 ± 0.0115**

To further illustrate the proficiency of the GCN model in accurately classifying Aβ-positive and -negative samples within a population graph, we offer a visualization of the final embedding of each GCN model in Figure 3. Notably, the GCN model with randomly assigned edges (GCN-random) demonstrated difficulty differentiating between the two classes across all three groups: cognitively normal (CN), mild cognitive impairment (MCI), and the combined CN + MCI group.

In contrast, the correlation-based population graph model (GCN-corr) effectively distinguished between the two classes across all sample groups, emphasizing the significant contribution of our proposed correlation-based population graph to the enhancement of Aβ positivity prediction using the GCN model.

The strength of these findings lends considerable support to our research hypothesis. It suggests the existence of shared biomarkers within groups of individuals whose demographic and neuroimaging features strongly correlate, thus positively influencing Aβ positivity prediction and enriching our understanding of AD prognosis. These results also suggest variability in AD risk factors across different individual groups, hinting at the potential benefits of tailoring prediction models and preventative strategies.

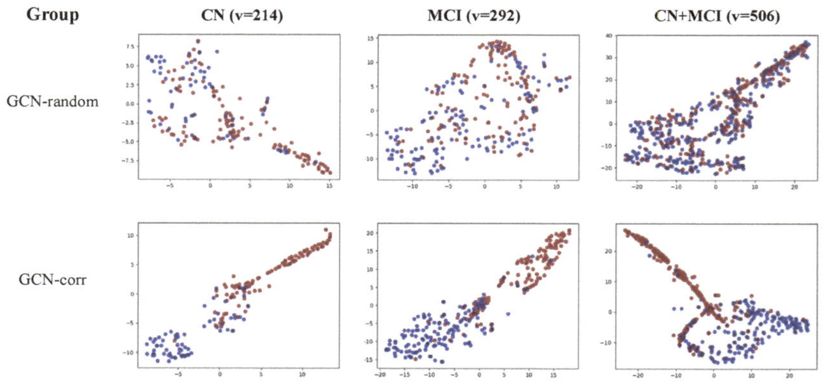

Figure 3. Visual representation of the final node embedding of the GCN-random and GCN-corr models across three groups: cognitively normal (CN), mild cognitive impairment (MCI), and a combined CN + MCI group. Nodes representing Aβ-positive samples are colored blue, while those representing Aβ-negative samples are in red. v represents the number of vertices (individuals).

3.3. Interpreting Predictions of Graph Neural Networks

In our endeavor to understand the intricate interplay between demographic and neuroimaging features, we utilized GNNExplainer. This tool helped us identify key biomarkers that significantly contribute to the prediction of Aβ positivity within GCN models. This investigation deepens our understanding of the model's decision-making process and bolsters our confidence in its predictive capabilities.

Figure 4 offers a visual representation of these prioritized biomarkers and their associated importance scores for each individual. For an improved visualization, we randomly selected 50 individuals from the test set that delivered the most accurate predictions across all cross-validation splits. Notably, these samples were correctly classified, which facilitated the identification of the most influential biomarkers for accurate predictions. To further enhance data interpretability, we employed both feature-wise and sample-wise clustering. This analysis led to the identification of four distinct groups (A, B, C, and D), each distinguished by the significance of their biomarkers. We delineated the top 10 biomarkers based on their averaged importance scores, uncovering unique patterns of significance across the four groups, as detailed in Table S1.

In group A, the left precentral gyrus, known for its involvement in motor function, emerged as the most significant biomarker with an average score of 0.9954. Additional notable biomarkers included the right precentral gyrus, the APOEe4 gene variant—associated with an increased risk of AD—and the left caudal middle frontal gyrus, all scoring above 0.89. In group B, the left precentral gyrus was also the top biomarker. However, the demographic feature of age was highlighted, receiving an average score of 0.7768. This underlines the well-established link between advancing age and increased risk of AD. In group C, the left precuneus, a brain region involved in episodic memory, stood out as the most significant biomarker with a score of 0.7137. This group also prioritized demographic features, such as the level of education (years) and the individual's sex, with scores of 0.6725 and 0.6160, respectively. These results may indicate a potential influence of educational attainment and biological sex on the disease's onset and progression. Group D highlighted the value of education (years) as the top biomarker, scoring a near-perfect 0.9970. This aligns with the cognitive reserve theory, which suggests that higher levels of education may offer a protective effect against cognitive decline. Neuroimaging biomarkers, such as the right precuneus and left pars orbitalis, also held high priority in this group.

Taken together, these findings underscore the complex interaction between structural brain changes and demographic factors in predicting Aβ positivity. They shed light on the heterogeneity of the disease, revealing different progression patterns across unique groups.

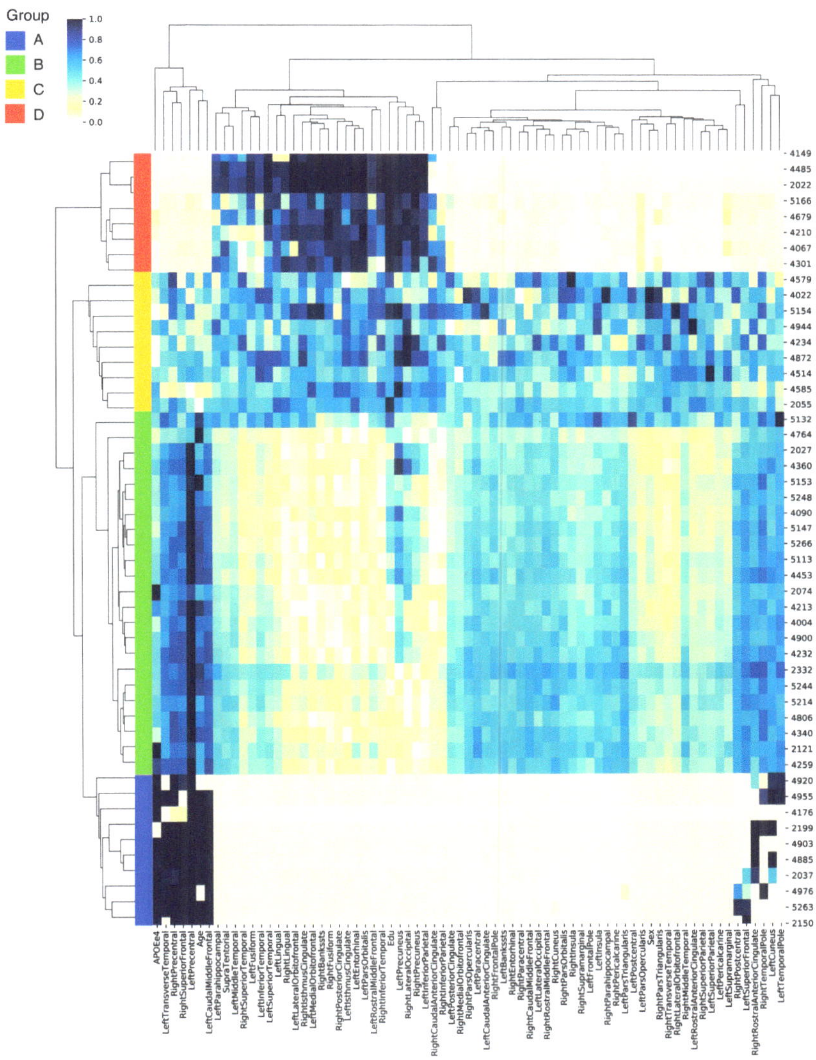

Figure 4. Heatmap visualization of the prioritized biomarkers derived from the GCN model with a correlation-based population graph. Each row represents an individual, and each column represents a biomarker. The color intensity indicates the importance score of each biomarker for Aβ-positivity prediction. The individuals and biomarkers are clustered feature-wise and sample-wise, revealing four distinct groups (A, B, C, and D) based on their important biomarkers.

4. Discussion

Our study illuminates the potential value of applying GCNs within a correlation-based population graph for enhanced AD prognosis. The superior performance of this model underscores the efficacy of harnessing the inherent structure and correlations in the data to augment predictive accuracy.

Our findings comprehensively demonstrate that a population graph built on correlations rather than random assignment can significantly elevate the model's ability to discern patterns and predict accurately. This suggests that the interplay between demographic and neuroimaging features is not random; instead, it exhibits a specific correlation structure that is instrumental for the prediction task.

Moreover, the improved performance lends credence to the hypothesis of the existence of common biomarkers among groups of individuals who exhibit high correlations in their demographic and neuroimaging features. This implies that different groups may possess distinctive sets of biomarkers that are particularly indicative of Aβ positivity, thereby illuminating the heterogeneity of AD progression.

Thus, our research underscores the potential of utilizing correlation-based population graphs in tandem with GCN models for a more nuanced and effective AD prognosis. This approach could potentially inform the development of more personalized, precise therapeutic strategies.

Future research could build upon our findings and explore several promising directions. This could include enriching the GCN model by incorporating diverse data sources, such as genetic, proteomic, and lifestyle factors, alongside demographic and neuroimaging features, to enhance the predictive performance. Transitioning from a cross-sectional model to one that exploits longitudinal data could offer a more dynamic understanding of AD progression. A deeper exploration of model interpretability and biomarker validation could provide greater transparency for the decision-making process, ensuring the identified biomarkers' relevance. Understanding the unique contributions of biomarkers to AD progression in different groups could inform the development of personalized treatment strategies. Lastly, the proposed framework in this study could be expanded to other neurological disorders or complex diseases, potentially offering valuable insights into disease mechanisms and therapeutic targets.

Supplementary Materials: The following supporting information can be downloaded at: https://www.mdpi.com/article/10.3390/bioengineering10060701/s1, Figure S1: Performances of the GCN-corr and GCN-random models as network density varies from a sparse 1% connectivity to a fully connected 100% connectivity in increments of 10%. Performance metrics were calculated as the mean area under the ROC curve (AUC), with a 95% confidence interval, derived from 10 repetitions of five-fold cross-validation. The confidence interval's upper and lower bounds are visually represented as a shaded area surrounding the line; Figure S2: Visual representation of the final node embedding of the GCN-random and GCN-corr models at selected network densities of 1%, 10%, and 50%. These densities were chosen to provide a clear comparison across varying levels of network densities. Nodes representing Aβ-positive samples are colored blue, while those representing Aβ-negative samples are in red; Table S1: The top 10 prioritized biomarkers from demographic and neuroimaging features, along with their corresponding averaged feature importance scores, listed in descending order. The biomarkers are divided into four groups (A–D) based on the results of heatmap clustering.

Funding: This work was supported by a National Research Foundation of Korea (NRF) grant funded by the Korean government (MSIT) (NRF-2022R1C1C1012060, NRF-2020R1I1A1A01071671).

Institutional Review Board Statement: Not applicable.

Informed Consent Statement: Not applicable.

Data Availability Statement: Data used in the preparation of this article were obtained from the Alzheimer's Disease Neuroimaging Initiative (ADNI) database (adni.loni.usc.edu, accessed on 21 May 2023). As such, the investigators within the ADNI contributed to the design and implementation of the ADNI and/or provided data but did not participate in analysis or writing of this report. A complete listing of ADNI investigators can be found at: http://adni.loni.usc.edu/wp-content/uploads/how_to_apply/ADNI_Acknowledgement_List.pdf (accessed on 21 May 2023).

Acknowledgments: Data collection and sharing for this project were funded by the Alzheimer's Disease Neuroimaging Initiative (ADNI) (National Institutes of Health Grant U01 AG024904) and DOD ADNI (Department of Defense award number W81XWH-12-2-0012). ADNI is funded by the National Institute on Aging and the National Institute of Biomedical Imaging and Bioengineering and through generous contributions from the following: AbbVie, Alzheimer's Association; Alzheimer's Drug Discovery Foundation; Araclon Biotech; BioClinica, Inc.; Biogen; Bristol-Myers Squibb Company; CereSpir, Inc.; Cogstate; Eisai Inc.; Elan Pharmaceuticals, Inc.; Eli Lilly and Company; EuroImmun; F. Hoffmann-La Roche Ltd and its affiliated company Genentech, Inc.; Fujirebio; GE Healthcare; IXICO Ltd.; Janssen Alzheimer Immunotherapy Research & Development, LLC.; Johnson & Johnson Pharmaceutical Research & Development LLC.; Lumosity; Lundbeck; Merck & Co., Inc.; Meso Scale Diagnostics, LLC.; NeuroRx Research; Neurotrack Technologies; Novartis Pharmaceuticals Corporation; Pfizer Inc.; Piramal Imaging; Servier; Takeda Pharmaceutical Company; and Transition Therapeutics. The Canadian Institutes of Health Research provides funds to support ADNI clinical sites in Canada. Private sector contributions are facilitated by the Foundation for the National Institutes of Health (www.fnih.org, accessed on 21 May 2023). The grantee organization was the Northern California Institute for Research and Education, and the study was coordinated by the Alzheimer's Therapeutic Research Institute at the University of Southern California. ADNI data are disseminated by the Laboratory for Neuro Imaging at the University of Southern California.

Conflicts of Interest: The authors declare no conflict of interest.

References

1. Cullen, N.C.; Leuzy, A.; Janelidze, S.; Palmqvist, S.; Svenningsson, A.L.; Stomrud, E.; Dage, J.L.; Mattsson-Carlgren, N.; Hansson, O. Plasma biomarkers of Alzheimer's disease improve prediction of cognitive decline in cognitively unimpaired elderly populations. *Nat. Commun.* **2021**, *12*, 3555. [CrossRef] [PubMed]
2. Jeremic, D.; Jiménez-Díaz, L.; Navarro-López, J.D. Past, present and future of therapeutic strategies against amyloid-β peptides in Alzheimer's disease: A systematic review. *Ageing Res. Rev.* **2021**, *72*, 101496. [CrossRef] [PubMed]
3. Nakamura, A.; Kaneko, N.; Villemagne, V.L.; Kato, T.; Doecke, J.; Doré, V.; Fowler, C.; Li, Q.X.; Martins, R.; Rowe, C.; et al. High performance plasma amyloid-β biomarkers for Alzheimer's disease. *Nature* **2018**, *554*, 249–254. [CrossRef] [PubMed]
4. Selkoe, D.J.; Schenk, D. Alzheimer's disease: Molecular understanding predicts amyloid-based therapeutics. *Annu. Rev. Pharmacol. Toxicol.* **2003**, *43*, 545–584. [CrossRef]
5. Oddo, S.; Caccamo, A.; Kitazawa, M.; Tseng, B.P.; LaFerla, F.M. Amyloid deposition precedes tangle formation in a triple transgenic model of Alzheimer's disease. *Neurobiol. Aging* **2003**, *24*, 1063–1070. [CrossRef]
6. Shirbandi, K.; Khalafi, M.; Mirza-Aghazadeh-Attari, M.; Tahmasbi, M.; Shahvandi, H.K.; Javanmardi, P.; Rahim, F. Accuracy of deep learning model-assisted amyloid positron emission tomography scan in predicting Alzheimer's disease: A systematic review and meta-analysis. *Inform. Med. Unlocked* **2021**, *25*, 100710. [CrossRef]
7. Ezzati, A.; Harvey, D.J.; Habeck, C.; Golzar, A.; Qureshi, I.A.; Zammit, A.R.; Hyun, J.; Truelove-Hill, M.; Hall, C.B.; Davatzikos, C.; et al. Predicting amyloid-β levels in amnestic mild cognitive impairment using machine learning techniques. *J. Alzheimer's Dis.* **2020**, *73*, 1211–1219. [CrossRef]
8. Vergallo, A.; Mégret, L.; Lista, S.; Cavedo, E.; Zetterberg, H.; Blennow, K.; Vanmechelen, E.; De Vos, A.; Habert, M.O.; Potier, M.C.; et al. Plasma amyloid β 40/42 ratio predicts cerebral amyloidosis in cognitively normal individuals at risk for Alzheimer's disease. *Alzheimer's Dement.* **2019**, *15*, 764–775. [CrossRef]
9. Kim, S.; Lee, P.; Oh, K.T.; Byun, M.S.; Yi, D.; Lee, J.H.; Kim, Y.K.; Ye, B.S.; Yun, M.J.; Lee, D.Y.; et al. Deep learning-based amyloid PET positivity classification model in the Alzheimer's disease continuum by using 2-[18F] FDG PET. *EJNMMI Res.* **2021**, *11*, 56. [CrossRef]
10. Chang, C.H.; Lin, C.H.; Lane, H.Y. Machine learning and novel biomarkers for the diagnosis of Alzheimer's disease. *Int. J. Mol. Sci.* **2021**, *22*, 2761. [CrossRef]
11. Basheera, S.; Ram, M.S.S. Deep learning based Alzheimer's disease early diagnosis using T2w segmented gray matter MRI. *Int. J. Imaging Syst. Technol.* **2021**, *31*, 1692–1710. [CrossRef]
12. Venugopalan, J.; Tong, L.; Hassanzadeh, H.R.; Wang, M.D. Multimodal deep learning models for early detection of Alzheimer's disease stage. *Sci. Rep.* **2021**, *11*, 3254. [CrossRef] [PubMed]
13. Houria, L.; Belkhamsa, N.; Cherfa, A.; Cherfa, Y. Multi-modality MRI for Alzheimer's disease detection using deep learning. *Phys. Eng. Sci. Med.* **2022**, *45*, 1043–1053. [CrossRef]
14. Ramzan, F.; Khan, M.U.G.; Rehmat, A.; Iqbal, S.; Saba, T.; Rehman, A.; Mehmood, Z. A deep learning approach for automated diagnosis and multi-class classification of Alzheimer's disease stages using resting-state fMRI and residual neural networks. *J. Med. Syst.* **2020**, *44*, 37. [CrossRef]
15. Zhang, T.; Liao, Q.; Zhang, D.; Zhang, C.; Yan, J.; Ngetich, R.; Zhang, J.; Jin, Z.; Li, L. Predicting MCI to AD conversation using integrated sMRI and rs-fMRI: machine learning and graph theory approach. *Front. Aging Neurosci.* **2021**, *13*, 688926. [CrossRef] [PubMed]

16. Grassi, M.; Rouleaux, N.; Caldirola, D.; Loewenstein, D.; Schruers, K.; Perna, G.; Dumontier, M.; Initiative, A.D.N. A novel ensemble-based machine learning algorithm to predict the conversion from mild cognitive impairment to Alzheimer's disease using socio-demographic characteristics, clinical information, and neuropsychological measures. *Front. Neurol.* **2019**, *10*, 756. [CrossRef] [PubMed]
17. Spasov, S.; Passamonti, L.; Duggento, A.; Lio, P.; Toschi, N.; Alzheimer's Disease Neuroimaging Initiative. A parameter-efficient deep learning approach to predict conversion from mild cognitive impairment to Alzheimer's disease. *Neuroimage* **2019**, *189*, 276–287. [CrossRef]
18. Lee, G.; Nho, K.; Kang, B.; Sohn, K.A.; Kim, D. Predicting Alzheimer's disease progression using multi-modal deep learning approach. *Sci. Rep.* **2019**, *9*, 1952. [CrossRef]
19. Ocasio, E.; Duong, T.Q. Deep learning prediction of mild cognitive impairment conversion to Alzheimer's disease at 3 years after diagnosis using longitudinal and whole-brain 3D MRI. *PeerJ Comput. Sci.* **2021**, *7*, e560. [CrossRef] [PubMed]
20. Li, J.; Bur, A.M.; Villwock, M.R.; Shankar, S.; Palmer, G.; Sykes, K.J.; Villwock, J.A. Olfactory phenotypes differentiate cognitively unimpaired seniors from Alzheimer's disease and mild cognitive impairment: A combined machine learning and traditional statistical approach. *J. Alzheimer's Dis.* **2021**, *81*, 641–650. [CrossRef]
21. Masters, C.L.; Selkoe, D.J. Biochemistry of amyloid β-protein and amyloid deposits in Alzheimer disease. *Cold Spring Harb. Perspect. Med.* **2012**, *2*, a006262. [CrossRef]
22. Engler, H.; Forsberg, A.; Almkvist, O.; Blomquist, G.; Larsson, E.; Savitcheva, I.; Wall, A.; Ringheim, A.; Långström, B.; Nordberg, A. Two-year follow-up of amyloid deposition in patients with Alzheimer's disease. *Brain* **2006**, *129*, 2856–2866. [CrossRef]
23. Leffa, D.T.; Ferrari-Souza, J.P.; Bellaver, B.; Tissot, C.; Ferreira, P.C.; Brum, W.S.; Caye, A.; Lord, J.; Proitsi, P.; Martins-Silva, T.; et al. Genetic risk for attention-deficit/hyperactivity disorder predicts cognitive decline and development of Alzheimer's disease pathophysiology in cognitively unimpaired older adults. *Mol. Psychiatry* **2023**, *28*, 1248–1255. [CrossRef]
24. Gottesman, R.F.; Schneider, A.L.; Zhou, Y.; Chen, X.; Green, E.; Gupta, N.; Knopman, D.S.; Mintz, A.; Rahmim, A.; Sharrett, A.R.; et al. The ARIC-PET amyloid imaging study: Brain amyloid differences by age, race, sex, and APOE. *Neurology* **2016**, *87*, 473–480. [CrossRef] [PubMed]
25. Wu, K.Y.; Hsiao, I.T.; Chen, C.S.; Chen, C.H.; Hsieh, C.J.; Wai, Y.Y.; Chang, C.J.; Tseng, H.J.; Yen, T.C.; Liu, C.Y.; et al. Increased brain amyloid deposition in patients with a lifetime history of major depression: Evidenced on 18 F-florbetapir (AV-45/Amyvid) positron emission tomography. *Eur. J. Nucl. Med. Mol. Imaging* **2014**, *41*, 714–722. [CrossRef]
26. Genin, E.; Hannequin, D.; Wallon, D.; Sleegers, K.; Hiltunen, M.; Combarros, O.; Bullido, M.J.; Engelborghs, S.; De Deyn, P.; Berr, C.; et al. APOE and Alzheimer disease: A major gene with semi-dominant inheritance. *Mol. Psychiatry* **2011**, *16*, 903–907. [CrossRef] [PubMed]
27. Meyer, M.R.; Tschanz, J.T.; Norton, M.C.; Welsh-Bohmer, K.A.; Steffens, D.C.; Wyse, B.W.; Breitner, J. APOE genotype predicts when—not whether—one is predisposed to develop Alzheimer disease. *Nat. Genet.* **1998**, *19*, 321–322. [CrossRef] [PubMed]
28. Strittmatter, W.J.; Roses, A.D. Apolipoprotein E and Alzheimer disease. *Proc. Natl. Acad. Sci. USA* **1995**, *92*, 4725–4727. [CrossRef] [PubMed]
29. Lerch, J.P.; Pruessner, J.C.; Zijdenbos, A.; Hampel, H.; Teipel, S.J.; Evans, A.C. Focal decline of cortical thickness in Alzheimer's disease identified by computational neuroanatomy. *Cereb. Cortex* **2005**, *15*, 995–1001. [CrossRef]
30. Lerch, J.P.; Pruessner, J.; Zijdenbos, A.P.; Collins, D.L.; Teipel, S.J.; Hampel, H.; Evans, A.C. Automated cortical thickness measurements from MRI can accurately separate Alzheimer's patients from normal elderly controls. *Neurobiol. Aging* **2008**, *29*, 23–30. [CrossRef]
31. Sun, Z.; Yin, H.; Chen, H.; Chen, T.; Cui, L.; Yang, F. Disease prediction via graph neural networks. *IEEE J. Biomed. Health Inform.* **2020**, *25*, 818–826. [CrossRef] [PubMed]
32. Cui, H.; Dai, W.; Zhu, Y.; Li, X.; He, L.; Yang, C. Brainnnexplainer: An interpretable graph neural network framework for brain network based disease analysis. *arXiv* **2021**, arXiv:2107.05097.
33. Zhang, X.M.; Liang, L.; Liu, L.; Tang, M.J. Graph neural networks and their current applications in bioinformatics. *Front. Genet.* **2021**, *12*, 690049. [CrossRef]
34. Bessadok, A.; Mahjoub, M.A.; Rekik, I. Graph neural networks in network neuroscience. *IEEE Trans. Pattern Anal. Mach. Intell.* **2022**, *45*, 5833–5848. [CrossRef]
35. Shen, Z.A.; Luo, T.; Zhou, Y.K.; Yu, H.; Du, P.F. NPI-GNN: Predicting ncRNA—Protein interactions with deep graph neural networks. *Briefings Bioinform.* **2021**, *22*, bbab051. [CrossRef]
36. Lv, G.; Hu, Z.; Bi, Y.; Zhang, S. Learning unknown from correlations: Graph neural network for inter-novel-protein interaction prediction. *arXiv* **2021**, arXiv:2105.06709.
37. Jha, K.; Saha, S.; Singh, H. Prediction of protein—Protein interaction using graph neural networks. *Sci. Rep.* **2022**, *12*, 8360. [CrossRef]
38. Parisot, S.; Ktena, S.I.; Ferrante, E.; Lee, M.; Guerrero, R.; Glocker, B.; Rueckert, D. Disease prediction using graph convolutional networks: Application to autism spectrum disorder and Alzheimer's disease. *Med. Image Anal.* **2018**, *48*, 117–130. [CrossRef]
39. Gu, Y.; Yang, X.; Tian, L.; Yang, H.; Lv, J.; Yang, C.; Wang, J.; Xi, J.; Kong, G.; Zhang, W. Structure-aware siamese graph neural networks for encounter-level patient similarity learning. *J. Biomed. Inform.* **2022**, *127*, 104027. [CrossRef] [PubMed]

40. Quan, Z.; Guo, Y.; Lin, X.; Wang, Z.J.; Zeng, X. Graphcpi: Graph neural representation learning for compound-protein interaction. In Proceedings of the 2019 IEEE International Conference on Bioinformatics and Biomedicine (BIBM), San Diego, CA, USA, 18–21 November 2019; pp. 717–722.
41. Zhang, Z.; Chen, L.; Zhong, F.; Wang, D.; Jiang, J.; Zhang, S.; Jiang, H.; Zheng, M.; Li, X. Graph neural network approaches for drug-target interactions. *Curr. Opin. Struct. Biol.* **2022**, *73*, 102327. [CrossRef]
42. Qin, Z.; Liu, Z.; Zhu, P. Aiding Alzheimer's Disease Diagnosis Using Graph Convolutional Networks Based on rs-fMRI Data. In Proceedings of the 2022 15th International Congress on Image and Signal Processing, BioMedical Engineering and Informatics (CISP-BMEI), Beijing, China, 5–7 November 2022; pp. 1–7.
43. Kolahkaj, S.; Zare, H. A connectome-based deep learning approach for Early MCI and MCI detection using structural brain networks. *Neurosci. Inform.* **2023**, *3*, 100118. [CrossRef]
44. Song, T.A.; Chowdhury, S.R.; Yang, F.; Jacobs, H.; El Fakhri, G.; Li, Q.; Johnson, K.; Dutta, J. Graph convolutional neural networks for Alzheimer's disease classification. In Proceedings of the 2019 IEEE 16th international symposium on biomedical imaging (ISBI 2019), Venice, Italy, 8–11 April 2019; pp. 414–417.
45. Kim, M.; Kim, J.; Qu, J.; Huang, H.; Long, Q.; Sohn, K.A.; Kim, D.; Shen, L. Interpretable temporal graph neural network for prognostic prediction of Alzheimer's disease using longitudinal neuroimaging data. In Proceedings of the 2021 IEEE International Conference on Bioinformatics and Biomedicine (BIBM), Virtual, 9–12 December 2021; pp. 1381–1384.
46. Ma, J.; Zhu, X.; Yang, D.; Chen, J.; Wu, G. Attention-guided deep graph neural network for longitudinal Alzheimer's disease analysis. In Proceedings of the Medical Image Computing and Computer Assisted Intervention—MICCAI 2020: 23rd International Conference, Lima, Peru, 4–8 October 2020; pp. 387–396.
47. Kipf, T.N.; Welling, M. Semi-supervised classification with graph convolutional networks. *arXiv* **2016**, arXiv:1609.02907.
48. Selvaraju, R.R.; Cogswell, M.; Das, A.; Vedantam, R.; Parikh, D.; Batra, D. Grad-cam: Visual explanations from deep networks via gradient-based localization. In Proceedings of the IEEE International Conference on Computer Vision, Venice, Italy, 22–29 October 2017; pp. 618–626.
49. Pope, P.E.; Kolouri, S.; Rostami, M.; Martin, C.E.; Hoffmann, H. Explainability methods for graph convolutional neural networks. In Proceedings of the IEEE/CVF Conference on Computer Vision and Pattern Recognition, Long Beach, CA, USA, 15–20 June 2019; pp. 10772–10781.
50. Luo, D.; Cheng, W.; Xu, D.; Yu, W.; Zong, B.; Chen, H.; Zhang, X. Parameterized explainer for graph neural network. *Adv. Neural Inf. Process. Syst.* **2020**, *33*, 19620–19631.
51. Vu, M.; Thai, M.T. Pgm-explainer: Probabilistic graphical model explanations for graph neural networks. *Adv. Neural Inf. Process. Syst.* **2020**, *33*, 12225–12235.
52. Yuan, H.; Tang, J.; Hu, X.; Ji, S. Xgnn: Towards model-level explanations of graph neural networks. In Proceedings of the 26th ACM SIGKDD International Conference on Knowledge Discovery & Data Mining, Virtual, 23–27 August 2020; pp. 430–438.
53. Ying, Z.; Bourgeois, D.; You, J.; Zitnik, M.; Leskovec, J. Gnnexplainer: Generating explanations for graph neural networks. *Adv. Neural Inf. Process. Syst.* **2019**, *32*, 9240–9251.
54. Van der Velden, B.H.; Kuijf, H.J.; Gilhuijs, K.G.; Viergever, M.A. Explainable artificial intelligence (XAI) in deep learning-based medical image analysis. *Med. Image Anal.* **2022**, *79*, 102470. [CrossRef] [PubMed]
55. Singh, A.; Sengupta, S.; Lakshminarayanan, V. Explainable deep learning models in medical image analysis. *J. Imaging* **2020**, *6*, 52. [CrossRef]
56. Borys, K.; Schmitt, Y.A.; Nauta, M.; Seifert, C.; Krämer, N.; Friedrich, C.M.; Nensa, F. Explainable AI in Medical Imaging: An overview for clinical practitioners–Saliency-based XAI approaches. *Eur. J. Radiol.* **2023**, *162*, 110787. [CrossRef]
57. Jiménez-Luna, J.; Grisoni, F.; Schneider, G. Drug discovery with explainable artificial intelligence. *Nat. Mach. Intell.* **2020**, *2*, 573–584. [CrossRef]
58. Preuer, K.; Klambauer, G.; Rippmann, F.; Hochreiter, S.; Unterthiner, T. Interpretable deep learning in drug discovery. In *Explainable AI: Interpreting, Explaining and Visualizing Deep Learning*; Springer: Cham, Switzerland, 2019; pp. 331–345.
59. Dey, S.; Luo, H.; Fokoue, A.; Hu, J.; Zhang, P. Predicting adverse drug reactions through interpretable deep learning framework. *BMC Bioinform.* **2018**, *19*, 476. [CrossRef]
60. Karimi, M.; Wu, D.; Wang, Z.; Shen, Y. DeepAffinity: Interpretable deep learning of compound–protein affinity through unified recurrent and convolutional neural networks. *Bioinformatics* **2019**, *35*, 3329–3338. [CrossRef] [PubMed]
61. Clark, C.M.; Schneider, J.A.; Bedell, B.J.; Beach, T.G.; Bilker, W.B.; Mintun, M.A.; Pontecorvo, M.J.; Hefti, F.; Carpenter, A.P.; Flitter, M.L.; et al. Use of florbetapir-PET for imaging β-amyloid pathology. *JAMA* **2011**, *305*, 275–283. [CrossRef] [PubMed]
62. Joshi, A.D.; Pontecorvo, M.J.; Clark, C.M.; Carpenter, A.P.; Jennings, D.L.; Sadowsky, C.H.; Adler, L.P.; Kovnat, K.D.; Seibyl, J.P.; Arora, A.; et al. Performance characteristics of amyloid PET with florbetapir F 18 in patients with Alzheimer's disease and cognitively normal subjects. *J. Nucl. Med.* **2012**, *53*, 378–384. [CrossRef] [PubMed]

bioengineering

MDPI

Article

GNN-surv: Discrete-Time Survival Prediction Using Graph Neural Networks

So Yeon Kim [1,2]

[1] Department of Artificial Intelligence, Ajou University, Suwon 16499, Republic of Korea; jebi1771@ajou.ac.kr
[2] Department of Software and Computer Engineering, Ajou University, Suwon 16499, Republic of Korea

Abstract: Survival prediction models play a key role in patient prognosis and personalized treatment. However, their accuracy can be improved by incorporating patient similarity networks, which uncover complex data patterns. Our study uses Graph Neural Networks (GNNs) to enhance discrete-time survival predictions (GNN-surv) by leveraging relationships in these networks. We build these networks using cancer patients' genomic and clinical data and train various GNN models on them, integrating Logistic Hazard and PMF survival models. GNN-surv models exhibit superior performance in survival prediction across two urologic cancer datasets, outperforming traditional MLP models. They maintain robustness and effectiveness under varying graph construction hyperparameter μ values, with performance boosts of up to 14.6% and 7.9% in the time-dependent concordance index and reductions in the integrated brier score of 26.7% and 24.1% in the BLCA and KIRC datasets, respectively. Notably, these models also maintain their effectiveness across three different types of GNN models, suggesting potential adaptability to other cancer datasets. The superior performance of our GNN-surv models underscores their wide applicability in the fields of oncology and personalized medicine, providing clinicians with a more accurate tool for patient prognosis and personalized treatment planning. Future studies can further optimize these models by incorporating other survival models or additional data modalities.

Keywords: discrete survival model; Graph Neural Networks; patient similarity network; survival prediction; time-to-event prediction

Citation: Kim, S.Y. GNN-surv: Discrete-Time Survival Prediction Using Graph Neural Networks. *Bioengineering* **2023**, *10*, 1046. https://doi.org/10.3390/bioengineering10091046

Academic Editors: Yan Pei and Jijiang Yang

Received: 2 August 2023
Revised: 31 August 2023
Accepted: 4 September 2023
Published: 6 September 2023

1. Introduction

The criticality of acknowledging censored observations in cancer research for accurate survival prediction is unquestionable [1]. Censored observations, such as patients lost to follow-up or outliving the study duration, often emerge in oncology studies. Overlooking these observations may cause significant bias in survival time and probability estimates, compromising the reliability of the study's findings [2]. Proper handling of censoring in survival analysis, therefore, is a cornerstone of oncology research, allowing a comprehensive and accurate portrayal of patient survival patterns, with significant implications for prognosis and clinical decision making.

Survival analysis involves a broad spectrum of continuous- and discrete-time survival models. The Cox proportional hazards regression model [3], a well-regarded continuous-time survival model, offers flexibility and interpretability. It estimates the hazard function based on a baseline hazard and an exponential function of linear predictors. The random survival forest model [4], an extension of the random forest model for right-censored survival time data, employs a decision tree ensemble trained on bootstrap data samples for robust survival time predictions. In contrast, discrete-time survival models like logistic regression models analyze hazard rates in discrete-time intervals [5,6], particularly beneficial when exact event times are unknown.

Recently, there has been an upsurge in using deep learning and machine learning for cancer survival prediction [7–17]. These advanced techniques excel at handling high-dimensional and heterogeneous data, unveiling complex patterns that traditional statistical models may miss. However, most of these studies focus predominantly on classifying patients into long-term or short-term survival groups, simplifying the multifaceted reality of patient survival times and often neglecting censored data.

To address the challenges, many researchers have developed survival models that incorporate neural networks with survival analysis. Kvamme et al. [12] combined the neural networks and Cox regression, offering a robust method to analyze survival data. Similarly, DeepSurv [13] proposed a Cox proportional hazards deep neural network, a state-of-the-art survival method to model the interplay between patients' covariates and treatment effectiveness to facilitate personalized treatment recommendations. Furthermore, DeepHit [14] introduced a deep neural network for survival analysis, specifically accounting for competing risks. This approach is particularly crucial for scenarios where multiple potential events of interest exist. In the domain of discrete-time survival models, a pioneering work by Brown [15] utilized indicator variables, which provided insights into how discrete markers can enhance survival model performance. Building on this, Nnet-survival [16] proposed a scalable approach to discrete-time survival modeling, which is designed to be used with neural networks. Collectively, these studies have advanced the integration of neural networks into survival analysis.

In the field of cancer research, Graph Neural Networks (GNNs) have achieved significant progress, uncovering intricate relationships often overlooked by traditional models. MGNN [18,19] provides a unified framework by building bipartite graphs between patients and multimodal data like gene expression and clinical information, demonstrating its efficacy in classifying short- and long-term survival across four cancer datasets. Qiu et al. [20] introduced an intratumor GNN model that leverages the spatial heterogeneity of multiple in situ biomarkers to reveal hidden prognostic value in breast cancer cases. The model's prognostic capability rivals that of conventional methods using routine biomarkers, advancing cancer prognosis. PathGNN [21] proposed a GNN model that can capture topological features in cancer pathways to predict long-term survival, identifying critical pathways linked to cancer outcomes. These studies underscore the versatility and effectiveness of GNNs in a variety of oncological research contexts. However, they still face challenges in appropriately considering censoring status in survival models, affecting the reliability of survival predictions. Although numerous studies have applied GNNs in survival prediction, their use in censoring-aware survival models remains largely unexplored. To refine and enhance the reliability of survival predictions, it is essential to integrate GNNs with survival models capable of effectively managing censoring. Our experiments with GNNs for discrete-time survival prediction models demonstrate this assumption, outperforming models that do not account for these relational structures.

In this study, we propose the hypothesis that distinct patient groups, characterized by similar genomic and clinical features, significantly influence their survivability and mortality rates. Recognizing these intergroup correlations in survival prediction can substantially improve the performance. We assume that relational structures exist within cancer patient data, contributing considerably to accurate survival predictions. We adopt patient similarity networks and GNNs to comprehend complex correlational structures and propose a GNN model specifically designed for discrete-time survival prediction (GNN-surv). As a proof-of-concept study, we conducted experiments on bladder and kidney cancer datasets. Our experiments demonstrated the effectiveness of the GNN-surv models in predicting discretized survival times, thus validating our hypothesis and research motivation. Our findings further highlight the importance of addressing the question of censoring in real-world scenarios and the potential for the broad applicability of GNN-surv models across diverse cancer datasets. The main contributions are summarized as follows.

- We design and construct a sophisticated cancer patient similarity network that integrates both genomic and clinical features, enabling a better understanding of patient characteristics and relationships.

- We propose GNN-surv, a novel GNN that incorporates discrete-time survival models. We demonstrate its broad applicability via experiments across two different survival models and three types of GNN layers.
- We empirically show the superior performance of GNN-surv in survival prediction for two urologic cancers, thereby showing its potential for broader application in oncological research.

2. Materials and Methods

2.1. Dataset

We obtained the RNA-Seq gene expression profiles of the TCGA bladder cancer (BLCA) and kidney clear cell carcinoma (KIRC) datasets. The gene expression data, comprising estimates for 20,530 genes, were measured using Illumina HiSeq 2000 RNA Sequencing, a level 3 data source from the TCGA data coordination center. We note that the same number of genes, 20,530, was present in both the BLCA and KIRC datasets. We retrieved these log-transformed RSEM normalized count data from the UCSC Xena platform [22].

The datasets included 400 patients for BLCA and 313 patients for KIRC. We excluded any patients with unrecorded or inaccurate clinical outcome variables, such as negative survival day values. For both cancer datasets, we used clinical variables such as overall survival (OS), event status, age, gender, and TNM stage. The event status was binary, with 1 indicating that an event occurred and 0 representing right-censored cases. The BLCA dataset contained 223 censored and 173 uncensored samples, yielding a censoring rate of 56.3%, whereas the KIRC dataset comprised 209 censored and 102 uncensored samples, yielding a censoring rate of 67.2%. In this study, we selected two types of urologic cancer as our subjects to illustrate the proof-of-concept. Although the TCGA dataset provides data for six types of urologic cancers, we decided to exclude certain types due to specific reasons. Testicular cancer (TGCT) and kidney chromophobe (KICH) were excluded due to their small sample sizes of 134 and 65 cases, respectively. Similarly, the kidney papillary cell carcinoma (KIRP) and prostate cancer (PRAD) datasets were not utilized. Despite their adequate sample sizes, these datasets were ruled out due to the excessive censoring rates of 85.1% and 98.2%, respectively. Such high censoring rates could introduce bias into our prediction model, thereby compromising the model's performance and the interpretation of the results [23].

We dichotomized the clinical variables, grouping ages into less than 65 years (0) and 65 years or older (1), T stages into T0-2 (0) and T3-4 (1), N stages into N0 and N1-3 (N+), and M stages into M0 and M1. Where pathologic stages were unknown, we filled in the gaps based on the American Joint Committee on Cancer (AJCC) staging system. Missing N or M stages were inferred from the number of positive lymph nodes or metastatic sites. For instance, cases with any positive lymph nodes were classified as N+, and if the only recorded metastatic site was 'lymph node only,' it was classified as M0. Notably, metastatic site data were only available in the BLCA dataset. We adopted this preprocessing strategy for clinical variables from the study [24]. The summary statistics of clinical features for both datasets are displayed in Table 1.

2.2. Patient Similarity Graph

We construct a patient similarity graph, denoted as $\mathcal{G} = (\mathcal{V}, \mathcal{E})$, where the vertex set $\mathcal{V} = (v_1, \ldots, v_n)$ symbolizes the cohort of cancer patients. Each patient, or vertex v_i, is characterized by a feature vector x_i, which combines their RNA-seq gene expression and clinical features. The set $\mathcal{E}$ defines undirected edges.

While clinical features are discretized into categories of 0 or 1, gene expression features are continuous variables. In joining genomic and clinical variables, we standardize gene expression features by subtracting the mean and scaling to unit variance.

We calculate patient similarity between patients v_i and v_j as $\mathcal{W}(i,j)$, derived from a scaled exponential similarity kernel, predicated on the probability density function of the normal distribution as follows:

$$W(i,j) = \frac{1}{\sqrt{2\pi\sigma^2}} \exp -\frac{\rho^2(x_i, x_j)}{2\sigma^2} \tag{1}$$

where $\rho(x_i, x_j)$ denotes the pairwise correlation distances between patients x_i and x_j. We compute the value σ as follows:

$$\sigma = \mu \frac{\overline{\rho}(x_i, N_i) + \overline{\rho}(x_j, N_j) + \rho(x_i, x_j)}{3} \tag{2}$$

Here, $\overline{\rho}(x_i, N_i)$ signifies the average value of the distances between x_i and its neighbors $N_{1...k}$, where k is the number of neighbors considered. μ is a hyperparameter modulating the extent to which we scale the similarity kernel W, and $\mu \in (0,1) \subset \mathbb{R}$

We define the adjacency matrix $A(i,j)$ as 1 if $W(i,j) > c$, where c is a correlation threshold, and otherwise $A(i,j) = 0$.

This study relies heavily on graph structures; thus, we utilize a sophisticated patient similarity network construction method proposed by [25]. A study's performance can be sensitive to the hyperparameter μ, also discussed in [25], leading us to empirically set μ in the range of [0.1, 1.0].

Table 1. Summary statistics of clinical features in the TCGA bladder cancer (BLCA) and kidney clear cell carcinoma (KIRC) data.

Feature		**BLCA**	**KIRC**
Number of Patients		**400**	**313**
Age	<65 years	147 (36.8%)	193 (61.7%)
	≥65 years	253 (63.2%)	120 (38.3%)
Gender	Male	295 (73.8%)	201 (64.2%)
	Female	105 (26.2%)	112 (35.8%)
Stage T (Primary tumor)	Negative (Stages T0–2)	148 (37.6%)	196 (62.6%)
	Positive (Stages T3–4)	246 (62.4%)	117 (37.4%)
Stage N (Regional lymph nodes)	Negative (Stage N0)	261 (67.4%)	244 (87.8%)
	Positive (Stage N1–3)	126 (32.6%)	34 (12.2%)
Stage M (Distant metastasis)	Negative (Stage M0)	340 (86.1%)	258 (82.7%)
	Positive (Stage M1)	55 (13.9%)	54 (17.3%)
Overall survival (OS)	Survival days (Mean ± SD [1])	810.5 ± 833.8	1310.3 ± 1062.7
	Uncensored patients	173 (43.7%)	102 (32.8%)
	Censored patients	223 (56.3%)	209 (67.2%)

[1] SD: Standard Deviation.

2.3. Graph Neural Networks for Survival Prediction

In this section, we detail our proposed architecture of Graph Neural Networks (GNNs), termed GNN-surv, tailored specifically for survival models. Our approach exploits GNNs on a patient similarity graph $\mathcal{G}$ to discern and learn from their correlational structures, which depict genomic and clinical similarities among patients. We employ an adjacency matrix A of the similarity graph $\mathcal{G}$ and a feature matrix X to train the GNN model. This model is designed to predict the discrete survival time while acknowledging the right-censored observations.

The GNN-surv architecture comprises multiple GNN layers, each succeeded by batch normalization, ReLU activation, and dropout for efficient learning and training. We also use dropout to regularize the model and prevent overfitting. The model is versatile, capable of using different types of GNN layers, such as GCNConv, SAGEConv, and GATConv. In the following section, we outline the functionality of each GNN layer.

2.3.1. Graph Convolutional Networks (GCN)

Introduced by Kipf and Welling [26], Graph Convolutional Networks (GCN) employ a convolution-based strategy that generates node embeddings by learning from the graph structure and node features, efficiently incorporating local neighborhood information into each node's embedding. Within each layer, nodes gather information from their immediate neighbors, apply a convolution operation on these features using a shared weight matrix, and pass through an activation function. The operation of GCN for a single layer is as follows:

$$h_v^{(l+1)} = \sigma\left(\frac{1}{\sqrt{D_v D_u}} \sum_{u \in N(v)} W^{(l)} h_u^{(l)}\right) \tag{3}$$

In this equation, $h_v^{(l)}$ and $h_v^{(l+1)}$ denote the feature vectors of node v at layers l and $l+1$, respectively; $N(v)$ is the set of neighboring nodes to node v; D_v and D_u are the degrees of node v and its neighboring node u, respectively, used for feature aggregation normalization; $W^{(l)}$ is the shared learnable weight matrix at layer l; and σ is an activation function, specifically a Rectified Linear Unit (ReLU) in this study.

2.3.2. GraphSAGE

Designed to generate node embeddings by sampling and aggregating features from a node's local neighborhood [27], GraphSAGE can operate on large graphs and generate embeddings for unseen nodes by leveraging node attribute information. Within each GraphSAGE layer, nodes aggregate information from their neighbors using various functions, such as mean, pooling, or LSTM, and subsequently use a learnable weight matrix to transform the aggregated information. The operation of a GraphSAGE layer is as follows:

$$h_v^{(l+1)} = \sigma\left(W^{(l)} \cdot \text{CONCAT}\left(h_v^{(l)}, \text{AGGREGATE}N(v)^{(l)}\left(hu^{(l)}\right)\right)\right) \tag{4}$$

Here, $h_v^{(l)}$ and $h_v^{(l+1)}$ represent the feature vectors of node v at layers l and $l+1$, respectively; $\text{AGGREGATE}_{N(v)}^{(l)}$ is an aggregation function that collects and processes features from the node's neighborhood $N(v)$ at layer l; $W^{(l)}$ is the learnable weight matrix at layer l; and σ is an activation function, specifically a Rectified Linear Unit (ReLU).

2.3.3. Graph Attention Networks (GAT)

Graph Attention Networks (GATs) are variants of GCNs, designed to compute node features by weighting the features of neighboring nodes with attention coefficients [28]. The attention mechanism allows the model to focus more on relevant neighbors and less on less significant ones, offering a level of flexibility that is absent in models like GCN or GraphSAGE. Within each GAT layer, each node calculates the attention coefficients with its neighbors, multiplies these coefficients with the neighbors' features, and subsequently aggregates this information. The operation of GAT for a single layer is as follows:

$$h_v^{(l+1)} = \sigma\left(\sum_{u \in N(v)} \alpha_{vu} W^{(l)} h_u^{(l)}\right) \tag{5}$$

In this equation, $h_v^{(l)}$ and $h_v^{(l+1)}$ represent the feature vectors of node v at layers l and $l+1$, respectively; α_{vu} are the attention coefficients that weigh the importance of node u's features to node v; $W^{(l)}$ is the learnable weight matrix at layer l; and σ is an activation function, specifically a LeakyReLU in this case. The attention coefficients α_{vu} are computed using a shared attention mechanism across all edges in the graph.

2.3.4. Discrete-Time Survival Models

We train the GNN-surv to minimize the loss using the ADAM optimizer, which in turn enables the learning of optimal parameters for the prediction task. The loss function and hazards are computed as defined in the discrete-time survival prediction models that we utilize. Specifically, we employ two discrete-time survival models, Logistic Hazard and the Probability Mass Function (PMF) model, both of which are implemented as per the methodology described in [29].

The Logistic Hazard method, first proposed in [15], is a discrete-time survival prediction model that has been enhanced by the capabilities of neural networks [29]. Also known as Partial Logistic Regression [30] or Nnet-survival[16], this method estimates discrete hazards—the probabilities of event occurrence within discrete-time intervals. By optimizing the survival likelihood, this method serves as a potent tool for survival analysis.

Complementarily, the PMF model, as detailed in [29], adopts a discrete-time survival prediction approach as well. It underlies methods such as DeepHit [14] and Multi-Task Logistic Regression (MTLR) [31]. These methods leverage neural networks to maximize the likelihood of right-censored time-to-event data within discrete time. The approach focuses on parameterizing PMF—the probabilities of discrete outcomes—and optimizing the survival likelihood. By probabilistically representing the distribution of event occurrence times, the PMF model provides a detailed understanding of survival analysis. Moreover, its integration with neural networks enables it to handle complex patterns in the data, thereby contributing to robust and reliable survival predictions.

2.4. *Performance Evaluation*

We evaluate the performance of our model using two metrics, namely the time-dependent concordance index (C^{td}) [32] and the integrated Brier score (IBS) [33,34]. Both metrics are utilized in accordance with the implementation described in [29].

In this study, we focus on a supervised node-level prediction problem, where the nodes represent cancer patients at risk. The primary objective of this model is to predict discrete-time survival while accounting for the patients' censoring status. As the discrete-time models necessitate the discretization of continuous survival time, we adopt the discretization scheme suggested in [29]. This scheme corresponds to either equidistant times or equidistant marginal survival probabilities. Moreover, it interpolates the discrete-time predictions, corresponding to either piecewise constant density functions or piecewise constant hazard rates.

The IBS metric accounts for both the discrimination and calibration of the survival estimates and also accommodates censored individuals by weighting the score inversely against the estimated censoring distribution. As the IBS is significantly influenced by the discretization scheme (which in turn depends on the number of output nodes), we compute the IBS over 100 equidistant points between the minimum and maximum observed times in the validation set during model training, as discussed in [29]. Conversely, the C^{td} only evaluates the discriminative capabilities of a method's predictions. It is informative to examine both metrics as they might indicate a trade-off between well-calibrated estimates and effective discriminative performance [29]. The best survival prediction performance is considered to be when the IBS is the lowest and the C^{td} is the highest.

For the purpose of our study, we randomly divide the entire sample into an 80% training set and a 20% test set. To further optimize and validate model training, we subdivide the training set into an 80% training subset and a 20% validation subset, resulting in a 64% training, 16% validation, and 20% test set. We repeat this process for 50 iterations of random splits and calculate the mean C^{td} and IBS of the total repetitions in the test set. When partitioning the data, we split the entire graph into three separate graphs: the training graph, the validation graph, and the test graph. This division signifies that GNN-surv possesses an inductive learning capacity, an important attribute as it allows the model to generalize and make predictions on unseen data, enhancing the robustness and utility of our model in real-world applications.

3. Results

3.1. Experimental Setting

In the process of constructing the patient similarity graph W, we carefully adjust the hyperparameter μ, as defined in Equation (2), within an empirical range of 0.1 to 1. This adjustment is made in increments of 0.1 due to the observed sensitivity of the model performance to variations in the hyperparameter μ [25]. The number of neighbors k within Equation (2) is set to 20 based on [25], indicating that model performance demonstrates relative insensitivity to changes in the number of neighbors. While creating the adjacency matrix A, we define a correlation threshold c of 0.5, adhering to the widely accepted consensus that correlations beneath this value are generally considered as low [35].

Our proposed model, GNN-surv, employs multiple GNN layers trained to predict discrete-time survival using discrete survival models. We utilize the PyTorch Geometric library (PyG) implementations [36] for each of the GNN layers: GCN, GraphSAGE, and GAT. To demonstrate the effectiveness of the proposed GNN-surv model, we incorporate a vanilla Multi-Layer Perceptron (MLP) layer with discrete survival models, designating this as MLP-surv in our experiments. It should be noted that the MLP layer does not utilize graph-structured data but is trained with a feature matrix $X \in \mathbb{R}^{N \times p}$, which represents p gene expression and clinical features across N samples.

We empirically set certain hyperparameters for both the GNN models and the MLP model. The GNN model consists of a three-layer network with a hidden layer size of 32. We train the model for a maximum of 500 epochs with a learning rate of 0.001, employing an early stopping scheme. In this training process, we incorporate a batch normalization layer and dropout. The batch size is determined to be 256, with the dropout rate set to 0.7 in the experiments. Additionally, we train with a batch of graphs to manage the complexity of graph data and set the batch size of these data to 32. For a fair comparison, we use the same hyperparameters for the MLP model as well.

3.2. Hyperparameter Optimization in Graph Construction

We evaluated and compared the performance of the MLP model and GNN models when paired with two discrete-time survival models: Logistic Hazard and PMF. To optimize the performance, we set the hyperparameter μ empirically during the construction of the patient similarity graph. We conducted an ablation study, detailed in Figures 1 and 2, to find the best μ value. In these studies, the time-dependent concordance index (C^{td}) was used as the performance metric.

We trained our GNN-surv models using a similarity network configured with the optimal μ value, which exhibited the best performance for each survival model. Interestingly, the scaling parameter μ demonstrated a negligible impact on the performance of the GNN-surv model in the BLCA dataset, as depicted in Figure 1.

For the KIRC dataset, the Logistic Hazard model maintained robust and enhanced performance. However, when μ was larger, GAT-surv's performance diminished compared to that of the MLP model in the PMF model, as shown in Figure 2. For this dataset, both GAT-surv and GCN-surv demonstrated relative instability in the PMF model, especially when contrasted with their performance in the BLCA dataset. Nevertheless, SAGE-surv consistently outperformed the MLP model in both survival models, underlining the efficacy of GraphSAGE when implemented in the GNN-surv model in the KIRC dataset.

For the bladder cancer (BLCA) data, we configured the μ parameter to 0.2 for the Logistic Hazard model and to 0.8 for the PMF model. In contrast, for the kidney cancer (KIRC) data, we set μ to 0.3 for the Logistic Hazard model and 0.2 for the PMF model.

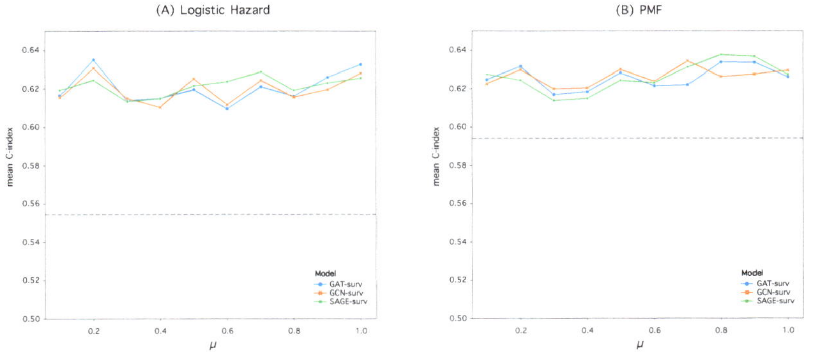

Figure 1. Performance variation of GNN-surv models in the BLCA dataset for different hyperparameter μ values. The discrete-time survival models are (**A**) Logistic Hazard and (**B**) PMF. The performance metrics are mean C^{td} values obtained from 50 random data splits. The gray dotted line represents the mean C^{td} of the MLP model, used as a baseline.

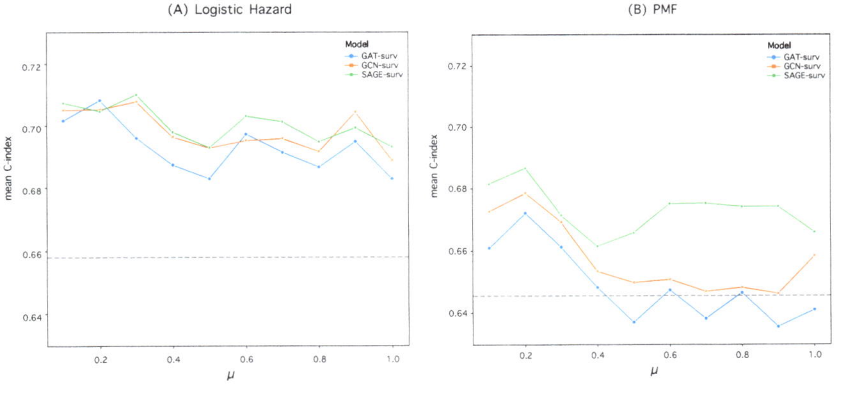

Figure 2. Performance variation of GNN-surv models in the KIRC dataset for different hyperparameter μ values. The discrete-time survival models are (**A**) Logistic Hazard and (**B**) PMF. The performance metrics are mean C^{td} values obtained from 50 random data splits. The gray dotted line represents the mean C^{td} of the MLP model, used as a baseline.

3.3. Survival Prediction Performance

The assessment of survival prediction performance was executed by evaluating the mean time-dependent concordance index (C^{td}) and the integrated Brier score (IBS), along with standard deviations across 50 data splits. The C^{td} is a metric used to evaluate the predictive accuracy of survival models over time. It measures the agreement between predicted and observed survival times, with a higher value indicating better model performance. The IBS quantifies the overall prediction error for a survival model across all time points, with lower values indicating better accuracy and calibration. Detailed results for the bladder cancer (BLCA) and kidney cancer (KIRC) datasets are provided in Tables 2 and 3, respectively.

The performance comparison of the GNN-surv models and the MLP model within the BLCA and KIRC datasets demonstrates a significant improvement when using GNN-surv models. In the BLCA dataset, the GAT-surv model outperforms the MLP-surv model, with an increase of approximately 14.6% and 7% in the C^{td} metric for the Logistic Hazard and PMF survival models, respectively. Concurrently, the GCN-surv and SAGE-surv models demonstrate superior performance, reducing the IBS by approximately 26.7% and 28.3%, respectively, for the same models. Upon examining the KIRC dataset, the SAGE-surv model emerges as the top performer, achieving a 7.9% and 6.4% increase in C^{td} for the Logistic Hazard and PMF models, respectively, while also reducing the IBS by 24.1% and 8.1%, respectively. Similar performance improvements are also exhibited by the GCN-surv and GAT-surv models.

These results underscore the superior performance of GNN-surv models in survival analysis within these datasets, revealing the benefits of integrating GNN models for discrete survival time-to-event predictions on a patient similarity graph. Given that the MLP model does not account for graph structures, this significant enhancement suggests the existence of a meaningful correlational structure within patient similarity graphs. By effectively utilizing these relationships, the proposed GNN-surv models substantially improve the survival prediction performance.

Upon examining the GNN-surv models, the differential performance between GCN-surv, SAGE-surv, and GAT-surv was relatively small, illustrating the adaptability and resilience of GNN-surv models across different GNN layers. However, a standout observation was the extraordinary performance of the SAGE-surv model within the KIRC dataset. This model demonstrated the highest performance metrics across both survival models, reinforcing the efficacy of the GraphSAGE model in survival prediction applications in the KIRC dataset. Interestingly, the majority of the GNN-surv models portrayed consistent and enhanced performance across both datasets, exhibiting insensitivity to variations in the μ parameter, generally for both survival models. This demonstrates their robustness and effectiveness and suggests that GNN-surv models could be seamlessly adapted to different GNN layers, providing a versatile framework for survival prediction.

Table 2. Performance comparison of GNN-surv models and MLP model within the BLCA dataset using two discrete-time survival models, Logistic Hazard and PMF. The metrics utilized for performance assessment include the mean C^{td} and IBS, along with their respective standard deviations, acquired from 50 random data splits. The highest performance for each metric and survival model is highlighted in bold text.

	Logistic Hazard ($\mu = 0.2$)		**PMF ($\mu = 0.8$)**	
Model	C^{td}	**IBS**	C^{td}	**IBS**
MLP-surv	0.5543 ± 0.0689	0.3183 ± 0.0497	0.5941 ± 0.0629	0.2324 ± 0.0222
GCN-surv	0.6309 ± 0.0481	**0.2331 ± 0.0358**	0.6265 ± 0.0493	**0.2130 ± 0.0231**
SAGE-surv	0.6247 ± 0.0505	**0.2331 ± 0.0389**	**0.6378 ± 0.0415**	0.2140 ± 0.0238
GAT-surv	**0.6352 ± 0.0520**	0.2341 ± 0.0365	0.6339 ± 0.0451	0.2154 ± 0.0229

Table 3. Performance comparison of GNN-surv models and MLP model within the KIRC dataset using two discrete-time survival models, Logistic Hazard and PMF. The metrics utilized for performance assessment include the mean C^{td} and IBS, along with their respective standard deviations, acquired from 50 random data splits. The highest performance for each metric and survival model is highlighted in bold text.

	Logistic Hazard ($\mu = 0.3$)		**PMF ($\mu = 0.2$)**	
Model	C^{td}	**IBS**	C^{td}	**IBS**
MLP-surv	0.6581 ± 0.0559	0.2577 ± 0.0902	0.6455 ± 0.0516	0.2022 ± 0.0174
GCN-surv	0.7077 ± 0.0373	0.1965 ± 0.0240	0.6785 ± 0.0464	0.1964 ± 0.0195
SAGE-surv	**0.7099 ± 0.0409**	**0.1955 ± 0.0269**	**0.6868 ± 0.047**	**0.1859 ± 0.0222**
GAT-surv	0.6962 ± 0.0362	0.2018 ± 0.0260	0.672 ± 0.05	0.1958 ± 0.0233

Furthermore, we delved into the exploration of the Brier scores relative to survival time (in days), as depicted in Figures 3 and 4. For the calculation of the IBS, the Brier score was obtained across 100 equidistant points, falling between the minimum and maximum observed times.

Reiterating the consistent superior performance of the GNN-surv models, these models continued to outperform the MLP model across all durations in the Logistic Hazard model within both datasets. Specifically, the MLP model showed noticeably poor performance during shorter durations across both survival models, in the BLCA data. When the PMF model was utilized, there were instances where SAGE-surv underperformed slightly compared to the MLP model within specific time durations—approximately around median time—in the BLCA data. Furthermore, the performance enhancements of the GNN-surv models, which were previously noted in the Logistic Hazard model, were relatively marginal and displayed instability for larger durations when applied to the PMF model in the KIRC data. However, the majority of the GNN-surv models substantially outperformed the MLP model and demonstrated stable performance across most durations.

The consideration of the integrated Brier score (IBS) revealed that, for both the MLP-surv and GNN-surv models, the PMF model could perform better than the Logistic Hazard. This was an interesting finding, given that no significant differences were observed between the survival models when the time-dependent concordance index (C^{td}) was considered. This points to the value of the IBS, as it takes into account both the discrimination and calibration of survival estimates, whereas the Logistic Hazard is primarily concerned with discriminative performance. In other words, the IBS not only considers the model's discriminative ability but also its calibration. Calibration refers to the accuracy of the model's predicted probabilities. A well-calibrated model's predicted probabilities of an event (such as survival in a certain time period) should match the actual proportion of that event in the observed data. In this context, the fact that the PMF model performs significantly better than the Logistic Hazard model when evaluated with the IBS (but not when evaluated with C^{td}) suggests that the PMF model's predicted probabilities may be more accurate (i.e., better calibrated) than those of the Logistic Hazard model.

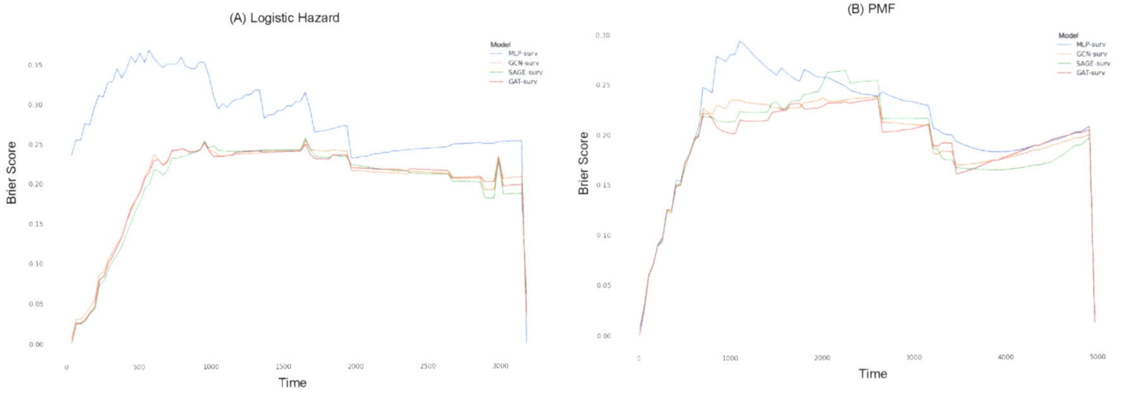

Figure 3. Brier scores over survival time (in days) for different GNN-surv models and MLP model in the BLCA dataset. Discrete-time survival models considered are (**A**) Logistic Hazard and (**B**) PMF.

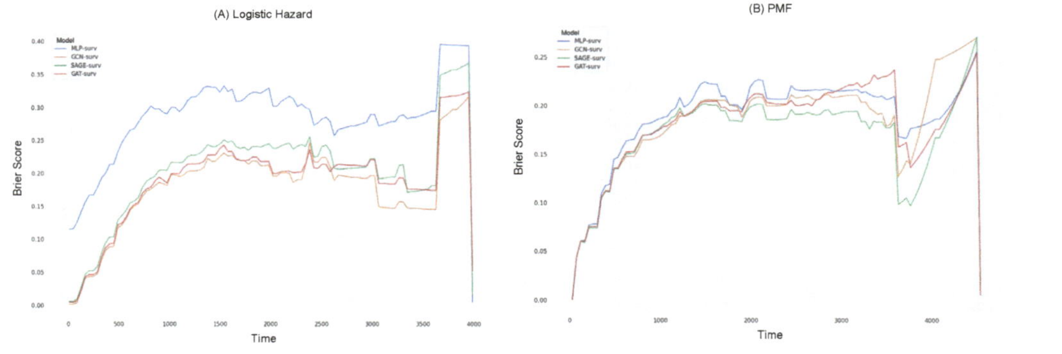

Figure 4. Brier scores over survival time (in days) for different GNN-surv models and MLP model in the KIRC dataset. Discrete-time survival models considered are (**A**) Logistic Hazard and (**B**) PMF.

4. Discussion

This study examined the potential of Graph Neural Networks (GNNs), specifically designed to enhance patient survival predictions for urologic cancers (GNN-surv). However, the research was limited by several key limitations. The scope of our investigation was limited to two types of urologic cancers, dictated by the need for a sufficient number of samples and balanced censoring rates. However, the potential to extend our method to a pan-cancer integrative analysis warrants further discussion. Our model's promising performance in predicting discrete survival outcomes for two urologic cancers hints at the possibility that it could be applicable across a wider range of cancer types. The foundation for such broad application lies in the inherent flexibility of the GNN model, which can adapt to different data types and structures.

Furthermore, the absence of feature selection in our analysis introduced potential challenges. Typically, survival prediction models involve feature selection to focus on statistically significant genes that correlate with survival outcomes. However, in this proof-of-concept study, we excluded feature selection to concentrate on the comparative effectiveness of the GNN model against a simple neural network model. This resulted in the inclusion of an excessive number of redundant genomic features, derived from over 20,000 genes, which may have impacted the overall performance of our models. Despite these limitations, our research provided several noteworthy findings. The superiority of the GNN-surv models over the MLP-surv model indicates that the former's ability to leverage the correlational structures within patient similarity graphs for survival prediction is beneficial. Furthermore, the enhanced performance of GNN-surv models, even with a excessive number of features and a relatively small sample size, suggests the potential to mitigate the effect of redundant features and extract meaningful patterns in a patient similarity graph for discrete survival prediction.

Despite the demonstrated efficacy of the discrete-time survival models employed in this study, there is potential for further enhancement through the incorporation of continuous-time survival models, such as the Cox Partial Hazard (Cox-PH) model, into GNN-surv models. This could allow for the consideration of time-varying covariates, providing a better understanding of patient survival probabilities over time. Furthermore, this integration could potentially enable the GNN-surv model to capture more intricate temporal dynamics within the data. This is particularly beneficial when dealing with longitudinal data or when a fine-grained temporal resolution is critical for clinical decision making. Moreover, by fusing both discrete- and continuous-time models, the GNN-surv framework could potentially serve as a unified platform for survival prediction, capable of handling a wide spectrum of clinical scenarios and data types.

Lastly, GNNs promise to revolutionize survival prediction models but face limitations, including reliance on robust graph structures and vulnerability to high-dimensionality in data. These challenges arise from the need for well-defined graph structures in the face of complex oncological data and high feature-to-sample ratios, common in biomedical datasets. Bypassing feature selection, GNNs can, however, prioritize informative structures over less significant ones. Thus, the elaborated graph construction methodology also presents an avenue for further exploration. Networks established using diverse types of omics profiles might capture the inherent patient heterogeneity more accurately, enhancing the precision of survival predictions. Future directions to address these constraints include structure learning to optimize graph structures, the development of explainable GNN models for comprehensive patient interaction analysis, and investigating feature-wise interactions such as gene–gene and gene–disease interactions to identify critical biomarkers for survival prediction. Moreover, the integration of other data modalities, such as pathological images, could provide a richer context to these networks, offering a more comprehensive picture of patients' health conditions.

5. Conclusions

To summarize, our research underscores the promising potential of GNN-surv models within the context of discrete-time survival prediction and patient similarity networks. Our findings reveal that GNN-surv models consistently outperformed traditional MLP models across various performance metrics in two urologic cancer datasets. The superior performance of these models could greatly assist clinicians by providing more accurate survival predictions, consequently guiding the formulation of personalized treatment strategies. The successful use of patient similarity graphs in our GNN-surv models also suggests the existence of valuable correlational structures within these networks, offering potential leverage for survival prediction. Despite the aforementioned limitations, our findings signal the potential for the wide applicability of GNN-surv models in survival prediction tasks. The continued exploration and refinement of these models, their application to diverse datasets, and the integration of various survival models could significantly enhance personalized treatment strategies in the realm of oncology research.

Funding: This work was supported by a National Research Foundation of Korea (NRF) grant funded by the Korean government (MSIT) [NRF-2022R1C1C1012060].

Institutional Review Board Statement: Not applicable.

Informed Consent Statement: Not applicable.

Data Availability Statement: The data used in this study come from the publicly accessible TCGA dataset and can be downloaded from UCSC Xena at: http://xena.ucsc.edu.

Acknowledgments: The authors gratefully acknowledge the TCGA Consortium and all its members within the TCGA Project initiative, for providing sample, tissues, and data processing and making data and results available. The results published here are in whole or part based upon data generated by The Cancer Genome Atlas pilot project established by the NCI and NHGRI. Information about TCGA and the investigators and institutions that constitute the TCGA research network can be found at http://cancergenome.nih.gov.

Conflicts of Interest: The authors declare no conflict of interest.

References

1. Leung, K.M.; Elashoff, R.M.; Afifi, A.A. Censoring issues in survival analysis. *Annu. Rev. Public Health* **1997**, *18*, 83–104. [CrossRef] [PubMed]
2. Van Wieringen, W.N.; Kun, D.; Hampel, R.; Boulesteix, A.L. Survival prediction using gene expression data: A review and comparison. *Comput. Stat. Data Anal.* **2009**, *53*, 1590–1603. [CrossRef]
3. Cox, D.R. Regression models and life-tables. *J. R. Stat. Soc. Ser. B* **1972**, *34*, 187–202. [CrossRef]
4. Ishwaran, H.; Kogalur, U.B.; Blackstone, E.H.; Lauer, M.S. Random survival forests. *Ann. Appl. Stat.* **2008**, *2*, 841–860. [CrossRef]
5. Efron, B. Logistic regression, survival analysis, and the Kaplan-Meier curve. *J. Am. Stat. Assoc.* **1988**, *83*, 414–425. [CrossRef]
6. Allison, P.D. Discrete-time methods for the analysis of event histories. *Sociol. Methodol.* **1982**, *13*, 61–98. [CrossRef]

7. Burke, H.B.; Goodman, P.H.; Rosen, D.B.; Henson, D.E.; Weinstein, J.N.; Harrell, F.E., Jr.; Marks, J.R.; Winchester, D.P.; Bostwick, D.G. Artificial neural networks improve the accuracy of cancer survival prediction. *Cancer* **1997**, *79*, 857–862. [CrossRef]
8. Ahmed, F.E. Artificial neural networks for diagnosis and survival prediction in colon cancer. *Mol. Cancer* **2005**, *4*, 29. [CrossRef]
9. Sun, D.; Wang, M.; Li, A. A multimodal deep neural network for human breast cancer prognosis prediction by integrating multi-dimensional data. *IEEE/ACM Trans. Comput. Biol. Bioinform.* **2018**, *16*, 841–850. [CrossRef]
10. Lundin, M.; Lundin, J.; Burke, H.; Toikkanen, S.; Pylkkänen, L.; Joensuu, H. Artificial neural networks applied to survival prediction in breast cancer. *Oncology* **1999**, *57*, 281–286. [CrossRef]
11. Lopez-Garcia, G.; Jerez, J.M.; Franco, L.; Veredas, F.J. Transfer learning with convolutional neural networks for cancer survival prediction using gene-expression data. *PLoS ONE* **2020**, *15*, e0230536. [CrossRef] [PubMed]
12. Kvamme, H.; Borgan, Ø.; Scheel, I. Time-to-event prediction with neural networks and Cox regression. *arXiv* **2019**, arXiv:1907.00825.
13. Katzman, J.L.; Shaham, U.; Cloninger, A.; Bates, J.; Jiang, T.; Kluger, Y. DeepSurv: Personalized treatment recommender system using a Cox proportional hazards deep neural network. *BMC Med. Res. Methodol.* **2018**, *18*, 24. [CrossRef] [PubMed]
14. Lee, C.; Zame, W.; Yoon, J.; Van Der Schaar, M. Deephit: A deep learning approach to survival analysis with competing risks. In Proceedings of the AAAI Conference on Artificial Intelligence, New Orleans, LA, USA, 2–7 February 2018; Volume 32.
15. Brown, C.C. On the use of indicator variables for studying the time-dependence of parameters in a response-time model. *Biometrics* **1975**, *31*, 863–872. [CrossRef] [PubMed]
16. Gensheimer, M.F.; Narasimhan, B. A scalable discrete-time survival model for neural networks. *PeerJ* **2019**, *7*, e6257. [CrossRef]
17. Khan, F.M.; Zubek, V.B. Support vector regression for censored data (SVRc): A novel tool for survival analysis. In Proceedings of the 2008 Eighth IEEE International Conference on Data Mining, Pisa, Italy, 15–19 December 2008; pp. 863–868.
18. Gao, J.; Lyu, T.; Xiong, F.; Wang, J.; Ke, W.; Li, Z. MGNN: A multimodal graph neural network for predicting the survival of cancer patients. In Proceedings of the 43rd International ACM SIGIR Conference on Research and Development in Information Retrieval, Xi'an, China, 25–30 July 2020; pp. 1697–1700.
19. Gao, J.; Lyu, T.; Xiong, F.; Wang, J.; Ke, W.; Li, Z. Predicting the survival of cancer patients with multimodal graph neural network. *IEEE/ACM Trans. Comput. Biol. Bioinform.* **2021**, *19*, 699–709. [CrossRef]
20. Qiu, L.; Kang, D.; Wang, C.; Guo, W.; Fu, F.; Wu, Q.; Xi, G.; He, J.; Zheng, L.; Zhang, Q.; et al. Intratumor graph neural network recovers hidden prognostic value of multi-biomarker spatial heterogeneity. *Nat. Commun.* **2022**, *13*, 4250. [CrossRef]
21. Liang, B.; Gong, H.; Lu, L.; Xu, J. Risk stratification and pathway analysis based on graph neural network and interpretable algorithm. *BMC Bioinform.* **2022**, *23*, 394. [CrossRef]
22. Goldman, M.J.; Craft, B.; Hastie, M.; Repečka, K.; McDade, F.; Kamath, A.; Banerjee, A.; Luo, Y.; Rogers, D.; Brooks, A.N.; et al. Visualizing and interpreting cancer genomics data via the Xena platform. *Nat. Biotechnol.* **2020**, *38*, 675–678. [CrossRef]
23. Zhu, X.; Zhou, X.; Zhang, Y.; Sun, X.; Liu, H.; Zhang, Y. Reporting and methodological quality of survival analysis in articles published in Chinese oncology journals. *Medicine* **2017**, *96*, e9204. [CrossRef]
24. Kim, S.Y.; Choe, E.K.; Shivakumar, M.; Kim, D.; Sohn, K.A. Multi-layered network-based pathway activity inference using directed random walks: Application to predicting clinical outcomes in urologic cancer. *Bioinformatics* **2021**, *37*, 2405–2413. [CrossRef] [PubMed]
25. Wang, B.; Mezlini, A.M.; Demir, F.; Fiume, M.; Tu, Z.; Brudno, M.; Haibe-Kains, B.; Goldenberg, A. Similarity network fusion for aggregating data types on a genomic scale. *Nat. Methods* **2014**, *11*, 333–337. [CrossRef] [PubMed]
26. Kipf, T.N.; Welling, M. Semi-supervised classification with graph convolutional networks. *arXiv* **2016**, arXiv:1609.02907.
27. Hamilton, W.; Ying, Z.; Leskovec, J. Inductive representation learning on large graphs. In Proceedings of the 31st Conference on Neural Information Processing Systems, Long Beach, CA, USA, 4–9 December 2017; Volume 30
28. Veličković, P.; Cucurull, G.; Casanova, A.; Romero, A.; Liò, P.; Bengio, Y. Graph Attention Networks. In Proceedings of the 6th International Conference on Learning Representations, Vancouver, BC, Canada, 30 April–3 May 2018.
29. Kvamme, H.; Borgan, Ø. Continuous and discrete-time survival prediction with neural networks. *Lifetime Data Anal.* **2021**, *27*, 710–736. [CrossRef] [PubMed]
30. Biganzoli, E.; Boracchi, P.; Mariani, L.; Marubini, E. Feed forward neural networks for the analysis of censored survival data: A partial logistic regression approach. *Stat. Med.* **1998**, *17*, 1169–1186. [CrossRef]
31. Yu, C.N.; Greiner, R.; Lin, H.C.; Baracos, V. Learning patient-specific cancer survival distributions as a sequence of dependent regressors. In Proceedings of the 25th Conference on Neural Information Processing Systems, Granada, Spain, 12–14 December 2011; Volume 24
32. Antolini, L.; Boracchi, P.; Biganzoli, E. A time-dependent discrimination index for survival data. *Stat. Med.* **2005**, *24*, 3927–3944. [CrossRef]
33. Graf, E.; Schmoor, C.; Sauerbrei, W.; Schumacher, M. Assessment and comparison of prognostic classification schemes for survival data. *Stat. Med.* **1999**, *18*, 2529–2545. [CrossRef]
34. Gerds, T.A.; Schumacher, M. Consistent estimation of the expected Brier score in general survival models with right-censored event times. *Biom. J.* **2006**, *48*, 1029–1040. [CrossRef]
35. Hinkle, D.E.; Wiersma, W.; Jurs, S.G. *Applied Statistics for the Behavioral Sciences*; Houghton Mifflin College Division: Boston, MA, USA, 2003; Volume 663.
36. Fey, M.; Lenssen, J.E. Fast Graph Representation Learning with PyTorch Geometric. In Proceedings of the ICLR Workshop on Representation Learning on Graphs and Manifolds, New Orleans, LA, USA, 6–9 May 2019.

bioengineering

MDPI

Article

Prediction of Cognitive Load from Electroencephalography Signals Using Long Short-Term Memory Network

Gilsang Yoo [1], Hyeoncheol Kim [2] and Sungdae Hong [3,*]

[1] Creative Informatics and Computing Institute, Korea University, Seoul 02841, Republic of Korea
[2] College of Informatics, Korea University, Seoul 02841, Republic of Korea
[3] Division of Design, Seokyeong University, Seoul 02713, Republic of Korea
* Correspondence: sungdaehong@gmail.com

Abstract: In recent years, the development of adaptive models to tailor instructional content to learners by measuring their cognitive load has become a topic of active research. Brain fog, also known as confusion, is a common cause of poor performance, and real-time detection of confusion is a challenging and important task for applications in online education and driver fatigue detection. In this study, we propose a deep learning method for cognitive load recognition based on electroencephalography (EEG) signals using a long short-term memory network (LSTM) with an attention mechanism. We obtained EEG signal data from a database of brainwave information and associated data on mental load. We evaluated the performance of the proposed LSTM technique in comparison with random forest, Adaptive Boosting (AdaBoost), support vector machine, eXtreme Gradient Boosting (XGBoost), and artificial neural network models. The experimental results demonstrated that the proposed approach had the highest accuracy of 87.1% compared to those of other algorithms, including random forest (64%), AdaBoost (64.31%), support vector machine (60.9%), XGBoost (67.3%), and artificial neural network models (71.4%). The results of this study support the development of a personalized adaptive learning system designed to measure and actively respond to learners' cognitive load in real time using wireless portable EEG systems.

Keywords: electroencephalography; long short-term memory network; attention mechanism; cognitive load; deep learning

Citation: Yoo, G.; Kim, H.; Hong, S. Prediction of Cognitive Load from Electroencephalography Signals Using Long Short-Term Memory Network. *Bioengineering* **2023**, *10*, 361. https://doi.org/10.3390/bioengineering10030361

Academic Editors: Yan Pei, Jijiang Yang and Larbi Boubchir

Received: 9 February 2023
Revised: 11 March 2023
Accepted: 14 March 2023
Published: 15 March 2023

1. Introduction

The COVID-19 pandemic has caused significant disruptions to traditional classroom education worldwide, resulting in a surge in distance learning methods [1,2]. The rapid development of information technology (IT) has facilitated this transition by allowing students to continue their education from a distance. Consequently, traditional classroom education has gradually integrated online and distance learning methods, with distance learning emerging as a new trend in education [3,4]. Distance learning offers learners the flexibility to create a learning environment that transcends spatial and temporal constraints.

During the pandemic, many people were forced to work and study remotely, which has increased interest in developing methods for monitoring cognitive load levels in these settings. A recent issue related to cognitive load recognition has been its application to remote work and online learning. In this context, the challenge is the lack of face-to-face interactions, which makes it difficult to detect non-verbal cues that indicate cognitive load levels. Consequently, researchers have been exploring the use of physiological signals, such as EEG and eye tracking, to monitor cognitive load levels in real time.

Cognitive load is a measure of the mental effort required to complete a task and can be used to predict performance, fatigue, and stress levels. Cognitive load recognition is designed to improve human performance by identifying and monitoring cognitive load levels in real time. Therefore, cognitive load detection has numerous applications in various domains, such as healthcare, education, and aviation.

Driven by advances in computational neuroscience, research has been conducted to measure learners' cognitive load based on the cognitive load theory. Cognitive load is one of the main causes of poor performance in a wide variety of tasks, including learning processes and associated thinking or reflection. If the degree of cognitive load of learners in learning or work processes is reflected, it can be used to develop adaptive instructional designs. However, most existing studies have focused on methods to estimate learners' degree of cognitive load during the learning outcome stage [5,6]. A study investigating prefrontal cortex (PFC) hemodynamics using functional near-infrared spectroscopy (fNIRS) while performing n-back and random number generation (RNG) tasks with multiple cognitive loads suggested a relationship between subjective workload and brain activity [7]. In attempting to quantify the cognitive load, cognitive load modeling techniques using deep learning are also being studied, considering workload mechanisms and their impact on human performance [8].

Measuring the degree of cognitive load of learners after completing a learning experience has certain limitations. However, the measurement of these qualities during learning presents several challenges. To overcome these limitations, the development of adaptive models that provide instructional control to learners by measuring their cognitive load in real time has emerged as a promising approach. Real-time teaching feedback could facilitate active support that reflects changes in learning status and real-time applications to help develop adaptive instructional materials. These materials based on real-time measurement can manage learners' cognitive load and participation in learning at an appropriate level during learning and help educators identify learners' difficulties. To realize such real-time adaptive teaching, a method to measure learners' cognitive load in real time during their learning experiences must be developed as a technical prerequisite.

In this study, we consider that learners' cognitive load can be measured in real time using data on their physiological and psychological responses. Electroencephalography (EEG) is commonly used to collect these data. Hence, measuring cognitive load by collecting learners' physiological data does not interfere with their learning experience. Cognitive load was measured using EEG analysis. EEG is the flow of electricity generated by signal transmission between brain nerves, and EEG analysis analyzes the frequency change in the EEG. Because EEGs exhibit different frequency wavelengths depending on mental activity, the degree of cognitive load can be measured by EEG analysis [9,10]. However, the generation of brainwaves is greatly affected by physical exercise and by differences in individual cognitive abilities. Noise in the signals may also pose some difficulties in interpreting information. Moreover, some authors have noted that EEG readings can be affected by other mental activities and that the continuous nature of the collected data poses notable difficulties in determining a person's degree of cognitive load.

However, this approach can be used to develop a model to predict specific results using learner information by applying artificial intelligence-based methods such as machine learning. Friedman et al. explored various cognitive load prediction models based on machine learning using learners' EEG measurement data. They compared and analyzed four machine learning algorithms (XGBoost, random forest, artificial neural network, and simple linear regression models) and reported that the XGBoost algorithm exhibited the highest predictive accuracy [11]. Machine learning algorithms may vary in prediction accuracy owing to variables such as the size of the training dataset. Similarly, the accuracy of artificial neural network algorithms varies with the number of hidden layers implemented in different models. Hence, a comparative analysis of the various algorithms is required. Therefore, we trained several machine learning models to predict cognitive load based on EEG data to compare their predictive performance.

In this study, we aimed to develop a model to measure learners' cognitive load based on their neurophysiological reactions. Additionally, we are interested in creating personalized models that can account for individual differences in cognitive load responses. To this end, we developed a long short-term memory (LSTM)-based machine learning model to predict the degree of cognitive load using EEG data. To induce a measurable difference in cognitive load, we presented participants with video learning tasks of different difficulty levels and collected EEG data to compare the degree of understanding of the content that the participants showed during the tasks. Based on these data, we applied support vector machines, K-nearest neighbors, artificial neural networks, convolutional neural networks, deep belief networks, (recurrent neural network) RNN-LSTM, bidirectional LSTM, and bidirectional LSTM attention models to compare their performance in handling data most predictably and efficiently. A recent issue related to cognitive load recognition has been its application to remote work and online learning and the development of more accurate and personalized models to monitor cognitive load levels in these settings.

The remainder of this paper is organized as follows. Section 2 introduces previous studies related to cognitive load, and Section 3 explains the implementation of the proposed bidirectional LSTM combination model. Section 4 explains the analysis and results of the study, and Section 5 presents conclusions and future research directions.

2. Related Work

2.1. Cognitive Load

Methods of measuring mental workload include subjective methods using response forms filled out by participants, and objective methods, including the use of psychophysiological measurements [12,13]. One of the best ways to measure mental workload with a high temporal resolution is to utilize EEG data [11–13]. In this study, we propose an algorithm to explore the mental workload associated with multitasking activities using EEG measurements and to recognize different levels of mental workload.

According to the theory of cognitive load, learners' management of cognitive resources is considered important for effective learning. Cognitive load theory argues that information processing that occurs in the learning process must be implemented within a limited capacity of working memory and that cognitive overload occurs if mental activity exceeds this limit [14]. The total cognitive load is composed of the sum of the extrinsic and intrinsic loads, of which the extrinsic load is considered to be lowered through efficient instructional design because it is a negative load owing to an incorrect design [15]. In contrast, because the intrinsic load is considered a positive load that helps form cognitive schemas, the total amount of cognitive load must be low for it to not be positive. For successful learning, appropriate teaching controls should be provided depending on learners' individual characteristics to avoid imposing either an excessively high or low cognitive load for a given learning situation [16]. This argument of the cognitive load theory is related to the need for adaptive teaching. Teaching should be adjusted according to the level and characteristics of each learner. In particular, the expertise reversal effect of teaching guidance that does not meet the needs of learners can act as an unnecessary cognitive load, highlighting the need for adaptive teaching considering learners' individual levels of knowledge [5]. Beginners can learn more effectively if they are provided with sufficient instructional guidance because they do not form mental schemas for certain learning topics. In contrast, for learners who have sufficiently developed a related schema, providing excessive instructional guidance hinders learning [17]. In other words, the application of adaptive teaching can optimize learners' cognitive load to achieve more positive learning outcomes. Studies applying adaptive teaching have reported that groups presented with adaptive teaching methods showed significantly higher knowledge acquisition, shorter learning time, and higher teaching efficiency compared to groups that did not.

Various methods have been proposed to measure cognitive load; however, there is no absolute method. Brünken, Plass, and Leutner divided cognitive load measurement methods into two categories: distinctions between subjective and objective methods and those between direct and indirect methods [18]. In the subjective-direct method, the level of stress perceived by learners and the degree of task difficulty were measured using a questionnaire. In contrast, the subjective-indirect method adopts a self-reporting approach to evaluate the degree of mental effort that learners experience through a written questionnaire. Objective-direct methods include electroencephalogram measurements or double-task response time measurements, whereas objective-indirect methods include physiological characteristics or behavioral measurements. According to Brünken et al.'s classification, EEG measurements consider cognitive load directly, whereas physiological signals (sweat, pupil, etc.) indirectly measure cognitive load. In general, subjective questionnaires, double-task reaction time measurements, and physiological signal measurement methods have been used [19]. A representative subjective method involves asking the participants to respond to subjective questionnaires. This method involves self-reporting the difficulty of the task by learners based on their subjective experience with the mental effort they put in through a questionnaire. This approach is the most commonly used and provides a relatively simple measurement of the degree of cognitive load without requiring special equipment. However, it has a disadvantage in that changes in the degree of cognitive load that occur during learning cannot be observed, and it relies on subjective reports provided after the end of learning. In addition, previous studies have not reached a clear consensus on which aspects should be measured by subjective perception, such as task difficulty. Similarly, care should be taken when interpreting results according to various learning contexts. For example, in the case of task difficulty, Kalyuga and Sweller measured the total cognitive load, while DeLeeuw and Mayer (2008) argued that it was related to the essential load [5,14,17,20]. Objective methods include the double-task response time and physiological signal measurement methods. First, the dual-task response time method measures the cognitive load based on the speed at which learners respond to additional tasks presented while performing a given task [21]. In general, if the response rate to an additional task is high, the level of cognitive load involved in processing the initial task is low. The physiological signal measurement method checks the cognitive load by measuring learners' physiological responses [22,23]. Because the physiological signal measurement method is based on objective data, it can be used to collect information in a relatively accurate and real-time manner without affecting task performance [24].

State-of-the-art works in cognitive load recognition involve using various physiological signals such as EEG, fNIRS, and ECG in developing models for predicting cognitive load. Researchers have used machine learning and deep learning algorithms to process these signals and classify cognitive load levels. However, despite the significant progress made in this field, challenges remain, such as high individual variability, noise, and poor generalization of models. Therefore, it is necessary to develop more accurate and robust models for cognitive load recognition.

The proposed method for cognitive load detection uses deep learning techniques and is motivated by the need for a more accurate and robust model that can address some of the challenges encountered by existing methods. Random forest, AdaBoost, SVM, XGBoost, and ANN are all traditional machine learning models that operate on fixed-length feature vectors. These models are often trained using labeled data and can predict unseen examples based on the patterns learned during training. Bi-LSTM, on the other hand, is a type of deep learning model that operates on sequential data, such as text or speech. Bi-LSTM is a variant of the long short-term memory (LSTM) network and is a type of recurrent neural network (RNN). Bi-LSTM has been successful in many natural language processing tasks, such as sentiment analysis, machine translation, and speech recognition. Unlike traditional machine learning models, Bi-LSTM can learn from the temporal relationships between inputs, which makes it well suited for tasks that involve sequential data. Bi-LSTM attention models can handle variable-length inputs and automatically extract relevant features from the input sequence, enabling them to capture complex patterns in the data.

Accordingly, the proposed method has the potential to improve the accuracy of cognitive load detection and can be applied to various real-world scenarios.

2.2. Preprocessing and Feature Extraction

Neural oscillations or brainwaves are electrical reactions that occur in the interaction between brain nerves and human mental activity, and these oscillations serve as indicators that reflect brain activity. EEG analysis considers changes in the intensity of electrical signals generated in the brain by frequency and is often used as a physiological signal measurement method to measure cognitive load. EEG measurements were made using the potential differences between the electrodes attached to the head. Electrodes can be attached to specific locations on the head to measure the EEG data in specific brain regions. EEG analysis generally analyzes the frequency of collected EEG signals by applying a Fast Fourier Transform (FFT). Brainwaves are divided into delta (0–4 Hz), theta (4–8 Hz), alpha (8–12 Hz), low beta (12–16 Hz), high beta (16–25 Hz), and gamma (25–50 Hz) waves, depending on their frequency. The intensity of brainwaves varies according to the state of human mental activity, and the degree of human cognitive load can be estimated based on this measurement. For example, alpha waves appear mainly in relaxation, beta waves appear in problem solving, and gamma waves appear mainly in more complex mental functions [25]. Considering these characteristics, we measured the constituent factors of cognitive load separately in terms of the degree of activation of the brainwaves by frequency.

The institutional review board (IRB) protects the rights and well-being of the subjects in life-oriented research. The proposed study, first, does not involve invasive behavior, such as drug administration and blood collection. Second, data were collected using only simple contact-measuring equipment that did not follow physical changes. Therefore, it corresponds to the IRB review that is not required in accordance with the regulations of the National Bioethics Policy Institute of Korea. The non-copyright dataset used in the experiment was obtained from Kaggle [26]. EEG signals were collected from 10 college students while they watched video footage. A total of 20 videos were provided, including 10 with and 10 without a mental load. Students wore single-channel wireless headsets (MindSet) to obtain EEG signals, which were measured on a 7-point scale from 1 to 7. The MindSet device measured the voltage between a forehead electrode and two electrodes (one for the ground and the other for reference) in contact with the ear. It provides an output of 0 for mental states and 1 for nonmental states. While the students watched a two-minute video, the EEG device emitted various previously listed signals. If the student was not ready at the beginning of the video, we removed the first and last 30 s of the video and analyzed only the middle 60 s of the EEG signal. The average of each firing interval was calculated to characterize the overall values. Several features were calculated to characterize the time profile of the EEG signal. Some of these distributions are typically used to measure the shape (minimum, maximum, variance, skewness, and kurtosis) of statistical distributions rather than time series. However, the small number of data samples (100 data points for 10 subjects who watched 10 videos each), including the aforementioned features, can overfit the training data and degrade the performance of the classification models. Accordingly, we used only the mean as a feature of the classifier.

Table 1 shows the structure of the EEG dataset used for deep learning, including the number of samples, the number of channels, and the range of values for the maximum and minimum amplitudes. We preprocessed 11,388 data points and partitioned them into separate training and validation sets. Specifically, we allocated 75% (8541 data points) for training and 25% (2847 data points) for validation.

Table 1. EEG data set.

	Feature	Count	Max	Min
0	Attention	11,388	100.0	1.0
1	Mediation	11,388	100.0	1.0
2	Raw	11,388	1440.0	−2048.0
3	Delta	11,388	3,964,663.0	448.0
4	Theth	11,388	2,567,643.0	17.1
5	Alpha1	11,388	1,369,955.0	2.0
6	Alpha2	11,388	1,016,913.0	2.0
7	Beta1	11,388	840,994.0	3.0
8	Beta2	11,388	1,083,461.0	2.0
9	Gamma1	11,388	658,008.0	1.0
10	Gamma2	11,388	283,517.0	2.0
11	User-defined label	11,388	1.0	0.0
12	Age	11,388	31	24
13	Ethnicity	11,388	Han Chinese	Bengali
14	Sex	11,388	M	F

2.3. LSTM-Based Recurrent Neural Network

In contrast to CNN models, LSTM architectures are incapable of large-scale parallel processing. Unlike RNNs, they include input, output, and forget gates that can control the flow of data in the network at any time. The gates of the LSTM architecture can place memory blocks on hidden nodes to solve the long-term dependency problem of CNN models, although the memory block cannot remember all data. Moreover, when LSTMs are used in the pooling layer of a CNN, spatial and temporal features can be considered simultaneously, owing to the end-to-end structure. The LSTM layer compensates for the long-term dependence problem of the CNN. LSTMs are used to recognize the characteristics of sequential data and store them in memory using a variable called the cell state. As shown in Figure 1, the LSTM architecture includes input, output, and forget gates, which enable it to be variably controlled according to the characteristics of the input data.

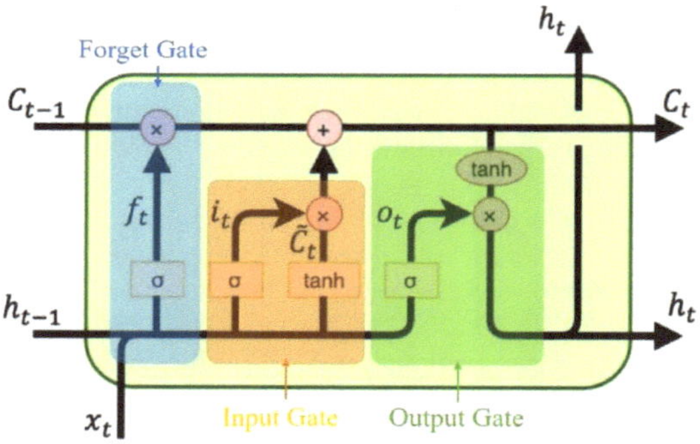

Figure 1. LSTM architecture comprises four components.

The LSTM architecture consists of four components: input gate, forget gate, cell state, and output gate. The purpose of the input gate is to obtain new information using two features referred to as Rt and dt. Rt combines the previous hidden vector h_{t-1} with the new information x_t. In other words, we multiply $[h_{t-1}, x_t]$ by the new matrix W_r and add the noise vector br. Then, we perform the same procedure for dt. Rt and dt multiply elements by element and import them into cell state ct. The slope of the forget gate is similar to that of the input gate; this component controls the limits of the values retained in memory. The cell state calculates the element multiplication between the previous cell states C_{t-1} and the forget ft. Then, we add the input gate rt multiplied by dt. The output gate is a symbol representing the output gate at t, and W_0 and b_0 are the weight and bias of the output gate, respectively. The hidden layer h_t is moved to the next point or output y_t.

2.4. Bi-LSTM

Bi-LSTM represents bidirectional long short-term memory. Bi-LSTM is a type of recurrent neural network (RNN) that is widely used for modeling sequential data. Unlike traditional RNNs that process input data in only one direction, Bi-LSTM models can process input data in both forward and backward directions simultaneously [27]. This makes them particularly useful for tasks such as natural language processing, speech recognition, and handwriting recognition, in which the context of each input data point depends on both past and future data points. The architecture of a Bi-LSTM model consists of two long short-term memory (LSTM) layers: one that processes input data in the forward direction and one that processes input data in the backward direction. Each LSTM layer has a series of memory cells that can store information over time and a series of gates that control the flow of information into and out of the memory cells. The gates are composed of sigmoid and tanh activation functions that determine the amount of information to be retained or discarded based on the relevance of the input data. By processing data in both directions, Bi-LSTM networks can capture both past and future contexts of a sequence, allowing them to better model complex dependencies and relationships within the data.

During training, the Bi-LSTM model was fed with the input sequences, and the weights of the network were updated using a backpropagation algorithm. The final output of the model was generated by concatenating the outputs of both LSTM layers, allowing the model to capture both the past and future contexts of the input sequence.

Overall, the Bi-LSTM model showed promising results in various applications, demonstrating its effectiveness in capturing long-term dependencies and improving the performance of sequential data-processing tasks.

3. Materials and Methods

Deep Learning-Based Cognitive Load Analysis Model

The model proposed in this study comprises a bidirectional LSTM and an attention mechanism to extract the positive and negative characteristics of the mental load. Contrary to conventional machine learning techniques, LSTM models are not capable of large-scale parallel processing, unlike CNNs. Instead, they utilize input, output, and forget gates to process the data. The gates have the advantage of being able to place a memory block on a hidden node. This can solve the long-term dependency problem of CNN models, although the memory block cannot remember all data. Moreover, when LSTM is used in the pooling layer of a CNN, spatial and temporal features can be considered simultaneously owing to its end-to-end structure. In addition, LSTM models can exhibit improved accuracy because they can equally model sequence vectors when predicting words. LSTMs provide sequential data characteristics and store them in memory using a variable called the cell state. This specialized architecture enables the data to be processed differently according to different situations by controlling the calculation process. Next, a single value is outputted using the sigmoid function in a fully connected layer called the dense layer.

A typical BCI system utilizes data preprocessing processes to remove noise, extract features, and classify the data to reflect characteristics and extract meaningful data from unprocessed brainwaves [28]. The classified information may be used as an instruction for device control or provided to the user. Because brainwaves are characterized by nonlinearity and high variability between individuals and situations, the implementation of stable and reliable BCI systems is challenging. In this study, we used an LSTM model to extract and classify cognitive load and related brainwave characteristics. The collected data were used as input to the LSTM model. The data to which the output value was assigned underwent a conversion process to make it suitable as input to the LSTM model.

In this study, we adopted a one-way LSTM layer followed by an attention mechanism to model the effect of the mental load generated at a given time on overall emotion. The attention mechanism is a learning method that weighs a part of the input that affects the output the most. Bidirectional LSTM layers are generally known to perform better when considering both the front and rear concealed states than unidirectional LSTM layers when using an attention mechanism. The overall structure of the proposed model is shown in Figure 2.

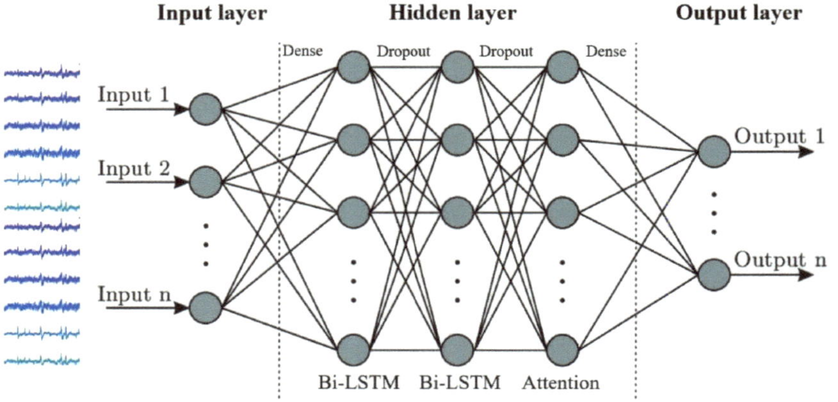

Figure 2. LSTM network structure with an attention mechanism.

The first hidden layer included 128 neurons and used a bidirectional long short-term memory (LSTM) layer with a rectified linear unit activation function. Then, to avoid overfitting, we included a dropout with a probability of 0.2 and passed the data through a second bidirectional LSTM layer with 64 neurons and the ReLU activation function. After calculating the attention weight with the hidden state, which is the output of the second layer, the dimensionality of the data was reduced by passing through a layer with an output size of 16 and the first dense layer using the ReLU activation function. The final output is obtained by passing the data through the second dense layer, which has an output size of 1 and using the sigmoid activation function in the second classification and the softmax function in the third. In the second-stage classification, the output values are low or high for valence or arousal, and in the third-stage classification, the results are classified as low, middle, or high. The attention weights were processed in the following order: This method calculates the attention weight of the part of the input that affects the output; the higher the weight of the input part, the greater the value when the network is trained. The order of calculation is as follows: The hidden state vector calculated via the second bidirectional LSTM layer is multiplied by a randomly initialized attention weight, whose length is equal to the length of the hidden state vector. The output size of the second bidirectional LSTM layer was 64, with a total of 128, owing to the bidirectional architecture. Thus, the length of the attention weight is 128. The resulting value from this calculation was converted into a probability value through the softmax activation layer, and the transformed attention vector was combined with the first calculated hidden state vector to be calculated as the final attention output. Dense layer 1, which is connected to the attention layer, receives the corresponding attention output as input and reflects the part of the weight that is most important for future learning to produce more accurate results. The Adam optimizer was used, with a learning rate of 0.001. A cross-entropy loss function suitable for binary classification was also used. To measure accuracy, we adopted the Stratified *K*-fold cross-validation method with four iterations. Using this method, labels were distributed in a balanced manner for each fold; 75% of the data were used as the training set and 25% as the testing set for each iteration. Each iteration was trained for 30 epochs with an input batch size of 32. Each hyperparameter was optimized experimentally.

Figure 3 shows the results of the analysis of the correlation between variables. In the heatmap, the *X*- and *Y*-axes were set to the same variable and plotted as points. Consequently, we observed a suitable correlation between Gamma1 and Beta2.

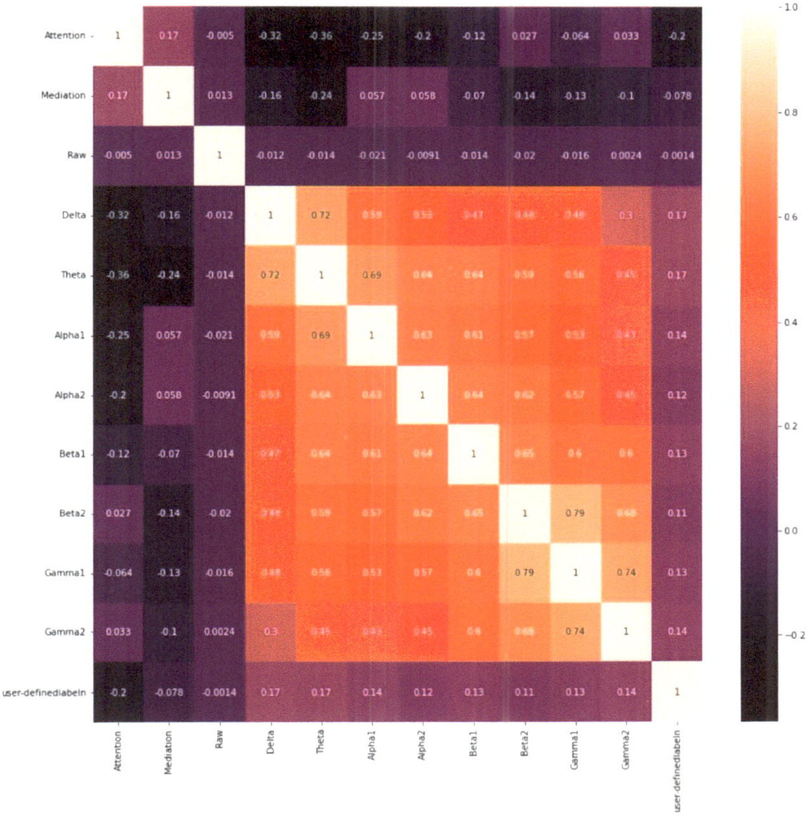

Figure 3. Results of the correlation analysis between variables.

Table 2 summarizes the proposed LSTM model. The structure of the model was the same as that of the model implemented in Python. The ReLu activation function was used with a dropout of 0.2, fork of 100, and batch size of 10, and a sigmoid activation function was used in the last dense layer.

Table 2. The proposed LSTM model.

Layer Num	Type	Output Shape	Parameters
Layer 1	Input Layer	(None, 16, 1)	0
Layer 2	Dense	(None, 16, 64)	128
Layer 3	Dense	(None, 16, 128)	8320
Layer 4	Bidirectional LSTM	(None, 16, 512)	788,480
Layer 5	Dropout	(None, 16, 512)	0
Layer 6	Bidirectional LSTM	(None, 16, 512)	1,574,912
Layer 7	Dropout	(None, 16, 512)	0
Layer 8	Attention	(None, 16, 512)	528
Layer 9	Dense	(None, 16, 128)	65,664
Layer 10	Dense	(None, 16, 1)	129

4. Results

The performance of the proposed Bi-LSTM model has a decisive influence on the overall system because control is performed based on accurate EEG feature extraction and classification according to the learning process. Loss and accuracy were used to verify the performance of the proposed LSTM model. The smaller the loss, the better because it indicates a difference from the result value. A categorical cross-entropy loss function was also used. Accuracy is the ratio of the total number of positive recognitions to that of negative recognitions. Thus, accuracy values closer to 1 indicate better performance. For the best training data, the proposed LSTM model exhibited a loss value of 0.08 and an accuracy of 96.9% over 76 epochs. For the verification data, the loss value was 0.3636, the accuracy was 84.8%, the loss value was 0.8289, and the accuracy was 87.11%. Figure 4 shows the loss and accuracy of the LSTM model according to data type. The x-axis represents the number of epochs, and the y-axis represents the accuracy according to the loss.

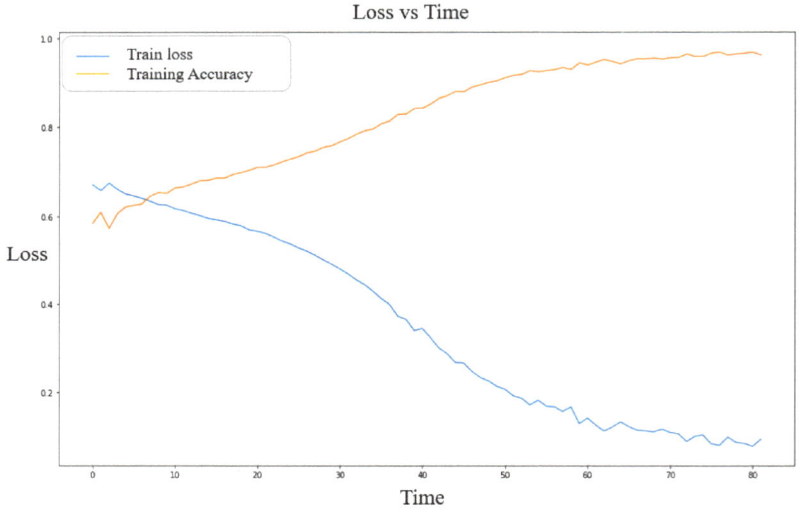

Figure 4. Loss and accuracy of the LSTM model.

Table 3 presents the results of the comparison between the proposed method and the other algorithms. We evaluated the accuracy and F1-score of each model using both EEG training and test datasets. Our experimental results allowed us to identify the best model based on its accuracy and bias-variance balance. We can clearly identify the models with the highest scores. Specifically, the SVM model performed the worst, whereas the AdaBoost and random forest models performed similarly, with a performance 15.68% better than that of the most accurate ANN algorithm with an average accuracy of 0.871. The accuracy results of the models are as follows: two-way LSTM attention (0.871), ANN (0.714), RNN-LSTM (0.69), Bi-LSTM (0.6743), XGBoost (0.6733), AdaBoost (0.6431), and vector machine (0.6094).

As shown in Table 4, we employed a technique to determine the optimal hyperparameters for all the models. By defining a grid of possible hyperparameter values and training the models with each combination of hyperparameters, the values that yielded the best performance in the validation set were identified.

Table 3. Classification result of comparison of the proposed method with other algorithms.

Classification Methods	Average Accuracy	F1-Score
Random Forest	0.6416	0.657
AdaBoost	0.6431	0.660
Support Vector Machine	0.6094	0.629
XGBoost	0.6733	0.686
ANN	0.7142	0.710
RNN-LSTM	0.6900	0.690
Bidirectional LSTM	0.6743	0.670
Bidirectional LSTM Attention	0.8710	0.870

Table 4. Grid search results for the best combination of parameters.

Models	Parameters (Grid Search)	Best Params
Random Forest	'max_depth': list (range (10, 20, 5)),	15
	'n_estimators': [50,100]	100
AdaBoost	'algorithm': ['SAMME','SAMME.R']	'SAMME.R'
	'n_estimators': [10,40,60,100,120,130,140]	120
SVC	'kernel': ['rbf']	'rdf'
	'C': list (np.arange (0.5, 1.5, 0.1))	0.7
	'gamma': ['scale', 'auto']	'scale', 'auto'
XGBoost	'base_score': list (np.arange (0.2, 0.5, 0.1))	0.4
	'n_estimators': [10,40,60,100,120,130,140]	60
	'objective': ['binary:logistic']	'logistic'
ANN	Model hidden layer	{32, 16, 16}
	Dense (activation = 'sigmoid')	'sigmoid'
	compile (loss = 'binary_crossentropy')	'binary_crossentropy'
	optimizer = 'adam', metrics = ['accuracy'])	'adam'
	Dense (activation = 'relu',kernel_regularizer = 'l2')	'relu', 12

A confusion matrix is a table for comparing predicted and actual values to measure prediction performance achieved through training [29,30]. As shown in Figure 5, the rows represent the correct answer class, and the column represents the predicted class. The confusion matrix, with 2847 data points, had a true negative, false positive, false negative, and true positive of 1058, 277, 124, and 1388, respectively.

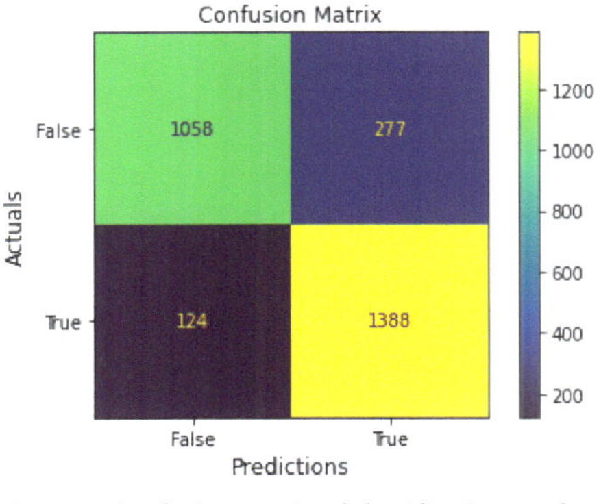

Figure 5. Confusion matrix of classification results.

Table 5 shows the precision, reproduction rate, and detailed classification report of the F1-score of the proposed model. Precision refers to the ratio of the number of samples belonging to the positive class among the samples shown to belong to the positive class, indicating a high precision of 0.918. Recall refers to the ratio of the number of samples detected to belong to the positive class among the samples in the actual positive class. The weight harmonic average of precision and recall is called the f-score, and the best result is false positive and false negative values close to 1. The classification report in Table 5 shows that 92% of the data predicted at 0 (mental load) were actually 0, and 83% of the data predicted at 1 (nonmental load) were actually 1. In addition, 80% of the actual cognitive load data was predicted to be cognitive load, and 93% of the non-cognitive load data was predicted to be non-cognitive load.

Table 5. Model performance of Bi-LSTM attention.

Bi-LSTM Attention	Precision	Recall	F1-Score	Support
0—Mental load	0.92	0.80	0.85	1360
1—Not mental load	0.83	0.93	0.88	1487
Accuracy	-	-	0.87	2847
Macro average	0.88	0.87	0.87	2847
Weighted average	0.87	0.87	0.87	2847

From the experimental results on algorithm comparisons, it was determined that traditional machine learning models, such as random forest, AdaBoost, SVM, XGBoost, and ANN, are best suited for tasks in which the input is a fixed-length feature vector. By contrast, Bi-LSTM is ideal for tasks that involve sequential data because it can learn from the temporal relationships between inputs, making it well suited for tasks such as EEG processing and cognitive load prediction.

The limitations of this study are as follows: First, the data used were insufficient to clearly reveal the difference in the degree of cognitive load in the composition of a given video based on participants' understanding of the online learning video. We can consider a difference in the degree of cognitive load calculated by dividing the difficulty according to the understanding of the video; however, factors other than learning difficulty may have affected this value. Considering these limitations, the results of this study only suggest the possibility of determining the degree of cognitive load through machine learning using brainwave data. Accordingly, subsequent studies should clearly determine the differences in the difficulty of the learning tasks given to the experimental participants. Second, it is difficult to conclude that the accuracy of the model was represented well for all situations because the amount of student data collected was relatively small. Although the machine learning model was trained by integrating the data extracted from each experimental participant, the trained machine learning model would likely be unable to determine the degree of cognitive load universally because the number of participants was only 10. The accuracy of machine learning models depends on the amount of data available for training and verification. In particular, in the case of the ANN model used in this study, the accuracy was relatively high, and the possibility of overfitting was suspected. To confirm overfitting, data collected from additional participants were required. Third, the participants of the experiment were not evenly distributed. Accordingly, the ratio of the number of data samples used to train the artificial intelligence model is not equal. Because the training data are the basis for the model to determine the level of cognitive load, the uneven proportion of training data may have caused the accuracy of the model's judgments to be inconsistent. Fourth, a real-time adaptive teaching model must be developed. In this study, we have presented EEG wavelength and electrode locations with a relatively large impact on EEG-based cognitive load determination and proposed an appropriate machine learning algorithm for the development of a cognitive load discrimination model. These research results only confirm the level of learners' cognitive load, and it is difficult to confirm what support should be provided to learners from these data. To apply this to the educational field, an adaptive teaching model that provides appropriate teaching support according to learners' cognitive load levels must be developed.

5. Discussion

The main reason for measuring the mental workload is to quantify the cognitive load that performs tasks to predict human performance. However, the existing method of evaluating mental workload presents a relationship between subjective workload and brain activity, making objective verification difficult. For instance, Agbangla et al. suggested a relationship between subjective workload and brain activity through PFC hemodynamics using fNIRS while performing n-back and RNG tasks with multiple cognitive loads, while Longo et al. suggested the possibility of mental workload modeling in EEG data using deep learning [7,8]. In this study, we trained an artificial intelligence model to determine learners' levels of cognitive load using EEG data and confirmed the influence of different variables on cognitive load determination and the accuracy of the model with different machine learning algorithms. Applying bidirectional LSTM cyclic neural networks to classify student confusion regarding online course videos with EEG data showed that the bidirectional LSTM model achieved state-of-the-art performance compared to other machine learning approaches and showed suitable robustness as evaluated by cross-validation. As a result, gamma and alpha waves significantly influenced the determination of the discriminant model, and the bidirectional LSTM attention and ANN models exhibited the highest accuracy.

In this study, we propose a two-way LSTM recurrent neural network framework to detect a student's mental load when watching online course videos. We implemented an attention-based LSTM deep learning model that effectively classifies cognitive load models by applying an attention mechanism, which is a state-of-the-art technology suitable for the mental load. The proposed model achieved an accuracy of 87.1% using EEG signals without a separate feature-extraction process. The results of a comparative analysis with other algorithms also showed that the accuracy of the proposed model outperformed that of other machine learning approaches, including a tomography LSTM model. The architecture of the bidirectional LSTM model helps leverage time-series capabilities for improved performance. An analysis of the contributions of each function to the model also confirmed that gamma and beta values are the most important for the cognitive load. In the future, the model should be trained with more EEG datasets, and the experimental results can be applied not only to learning but also to other EEG-related tasks, such as task evaluation and detection of drowsy driving.

In future studies, we intend to improve the accuracy of measuring cognitive load even in the lower class by applying a method to solve the data imbalance problem. In addition, for continuous cognitive load models, the degree of the mental load is important. Hence, we plan to apply a regression model to the last stage of the deep learning-based cognitive load model to analyze it in various ways.

Author Contributions: Conceptualization, G.Y. and S.H.; methodology, G.Y.; software, H.K.; validation, G.Y., H.K. and S.H.; formal analysis, G.Y.; investigation, G.Y.; resources, G.Y.; data curation, G.Y.; writing—original draft preparation, G.Y.; writing—review and editing, S.H.; visualization, G.Y.; supervision, H.K.; project administration, H.K.; funding acquisition, S.H. All authors have read and agreed to the published version of the manuscript.

Funding: This research was supported by the Basic Science Research Program through the National Research Foundation of Korea (NRF), funded by the Ministry of Education (2020R1I1A1A01064580).

Institutional Review Board Statement: Not applicable.

Informed Consent Statement: Not applicable.

Data Availability Statement: The data presented in this study are available upon request from the corresponding author.

Conflicts of Interest: The authors declare no conflict of interest.

References

1. Salta, K.; Paschalidou, K.; Tsetseri, M.; Koulougliotis, D. Shift from a traditional to a distance learning environment during the COVID-19 pandemic: University students' engagement and interactions. *Sci. Educ.* **2022**, *31*, 93–122. [CrossRef] [PubMed]
2. Singh, J.; Singh, L.; Matthees, B. Establishing social, cognitive, and teaching presence in online learning—A panacea in COVID-19 pandemic, post vaccine and post pandemic times. *J. Educ. Technol. Syst.* **2022**, *51*, 28–45. [CrossRef]
3. O'Keefe, R.; Auffermann, K. Exploring the effect of COVID-19 on graduate nursing education. *Acad. Med.* **2022**, *97*, S61. [CrossRef] [PubMed]
4. Cui, Y.; Ma, Z.; Wang, L.; Yang, A.; Liu, Q.; Kong, S.; Wang, H. A survey on big data-enabled innovative online education systems during the COVID-19 pandemic. *J. Innov. Knowl.* **2023**, *8*, 100295. [CrossRef]
5. Likourezos, V.; Kalyuga, S.; Sweller, J. The variability effect: When instructional variability is advantageous. *Educ. Psychol. Rev.* **2019**, *31*, 479–497. [CrossRef]
6. Liu, J.C.; Li, K.A.; Yeh, S.L.; Chien, S.Y. Assessing perceptual load and cognitive load by fixation-related information of eye movements. *Sensors* **2022**, *22*, 1187. [CrossRef]
7. Agbangla, N.F.; Audiffren, M.; Pylouster, J.; Albinet, C.T. Load-dependent prefrontal cortex activation assessed by continuous-wave near-infrared spectroscopy during two executive tasks with three cognitive loads in young adults. *Brain Sci.* **2022**, *12*, 1462. [CrossRef]
8. Longo, L. Modeling cognitive load as a self-supervised brain rate with electroencephalography and deep learning. *Brain Sci.* **2022**, *12*, 1416. [CrossRef]
9. Liu, H.; Shi, R.; Liao, R.; Liu, Y.; Che, J.; Bai, Z.; Cheng, N.; Ma, H. Machine learning based on event-related EEG of sustained attention differentiates adults with chronic high-altitude exposure from healthy controls. *Brain Sci.* **2022**, *12*, 1677. [CrossRef]
10. Nafea, M.S.; Ismail, Z.H. Supervised machine learning and deep learning techniques for epileptic seizure recognition using EEG signals—A systematic literature review. *Bioengineering.* **2022**, *9*, 781. [CrossRef]
11. Friedman, N.; Fekete, T.; Gal, K.; Shriki, O. EEG-based prediction of cognitive load in intelligence tests. *Front. Hum. Neurosci.* **2019**, *13*, 191. [CrossRef]
12. Rubio, S.; Díaz, E.; Martín, J.; Puente, J.M. Evaluation of subjective mental workload: A comparison of SWAT, NASA-TLX, and workload profile methods. *Appl. Psychol.* **2004**, *53*, 61–86. [CrossRef]
13. Hogervorst, M.A.; Brouwer, A.M.; Van Erp, J.B. Combining and comparing EEG, peripheral physiology and eye-related measures for the assessment of mental workload. *Front. Neurosci.* **2014**, *8*, 322. [CrossRef] [PubMed]
14. Sweller, J.; Ayres, P.; Kalyuga, S. Acquiring information: The borrowing and reorganising principle and the randomness as genesis principle. In *Cognitive Load Theory. Explorations in the Learning Sciences, Instructional Systems and Performance Technologies*; Springer: New York, NY, USA, 2011; Volume 1. [CrossRef]
15. Paas, F.; Renkl, A.; Sweller, J. Cognitive load theory and instructional design: Recent developments. *Educ. Psychol.* **2003**, *38*, 1–4. [CrossRef]
16. Paas, F.; Renkl, A.; Sweller, J. *Cognitive Load Theory: A Special Issue of Educational Psychologist*; Routledge: New York, NY, USA, 2016.
17. Ericsson, K.A.; Hoffman, R.R.; Kozbelt, A.; Williams, A.M. (Eds.) *The Cambridge Handbook of Expertise and Expert Performance*; Cambridge University Press: Cambridge, UK, 2018; pp. 793–811.
18. Brünken, R.; Plass, J.L.; Leutner, D. Direct measurement of cognitive load in multimedia learning. *Educ. Psychol.* **2003**, *38*, 53–61. [CrossRef]
19. Schultheis, H.; Jameson, A. Assessing cognitive load in adaptive hypermedia systems: Physiological and behavioral methods. In *Adaptive Hypermedia and Adaptive Web-Based Systems*; De Bra, P.M.E., Nejdl, W., Eds.; Springer: Berlin/Heidelberg, Germany, 2004; Volume 3137, pp. 225–234. [CrossRef]
20. DeLeeuw, K.E.; Mayer, R.E. A comparison of three measures of cognitive load: Evidence for separable measures of intrinsic, extraneous, and germane load. *J. Educ. Psychol.* **2008**, *100*, 223–234. [CrossRef]
21. Nilsson, E.J.; Aust, M.L.; Engström, J.; Svanberg, B.; Lindén, P. Effects of cognitive load on response time in an unexpected lead vehicle braking scenario and the detection response task (DRT). *Transp. Res. F.* **2018**, *59*, 463–474. [CrossRef]
22. Krejtz, K.; Duchowski, A.T.; Niedzielska, A.; Biele, C.; Krejtz, I. Eye tracking cognitive load using pupil diameter and microsaccades with fixed gaze. *PLoS ONE* **2018**, *13*, e0203629. [CrossRef]
23. Kumar, N.; Kumar, J. Measurement of cognitive load in HCI systems using EEG power spectrum: An experimental study. *Procedia Comput. Sci.* **2016**, *84*, 70–78. [CrossRef]
24. Ikehara, C.S.; Crosby, M.E. Assessing cognitive load with physiological sensors. In Proceedings of the 38th Annual Hawaii International Conference on System Sciences, Big Island, HI, USA, 3–6 January 2005.
25. Liang, Y.; Liang, W.; Qu, J.; Yang, J. Experimental study on EEG with different cognitive load. In Proceedings of the IEEE International Conference on Systems, Man, and Cybernetics, Miyazaki, Japan, 7–10 October 2018; Volume 2018.
26. Confused Student EEG Brainwave Data. Available online: https://www.kaggle.com/datasets/wanghaohan/confused-eeg (accessed on 23 January 2023).
27. Algarni, M.; Saeed, F.; Al-Hadhrami, T.; Ghabban, F.; Al-Sarem, M. Deep learning-based approach for emotion recognition using electroencephalography (EEG) signals using bi-directional long short-term memory (Bi-LSTM). *Sensors* **2022**, *22*, 2976. [CrossRef]

28. Butt, A.M.; Alsaffar, H.; Alshareef, M.; Qureshi, K.K. AI prediction of brain signals for human gait using BCI device and FBG based sensorial platform for plantar pressure measurements. *Sensors* **2022**, *22*, 3085. [CrossRef] [PubMed]
29. Ahmed, A.A.; Ali, W.; Abdullah, T.A.A.; Malebary, S.J. Classifying cardiac arrhythmia from ECG signal using 1D CNN deep learning model. *Mathematics* **2023**, *11*, 562. [CrossRef]
30. Choudhary, A.; Ahlawat, S.; Urooj, S.; Pathak, N.; Lay-Ekuakille, A.; Sharma, N. A deep learning-based framework for retinal disease classification. *Healthcare* **2023**, *11*, 212. [CrossRef] [PubMed]

bioengineering

MDPI

Article

Elucidating Multimodal Imaging Patterns in Accelerated Brain Aging: Heterogeneity through a Discriminant Analysis Approach Using the UK Biobank Dataset

Lingyu Liu [1], Lan Lin [1,2,*], Shen Sun [1,2] and Shuicai Wu [1,2]

[1] Department of Biomedical Engineering, College of Chemistry and Life Science, Beijing University of Technology, Beijing 100124, China; lingyuliu98@sina.com (L.L.); sunshen@bjut.edu.cn (S.S.); wushuicai@bjut.edu.cn (S.W.)

[2] Intelligent Physiological Measurement and Clinical Translation, Beijing International Base for Scientific and Technological Cooperation, Beijing University of Technology, Beijing 100124, China

* Correspondence: lanlin@bjut.edu.cn

Abstract: Accelerated brain aging (ABA) intricately links with age-associated neurodegenerative and neuropsychiatric diseases, emphasizing the critical need for a nuanced exploration of heterogeneous ABA patterns. This investigation leveraged data from the UK Biobank (UKB) for a comprehensive analysis, utilizing structural magnetic resonance imaging (sMRI), diffusion magnetic resonance imaging (dMRI), and resting-state functional magnetic resonance imaging (rsfMRI) from 31,621 participants. Pre-processing employed tools from the FMRIB Software Library (FSL, version 5.0.10), FreeSurfer, DTIFIT, and MELODIC, seamlessly integrated into the UKB imaging processing pipeline. The Lasso algorithm was employed for brain-age prediction, utilizing derived phenotypes obtained from brain imaging data. Subpopulations of accelerated brain aging (ABA) and resilient brain aging (RBA) were delineated based on the error between actual age and predicted brain age. The ABA subgroup comprised 1949 subjects (experimental group), while the RBA subgroup comprised 3203 subjects (control group). Semi-supervised heterogeneity through discriminant analysis (HYDRA) refined and characterized the ABA subgroups based on distinctive neuroimaging features. HYDRA systematically stratified ABA subjects into three subtypes: SubGroup 2 exhibited extensive gray-matter atrophy, distinctive white-matter patterns, and unique connectivity features, displaying lower cognitive performance; SubGroup 3 demonstrated minimal atrophy, superior cognitive performance, and higher physical activity; and SubGroup 1 occupied an intermediate position. This investigation underscores pronounced structural and functional heterogeneity in ABA, revealing three subtypes and paving the way for personalized neuroprotective treatments for age-related neurological, neuropsychiatric, and neurodegenerative diseases.

Keywords: accelerated brain aging; advanced brain aging; subtypes; heterogeneity; structural MRI

Citation: Liu, L.; Lin, L.; Sun, S.; Wu, S. Elucidating Multimodal Imaging Patterns in Accelerated Brain Aging: Heterogeneity through a Discriminant Analysis Approach Using the UK Biobank Dataset. *Bioengineering* **2024**, *11*, 124. https://doi.org/10.3390/bioengineering11020124

Academic Editors: Richard Bayford, Yan Pei and Jijiang Yang

Received: 15 December 2023
Revised: 17 January 2024
Accepted: 24 January 2024
Published: 26 January 2024

1. Introduction

The brain aging process elicits intricate alterations in both the structure and function aspects of the brain. This phenomenon manifests in various forms of degeneration, encompassing cortical thinning [1], increased white-matter atrophy and lesions [2], and diminished functional connectivity [3]. Despite the profound impact of aging on the brain, current investigations into brain aging, specifically the categorization of brain aging subtypes based on neuroimaging features, are at an early stage of development [4,5]. Researchers are navigating the complexities of understanding the diverse patterns associated with brain aging. The complexity is further heightened by the interplay among genetic, lifestyle, and environmental factors, all of which contribute significantly to the observed diversity in the brain aging process [6–8]. This apparent heterogeneity intrinsic to brain aging emphasizes the imperative of studying brain-aging subtypes to unravel the underlying mechanisms and variations of brain aging within the aging population [9].

Traditional brain-aging research has categorized brain aging into distinct categories: resilient brain aging (RBA) [10–12], normal brain aging, and accelerated brain aging (ABA), occasionally referred to as advanced brain aging [13,14]. Such studies have revealed that RBA is associated with a greater resistance to the risk of neurodegenerative diseases like Alzheimer's disease (AD). Conversely, ABA represents a paradigm wherein aging processes within the cerebral domain proceed at an accelerated pace, surpassing the expected rate corresponding to an individual's chronological age. This acceleration is evidenced by the manifestation of cerebral characteristics that appear older than anticipated. Discernible disparities in the brain-aging biomarkers of ABA have been well-documented within the context of neuropsychiatric disorders. Conditions such as schizophrenia, post-traumatic stress disorder, anxiety disorders, and depression exhibit conspicuous deviations in some brain-aging biomarkers [15]. This observation underscores the intricate interplay between ABA and the underlying pathophysiological substrates of neuropsychiatric disorders. The implication arises that ABA processes may serve as potential contributors to the etiology and progression of such conditions. Expanding upon extensive databases of normative aging, the analysis of MRI data emerges as a pivotal avenue for scrutinizing ABA within the cerebral milieu. One such methodology, the Brain Age Gap Estimation (BrainAGE) method [16], leverages machine-learning techniques to discern individual-level variability in brain-aging dynamics. This involves the utilization of standard MRI sequences, wherein a prediction model, derived from a learning sample comprising neurologically healthy adults, is deployed to estimate the apparent biological age of a new individual's brain. In this process, the disparity between the estimated biological age and the subject's chronological age constitutes the brain-age "gap" (BAG), a metric quantifying the extent to which a given brain appears comparatively "older" or "younger" relative to the individual's chronological age. The BrainAGE method thus provides a sophisticated means of assessing and quantifying the accelerated aging phenomenon within the brain, offering insights into the individualized dynamics of cerebral aging beyond chronological timelines. Studies have established positive correlations between increased BrainAGE and numerous diseases, including obstructive sleep apnea [17], schizophrenia [18], AD [19], major depressive disorder [20], chronic poststroke language impairment [21], and Parkinson's disease [22]. Therefore, elucidating the heterogeneity of ABA is crucial for understanding the underlying pathophysiological processes in brain aging [23].

Within the realm of neurological disorders, the application of unsupervised clustering algorithms stands as a pervasive methodology for conducting ABA subtype analyses. In a seminal study by Wrigglesworth et al. [24], 167 individuals exhibiting ABA were identified from a cohort of 326 community-dwelling older adults based on their BrainAGE metrics. The investigators proceeded to employ latent class analysis (LCA) on the ABA subjects, incorporating a comprehensive array of cognitive, lifestyle, and health measures. The results of the LCA revealed the presence of two distinct ABA subtypes. The first subtype exhibited a low prevalence of obesity, a diminished likelihood of low general cognitive status, a smaller probability of low mental quality of life (QoL), and a reduced likelihood of low physical QoL. In contrast, the second subtype was characterized by a higher prevalence of hypertension, a lower probability of high general cognitive status, moderate scores in mental QoL, and a diminished likelihood of high physical QoL. These findings underscore the utility of unsupervised clustering in unraveling nuanced health-related subtypes within the context of ABA. Unlike unsupervised learning, semi-supervised learning methods utilize labeled and unlabeled data to train a base classifier to distinguish between target and control groups, which is then updated in an unsupervised manner to discover the heterogeneity of the target group. This approach leads to more accurate predictions and a deeper understanding of the disease. Eavani et al.'s study [25] in ABA, utilizing the Mixture of Experts (MOEs) method on MRI data from 400 participants aged 50 to 96, identifies 5 distinct ABA phenotypes. This research underscores the importance of capturing the heterogeneity and subtypes of ABA rather than seeking a single signature, providing insights for future studies in understanding the neurobiological underpinnings of ABA. However, this study confronts two notable challenges. First, the limited dataset, encompassing approximately 261 subjects displaying ABA that were derived from around 400 participants, may raise concerns about the generalizability of the results. Additionally, while the MOE framework amalgamates the unsupervised modeling of mixtures of distributions with the supervised learning of classifiers, bestowing it with commendable merits in subgroup identification and multivariate pattern discrimination, it is not without its shortcomings. The MOE method's integration of classification with clustering strategies leads it to inherit the limitations inherent in traditional clustering methods, particularly in the context of high-dimensional data where sparsity and dimensionality challenges prevail. With escalating dimensionality, the notion of distance between data points loses its meaning and becomes increasingly inadequate for discerning inherent patterns. This predicament is further exacerbated by the high sparsity endemic to high-dimensional spaces, yielding clustering outcomes that are marked by instability and unreliability. In this context, even minor perturbations in data points can yield entirely disparate cluster assignments, thereby compromising the robustness and consistency of the clustering results. In stark contrast, the recently developed heterogeneity through discriminant analysis (HYDRA) [26] was used for this study. A multiple max-margin discriminative analysis framework algorithm offers a promising and innovative solution. HYDRA's prowess lies in its remarkable capacity to effectively capture neuroanatomical subtypes by utilizing multiple linear hyperplanes to create a convex polytope that distinctly separates various subgroups. Notably, HYDRA leverages the modeling capabilities of linear support vector machines (SVMs) to discriminate between homogeneous classes, even within high-dimensional data spaces. Moreover, HYDRA adopts a sophisticated two-pronged approach to improve upon its predecessor. Firstly, it meticulously initializes the iterative algorithm with great care, with a specific emphasis on promoting clustering solutions that exhibit diversity related to disease characteristics. This is achieved through the application of determinantal point processes (DPPs) to sample a wide array of aging directions, thus refining the initial clustering assignments. Secondly, HYDRA acknowledges the variability inherent in estimated solutions, particularly in non-convex settings, and skillfully employs a multi-initialization strategy in tandem with a fusion step. This comprehensive approach results in the production of robust and consistent results that accentuate the underlying group structure while simultaneously minimizing the impact of noisy perturbations. Overall, the innate advantages and advanced methodologies of HYDRA position it as a compelling and superior choice for heterogeneous analysis. Consequently, HYDRA has garnered widespread adoption in disease subtype analysis and recognition as the preeminent heterogeneous analysis algorithm in current practice [26–30].

In the course of this scientific investigation, BrainAGE functions as the pivotal tool for the stratification of the aging population into ABA and RBA subpopulations. Subsequent to this stratification, our study delves into the nuanced task of estimating diverse aging trajectories within the ABA population relative to the RBA. This intricate analysis is facilitated through the utilization of multimodal MRI image features. This research methodology signifies a departure from previous studies as it entails the examination of a notably expansive dataset comprising 5152 subjects. Within this dataset, 1949 subjects are representative of the ABA subpopulation, while 3203 represent the RBA subpopulation. The data were meticulously sourced from the UK Biobank (UKB) database, a reservoir of information that spans multiple distinct imaging modalities. Specifically, the brain imaging-derived phenotypes (IDPs) encompass 207 features derived from structural magnetic resonance imaging (sMRI), 144 features from diffusion magnetic resonance imaging (dMRI), and 210 features from resting-state functional magnetic resonance imaging (rsfMRI). This comprehensive approach substantially fortifies the robustness of our analysis, enabling a more nuanced understanding of the intricate interplay between ABA subpopulations and their corresponding neuroimaging profiles. By meticulously dissecting these multimodal MRI features, our investigation aims to contribute novel insights into the complex landscape of

ABA, thus advancing our comprehension of the underlying mechanisms at play within the aging process.

In this study, we make three crucial contributions. Firstly, our investigation conducts a thorough examination of ABA subtypes through the application of multimodal neuroimaging techniques. To our knowledge, this study stands as the first-ever exploration into ABA heterogeneity utilizing multimodal neuroimaging on a significant scale, drawing from a substantial cohort of healthy volunteers (n = 31,621). This addresses challenges associated with limited sample sizes in previous research, ensuring a more comprehensive understanding of the diverse ABA population. Secondly, the high dimensionality (n = 561) poses challenges in heterogeneity analysis algorithms, prompting the introduction of the innovative HYDRA method for scrutinizing brain-aging heterogeneity. This approach effectively addresses inherent limitations in traditional methods applied to the analysis of ABA heterogeneity, showcasing its potential to unravel intricate patterns within high-dimensional neuroimaging datasets. This promises a fruitful avenue for future research into brain-aging heterogeneity. Beyond these advancements, the study's contributions extend to the broader significance of understanding brain aging. Stratifying ABA subjects into three subtypes establishes a foundation for personalized prevention approaches against conditions like dementia. In essence, this study not only propels the methodological landscape of neuroimaging research forward but also holds profound implications for translational applications in the realms of personalized medicine and preventative neurology.

The organization of the remainder of this paper is as follows. In Section 2, a comprehensive introduction unfolds, elucidating essential facets such as the UKB data, the neuroimaging processing pipeline, the machine-learning model employed for brain-age prediction and the identification of ABA subgroups, and a meticulous overview of the statistical procedures that underpin this study. Section 3 meticulously unveils the study results, with a particular focus on the nuanced analysis of ABA subtypes. Following the presentation of results, Section 4 engages in a comprehensive discussion that contextualizes the findings within the broader landscape of neuroimaging research and the understanding of ABA, while a concise summary is encapsulated in Section 5.

2. Materials and Methods

2.1. Participants

The data utilized in this investigation emanated from a population-based prospective cohort study, namely the UKB [31], which is accessible at www.ukbiobank.ac.uk (accessed on 11 January 2021). The UKB had previously secured ethical approval from the North West Multi-centre Research Ethics Committee (REC reference 11/NW/0382). Furthermore, the research initiative documented herein had received approval from the UKB, designated by application number 68,382. During in-person interviews, a standardized questionnaire was employed to systematically acquire an extensive array of lifestyle information from the study participants. Additionally, the cognitive status of the subjects was evaluated using a touch-screen questionnaire. The UKB encompassed a comprehensive cohort, comprising a total of over 500,000 individuals.

As part of the overarching UKB study, a subset of participants underwent neuroimaging procedures, resulting in the acquisition of brain imaging data. To ensure data homogeneity, each of the three imaging centers was equipped with identical scanners and fixed platforms, maintaining consistency by refraining from major software or hardware updates throughout the study. Specifically, each center utilized a 3T Siemens Skyra (Skyra 3T, Siemens Healthcare GmbH, Erlangen, Germany) with software platform VD13 and a 32-channel receive head coil dedicated to brain imaging. The T1-weighted MRI employed a Magnetization-Prepared Rapid Gradient-Echo (MPRAGE) sequence characterized by a high spatial resolution of $1 \times 1 \times 1$ mm, an image matrix of $208 \times 256 \times 256$ mm^3, and inversion time (TI)/repetition time (TR) of 880/2000 ms. DMRI data acquisition encompassed two b values (b = 1000, 2000 s/mm^2) with a spatial resolution of $2 \times 2 \times 2$ mm, covering a comprehensive set of 100 distinct directions. This protocol incorporated a multiband acceleration factor of 3. RsfMRI was executed with specific acquisition parameters, featuring a spatial resolution of 2.4 × 2.4 × 2.4 mm, a TR of 0.735 s, an echo time (TE) of 39 ms, and a multiband acceleration factor of 8. These standardized imaging protocols contribute to the reliability and consistency of the acquired data across the study cohort.

The selection process, detailed in the accompanying Figure 1, adhered to rigorous criteria aimed at ensuring data quality. Exclusion was based on the International Classification of Diseases, Tenth Revision (ICD-10), diagnostic classification system. Individuals diagnosed with malignant tumors of the eye, brain, and other parts of the central nervous system; cerebrovascular disease; psychiatric and behavioral disorders; neurological disorders; and other disorders affecting brain health were excluded from the analysis. This screening procedure led to the inclusion of 388,721 subjects between the ages of 45 and 83. Subsequently, 31,621 subjects possessing comprehensive sMRI, dMRI, and rsfMRI data were selected. This subset was then randomly divided into a training set (40%) and a test set (60%) to implement the brain-age prediction model. Within the test set, individuals exhibiting characteristics of ABA or RBA were identified using the brain-age prediction model. Specifically, individuals demonstrating a positive BrainAGE across all three imaging modalities were assigned to the ABA group. Conversely, those displaying consistently negative values in all three imaging modalities were assigned to the RBA group. Furthermore, participants who had not completed all nine cognitive tests, as well as those with incomplete data on covariates, were eliminated from the final analysis. As a result, the study ultimately comprised 1949 subjects classified within the RAB group, characterized by a mean age of 63.6 years with a standard deviation of 7.97. Concurrently, the ABA group encompassed 3203 subjects with a mean age of 64.6 years and a standard deviation of 6.96.

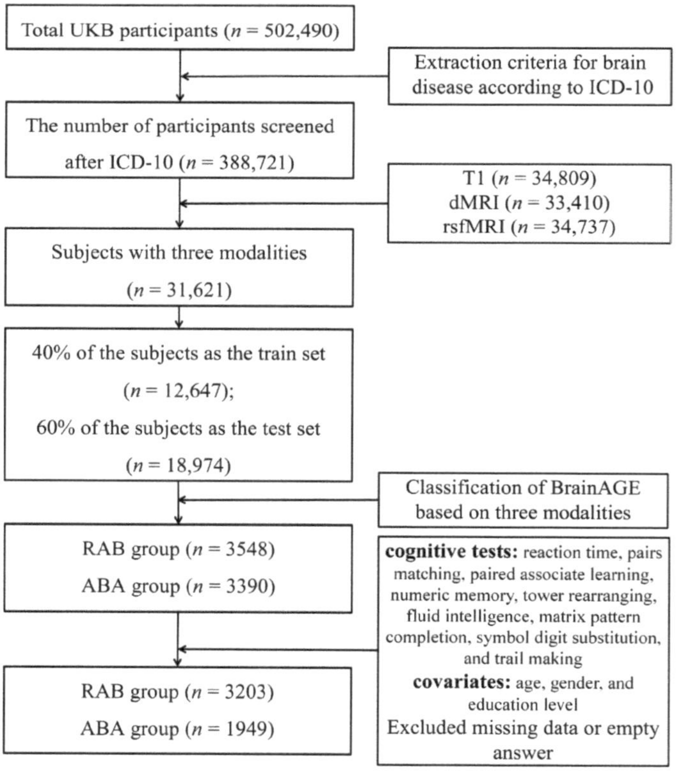

Figure 1. Flowchart depicting the subject screening process.

2.2. Imaging-Derived Phenotypes (IDPs)

The UKB presents a diverse array of neuroimaging modalities [32]. Following the meticulous acquisition of data, a standardized methodology is applied for image preprocessing and preliminary analysis, leading to a comprehensive set of IDPs. The carefully curated IDPs serve as the foundation for capturing valuable insights into different aspects of the brain structure and function, facilitating a comprehensive investigation aligned with the study's objectives.

The T1 MRI, distinguished for its meticulous precision, stands as a structural modality acclaimed for its remarkable ability to intricately capture detailed brain anatomy at an impressive resolution. This imaging modality provides a potent contrast between gray and white matter, facilitating the accurate visualization of intricate brain structures. The quantification of volumes was meticulously conducted using the FMRIB software library (FSL, version 5.0.10), accessible at http://fsl.fmrib.ox.ac.uk/fsl (accessed on 16 February 2022). Employing the FMRIB's automated segmentation tool (FAST, version FAST3), a total of 139 IDPs were derived. This was achieved by aggregating partial volume estimations within 139 regions of interest (ROIs) (UKB ID: 25782-25920) established in the MNI152 space, amalgamating parcellations from various atlases, including the Harvard–Oxford cortical and subcortical atlases (https://fsl.fmrib.ox.ac.uk/fsl/fslwiki/Atlases, accessed on 16 February 2022) and the Diedrichsen cerebellar atlas (http://www.diedrichsenlab.org/imaging/propatlas.htm, accessed on 16 February 2022). The warp field, previously estimated to effectuate the transformation of subject data into a standardized space, underwent inversion and subsequent application to the ROIs. This process generated a version of the ROIs in the native space, facilitating precise masking within the segmentation framework. Extraction of cortical thickness from cortical regions involved the meticulous implementation of the established FreeSurfer parcellation scheme [33]. This scheme, grounded in the Desikan–Killiany atlas, comprehensively delineates cortical domains across both hemispheres, encompassing a total of 68 discrete regions (UKB ID: 25755-26788, 26856-26889).

DMRI serves as a crucial tool for evaluating water molecule movement within the local tissue environment. At the voxel level, local estimates of diffusion properties provide valuable insights into microstructural tissue integrity, encompassing diffusion tensor estimates. Furthermore, long-range estimates derived from tractography, which involves the meticulous tracing of brain pathways, offer comprehensive information about the structural connectivity between pairs of brain regions. In this study, we employed the DTIFIT tool (available at https://fsl.fmrib.ox.ac.uk/fsl/fdt, accessed on 16 February 2022), to fit a diffusion tensor at each voxel. This procedure yielded multiple diffusion measures, encompassing fractional anisotropy (FA) and mean diffusivity (MD) maps. These collective measures provide a comprehensive elucidation of the characteristics of water diffusion within the brain tissue. Moreover, the dMRI data underwent sophisticated processing leveraging using NODDI (Neurite Orientation Dispersion and Density Imaging). NODDI enables the estimation of crucial white-matter microstructural parameter isotropic water volume fraction (ISOVF).

To delve into the intricacies of the white-matter microstructure, we employed tract-based spatial statistics (TBSS). TBSS facilitates the alignment of the FA image onto a standard-space white-matter skeleton through high-dimensional FNIRT-based warping. This standardized-space warp is subsequently applied to all other dMRI measures. Each resulting skeletonized image for dMRI measures underwent averaging across 48 standard spatial tract masks, meticulously defined by Susumi Mori's group at Johns Hopkins University. This detailed averaging procedure produced a total of 144 distinctive IDPs (FA (UKB ID: 25056-25103), MDs (UKB ID: 25104-25151), and ISOVFs (UKB ID: 25440-25487)).

The analysis of rs-fMRI images was conducted using the MELODIC (Multivariate Exploratory Linear Decomposition into Independent Components) framework [34]. This processing pipeline integrated group principal component analysis and independent component analysis, culminating in the extraction of spatially orthogonal independent components (ICs) representing distinct resting-state neural networks. A low-dimensional group-independent component analysis approach was employed to obtain a population-level spatial map of the resting-state network. The functional images underwent pre-processing with 25 fractions (UKB ID: 25752), and a meticulous exclusion process eliminated 4 noise components, resulting in a set of 21 components of particular interest. Each of these components corresponded to unique resting-state networks, offering invaluable insights into the underlying neural activity patterns during rest. The online visualization of these ICs is facilitated through the Papaya viewer (https://www.fmrib.ox.ac.uk/ukbiobank/group_means/rfMRI_ICA_d25_good_nodes.html, accessed on 16 February 2022). This viewer, along with accompanying maps, provides an interactive and insightful platform for exploring and comprehending the spatial distribution of the ICs derived from rs-fMRI data. Moreover, a partial correlation matrix derived from rsfMRI data was utilized to represent the number of network connections, totaling 210 values. This was calculated by multiplying the 21 networks by 20 (excluding identity correlations) and dividing by 2, considering the matrix's diagonal symmetry. The implementation of partial correlation aimed to enhance the precision of estimating direct "connections" between networks compared to full correlation.

2.3. Brain-Age Prediction Model

Lasso, short for "Least Absolute Shrinkage and Selection Operator," is a statistical regularization technique in machine learning. It adds a penalty term to the regression equation, constraining the absolute size of the coefficients and effectively promoting sparsity by forcing some coefficients to be exactly zero. Lasso is widely employed in predictive modeling, particularly when dealing with high-dimensional datasets. Prior investigations into brain-age prediction [35,36] have consistently demonstrated the superior performance of the Lasso model when compared to other machine-learning models. Given these compelling findings, we have chosen the Lasso model as the method of choice for brain-age prediction in our study.

Within the Lasso model, the penalty regularization parameter, denoted as alpha, assumes a pivotal role in determining the intensity of the penalty applied to model parameters. The magnitude of alpha directly influences the strength of the penalties assigned to each parameter, resulting in varying degrees of model shrinkage. In the context of this study, we meticulously defined the grid search space for the alpha parameter as (0.001, 0.01, 0.1, 1, 10, 100). This specific range was chosen to efficiently explore the parameter space and identify the optimal alpha value that would maximize model performance.

BrainAGE [37] is a neuroimaging-based metric designed to quantify the difference between an individual's actual chronological age and the predicted age of their brain. This innovative approach leverages structural brain imaging data to provide insights into the aging process at the neural level. The fundamental premise behind BrainAGE is to assess the extent to which the brain either accelerates or decelerates in comparison to the individual's chronological age. BrainAGE also has a strong correlation with brain maintenance (BM) [38]. The brain-age prediction model entails the application of machine-learning techniques to brain imaging data, enabling the development of a predictive model for estimating the "age" of the brain through its imaging features. The BrainAGE score is derived by calculating the difference between the age predicted by the model and an individual's chronological age (Equation (1)). A positive BrainAGE score suggests that the brain is aging at a faster rate than expected, potentially indicating accelerated aging or suboptimal BM. Conversely, a negative BrainAGE score implies a more youthful-appearing brain, indicative of better-preserved structural characteristics than what would be expected based on chronological age.

$$\text{BrainAGE} = \text{Predicted age} - \text{Chronological age} \tag{1}$$

Recent studies have underscored the presence of a proportional bias in the computation of brain age, where the disparity between chronological age and predicted brain age exhibits a negative correlation with chronological age. This phenomenon is attributed to the well-documented effect of regression toward the mean [35,39]. This phenomenon has the potential to introduce bias in the prediction of age, which may be overestimated in younger subjects and underestimated in older subjects compared to their respective chronological ages. Given the inherent age-related bias, the imperative arises for the implementation of an age-bias correction procedure, as outlined in Equation (2).

$$\text{Predicted age}_{\text{corrected}} = \text{Predicted age}_{\text{raw}} - \alpha - \beta \times \text{Chronological age} \quad (2)$$

where Predicted age_{raw} indicates brain age predicted by the Lasso model, and α and β represent the intercept and slope of the regression line between chronological age and predicted age in the training set.

Subsequently, subjects in the test sets were systematically categorized based on their BrainAGE. Individuals displaying positive BrainAGE values across all three modalities were categorized into the ABA group, indicating that their predicted brain age exceeded their chronological age. Conversely, subjects with negative BrainAGE values across the three imaging modalities were assigned to the RBA group, indicating a favorable condition where the predicted brain age suggested a structure and function younger than their actual age. This stratification provides a nuanced understanding of age-related deviations in brain structure and function, fostering a comprehensive characterization of individual differences in brain aging within the study cohort.

2.4. Non-Imaging Derived Phenotypes (Non-IDPs)

Throughout their active engagement in the UKB study, subjects were diligently queried to furnish comprehensive insights into their lifestyle and physical health using diverse methodologies. The amalgamation of this wealth of information culminated in the creation of non-imaging derived phenotypes (Non-IDPs), which serve as integral components of the broader analytical framework. The study comprehensively examined six Non-IDPs intricately associated with lifestyle and physical health. These variables included systolic blood pressure (UKB ID: 4080), time spent driving (UKB ID: 1090), hand grip strength (UKB ID: 46, 47), usual walking pace (UKB ID: 924), and diabetes diagnosed by a doctor (UKB ID: 2443).

2.5. Neuropsychological Tests

The neuropsychological battery, consisting of nine cognitive domains [40], served as the foundation for cognitive evaluation in this study. Specifically, two cognitive scales within the scope of our investigation—reaction time (UKB ID: 20023) and trail-making (UKB ID: 6350)—both incorporating time as a test outcome, underwent a log transformation to enhance their analytical robustness. A detailed overview of the neuropsychological tests is provided in Table 1.

Table 1. Cognitive domain, neuropsychological tests, and test descriptions.

Testing	Description	Cognitive Domain	UKB ID
Pairs matching	Number of incorrect matches made in round	Visual declarative memory	399
Numeric memory	Maximum number of digits remembered correctly	Working memory	4282
Fluid intelligence	Fluid intelligence score assessment	Verbal and numerical reasoning	20016
Paired associate learning	Number of correctly associated word pairs	Verbal declarative memory	20197
Matrix pattern completion	Number of correctly solved puzzles	Non-verbal reasoning	6373
Reaction time	Mean time taken to correctly identify matches	Processing speed	20023
Symbol digit substitution	Number of correct symbol digit matches made	Processing speed	23324

Table 1. *Cont.*

Testing	Description	Cognitive Domain	UKB ID
Tower rearranging	Number of correctly solved puzzles	Executive function	21004
Trail-making	Duration to complete alphanumeric path	Executive function	6350

2.6. Identification of ABA Subgroups Using HYDRA

Leveraging the information derived from IDPs, we employed the HYDRA algorithm to discern distinct ABA subtypes [26]. HYDRA is a semi-supervised machine-learning algorithm tailored for unraveling the intricacies of disease heterogeneity. In this study, this algorithm achieves ABA heterogeneity by partitioning ABA subjects, discerning patterns or transformations between subpopulations within the ABA group and a reference group (i.e., RBA subjects). The partitioning process employs a convex polytope, a construct amalgamating multiple linear max-margin classifiers. Notably, HYDRA demonstrates the capability to effectively regress out nuisance covariates, such as age and sex, enhancing its precision in discerning genuine patterns associated with brain aging. In its approach, HYDRA conceptualizes subjects as points within a high-dimensional space, aligning with the support vector machine (SVM) classification framework. Leveraging the discriminative power of linear SVMs in high-dimensional spaces, HYDRA extends this capability to the non-linear domain in a piecewise fashion. This extension involves the formation of a convex polytope through the combination of multiple hyperplanes, effectively segregating the two groups. Enclosed within this convex polytope are the RBA samples, while distinct faces of the polytope facilitate ABA subtyping. Each face encapsulates a distinct multivariate pattern of difference between the two groups, and hence a distinct accelerated aging process.

In the initial phase, HYDRA allocates different labels to the ABA and control groups (RBA subjects). Subsequently, the algorithm integrates multiple linear max-margin classifiers into a convex polyhedron by clustering the k-values, where k represents the number of clusters, effectively distinguishing control subjects from those exhibiting ABA. The assignment of ABA subjects to the nearest hyperplane within a single linear subclassifier results in the division of all ABA subjects into K clusters, with each polyhedron encapsulating the distinct characteristics of an ABA subtype. The optimization problem is systematically addressed through an iterative procedure, alternately assigning ABA samples to the faces of the polytope and estimating hyperplanes to maximize the overall margin. This iterative coupling between clustering and classification serves the dual purpose of segregating ABA subjects based on accelerated brain-aging effects, rather than a global anatomical perspective. For optimizing the identification of ABA subtypes, a systematic approach was employed, ranging from two to five clusters, with five-fold cross-validation. Covariates, including age, gender, and education level, were considered during the process. Of note, the educational level underwent a transformation into years of education, aligning with established practices in prior research [41]. The stability of clustering outcomes was quantified using the adjusted rand index (ARI) [26] in conjunction with five-fold cross-validation. The determination of the optimal number of clusters relied on the maximum ARI, ensuring the selection of the most reliable clusters. The comprehensive workflow is depicted in Figure 2.

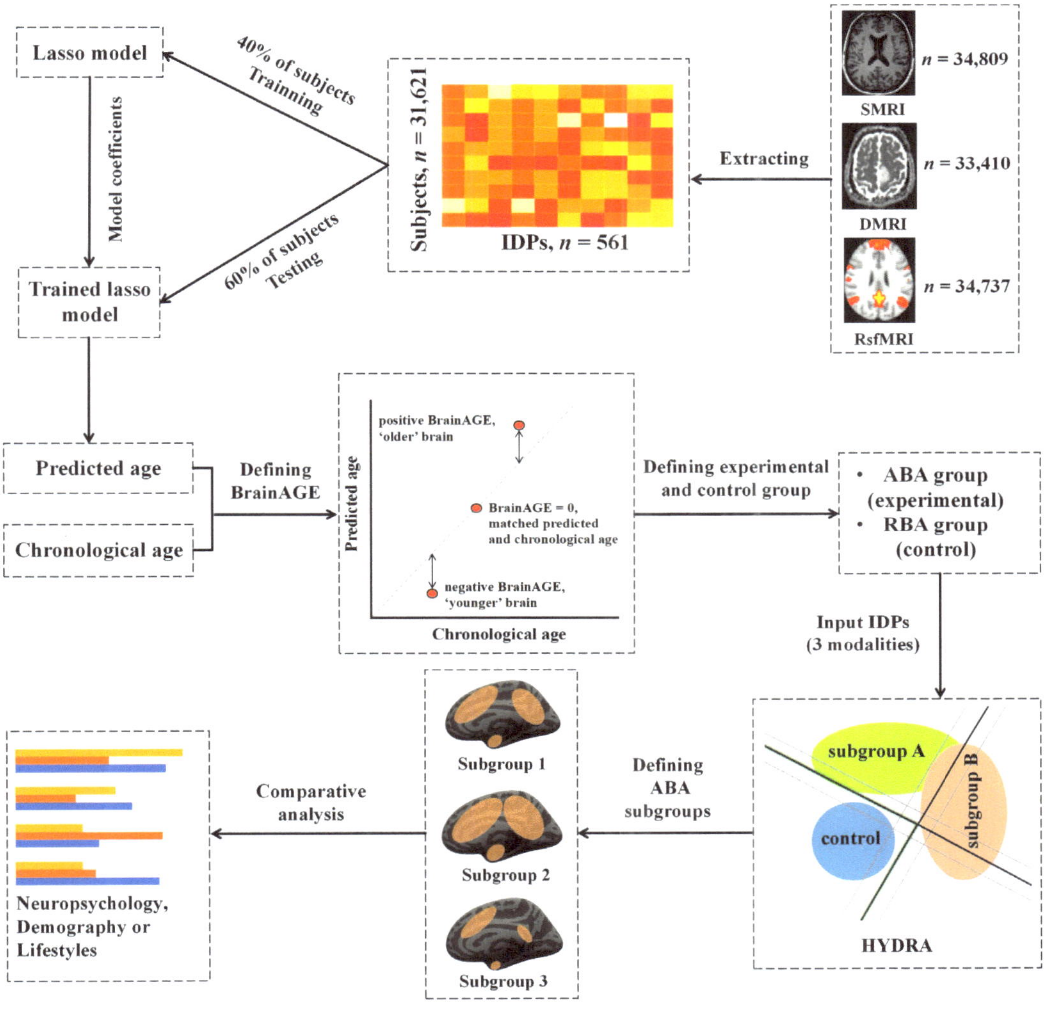

Figure 2. The comprehensive workflow of the present investigation.

2.7. Statistical Analysis

The study encompassed three primary sections delineating distinct characteristics: (1) Lifestyle and determinants, encompassing variables such as age, gender, years of education, and six lifestyle factors pertaining to physical health; (2) Neuropsychological exam, comprising a comprehensive battery of nine cognitive assessments; and (3) IDPs derived from T1, dMRI, and rsfMRI, totaling 561 IDPs. For sections (1) and (2), differences between matched subtypes were rigorously compared. Disparities in qualitative variables were assessed using the chi-square test, while quantitative variables underwent analysis of variance (ANOVA). Two-by-two comparisons were executed utilizing Dunnett's test, with a predefined statistical significance level set at $p < 0.05$. In section (3), the analytical framework encompassed a comparison of differences between subgroups and controls, employing ANOVA. To address the issue of multiple comparisons, the Bonferroni method was meticulously applied, imposing a stringent threshold of $q < 0.01$. All statistical analyses were conducted using SPSS 26 software, a widely acknowledged statistical package (SPSS, 1989; Apache Software Foundation, Chicago, IL, USA).

3. Results

3.1. Brain-Age Prediction

Within the scope of this investigation, Lasso regression analysis was selected as the preferred methodology for predicting brain age, with mean absolute error (MAE) serving as the metric for evaluating model performance. Interestingly, dMRI emerged as the modality with the highest predictive accuracy. The application of Lasso regression to dMRI data resulted in a remarkably low MAE of 4.03 years, indicating the effectiveness of this approach in estimating brain age. Moreover, the predictive accuracy based on T1 data, encompassing cortical thickness and gray-matter volume, resulted in an MAE of 4.17 years, while rsfMRI demonstrated an MAE of 5.28 years.

The categorization of ABA and RBA groups was contingent upon the consistency of positive or negative BrainAGE across the three modalities within the test set of brain-age prediction (n = 18,974). Specifically, if BrainAGE across all three modalities was positive, the subject was categorized as ABA; conversely, if BrainAGE was consistently negative, the subject was designated as RBA. This delineation led to the selection of 3203 subjects in the RBA group (mean age = 63.6 ± 7.97) and 1949 subjects in the ABA group (mean age = 64.6 ± 6.96).

3.2. Definition of ABA Subgroups

Within the confines of this investigation, the HYDRA framework was implemented to partition ABA heterogeneity, where the ABA population assumed the role of the experimental group, and the RBA population served as the control group. Subsequent to this partitioning, meticulous scrutiny of the fidelity of cluster assignment transpired. The examination involved systematically varying the cluster number from 1 to 5, employing the ARI as the metric for assessment. The ARI quantifies the similarity between true and predicted cluster assignments, offering a measure of clustering accuracy that accounts for chance. Notably, a monotonically increasing trend was observed within the range of 1 to 3 subtypes. However, as the subtype count extended to 4 and 5, a relative decline in the ARI values was discerned in comparison to the trinary configuration. This observation suggests that clustering efficacy may be optimized into three distinct subtypes (refer to Figure 3). It is imperative to note that HYDRA employed a robust five-fold cross-validation strategy. The delineation of optimal subtypes reflects the outcomes observed in the validation sets across these folds. Subsequent analysis delineated that, based on the cross-validation result, 783 individuals from the ABA cohort were allocated to SubGroup 1, while SubGroup 2 comprised 561 ABA subjects, and SubGroup 3 encompassed 605 ABA subjects.

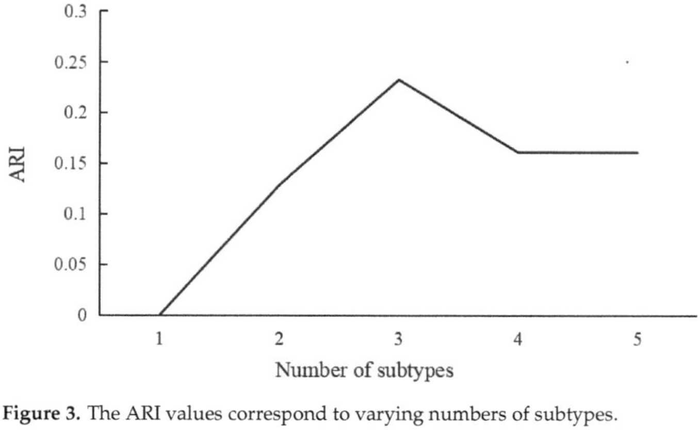

Figure 3. The ARI values correspond to varying numbers of subtypes.

The comprehensive delineation of demographic information, as meticulously presented in Table 2, highlights the nuanced distinctions within these demographic variables. Substantial statistical distinctions in age and sex distribution were evident within the tripartite classification of ABA subjects. Notably, there were no discernible differences in years of education. To mitigate the potential confounding effects stemming from these demographic variations, the Generalized Linear Model (GLM) for IDps incorporated three crucial demographic variables—namely, age, sex, and years of education—as covariates. Through this inclusion, their respective influences were systematically controlled and eliminated. The application of rigorously controlled covariate regressions serves to enhance the precision of subsequent analyses and facilitates a nuanced interpretation of the influence of specific subtypes on the observed outcomes.

Table 2. Demographics characteristics of RBA and ABA subgroups.

Characteristics	RBA Group	SubGroup 1	SubGroup 2	SubGroup 3	*p*-Values
n	3203	783	561	605	
Age (years)	64.62	61.57	66.78	63.40	<0.0001 [a,b,c]
Education (years)	16.17	15.46	15.54	15.75	0.527
Women, *n* (%)	1884 (58.8%)	326 (41.6%)	259 (46.2%)	302 (49.9%)	0.008 [c]

a: SubGroup 1 is significantly different from the SubGroup 2; b: SubGroup 2 is significantly different from the SubGroup 3; c: SubGroup 1 is significantly different from the SubGroup 3.

Following this, an ANOVA was employed to scrutinize discrepancies among RBA and distinct subgroups within the ABA cohort. In response to the inherent challenge of multiple comparisons, the Bonferroni method was applied with a stringent threshold ($q < 0.01$). Remarkably, this comprehensive examination revealed nuanced differences in the patterns of sMRI, dMRI, and rsfMRI features across the three subtypes. These findings underscore the intricate nature of neuroimaging alterations within distinct subtypes of ABA cohorts.

In the context of structural alterations discerned through sMRI, Figures 4 and S1 in Supplementary Materials have been meticulously crafted to provide comprehensive insights into the distinctions among the three identified ABA subgroups and the control group. SubGroup 1, consisting of 783 elderly subjects, exhibited diffuse cortical atrophy spanning the frontal, parietal, and temporal lobes bilaterally, with limited atrophy observed in the occipital and limbic lobes. This subgroup displayed extensive gray-matter volume reduction throughout the entire brain, emphasizing significant atrophy in key regions such as the Insula, Paracentral lobule, and Angular gyrus. SubGroup 2 demonstrated a comparable pattern of atrophy to SubGroup 1, with slight variations noted in the left cortex of the limbic lobe and Insula. In contrast, SubGroup 3 manifested small cortical and gray-matter volume atrophy, indicating regionally sparse and mild whole-brain atrophy. To further elucidate these morphological alterations, a graphical representation (Figure 5) of Z-values and their 95% confidence intervals for cortical thickness comparisons has been incorporated. This graphical representation unveils similar regions of atrophy across the three subgroups, yet discernible differences exist in the overall distribution pattern of atrophy. It is imperative to underscore that all Z-values were computed with respect to the mean and standard deviation of the RBA group, where Z-values for the RBA group serve as the baseline with a value of 0. SubGroup 3 exhibited the least pronounced atrophy, while SubGroup 2 showcased the most severe atrophy. These findings offer nuanced insights into structural distinctions among ABA subgroups, unraveling the intricacies of ABA-related morphological alterations.

Upon meticulous examination of white-matter microstructure using dMRI, SubGroup 2 emerged as a focal point characterized by substantial deviations from the control group, indicating pronounced alterations across nearly all scrutinized regions. In comparison to the control group, SubGroup 1 and 3 also manifested a comprehensive array of distinctions from controls, albeit with a noticeably lower magnitude than observed in SubGroup 2.

Graphical representations, as depicted in Figures 6–8, unravel the nuanced variations in Z-values and their 95% confidence intervals for FA, MD, and ISOVF. All Z-values are computed relative to the mean and standard deviation of the RBA group, serving as the baseline with an expected value of 0. In the context of white-matter integrity, lower FA values and elevated MD and ISOVF values are indicative of compromised microstructural integrity. Remarkably, SubGroup 2 exhibited lower Z-values in FA compared to the other two subgroups, accompanied by higher values in MD and ISOVF. This observation underscores a pronounced degradation of white-matter integrity in SubGroup 2. In contrast, the Z-values within SubGroups 1 and 3 exhibited comparable trends, displaying closer proximity to 0. Contrary to the control group, SubGroup 1 displayed a compromised white-matter microstructure akin to that observed in SubGroup 3, albeit with a milder impact.

Examining functional connectivity through rsfMRI, this study meticulously delineates the intricate connection strengths between distinct ABA subgroups and the control cohort. The categorization of connection strengths within the control group, discerned through positive and negative connections, facilitated the comparison of Z-values for the three subgroups, elegantly presented as a heatmap in Figure 9. The Z-values presented in the analysis are derived in relation to the mean and standard deviation of the RBA group, establishing 0 as the baseline for Z-values in the RBA group. Noteworthy observations emerge as groups 1 and 3 exhibit a more analogous pattern in both positive and negative connection strengths. However, SubGroup 3 stands out with notably more negatively linking nodes within the negative connection category. In stark contrast, SubGroup 2 presents a divergent pattern characterized by a smaller change in negative connection strengths in comparison to the control group.

3.3. Cognitive and Non-IDPs Characteristics between Matched Subtypes

Detailed cognitive characteristics among the three ABA subtypes and the RBA are elucidated in Table 3. SubGroup 2 prominently exhibited the most discernible cognitive impairment, notably differing from the other subtypes in reaction time, symbol digit substitution, and trail-making. In contrast, SubGroup 3 displayed superior cognitive performance across all tests, demonstrating significant differences, particularly in fluid intelligence and matrix pattern completion, compared to the other subgroups.

Table 3. Cognitive characteristics within the identified study subtypes.

Cognitive Function Test	UKB ID	RBA Group	SubGroup 1	SubGroup 2	SubGroup 3	*p*-Values
Pairs matching	399	3.577	3.664	3.814	3.540	0.307
Numeric memory	4282	6.819	6.554	6.452	6.688	0.089
Fluid intelligence	20016	6.820	6.307	6.435	6.927	<0.001 [b,c]
Paired associate learning	20197	7.234	6.670	6.445	6.854	0.097
Matrix pattern completion	6373	8.227	7.756	7.745	8.088	0.036 [b,c]
Reaction time	20023	2.764	2.764	2.784	2.769	<0.001 [a,c]
Symbol digit substitution	23324	19.634	18.654	17.633	18.832	0.003 [a,c]
Tower rearranging	21004	10.041	9.807	9.580	9.958	0.246
Trail-making	6350	2.711	2.733	2.765	2.718	<0.001 [a,c]

a: SubGroup 1 is significantly different from SubGroup 2 ($p < 0.05$); b: SubGroup 1 is significantly different from SubGroup 3 ($p < 0.05$); c: SubGroup 2 is significantly different from SubGroup 3 ($p < 0.05$).

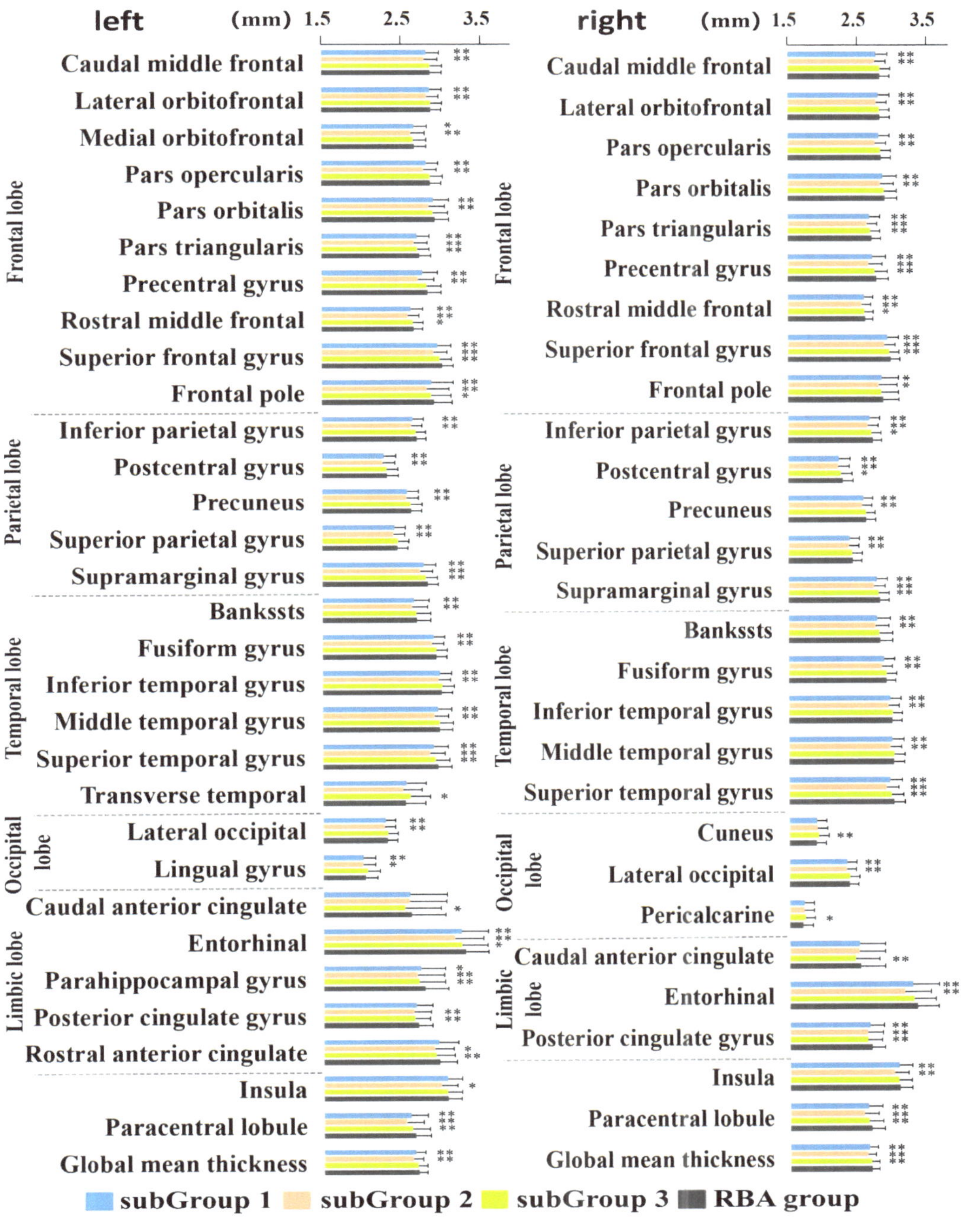

Figure 4. Comparative analysis of cortical thickness across the three ABA subgroups and the control group. Statistical significance denoted as * indicates $q < 0.01$, while ** signifies $q < 0.001$.

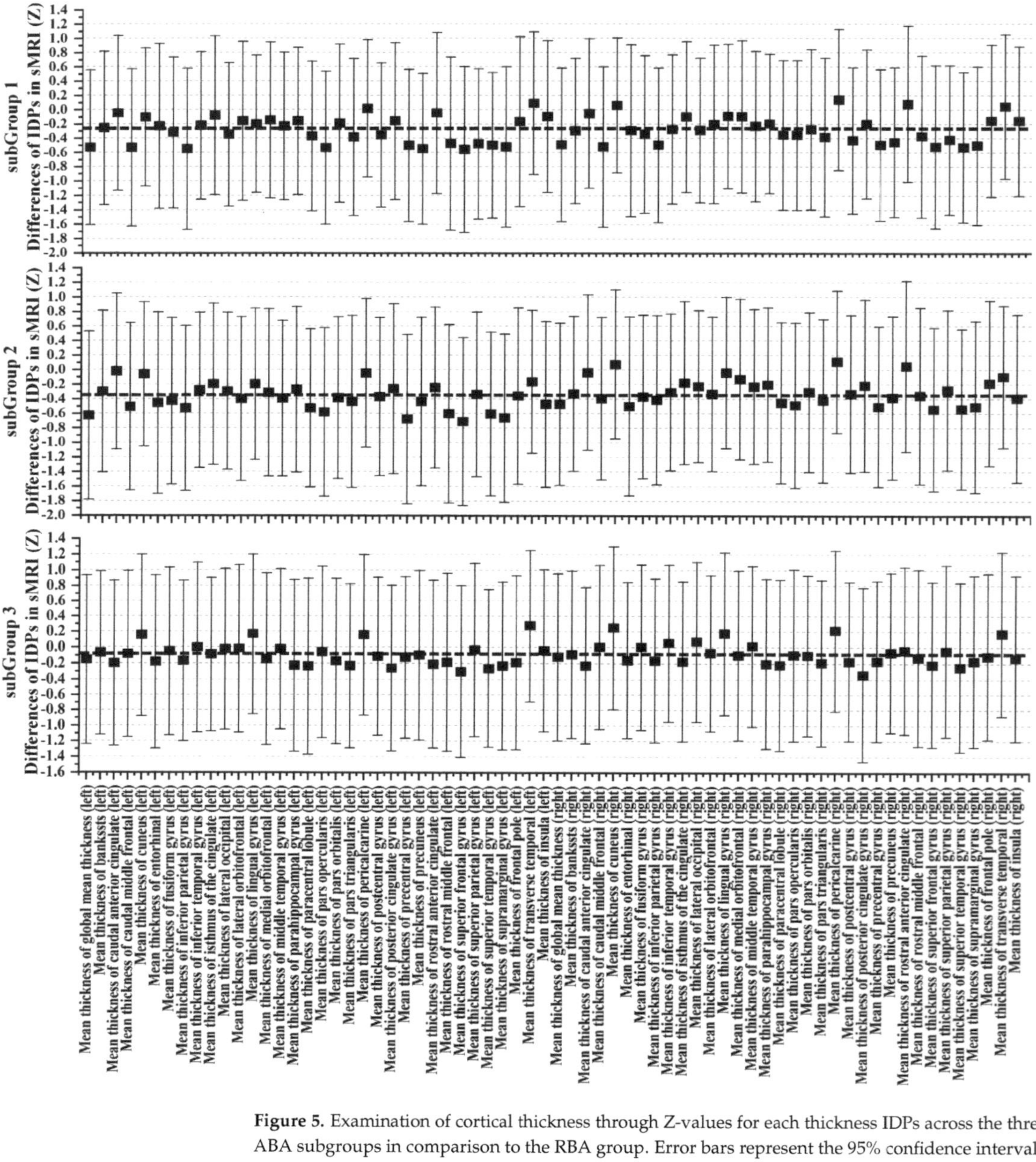

Figure 5. Examination of cortical thickness through Z-values for each thickness IDPs across the three ABA subgroups in comparison to the RBA group. Error bars represent the 95% confidence intervals. The black dotted line represents the mean Z-value for each subgroup.

Shifting the focus to Non-IDPs, differences among the subtypes are illustrated in Figures 10 and 11. SubGroup 2 exhibited the most pronounced distinctions compared to the other two ABA groups, featuring elevated blood pressure, diminished grip strength, a higher prevalence of confirmed diabetes, and a slower pace in usual walking. SubGroups 1 and 3 displayed relatively fewer differences, primarily diverging in the time spent driving and usual walking pace. Concurrently, notable distinctions were observed in blood pressure, confirmed diabetes prevalence, and usual walking pace between SubGroup 2 and the RBA group. However, no statistically significant differences were identified in grip strength values and driving time. In the case of SubGroup 1, marked disparities were evident in all Non-IDP variables as compared to the RBA group, except for the prevalence of diagnosed diabetes.

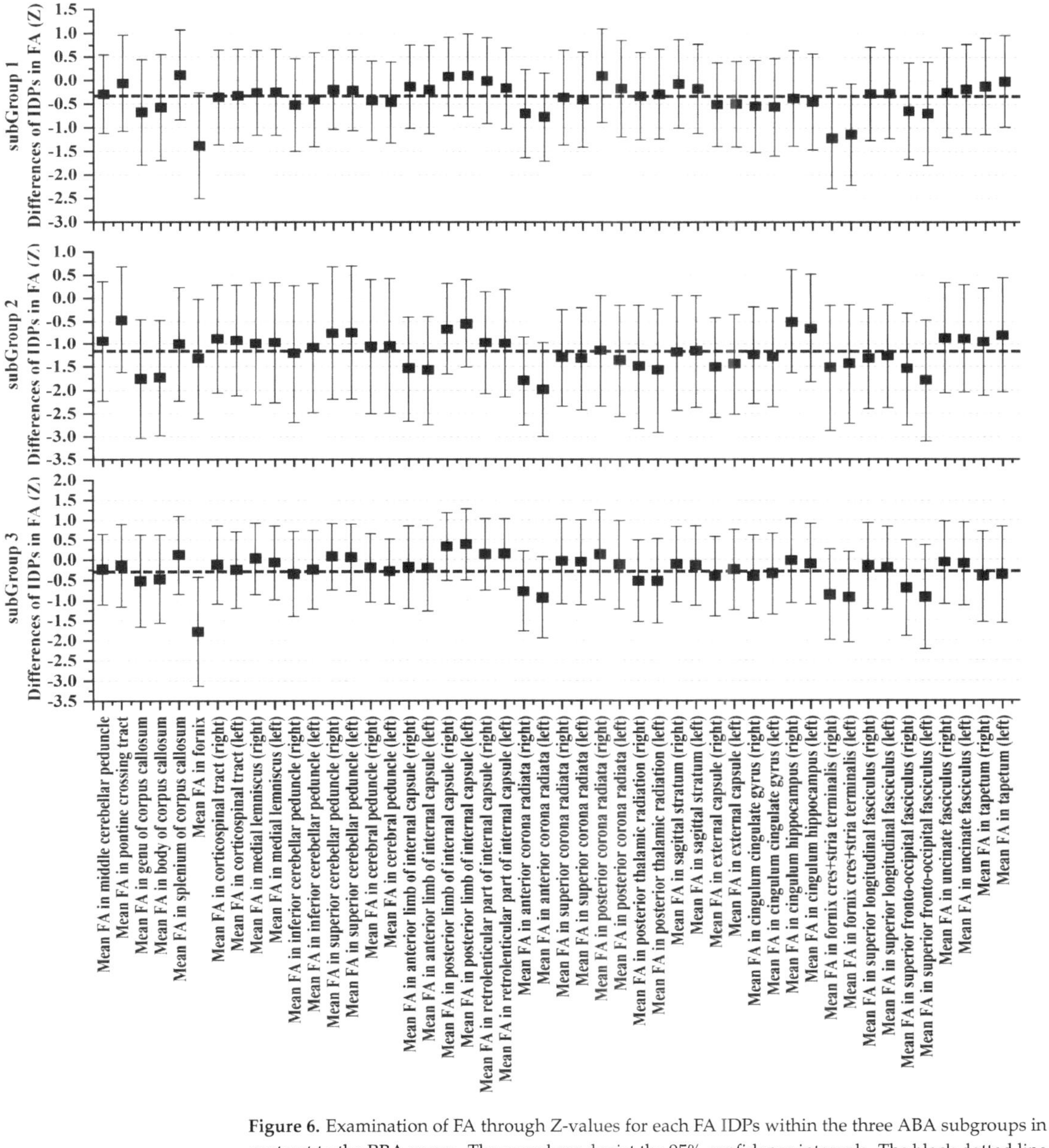

Figure 6. Examination of FA through Z-values for each FA IDPs within the three ABA subgroups in contrast to the RBA group. The error bars depict the 95% confidence intervals. The black dotted line represents the mean Z-value for each subgroup.

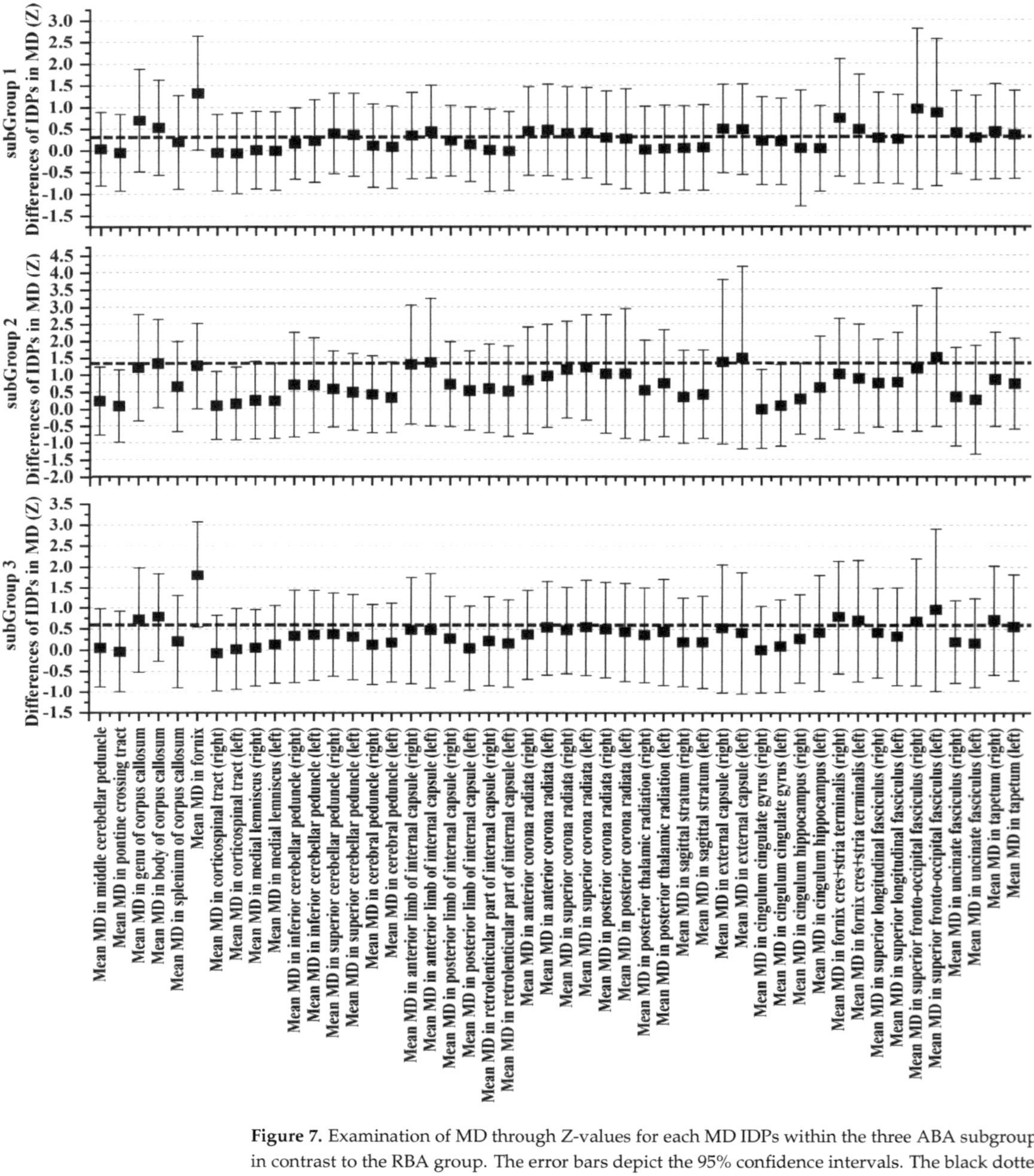

Figure 7. Examination of MD through Z-values for each MD IDPs within the three ABA subgroups in contrast to the RBA group. The error bars depict the 95% confidence intervals. The black dotted line represents the mean Z-value for each subgroup.

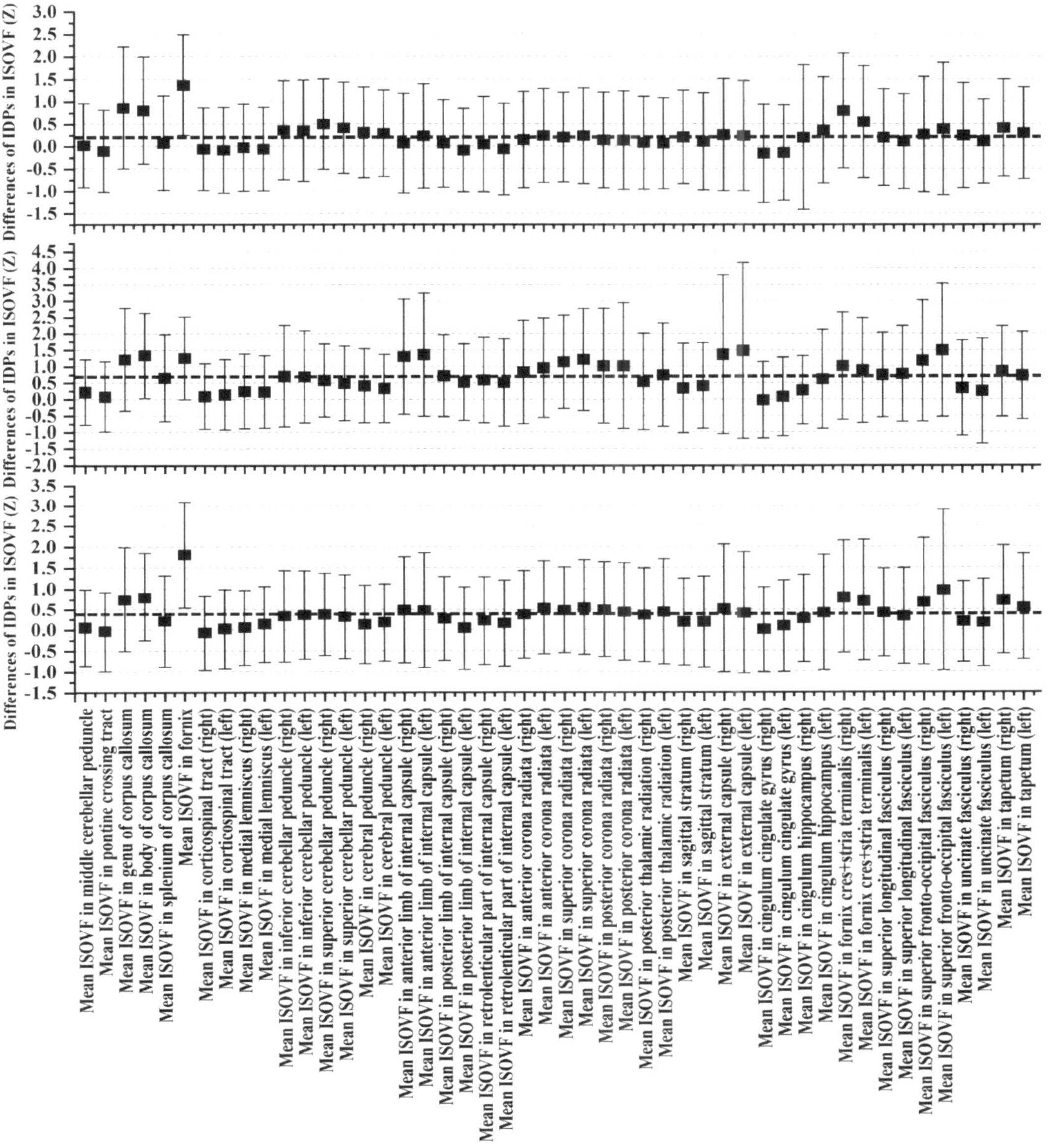

Figure 8. Examination of ISOVF through Z-values for each ISOVF IDPs within the three ABA subgroups in contrast to the RBA group. The error bars depict the 95% confidence intervals. The black dotted line represents the mean Z-value for each subgroup.

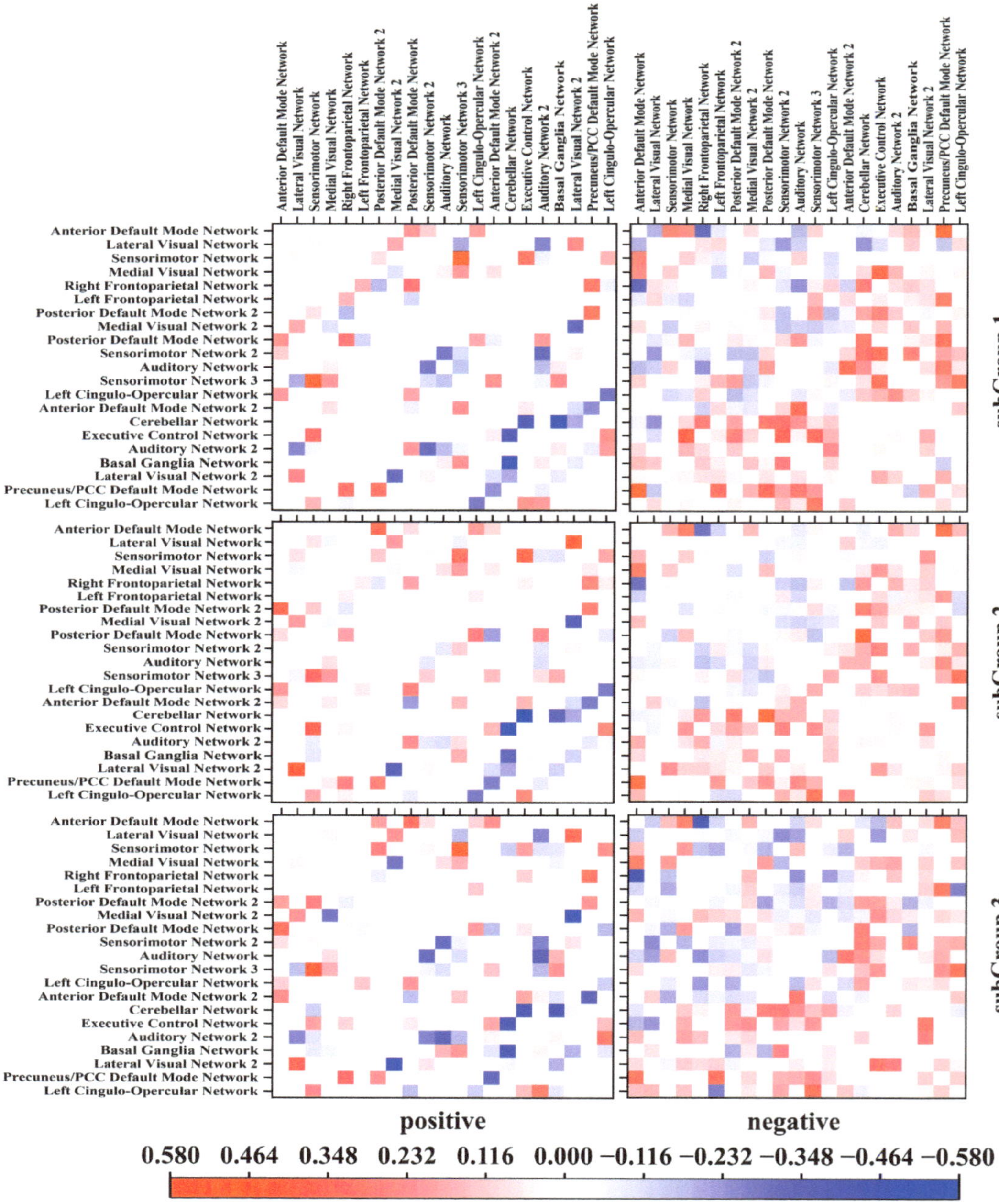

Figure 9. Heat map illustrating connection strength. The connection strengths within the three ABA groups were stratified based on the positive and negative phases of the RBA groups.

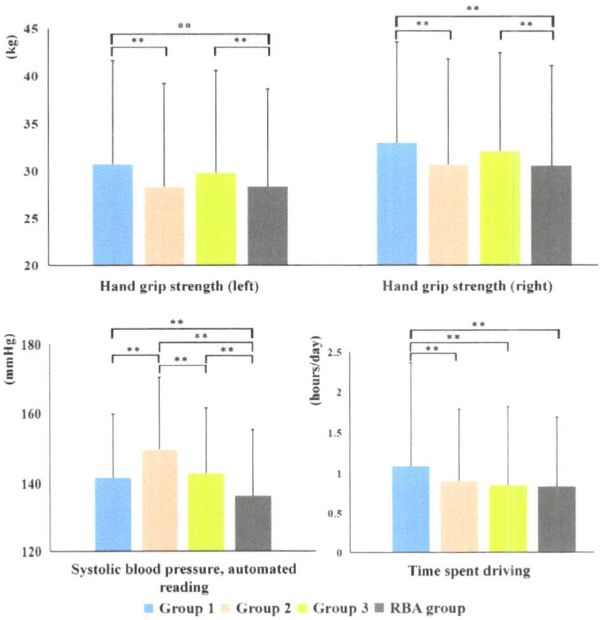

Figure 10. Quantitative analysis of Non-IDP variables among the three subtypes and the RBA group employed ANOVA, followed by pairwise comparisons using Dunnett's test. Significance is denoted by ** at $p < 0.001$.

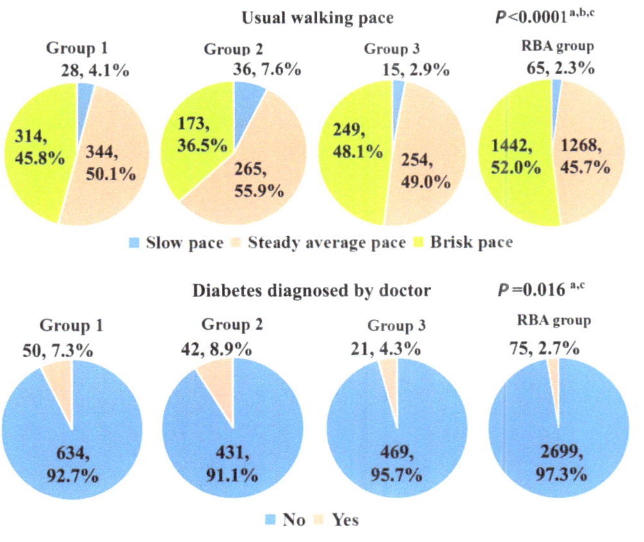

Figure 11. Comparative assessment of qualitative Non-IDP variables between the three subtypes and RBA group were then underwent by ANOVA, two-by-two comparisons were conducted employing Dunnett's test. a: SubGroup 2 is significantly different from SubGroup 3 ($p < 0.05$); b: SubGroup 1 is significantly different from RBA group ($p < 0.05$); c: SubGroup 2 is significantly different from RBA group ($p < 0.05$).

4. Discussion

Harnessing the capabilities of HYDRA in conjunction with the distinctive datasets provided by the UKB study, our research endeavors sought to scrutinize the existence of neuroimaging-defined subtypes within a cross-sectional sample of ABA. Our analyses discerned the presence of three discernible subtypes, each characterized by distinct neuroimaging profiles. These three subtypes manifest distinctive attributes of brain gray-matter structure, white-matter microstructure, and functional network connectivity. Notably, SubGroup 3 displayed the mildest atrophy, resembling SubGroup 1 in white-matter microstructure and functional connectivity strength. In contrast, SubGroup 2 exhibited no significant atrophy disparities compared to SubGroup 1; however, SubGroup 2 is characterized by the most impaired white-matter microstructural integrity and displays distinctive connectivity networks. This differentiation implies potential variations in underlying aging mechanisms, shedding light on the intricate heterogeneity inherent in the aging process.

4.1. Complex Landscape of ABA

Within the broader spectrum, age-related cognitive impairment seldom emerges as a consequence of a singular disease entity. Instead, it presents as a multifaceted interplay involving diverse factors, encompassing AD, various forms of dementia, and a range of health conditions like traumatic brain injury, stroke, depression, or developmental disabilities. The escalating apprehension regarding age-related cognitive decline arises from its widely recognized role as a pivotal determinant shaping the overall quality of life [42]. Given this backdrop, there is a heightened emphasis on the pursuit of biomarkers capable of assessing individual brain age and forecasting the trajectory of cognitive decline.

Methodologies deployed to ascertain brain age, grounded in neuroimaging data, are designed to elucidate deviations in age-related cerebral changes. This is accomplished through the establishment of robust reference curves for RBA and ABA, providing personalized metrics of brain age. Importantly, these approaches are tailored to accommodate the multidimensional patterns that characterize the aging process within the brain. Such sophisticated strategies hold considerable promise for advancing our understanding of cognitive aging and facilitating proactive interventions to enhance cognitive well-being in the aging population.

In the course of this comprehensive investigation, the ABA cohorts were meticulously characterized based on the discerning metric of BrainAGE, as detailed in Table 4. Within both sMRI and dMRI modalities, SubGroup 2 consistently exhibits the highest BrainAGE levels, indicative of the most pronounced accelerated aging. However, in the realm of rsfMRI, SubGroup 2 demonstrates the lowest BrainAGE, portraying a distinctive profile of accelerated aging within this specific modality. Shifting the focus to the domain of dMRI-defined BrainAGE, SubGroups 1 and 3 demonstrate comparable BrainAGE levels, both of which are lower than that of SubGroup 2. Delving deeper into the analysis of rsfMRI-defined BrainAGE, SubGroup 1 emerges as the category with the highest values, yet it exhibits proximity to SubGroup 3.

Table 4. BrainAGE of the ABA subtypes.

Group	sMRI	rsfMRI	dMRI
SubGroup 1	6.55 ± 4.51	11.51 ± 8.07	6.19 ± 4.55
SubGroup 2	7.85 ± 5.94	10.73 ± 7.79	7.19 ± 5.52
SubGroup 3	5.59 ± 4.39	11.41 ± 8.51	6.40 ± 4.64
RBA group	-6.19 ± 4.56	-10.85 ± 8.14	-5.91 ± 4.06

Numerous determinants intricately shape and modulate the trajectories of individual brain aging. The application of neuroimaging-based models in exploring brain aging has yielded compelling insights. Notably, robust correlations have been unveiled between ABA, AD severity, and the prospective decline in cognitive functions [43]. Additionally, associations have been established between ABA and mild cognitive impairment (MCI) [44], as well as the conversion to AD [45]. Furthermore, investigations have linked ABA to diverse factors such as traumatic brain injury [46], HIV [47], chronic pain [48], and type 2 diabetes mellitus [49]. ABA has proven indicative not only of diminished physical and mental fitness but also of heightened allostatic load and increased mortality [50]. Moreover, individual brain aging exhibits noteworthy connections with an array of health parameters, personal lifestyle choices, and drug utilization [19]. Education levels and engagement in physical activity have also emerged as significant determinants influencing the ABA process [51]. This intricate interplay underscores the multifaceted nature of brain aging, weaving a complex tapestry of connections with various health indicators, lifestyle elements, and physiological conditions. The dissection of underlying mechanisms expediting brain aging not only enables researchers to identify intervention and prevention targets but also sheds light on the heightened risk of individuals experiencing ABA for conditions such as AD, Parkinson's disease, and other neurodegenerative disorders.

4.2. ABA Subtype and Cognitive Reserve

In the realm of maintaining cognitive functioning amidst brain changes or insults, two pivotal forms of reserve come to the fore: brain reserve and cognitive reserve [52]. Brain age estimation serves as a valuable metric, providing a nuanced perspective on brain maintenance and reserves. Notably, ABA individuals, when compared to age-matched peers, exhibit compromised brain reserve capacities. This suggests that these individuals may face challenges in deploying alternative brain networks or cognitive strategies in the face of aging or insults. Cognitive reserve reflects the brain's adaptive capacity against insults or aging [53,54]. Educational attainment, commonly employed as a proxy for cognitive reserve [55–57], reveals that ABA subjects, across three subgroups, possess educational durations exceeding 15 years, signifying a population with high cognitive reserve. The neural implementation of cognitive reserve manifests in two distinct forms: neural reserve and neural compensation [58,59]. Neural reserve posits variability in primary brain networks or cognitive paradigms underlying task performance, thereby offering resilience against brain aging. On the other hand, neural compensation describes the utilization of non-normally engaged brain structures or networks to compensate for aging-induced changes. These mechanisms exemplify the brain's flexibility and adaptive strategies in the face of challenges. In the present study, the application of ICA facilitates the decomposition of fMRI data into distinctive brain networks. Positive connectivity within these networks signifies synchronized activity between networks, reflecting collaborative involvement in specific cognitive processes or tasks. The cooperative synergy inherent in positive connectivity is indispensable for the facilitation of streamlined information processing and the seamless execution of cognitive functions. Conversely, negative connectivity assumes a pivotal role in promoting cognitive flexibility, affording the brain the capacity to navigate between different cognitive states and alleviating interference among concurrent cognitive processes. Disparities in both positive and negative network connections observed between the ABA and RBA cohorts underscore a conspicuous neural compensation mechanism [60]. Specifically, the discernible augmentation in negative connectivity within ABA individuals suggests that, in the face of degeneration, the brain intensifies inhibitory interactions among disparate brain regions to counterbalance the disruptive effects of structural decline. In the specific context of ABA subtypes 1 and 2, despite structural similarities in neurodegeneration, nuanced differences in negative connectivity patterns are apparent. ABA subtype 1 prominently manifests a discernible proclivity towards cognitive compensation, indicating adaptive responses to the structural challenges inherent in neurodegeneration. Conversely, subtype 2 showcases a confluence of neural compensation and neural reserve. This observation underscores the inference that distinct strategies are employed by different subtypes within the ABA context, delineating nuanced approaches to addressing the intricacies of neurodegenerative processes.

4.3. Limitations

This study entails certain limitations that warrant careful consideration. Firstly, the exclusive utilization of data from the UKB introduces a notable limitation, as the subjects are predominantly of white ethnicity and hail from the United Kingdom. Consequently, the generalizability of the study findings to other countries or regions may be constrained. Secondly, in the implementation of HYDRA for semi-supervised learning, the RBA was deliberately chosen as the reference group. This decision stems from the discernible differences exhibited by the RBA when compared to the ABA cohort. Nevertheless, it is crucial to acknowledge that this choice may introduce potential bias into the subtype estimation. Thirdly, to validate the delineation of ABA subtypes, it is essential to broaden our experimental scope by integrating additional datasets and extending the spectrum of comparative analyses. However, a noteworthy limitation arises from the inherent inadequacies of outcomes derived from smaller datasets, which often lack the necessary representativeness. Moreover, the current state of the research landscape confronts a significant impediment characterized by a shortage of openly accessible datasets commensurate in magnitude to the UKB. This scarcity not only diminishes the depth of available data but also presents a formidable barrier to the facilitation of seamless cross-study comparisons. However, with increasing recognition from governments worldwide regarding the significance of large-scale neurobiological repositories in medical and clinical research [61,62], we anticipate a continual emergence of additional open-access large-scale biological databases.

5. Conclusions

Distinguishing itself from precedent investigations, this study capitalizes on considerable sample size and an extensive age spectrum, imparting significant robustness to the examination of brain variability within the ABA cohort. The utilization of the HYDRA methodology represents a notable methodological advancement, surpassing conventional heterogeneity analysis techniques used in ABA analysis. HYDRA not only discerns ABA subgroups but also enables the characterization of distinctions from the RBA group across multiple dimensions.

Looking ahead, the inclusion of subsequent follow-up waves from the UKB study promises a longitudinal exploration of the identified clusters. This longitudinal perspective is essential for unraveling the evolving nature of these clusters over time and elucidating their prognostic implications for brain and cognitive aging outcomes. The comprehensive insights derived from this study not only unveil inherent brain heterogeneity within ABA but also lay the groundwork for future analyses to deepen our understanding of the cognition and brain arising from the progressive ABA observed in UKB participants.

Supplementary Materials: The following supporting information can be downloaded at: https://www.mdpi.com/article/10.3390/bioengineering11020124/s1, Figure S1: Comparative analysis of gray matter volume across the three subgroups and the control group.

Author Contributions: Conceptualization, L.L. (Lingyu Liu), L.L. (Lan Lin) and S.W.; methodology, L.L. (Lingyu Liu) and L.L. (Lan Lin); software, L.L.(Lingyu Liu); validation, S.S.; resources, L.L. (Lingyu Liu) and L.L. (Lan Lin); data curation, S.S. and L.L. (Lingyu Liu); writing—original draft preparation, L.L. (Lingyu Liu); writing—review and editing, L.L. (Lingyu Liu) and L.L. (Lan Lin); visualization, L.L. (Lingyu Liu), L.L. (Lan Lin) and S.W. All authors have read and agreed to the published version of the manuscript.

Funding: This research was financially supported by grants from the National Natural Science Foundation of China (81971683) and the Natural Science Foundation of Beijing Municipality (L182010).

Institutional Review Board Statement: Not applicable.

Informed Consent Statement: Not applicable.

Data Availability Statement: The imaging datasets generated by the UK Biobank and analyzed in the present study can be accessed through the UK Biobank data access process, as outlined at http://www.ukbiobank.ac.uk/register-apply/ (accessed on 11 January 2021). The UK Biobank's Research

Access Administration Team impartially manages all data access requests from both academic and commercial researchers, without exhibiting any preference or exclusivity. Requests are diligently evaluated to ascertain their alignment with public health research interests, and if deemed supportive, are promptly approved. Comprehensive information regarding the available data from the UK Biobank can be obtained at http://www.ukbiobank.ac.uk (accessed on 11 January 2021). It is important to note that the precise count of participants with imaging data accessible in the UK Biobank may marginally differ from the figures delineated in this manuscript.

Acknowledgments: We express our gratitude to the UK Biobank for providing access to this invaluable resource and extend our appreciation to the participants of the UK Biobank for their commitment to contributing their time, thereby enabling the generation of this essential data.

Conflicts of Interest: The authors declare no conflicts of interest.

References

1. Salat, D.H.; Buckner, R.L.; Snyder, A.Z.; Greve, D.N.; Desikan, R.S.; Busa, E.; Morris, J.C.; Dale, A.M.; Fischl, B. Thinning of the cerebral cortex in aging. *Cereb. Cortex* **2004**, *14*, 721–730. [CrossRef]
2. Vernooij, M.W.; de Groot, M.; van der Lugt, A.; Ikram, M.A.; Krestin, G.P.; Hofman, A.; Niessen, W.J.; Breteler, M.M. White matter atrophy and lesion formation explain the loss of structural integrity of white matter in aging. *Neuroimage* **2008**, *43*, 470–477. [CrossRef]
3. Salami, A.; Pudas, S.; Nyberg, L. Elevated hippocampal resting-state connectivity underlies deficient neurocognitive function in aging. *Proc. Natl. Acad. Sci. USA* **2014**, *111*, 17654–17659. [CrossRef]
4. Planche, V.; Coupé, P.; Helmer, C.; Le Goff, M.; Amieva, H.; Tison, F.; Dartigues, J.F.; Catheline, G. Evolution of brain atrophy subtypes during aging predicts long-term cognitive decline and future Alzheimer's clinical syndrome. *Neurobiol. Aging* **2019**, *79*, 22–29. [CrossRef]
5. Lee, A.J.; Ma, Y.; Yu, L.; Dawe, R.J.; McCabe, C.; Arfanakis, K.; Mayeux, R.; Bennett, D.A.; Klein, H.U.; De Jager, P.L. Multi-region brain transcriptomes uncover two subtypes of aging individuals with differences in Alzheimer risk and the impact of APOEε4. *bioRxiv* **2023**. *preprint*. [CrossRef]
6. Mather, K.A. Genetic and Environmental Factors in Ageing and Age-Related Disease. *Genes* **2022**, *13*, 396. [CrossRef] [PubMed]
7. Pang, S.Y.; Ho, P.W.; Liu, H.F.; Leung, C.T.; Li, L.; Chang, E.E.S.; Ramsden, D.B.; Ho, S.L. The interplay of aging, genetics and environmental factors in the pathogenesis of Parkinson's disease. *Transl. Neurodegener.* **2019**, *8*, 23. [CrossRef] [PubMed]
8. Lin, L.; Xiong, M.; Jin, Y.; Kang, W.; Wu, S.; Sun, S.; Fu, Z. Quantifying Brain and Cognitive Maintenance as Key Indicators for Sustainable Cognitive Aging: Insights from the UK Biobank. *Sustainability* **2023**, *15*, 9620. [CrossRef]
9. Liu, L.; Sun, S.; Kang, W.; Wu, S.; Lin, L. A review of neuroimaging-based data-driven approach for Alzheimer's disease heterogeneity analysis. *Rev. Neurosci.* 2024, *in press*. [CrossRef] [PubMed]
10. Abellaneda-Pérez, K.; Cattaneo, G.; Cabello-Toscano, M.; Solana-Sánchez, J.; Mulet-Pons, L.; Vaqué-Alcázar, L.; Perellón-Alfonso, R.; Solé-Padullés, C.; Bargalló, N.; Tormos, J.M.; et al. Purpose in life promotes resilience to age-related brain burden in middle-aged adults. *Alzheimers Res. Ther.* **2023**, *15*, 49. [CrossRef] [PubMed]
11. Montine, T.J.; Cholerton, B.A.; Corrada, M.M.; Edland, S.D.; Flanagan, M.E.; Hemmy, L.S.; Kawas, C.H.; White, L.R. Concepts for brain aging: Resistance, resilience, reserve, and compensation. *Alzheimers Res. Ther.* **2019**, *11*, 22. [CrossRef] [PubMed]
12. Peeters, G.; Kok, A.; de Bruin, S.R.; van Campen, C.; Graff, M.; Nieuwboer, M.; Huisman, M.; van Munster, B.; van der Zee, E.A.; Kas, M.J.; et al. Supporting Resilience of Older Adults with Cognitive Decline Requires a Multi-Level System Approach. *Gerontology* **2023**, *69*, 866–874. [CrossRef] [PubMed]
13. Kozic, D.B.; Thurnher, M.M.; Boban, J.; Sundgren, P.C. Editorial: Accelerated Brain Aging: Different Diseases-Different Imaging Patterns. *Front. Neurol.* **2022**, *13*, 889538. [CrossRef] [PubMed]
14. Chaudhari, P.R.; Singla, A.; Vaidya, V.A. Early Adversity and Accelerated Brain Aging: A Mini-Review. *Front. Mol. Neurosci.* **2022**, *15*, 822917. [CrossRef]
15. Darrow, S.M.; Verhoeven, J.E.; Révész, D.; Lindqvist, D.; Penninx, B.W.; Delucchi, K.L.; Wolkowitz, O.M.; Mathews, C.A. The Association Between Psychiatric Disorders and Telomere Length: A Meta-Analysis Involving 14,827 Persons. *Psychosom. Med.* **2016**, *78*, 776–787. [CrossRef] [PubMed]
16. Franke, K.; Ziegler, G.; Klöppel, S.; Gaser, C. Estimating the age of healthy subjects from T1-weighted MRI scans using kernel methods: Exploring the influence of various parameters. *Neuroimage* **2010**, *50*, 883–892. [CrossRef]
17. Weihs, A.; Frenzel, S.; Wittfeld, K.; Obst, A.; Stubbe, B.; Habes, M.; Szentkirályi, A.; Berger, K.; Fietze, I.; Penzel, T.; et al. Associations between sleep apnea and advanced brain aging in a large-scale population study. *Sleep* **2021**, *44*, zsaa204. [CrossRef] [PubMed]
18. Klaus, F.; Nguyen, T.T.; Thomas, M.L.; Liou, S.C.; Soontornniyomkij, B.; Mitchell, K.; Daly, R.; Sutherland, A.N.; Jeste, D.V.; Eyler, L.T. Peripheral inflammation levels associated with degree of advanced brain aging in schizophrenia. *Front. Psychiatry* **2022**, *13*, 966439. [CrossRef]

19. Habes, M.; Janowitz, D.; Erus, G.; Toledo, J.B.; Resnick, S.M.; Doshi, J.; Van der Auwera, S.; Wittfeld, K.; Hegenscheid, K.; Hosten, N.; et al. Advanced brain aging: Relationship with epidemiologic and genetic risk factors, and overlap with Alzheimer disease atrophy patterns. *Transl. Psychiatry* **2016**, *6*, e775. [CrossRef]
20. Han, L.K.M.; Schnack, H.G.; Brouwer, R.M.; Veltman, D.J.; van der Wee, N.J.A.; van Tol, M.J.; Aghajani, M.; Penninx, B. Contributing factors to advanced brain aging in depression and anxiety disorders. *Transl. Psychiatry* **2021**, *11*, 402. [CrossRef]
21. Busby, N.; Wilmskoetter, J.; Gleichgerrcht, E.; Rorden, C.; Roth, R.; Newman-Norlund, R.; Hillis, A.E.; Keller, S.S.; de Bezenac, C.; Kristinsson, S.; et al. Advanced Brain Age and Chronic Poststroke Aphasia Severity. *Neurology* **2023**, *100*, e1166–e1176. [CrossRef]
22. Chen, C.L.; Kuo, M.C.; Wu, W.C.; Hsu, Y.C.; Wu, R.M.; Tseng, W.I. Advanced brain aging in multiple system atrophy compared to Parkinson's disease. *NeuroImage Clin.* **2022**, *34*, 102997. [CrossRef] [PubMed]
23. Whitwell, J.L.; Petersen, R.C.; Negash, S.; Weigand, S.D.; Kantarci, K.; Ivnik, R.J.; Knopman, D.S.; Boeve, B.F.; Smith, G.E.; Jack, C.R., Jr. Patterns of atrophy differ among specific subtypes of mild cognitive impairment. *Arch. Neurol.* **2007**, *64*, 1130–1138. [CrossRef] [PubMed]
24. Wrigglesworth, J.; Ryan, J.; Ward, P.G.D.; Woods, R.L.; Storey, E.; Egan, G.F.; Murray, A.; Espinoza, S.E.; Shah, R.C.; Trevaks, R.E.; et al. Health-related heterogeneity in brain aging and associations with longitudinal change in cognitive function. *Front. Aging Neurosci.* **2022**, *14*, 1063721. [CrossRef]
25. Eavani, H.; Habes, M.; Satterthwaite, T.D.; An, Y.; Hsieh, M.K.; Honnorat, N.; Erus, G.; Doshi, J.; Ferrucci, L.; Beason-Held, L.L.; et al. Heterogeneity of structural and functional imaging patterns of advanced brain aging revealed via machine learning methods. *Neurobiol. Aging* **2018**, *71*, 41–50. [CrossRef]
26. Varol, E.; Sotiras, A.; Davatzikos, C. HYDRA: Revealing heterogeneity of imaging and genetic patterns through a multiple max-margin discriminative analysis framework. *Neuroimage* **2017**, *145*, 346–364. [CrossRef]
27. Zhang, H.; Sun, H.; Li, J.; Fan, Y.; Jülich, S.T.; Lei, X. Subtypes of insomnia revealed by the heterogeneity of neuroanatomical patterns: A structural MRI study. *Biol. Psychol.* **2023**, *180*, 108591. [CrossRef]
28. Liu, H.; Zheng, R.; Zhang, Y.; Zhang, B.; Hou, H.; Cheng, J.; Han, S. Two distinct neuroanatomical subtypes of migraine without aura revealed by heterogeneity through discriminative analysis. *Brain Imaging Behav.* **2023**, *17*, 715–724. [CrossRef] [PubMed]
29. Han, S.; Xu, Y.; Guo, H.R.; Fang, K.; Wei, Y.; Liu, L.; Cheng, J.; Zhang, Y.; Cheng, J. Two distinct subtypes of obsessive compulsive disorder revealed by heterogeneity through discriminative analysis. *Hum. Brain Mapp.* **2022**, *43*, 3037–3046. [CrossRef]
30. Liu, G.; Shi, L.; Qiu, J.; Lu, W. Two neuroanatomical subtypes of males with autism spectrum disorder revealed using semi-supervised machine learning. *Mol. Autism* **2022**, *13*, 9. [CrossRef]
31. Sudlow, C.; Gallacher, J.; Allen, N.; Beral, V.; Burton, P.; Danesh, J.; Downey, P.; Elliott, P.; Green, J.; Landray, M.; et al. UK biobank: An open access resource for identifying the causes of a wide range of complex diseases of middle and old age. *PLoS Med.* **2015**, *12*, e1001779. [CrossRef]
32. Miller, K.L.; Alfaro-Almagro, F.; Bangerter, N.K.; Thomas, D.L.; Yacoub, E.; Xu, J.; Bartsch, A.J.; Jbabdi, S.; Sotiropoulos, S.N.; Andersson, J.L.; et al. Multimodal population brain imaging in the UK Biobank prospective epidemiological study. *Nat. Neurosci.* **2016**, *19*, 1523–1536. [CrossRef]
33. Chung, B.S.; Park, J.S. Automatic segmentation of true color sectioned images using FMRIB Software Library: First trial in brain, gray matter, and white matter. *Clin. Anat.* **2020**, *33*, 1197–1203. [CrossRef]
34. Beckmann, C.F.; Smith, S.M. Probabilistic independent component analysis for functional magnetic resonance imaging. *IEEE Trans. Med. Imaging* **2004**, *23*, 137–152. [CrossRef] [PubMed]
35. Cole, J.H. Multimodality neuroimaging brain-age in UK biobank: Relationship to biomedical, lifestyle, and cognitive factors. *Neurobiol. Aging* **2020**, *92*, 34–42. [CrossRef]
36. Xiong, M.; Lin, L.; Jin, Y.; Kang, W.; Wu, S.; Sun, S. Comparison of Machine Learning Models for Brain Age Prediction Using Six Imaging Modalities on Middle-Aged and Older Adults. *Sensors* **2023**, *23*, 3622. [CrossRef]
37. Franke, K.; Gaser, C. Ten Years of BrainAGE as a Neuroimaging Biomarker of Brain Aging: What Insights Have We Gained? *Front. Neurol.* **2019**, *10*, 789. [CrossRef] [PubMed]
38. Anatürk, M.; Kaufmann, T.; Cole, J.H.; Suri, S.; Griffanti, L.; Zsoldos, E.; Filippini, N.; Singh-Manoux, A.; Kivimäki, M.; Westlye, L.T.; et al. Prediction of brain age and cognitive age: Quantifying brain and cognitive maintenance in aging. *Hum. Brain Mapp.* **2021**, *42*, 1626–1640. [CrossRef]
39. Smith, S.M.; Vidaurre, D.; Alfaro-Almagro, F.; Nichols, T.E.; Miller, K.L. Estimation of brain age delta from brain imaging. *Neuroimage* **2019**, *200*, 528–539. [CrossRef]
40. Fawns-Ritchie, C.; Deary, I.J. Reliability and validity of the UK Biobank cognitive tests. *PLoS ONE* **2020**, *15*, e0231627. [CrossRef] [PubMed]
41. Amin, V.; Fletcher, J.M.; Sun, Z.; Lu, Q. Higher educational attainment is associated with longer telomeres in midlife: Evidence from sibling comparisons in the UK Biobank. *SSM-Popul. Health* **2022**, *17*, 101018. [CrossRef]
42. Xu, X.; Lin, L.; Wu, S.; Sun, S. Exploring Successful Cognitive Aging: Insights Regarding Brain Structure, Function, and Demographics. *Brain Sci.* **2023**, *13*, 1651. [CrossRef] [PubMed]
43. Franke, K.; Gaser, C.; Initiative, F.S.D.N. Longitudinal Changes in Individual BrainAGE in Healthy Aging, Mild Cognitive Impairment, and Alzheimer's Disease. *Geriatr. Psychiatry* **2012**, *25*, 235–245. [CrossRef]
44. Ziegler, G.; Ridgway, G.R.; Dahnke, R.; Gaser, C. Individualized Gaussian process-based prediction and detection of local and global gray matter abnormalities in elderly subjects. *Neuroimage* **2014**, *97*, 333–348. [CrossRef]

45. Gaser, C.; Franke, K.; Klöppel, S.; Koutsouleris, N.; Sauer, H. BrainAGE in Mild Cognitive Impaired Patients: Predicting the Conversion to Alzheimer's Disease. *PLoS ONE* **2013**, *8*, e67346. [CrossRef] [PubMed]
46. Cole, J.H.; Leech, R.; Sharp, D.J. Prediction of brain age suggests accelerated atrophy after traumatic brain injury. *Ann. Neurol.* **2015**, *77*, 571–581. [CrossRef]
47. Cole, J.H.; Underwood, J.; Caan, M.W.; De Francesco, D.; van Zoest, R.A.; Leech, R.; Wit, F.W.; Portegies, P.; Geurtsen, G.J.; Schmand, B.A.; et al. Increased brain-predicted aging in treated HIV disease. *Neurology* **2017**, *88*, 1349–1357. [CrossRef] [PubMed]
48. Cruz-Almeida, Y.; Fillingim, R.B.; Riley, J.L., 3rd; Woods, A.J.; Porges, E.; Cohen, R.; Cole, J. Chronic pain is associated with a brain aging biomarker in community-dwelling older adults. *Pain* **2019**, *160*, 1119–1130. [CrossRef] [PubMed]
49. Franke, K.; Gaser, C.; Manor, B.; Novak, V. Advanced BrainAGE in older adults with type 2 diabetes mellitus. *Front. Aging Neurosci.* **2013**, *5*, 90. [CrossRef]
50. Cole, J.H.; Ritchie, S.J.; Bastin, M.E.; Valdés Hernández, M.C.; Muñoz Maniega, S.; Royle, N.; Corley, J.; Pattie, A.; Harris, S.E.; Zhang, Q.; et al. Brain age predicts mortality. *Mol. Psychiatry* **2018**, *23*, 1385–1392. [CrossRef]
51. Steffener, J.; Habeck, C.; O'Shea, D.; Razlighi, Q.; Bherer, L.; Stern, Y. Differences between chronological and brain age are related to education and self-reported physical activity. *Neurobiol. Aging* **2016**, *40*, 138–144. [CrossRef]
52. Stern, Y. What is cognitive reserve? Theory and research application of the reserve concept. *J. Int. Neuropsychol. Soc.* **2002**, *8*, 448–460. [CrossRef]
53. Corbo, I.; Marselli, G.; Di Ciero, V.; Casagrande, M. The Protective Role of Cognitive Reserve in Mild Cognitive Impairment: A Systematic Review. *J. Clin. Med.* **2023**, *12*, 1759. [CrossRef] [PubMed]
54. Panico, F.; Sagliano, L.; Magliacano, A.; Santangelo, G.; Trojano, L. The relationship between cognitive reserve and cognition in healthy adults: A systematic review. *Curr. Psychol.* **2022**, *42*, 24751–24763. [CrossRef]
55. Nogueira, J.; Gerardo, B.; Santana, I.; Simões, M.R.; Freitas, S. The Assessment of Cognitive Reserve: A Systematic Review of the Most Used Quantitative Measurement Methods of Cognitive Reserve for Aging. *Front. Psychol.* **2022**, *13*, 847186. [CrossRef] [PubMed]
56. Jin, Y.; Lin, L.; Xiong, M.; Sun, S.; Wu, S.C. Moderating effects of cognitive reserve on the relationship between brain structure and cognitive abilities in middle-aged and older adults. *Neurobiol. Aging* **2023**, *128*, 49–64. [CrossRef] [PubMed]
57. Lin, L.; Jin, Y.; Xiong, M.; Wu, S.; Sun, S. The Protective Power of Cognitive Reserve: Examining White Matter Integrity and Cognitive Function in the Aging Brain for Sustainable Cognitive Health. *Sustainability* **2023**, *15*, 11336. [CrossRef]
58. Stern, Y.; Barnes, C.A.; Grady, C.; Jones, R.N.; Raz, N. Brain reserve, cognitive reserve, compensation, and maintenance: Operationalization, validity, and mechanisms of cognitive resilience. *Neurobiol. Aging* **2019**, *83*, 124–129. [CrossRef] [PubMed]
59. Stern, Y.; Habeck, C.; Moeller, J.; Scarmeas, N.; Anderson, K.E.; Hilton, H.J.; Flynn, J.; Sackeim, H.; van Heertum, R. Brain networks associated with cognitive reserve in healthy young and old adults. *Cereb. Cortex* **2005**, *15*, 394–402. [CrossRef] [PubMed]
60. Geng, H.; Wu, F.; Kong, L.; Tang, Y.; Zhou, Q.; Chang, M.; Zhou, Y.; Jiang, X.; Li, S.; Wang, F. Disrupted Structural and Functional Connectivity in Prefrontal-Hippocampus Circuitry in First-Episode Medication-Naïve Adolescent Depression. *PLoS ONE* **2016**, *11*, e0148345. [CrossRef]
61. Toccaceli, V.; Brescianini, S.; Fagnani, C.; Gigantesco, A.; D'Abramo, F.; Stazi, M.A. What Potential Donors in Research Biobanking Want to Know: A Large Population Study of the Italian Twin Registry. *Biopreserv. Biobank.* **2016**, *14*, 456–463. [CrossRef] [PubMed]
62. Larsson, A. The Need for Research Infrastructures: A Narrative Review of Large-Scale Research Infrastructures in Biobanking. *Biopreserv. Biobank.* **2017**, *15*, 375–383. [CrossRef] [PubMed]

bioengineering MDPI

Article

RGGC-UNet: Accurate Deep Learning Framework for Signet Ring Cell Semantic Segmentation in Pathological Images

Tengfei Zhao [1], Chong Fu [1,2,3,*], Wei Song [1] and Chiu-Wing Sham [4]

1 School of Computer Science and Engineering, Northeastern University, Shenyang 110819, China
2 Engineering Research Center of Security Technology of Complex Network System, Ministry of Education, Shenyang 110819, China
3 Key Laboratory of Intelligent Computing in Medical Image, Ministry of Education, Shenyang 110819, China
4 School of Computer Science, The University of Auckland, Auckland 1142, New Zealand
* Correspondence: fuchong@mail.neu.edu.cn

Abstract: Semantic segmentation of Signet Ring Cells (SRC) plays a pivotal role in the diagnosis of SRC carcinoma based on pathological images. Deep learning-based methods have demonstrated significant promise in computer-aided diagnosis over the past decade. However, many existing approaches rely heavily on stacking layers, leading to repetitive computational tasks and unnecessarily large neural networks. Moreover, the lack of available ground truth data for SRCs hampers the advancement of segmentation techniques for these cells. In response, this paper introduces an efficient and accurate deep learning framework (RGGC-UNet), which is a UNet framework including our proposed residual ghost block with ghost coordinate attention, featuring an encoder-decoder structure tailored for the semantic segmentation of SRCs. We designed a novel encoder using the residual ghost block with proposed ghost coordinate attention. Benefiting from the utilization of ghost block and ghost coordinate attention in the encoder, the computational overhead of our model is effectively minimized. For practical application in pathological diagnosis, we have enriched the DigestPath 2019 dataset with fully annotated mask labels of SRCs. Experimental outcomes underscore that our proposed model significantly surpasses other leading-edge models in segmentation accuracy while ensuring computational efficiency.

Keywords: semantic segmentation; signet ring cell; residual ghost block; ghost coordinate attention

Citation: Zhao, T.; Fu, C.; Song, W.; Sham, C.-W. RGGC-UNet: Accurate Deep Learning Framework for Signet Ring Cell Semantic Segmentation in Pathological Images. *Bioengineering* **2024**, *11*, 16. https://doi.org/10.3390/bioengineering11010016

Academic Editors: Andrea Cataldo, Yan Pei and Jijiang Yang

Received: 10 November 2023
Revised: 20 December 2023
Accepted: 22 December 2023
Published: 23 December 2023

1. Introduction

Signet ring cell carcinoma (SRCC) represents a relatively uncommon subtype of profoundly aggressive adenocarcinoma [1]. Predominantly encountered within the gastric glandular cells, primary SRCCs exhibit a notable association with gastric malignancies [2]. In the SRCC, a signet ring cell (SRC) contains a lot of mucins that push the nucleus to the periphery [3]. Moreover, SRCC has the highest malignancy and poorest prognosis in advanced gastric cancer. The prompt and precise diagnosis followed by timely intervention of SRCs in the gastric region can substantially enhance patients' survival rates. In the realm of the digestive system, the gold standard for diagnosing SRCC is the examination of pathological images [4]. Therefore, detecting the SRCs in pathological images is essential for diagnosing SRCC. Nevertheless, the conventional manual segmentation of signet ring cells is susceptible to time-consuming processes and human error. Automatic segmentation methods have, therefore, been devised to enhance both accuracy and efficiency. These methods typically involve using image processing and machine learning algorithms to identify and segment signet ring cells from surrounding tissue or other types of cells. By automating the segmentation process, medical professionals can quickly and accurately analyze large amounts of data, leading to earlier detection and improved treatment of cancer. Hence, the computer-aided diagnosis-based analysis of SRC, serving as a supplementary investigation, holds significant promise and is in high demand.

The Digestive-System Pathological Detection and Segmentation Challenge of 2019, abbreviated as DigestPath 2019, marks the inaugural competition and open dataset dedicated to the detection of signet ring cells (SRCs) within pathological images [5]. Automatic SRC detection algorithms had not been thoroughly investigated prior to this challenge. As a result, the DigestPath 2019 challenge has driven research into the SRC detection algorithms. Unfortunately, only a portion of the data has been annotated, and the algorithms for this research are all based on semi-supervised object detection methods [4,6–10]. Therefore, existing semi-supervised detection labels in the DigestPath 2019 dataset have not led to an increase in the network performance, limiting the application in practical medicine.

In recent years, deep learning methods [11–18] have achieved success in medical image analysis, such as biomedical segmentation and nuclei instance segmentation [19–24]. Most of this research is based on convolutional neural networks (CNNs) and has performed well in diverse biomedical segmentation applications. As an illustration, Lu et al. [20] presented an enhanced algorithm that employs a collaborative optimization approach involving multiple-level set functions. This method is designed for the segmentation of cytoplasm and nuclei in cases where cervical cells overlap and form clumps. Chen et al. [21] introduced the concept of deep contour-aware networks for precise gland segmentation, abbreviated as DCAN. This framework generates precise probability maps for glands while simultaneously delineating accurate contours, enabling effective separation of clustered objects and thereby enhancing gland segmentation performance. Naylor et al. [22] introduced an innovative approach involving fully convolutional networks designed for the automated segmentation of nuclei within histopathology data stained with hematoxylin and eosin (H&E). Their methods address the challenge of segmenting touching nuclei by treating the problem as a regression task for distance maps, thereby providing a solution to this segmentation issue. Simon et al. [23] introduced the HoVer-Net, a novel approach designed for both simultaneous nuclei segmentation and classification. This method harnesses the wealth of information embedded in the vertical and horizontal distances from nuclear pixels to their respective centers of mass. Zhou et al. [25] proposed the CIA-Net, which incorporates a multi-level information aggregation module between two task-specific decoders. This approach exploits the advantages of spatial and texture dependencies between nuclei and contours by bidirectionally aggregating task-specific features. Unfortunately, these methods suffer from model redundancy, resulting in low efficiency.

Hence, lightweight deep learning frameworks have become another topic of study. In particular, lightweight deep learning frameworks have been applied to medical image analysis. For instance, Zhang et al. [26] proposed a lightweight hybrid convolutional network for liver tumor segmentation. Zhao et al. [27] introduced a streamlined feature attention network to segment both nucleus and cytoplasm regions within cervical images. Unfortunately, the above methods suffer from insufficient model expression ability, resulting in low accuracy. In addition, these methods cannot be used directly in actual medical scenarios because they are low in efficiency and accuracy.

The segmentation of SRCs poses a challenge that remains unaddressed in current research, primarily because of the absence of reliable ground truth for SRCs. This deficiency has notably hampered advancements in the field of SRC segmentation. In the clinical diagnosis, pathologists rely on the presence of a substantial number of SRCs within pathological Whole Slide Images (WSI) as a key indicator suggesting a higher likelihood of the WSI being of the SRCC type. In this paper, we introduce an efficient and accurate deep learning framework tailored for the semantic segmentation of SRCs in pathology images. In particular, we have fully annotated the mask labels for SRC in the DigestPath 2019 SRC detection dataset [6]. In our approach, we employ an encoder-decoder architecture that incorporates a residual ghost block featuring ghost coordinate attention (GCA). In addition, our proposed encoder enhances the extraction of the features of the SRC boundary region. Our main contributions are summarized as follows.

- We propose an efficient and accurate deep learning framework for signet ring cell semantic segmentation in pathological images.

- We design a novel encoder that not only refines the network's capability but also notably enhances its performance in segregating overlapping and clustered cells.
- We propose ghost coordinate attention, which can efficiently capture the long-range dependencies.
- We provide full mask labels of SRC on the DigestPath 2019 dataset, referred to as the SRC dataset.
- Our experimental findings validate that the network proposed in this study attains superior evaluation scores and generates more refined segmentation outcomes when compared to other state-of-the-art methods for SRC segmentation.

The structure of this paper is as follows: Section 2 provides an introduction to the proposed method. In Section 3, we present the dataset, evaluation metrics and implementation details related to the experiment. In Section 4, the experimental results are the discussion and analysis. Lastly, Section 5 offers a summary of our work and a brief discussion on potential future research directions.

2. Methods

Figure 1 provides an overview of our proposed efficient and accurate deep learning framework for SRC semantic segmentation in pathology images. In this study, we begin with $128 \times 128 \times 3$ image patches, which are generated using dense cropped methods to extract relevant features from the original images. Detailed descriptions will be presented in the following subsections.

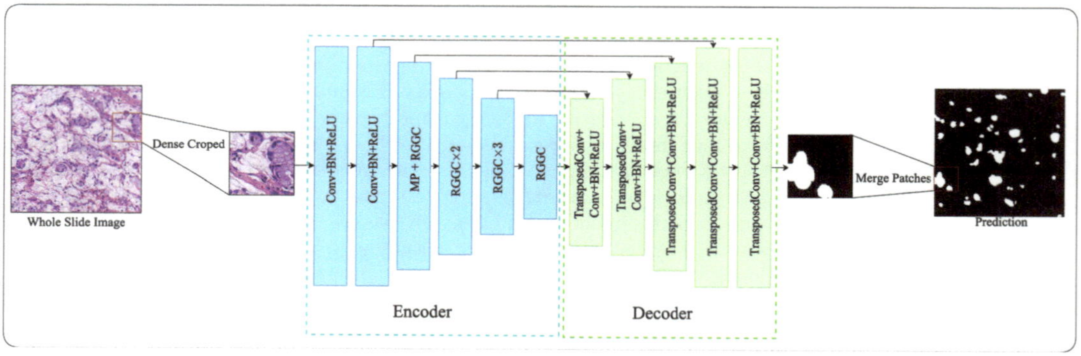

Figure 1. Overview of RGGC-UNet.

2.1. Network Architecture

Figure 2 provides a comprehensive depiction of the intricate architecture of the proposed RGGC-UNet. Our proposed network is an adaptation of the UNet framework, comprising an encoder and a decoder designed for the segmentation of SRCs. The encoder is proficient at extracting a highly effective set of features. Meanwhile, the decoder incorporates transposed convolution and 1×1 convolution operations.

In the encoder, we incorporate the ghost block with ghost coordinate attention, which is extensively discussed in Section 2.2. Detailed explanations of ghost coordinate attention mechanisms are presented in Section 2.3. Additionally, we delve into the RGGC block in Section 2.4 and the decoder in Section 2.5. The utilization of deep supervision is addressed in Section 2.6. We introduce loss function in Section 2.7.

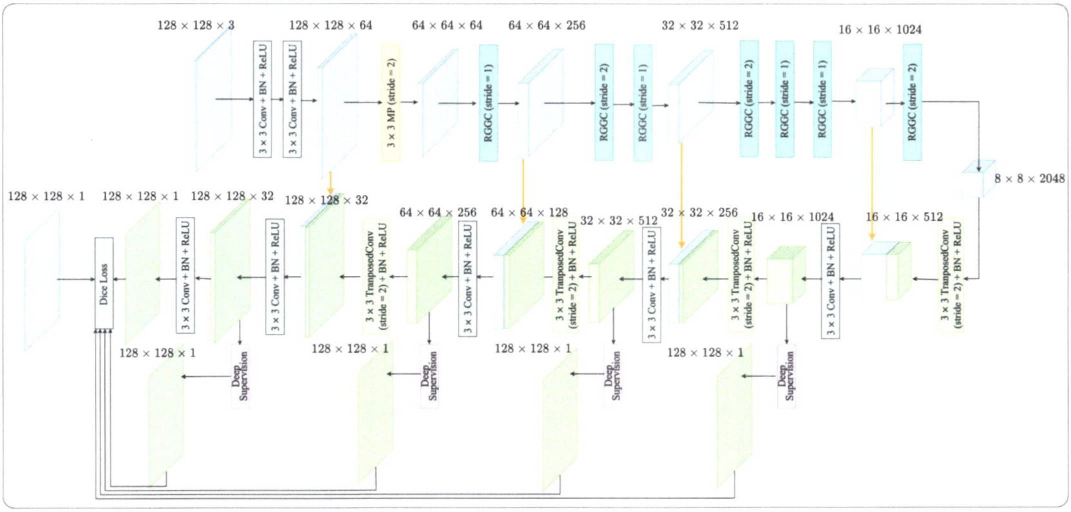

Figure 2. Detailed architecture of RGGC-UNet.

2.2. Encoder

In order to derive a valid set of features from the SRC, we introduce an innovative downsampling mechanism as an integral component of the encoder. The encoder primarily employs a sequence of residual ghost blocks with ghost coordinate attention (RGGC) for the downsampling process.

Our network comprises four downsampling modules, each incorporating a variable number of ghost blocks with ghost coordinate attention (GGC). As illustrated in Figure 2, the initial downsampling module utilizes a 3×3 max pooling (MP) operation followed by an RGGC block. Subsequently, the second and third downsampling modules incorporate two and three stacked RGGC blocks with stride = 2 where an RGGC block performs the downsampling operation, respectively. Meanwhile, the fourth downsampling module solely relies on a RGGC block for the downsampling operation.

Through the utilization of ghost blocks, our network is capable of generating feature-rich maps with significantly fewer input features compared to conventional convolution methods, thus enhancing the computational efficiency of our encoder. Particularly noteworthy is the advantage conferred by ghost coordinate attention (GCA), which empowers our proposed encoder to effectively capture dependence between long-range pixels.

2.3. Ghost Coordinate Attention

Figure 3 depicts the ghost block [28], a novel component in our study. It is well-established that the inclusion of ghost blocks can significantly enhance the feature generation capabilities of a convolutional neural network while maintaining a remarkably lower computational overhead.

This enhancement is achieved through a two-step process within the ghost block. Initially, it generates a set of intrinsic features utilizing a 1×1 point-wise convolution operation. Subsequently, it employs computationally economical operations to further expand the feature set based on these intrinsic features. The resultant feature sets are then concatenated along the channel dimension.

It is worth noting that the computational cost associated with linear operations on feature maps within the ghost block is substantially lower when compared to traditional convolutional techniques, thereby surpassing the efficiency of other existing approaches.

Mathematically, ghost block is defined by

$$Y = Concat([X * F_{1\times1}, (X * F_{1\times1}) * F_{dp}]), \quad (1)$$

where $*$ denote convolution operation, and $X \in \mathbb{R}^{H\times W\times C}$ with height H, width W and channel's number C is the input feature. $F_{1\times1}$ and F_{dp} are the 1×1 point-wise and 3×3 depth-wise convolutional filter, respectively. $Y \in \mathbb{R}^{H\times W\times C_{out}}$ is the output feature.

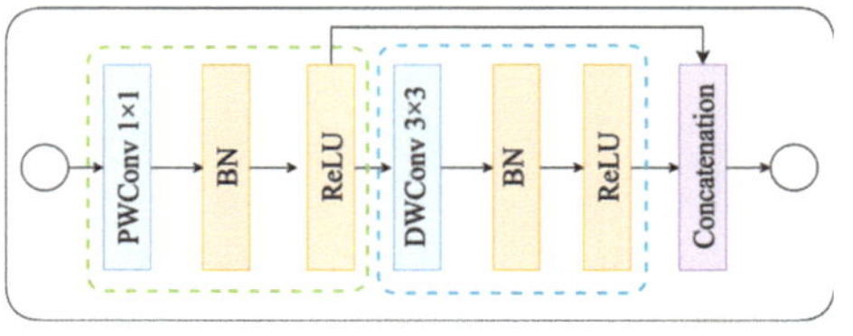

Figure 3. Diagram of ghost block. The green dash box represents identity operation. The blue dash box represents the efficiency operation [28].

Unfortunately, as evident from Equation (1), it becomes apparent that the spatial information is exclusively captured by the cost-effective operations for merely half of the features. The residual features, generated solely through 1×1 point-wise convolutions, lack any form of interaction with neighboring pixels. Consequently, this limited capacity to capture spatial information could potentially hinder the further enhancement of performance.

As aforementioned, the ghost block has previously been identified as having limitations due to its weak ability to capture spatial information, which may negatively impact its performance. However, the proposed ghost coordinate attention (GCA) solves this problem. Our GCA adopts the advantage of coordinate attention [29] and ghost block. While channel attention converts a feature tensor into a single feature vector through 2D global pooling, ghost coordinate attention takes a different approach by breaking down channel attention into two distinct 1D feature encoding processes. These processes aggregate features along two spatial directions separately. As a result of this approach, long-range dependencies are captured effectively along one spatial direction, and, at the same time, precise positional information is carefully preserved along the other spatial directions. The outcome of this process is two separate sets of encoded feature maps, each characterized by its direction awareness and sensitivity to positional information. These feature maps can be applied in a complementary manner to the input feature map, thereby enhancing the representations of the objects of interest.

In Figure 4, the blue dashed square denotes a comprehensive elucidation of the ghost coordinate attention mechanism. This mechanism adeptly encapsulates both channel interrelations and long-range dependencies, enables a global receptive field, and encodes precise positional information.

Global pooling is a frequently utilized technique in channeling attention to encode spatial information on a broad scale. However, its method of compressing global spatial information into a channel descriptor makes the preservation of positional information challenging. Such positional information is crucial for recognizing spatial structures in vision-related tasks. Attention blocks efficiently capture long-range interactions with accurate positional information. Unlike conventional methods, the X adaptive average pool and Y adaptive average pool aggregate features in two spatial directions. This approach diverges significantly from the squeeze operation seen in channel attention methods, which usually yield a singular feature vector. These transformations facilitate the attention block in encoding long-range dependencies in one spatial direction while maintaining precise positional information in the other. This dual-action allows networks to pinpoint the objects of interest with heightened accuracy.

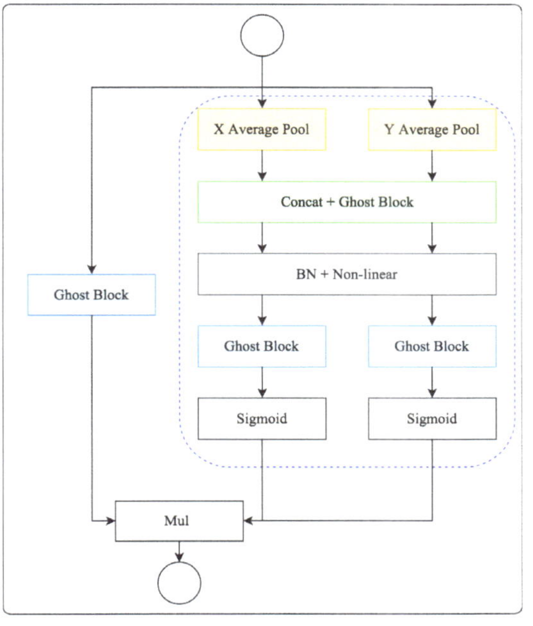

Figure 4. The diagram of GGC block. The blue dash square denotes the ghost coordinate attention (GCA).

As explained earlier, the X adaptive average pool and Y adaptive average pool allow for a global receptive field and encapsulate precise positional information. To leverage the high-level representations derived, a method coined as coordinate attention generation is introduced as a subsequent transformation. Specifically, the feature maps amalgamated by the X adaptive average pool and Y adaptive average pool are first concatenated and then subjected to a shared ghost block. The resulting feature map is then divided along the spatial dimension into distinct tensors and dispatched to two separate ghost blocks and sigmoid functions.

In contrast to channel attention, which prioritizes re-calibrating the significance of varied channels, the ghost coordinate attention block also aspires to integrate spatial information. The concurrent application of attention along both horizontal and vertical directions to the input tensor enables each element in the attention maps to signify the presence of an object of interest in the corresponding row and column. Especially, our proposed GCA can enhance the feature generation capability through using ghost blocks. This intricate encoding mechanism empowers the ghost coordinate attention to precisely discern the exact locations of objects of interest, enhancing the model's overall representation capabilities.

2.4. Residual Ghost Block with Ghost Coordinate Attention

The residual ghost block with ghost coordinate attention (RGGC), which incorporates the ghost block and GCA is illustrated in Figure 5. A RGGC comprises the residual block consisting of a GGC block and a ghost block. As shown in Figure 4, the GGC block generates expanded features with more channels, while the ghost block reduces the channel count to produce output features. Importantly, the GCA can help a ghost block to preserve information along one spatial direction while precise positional information can be preserved along the other spatial direction.

Figure 4 also shows that the GGC block consists of two parallel branches, a ghost block, and a GCA branch, which extract information from different perspectives. As mentioned earlier, the GCA branch can help the ghost block branch to enhance its representation ability. In the GGC block, the GCA branch operates in parallel with the ghost block branch to enhance the expanded features. Then the output features from GGC block are sent to another ghost block for producing output features. This allows the RGGC block to capture long-range dependence between pixels in different spatial locations and enhance the model's expressiveness.

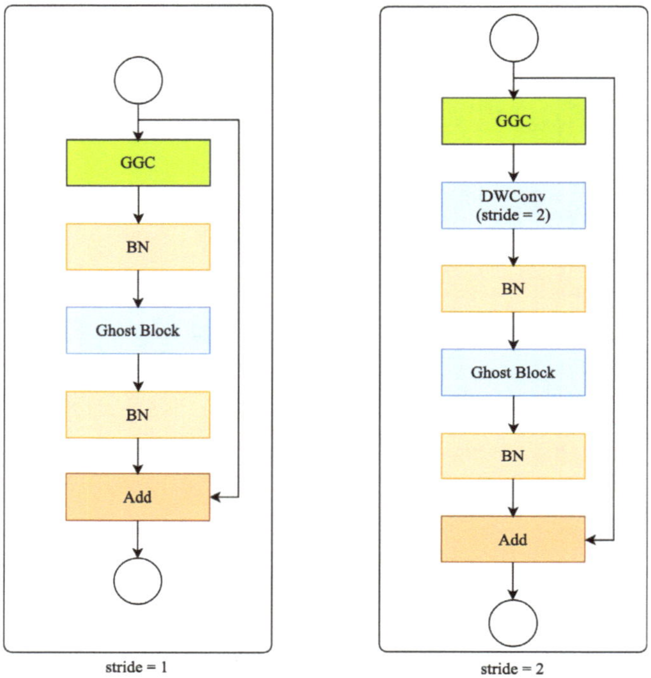

Figure 5. Diagram of an RGGC block.

2.5. Decoder

As depicted in Figure 2, the decoder is constructed with four upsampling modules, employing a combination of transposed convolution and 1×1 convolution with Rectified Linear Unit (ReLU) activation. This configuration effectively doubles the spatial resolution of the input data.

The concatenation operation plays a pivotal role in this process by merging the skip and output features of the TransposedConv-ReLU modules. This operation seamlessly integrates the low-level features from the encoder, located at the same level, directly into the decoder at that level. Consequently, it augments the granularity of information within the target region under evaluation. This enhancement in information granularity leads to an improvement in the segmentation performance of the model.

2.6. Deep Supervision

To enhance back-propagation and ensure greater stability in the decoder, we implement deep supervision (DS) across all four stages of the decoding process, as shown in Figure 2. Figure 6 shows the detailed construction. Our deep supervision block comprises a residual block, two 1×1 convolution layers, and an upsampling layer with bilinear operation for enlarging the feature map. Deep supervision effectively directs the learning of features in the intermediate layers, guided directly by loss functions and corresponding labels. We perform upsampling on features from the initial four hidden stages, aligning them with the dimensions of the final prediction stage. Subsequently, we use the Dice loss functions to supervise these stages. After decoding, the final output is rescaled to match the original input dimensions. This rescaled output is then processed through a softmax layer to generate the distribution of class probabilities. It is important to note that deep supervision is not employed during the inference stage. In this phase, only the last layer of the decoder is utilized to generate the segmentation prediction.

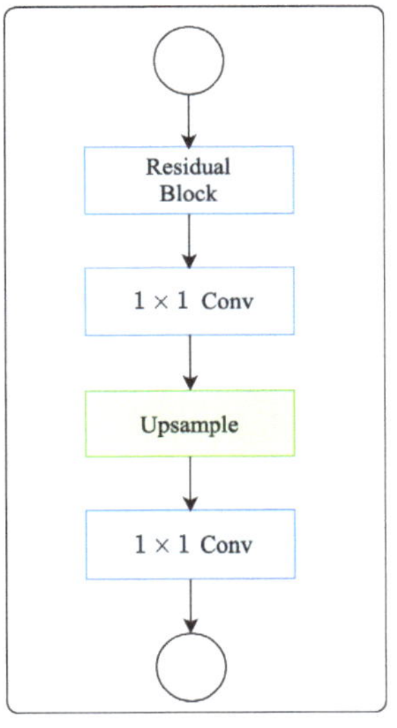

Figure 6. Diagram of deep supervision.

2.7. Loss Function

The Dice loss serves as a conventional loss function in image segmentation tasks, quantifying the disparity between the predicted mask and the ground-truth mask, as established in [30]. However, certain limitations persist when employing this function. Notably, in the absence of a segmentation target, the Dice loss yields a score of 0. This signifies that the Dice loss function does not penalize false positives.

To address this issue, we employ the enhanced class-wise Dice loss function to compute Dice Similarity Coefficients (DSCs) for background and SRC segmentation in benign and malignant images, respectively, as detailed in [31]. This refined loss function effectively mitigates false positives, underscoring its practical utility in clinical applications. The enhanced class-wise dice loss (CDL) function is detailed by:

$$L_{CDL} = 1 - \sum_{i}^{N}(y_p \frac{y_i \hat{y}_i}{y_i + \hat{y}_i} + \frac{(1-y_p)(1-y_i)(1-\hat{y}_i) + \epsilon}{(1-y_i) + (1-\hat{y}_i) + \epsilon}), \tag{2}$$

where y_i represents the binary label for pixel i, $\hat{y}_i$ corresponds to the predicted probability, and N denotes the total pixel count within a patch. The parameter ϵ is introduced as a small value to prevent division by zero.

The assignment of a patch label (y_p) hinges on the presence or absence of a lesion area. The employment of the L_{CDL} loss function effectively mitigates pixel-level class imbalance, leading to the generation of an all-zero mask during training for negative samples.

3. Experiments

This section describes our experiments designed to assess and appraise the segmentation performance of the proposed approach. In particular, we provide an elaborate account of our SRC dataset, evaluation metrics, and implementation specifics.

3.1. Dataset

In our experiments, we employed the SRC dataset to train and validate our model sourced from two organs: the gastric mucosa and intestine. Our dataset was comprised of 308 high-resolution images, with 77 positive and 231 negative samples. These positive samples were cropped from 20 whole slide images (WSIs), all of which are comprehensively annotated. Each WSI was stained with H&E, scanned at a $\times 40$ magnification and sourced from two organs: the gastric mucosa and intestine. Experienced pathologists identified and labeled each signet ring cell using the labelme, ensuring accuracy with a precise ground truth surrounding each cell. For our proposed model, we selected 62 positive and 186 negative samples from our dataset for training. During the training process, we also used 7 positive and 21 negative samples for validation. To assess the effectiveness of our model, we employed 8 negative and 24 positive samples as test data.

To demonstrate our proposed method's generalizability and its performance in different contexts, we used the GlaS dataset to verify the network. Glands represent pivotal histological structures found across various organ systems, primarily responsible for the secretion of proteins and carbohydrates. Adenocarcinomas, malignant tumors originating from glandular epithelium, stand out as the most prevalent form of cancer. Pathologists routinely rely on gland morphology to assess the malignancy levels of various adenocarcinomas in organs such as the prostate, breast, lung, and colon. Accurate gland segmentation is imperative for acquiring dependable morphological data. However, this task can be challenging due to the diverse glandular morphologies present across different histological grades. The GlaS dataset comprises a total of 165 tissue sections, encompassing both positive and negative samples. Within this dataset, our training subset contained 85 samples, with an additional 17 samples reserved for the validation set. Furthermore, the GlaS dataset offers two distinct test sets, denoted as testA and testB, consisting of 60 and 20 samples, respectively. We employed the validation set to identify the optimal model, conducting all performance evaluations on the combined results from testA and testB.

Two examples from the SRC and GlaS datasets are illustrated in Figure 7. Notably, most previous studies have concentrated on gland segmentation within either healthy or benign samples, often overlooking intermediate or high-grade cancers. Consequently, these studies frequently tailor their methods to specific datasets.

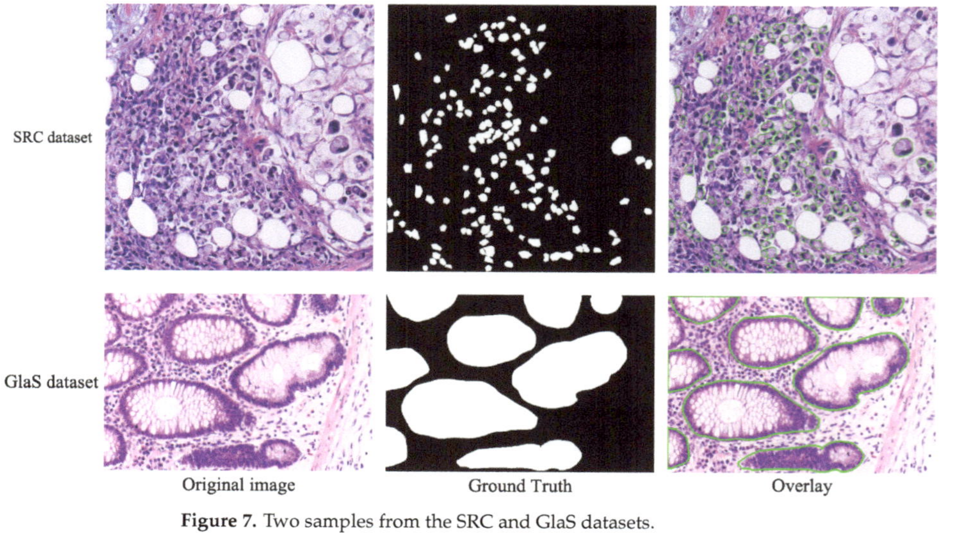

Figure 7. Two samples from the SRC and GlaS datasets.

3.2. Evaluation Metrics

In the context of evaluating segmented models, pixel-based metrics are often employed for assessing accuracy. We use a variety of metrics to evaluate the performance of our network, including the Dice similarity coefficient (DSC), Jaccard index,precision, and recall.

While both DSC and Jaccard are used to measure the similarity between predicted and labeled images, they have distinct focuses. Jaccard measures the consistency of extracted features and is suitable for comparing similarities and differences between limited sample sets. In contrast, DSC is more sensitive to the inner padding of the mask and is primarily used to calculate the similarity of two sets, making it our primary performance indicator.

In addition to DSC and Jaccard, we also employ precision and recall to evaluate our network's performance. Precision measures the proportion of predicted targets that are accurately identified, while recall represents the number of actual targets correctly identified based on predicted results.

Overall, these metrics allow us to comprehensively assess the accuracy of our segmentation network in identifying and classifying targets in the SRC dataset. These metrics are formulated as follows:

$$DSC = \frac{2TP}{FP + 2TP + FN}, \tag{3}$$

$$Jaccard = \frac{TP}{FP + TP + FN}, \tag{4}$$

$$Precision = \frac{TP}{FP + TP}, \tag{5}$$

$$Recall = \frac{TP}{TP + FN}, \tag{6}$$

where *TP*, *FP*, and *FN* correspond to the true positive predictions, false positive predictions, and false negative predictions, respectively.

3.3. Implementation Details

Our proposed method was implemented using PyTorch 1.8.0 and trained on a single NVIDIA GeForce RTX 3090 GPU. The initial learning rate was set to 1.0×10^{-4}. We employed the Adam optimizer for training the algorithm on the SRC dataset, with momentum and weight decay values of 0.99 and 1×10^{-8}, respectively.

For our SRC dataset, input images were densely cropped into patches with 128×128 pixels. The training process consisted of 2000 epochs with a batch size of 4. Data augmentation techniques included Gaussian blur, hue and saturation adjustments, affine transformations, as well as horizontal and vertical flips.

4. Discussion and Analysis

4.1. Discussion on Different Blocks

Table 1 presents the outcomes of an ablation study, illustrating the improvements in performance resulting from the integration of various blocks into the UNet architecture. These integrated blocks include ResGhost, GCA, and DS. It is evident that ResGhost, GCA, and DS all contribute to the enhancement of model performance. Our proposed RGGC-UNet, in particular, achieves the highest DSC. Furthermore, we conduct a detailed analysis of the performance of different discriminators in the context of the RGGC-UNet architecture. The corresponding results are provided in Table 2.

Table 1. Performance gain by integrating different blocks into UNet on the SRC dataset. The best results are indicated in bold.

UNet	ResGhost	GCA	DS	DSC
✓				0.5298
✓			✓	0.5621
✓	✓			0.5635
✓	✓		✓	0.5827
✓	✓	✓		0.7231
✓	✓	✓	✓	**0.7852**

Table 2. Comparative results for signet ring cell segmentation on the proposed dataset. The best results are indicated in bold.

Method	DSC	Jaccard	Precision	Recall
UNet(Baseline) [19]	0.5621	0.4007	0.5160	0.6434
UNet(Backbone: Vgg11) [32]	0.5771	0.4160	0.5530	0.6271
UNet(Backbone: Vgg16) [33]	0.5817	0.4191	0.5599	0.6304
UNet(Backbone: Vgg19) [31]	0.5850	0.4232	0.5930	0.6036
UNet(Backbone: ResNet50) [34]	0.5531	0.3943	0.6512	0.5316
DeepLabV3(Backbone:Mobilenet) [35]	0.4620	0.3098	0.3320	0.7804
DeepLabV3(Backbone: Drn) [35]	0.4564	0.3035	0.3361	0.7340
DeepLabV3(Backbone: ResNet50) [35]	0.5200	0.3576	0.4210	0.6916
DeepLabV3(Backbone: Xception) [35]	0.5227	0.3599	0.4020	0.7572
GCN [36]	0.4574	0.3026	0.3691	0.6270
SegNet [12]	0.4728	0.3198	0.4084	0.5867
Proposed	**0.7852**	**0.6482**	**0.7800**	**0.7964**

4.2. Comparison on SRC Dataset

Table 2 provides a comparative analysis of the performance between our proposed model and other popular models, using four metrics on our SRC dataset. The results clearly indicate that our proposed model achieves the highest scores in terms of DSC, Jaccard, recall, and precision. In all four metrics, our model outperforms the alternatives significantly.

Table 3 presents an overview of the computational complexity in terms of FLOPS and parameters. Although our proposed model may not boast the minimum number of FLOPS or parameters compared to other popular models, it effectively strikes a balance between computational load and model size. Consequently, our network represents an advantageous trade-off between accuracy and efficiency.

Table 3. Number of the FLOPS and parameters.

Model	GFLOPS	Params (M)
UNet (Baseline) [19]	16.70	14.50
UNet (Backbone: Vgg11) [32]	17.66	17.47
UNet (Backbone: Vgg16) [33]	22.79	22.96
UNet (Backbone: Vgg19) [31]	25.51	28.27
UNet (Backbone: ResNet50) [34]	55.87	59.04
DeepLabV3 (Backbone: Mobilenet) [35]	4.45	7.55
DeepLabV3 (Backbone: Drn) [35]	23.31	40.73
DeepLabV3 (Backbone: ResNet50) [35]	11.06	59.22
DeepLabV3 (Backbone: Xception) [35]	10.33	54.5
GCN [36]	7.64	58.25
SegNet [12]	20.06	29.44
Proposed	51.86	48.03

Figure 8 visually displays the segmentation results of various models, including ours and the findings of [12,19,31–36] on the SRC dataset. The visual evidence demonstrates that our model provides the most optimal alignment between its predictions and the ground truth. In comparison to other leading networks, our model excels in successfully segmenting SRCs. Overall, our proposed model excels at distinguishing between clustered and overlapping cells, achieving state-of-the-art accuracy in SRC segmentation tasks.

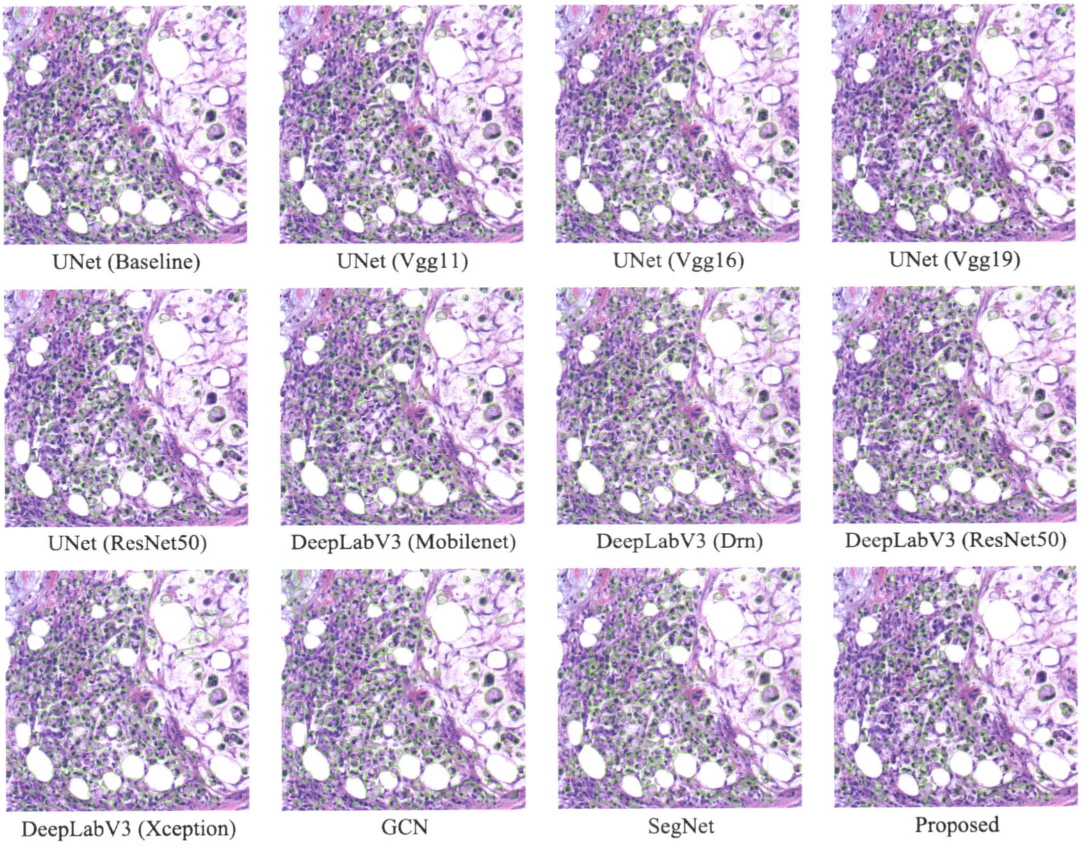

Figure 8. Segmentation results of various models on the SRC dataset.

4.3. Comparison on GlaS Dataset

To illustrate the generalizability of our proposed method and its performance under different scenarios, we also validate the network using the GlaS dataset. As demonstrated in Table 4, our proposed network consistently outperforms other methods in gland segmentation tasks, achieving the highest scores. Figure 9 visually presents the results of gland segmentation using various models on the test set. The visual evidence underscores that our proposed network effectively segments gland boundaries and attains superior DSC, Jaccard, precision, and recall. Our innovative approach has direct applicability in computer-aided pathological diagnosis systems, potentially alleviating the workload of pathologists.

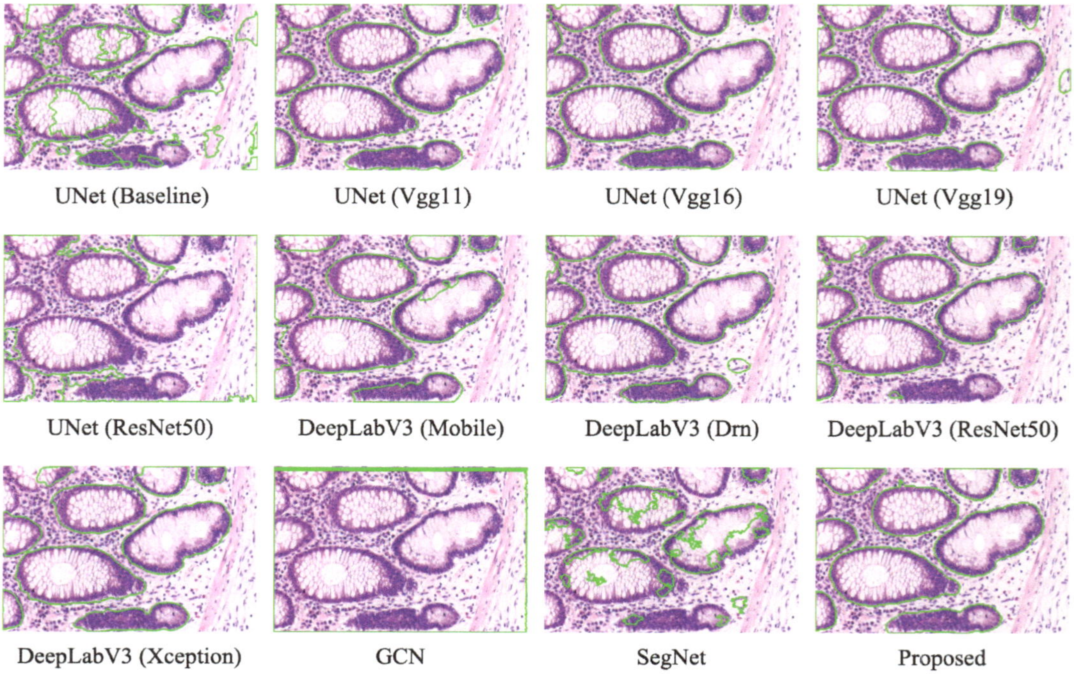

Figure 9. Segmentation results of different models on the GlaS dataset.

Table 4. Comparative results for gland segmentation on the Glas dataset. The best results are indicated in bold.

Method	DSC	Jaccard	Precision	Recall
UNet(Baseline) [19]	0.5132	0.3745	0.9285	0.3549
UNet(Backbone: Vgg11) [32]	0.7486	0.6195	0.9313	0.6268
UNet(Backbone: Vgg16) [33]	0.7324	0.6038	0.8375	0.6507
UNet(Backbone: Vgg19) [31]	0.7289	0.600	0.7928	0.6747
UNet(Backbone: ResNet50) [34]	0.6511	0.5065	0.9375	0.4985
DeepLabV3(Backbone:Mobilenet) [35]	0.6839	0.5410	0.9367	0.5388
DeepLabV3(Backbone: Drn) [35]	0.7367	0.6039	0.9375	0.6065
DeepLabV3(Backbone: ResNet50) [35]	0.6887	0.5503	0.9358	0.5203
DeepLabV3(Backbone: Xception) [35]	0.6867	0.5564	0.9342	0.5430
GCN [36]	0.5696	0.4220	0.7863	0.4464
SegNet [12]	0.5206	0.3799	0.9445	0.3592
Proposed	**0.9571**	**0.9190**	**0.9548**	**0.9611**

5. Conclusions

In this research, we have developed RGGC-UNet, an efficient and accurate deep learning framework specifically designed for the semantic segmentation of SRCs in pathological images. The central component of our model lies in its encoder-decoder architecture, where we have introduced an innovative encoder. This encoder is purposefully crafted to adeptly capture features, preserving the relationships between distant pixels. Particularly noteworthy is our introduction of the ghost coordinate attention mechanism, which inherits the advantages of coordinated attention. It adeptly models inter-channel relationships while simultaneously capturing long-range dependencies with precise positional information and ghost blocks.

To assess the effectiveness of RGGC-UNet, we conducted extensive experiments on a dataset that we curated. The results indicate that our proposed model can surpass leading models in terms of segmentation accuracy and efficiency, benefiting from ghost block and ghost coordinate attention. An important attribute of our proposed framework is its adaptability; it can seamlessly transition to other tasks related to pathological image analysis. Furthermore, the decoder structure we have presented exhibits flexibility and can be integrated into other deep convolutional neural networks dedicated to pathological image analysis.

Nonetheless, it is important to acknowledge certain limitations. We have yet to evaluate the performance of our model on natural images, leaving its effect in such contexts uncertain. Recognizing this as an existing challenge, our future research endeavors will involve an in-depth theoretical analysis to provide more robust insights.

Author Contributions: Conceptualization, T.Z., W.S. and C.F.; software, T.Z.; methodology, T.Z. and C.F.; validation, T.Z.; formal analysis, T.Z., W.S. and C.-W.S.; investigation, T.Z. and C.F.; resources, T.Z.; data curation, T.Z. and C.F.; writing—original draft preparation, T.Z.; writing—review and editing, T.Z., C.F. and C.-W.S.; visualization, T.Z.; supervision, T.Z.; project administration, T.Z.; funding acquisition, C.F. All authors have read and agreed to the published version of the manuscript.

Funding: This research is supported by the National Natural Science Foundation of China (No. 62032013), and the Fundamental Research Funds for the Central Universities (Nos. N2324004-12 and N2316010).

Institutional Review Board Statement: Not applicable.

Informed Consent Statement: Not applicable.

Data Availability Statement: The data that support the findings of this study are available from the corresponding author upon reasonable request.

Conflicts of Interest: The authors declare no conflict of interest.

References

1. Gnepp, D.R.; Henley, J.D.; Simpson, R.H.; Eveson, J. Chapter 6—Salivary and Lacrimal Glands. In *Diagnostic Surgical Pathology of the Head and Neck*, 2nd ed.; Gnepp, D.R., Ed.; W.B. Saunders: Philadelphia, PA, USA, 2009; pp. 413–562. [CrossRef]
2. Benesch, M.G.; Mathieson, A. Epidemiology of Signet Ring Cell Adenocarcinomas. *Cancers* **2020**, *12*, 1544. [CrossRef] [PubMed]
3. Hamilton, S.R.; Aaltonen, L.A. *Chapter 1—Tumours of the Oesophagus*; IARC Press: Lyon, France, 2000; Volume 2, pp. 9–30.
4. Ying, H.; Song, Q.; Chen, J.; Liang, T.; Gu, J.; Zhuang, F.; Chen, D.Z.; Wu, J. A semi-supervised deep convolutional framework for signet ring cell detection. *Neurocomputing* **2021**, *453*, 347–356. [CrossRef]
5. Da, Q.; Huang, X.; Li, Z.; Zuo, Y.; Zhang, C.; Liu, J.; Chen, W.; Li, J.; Xu, D.; Hu, Z.; et al. DigestPath: A benchmark dataset with challenge review for the pathological detection and segmentation of digestive-system. *Med. Image Anal.* **2022**, *80*, 102485. [CrossRef] [PubMed]
6. Li, J.; Yang, S.; Huang, X.; Da, Q.; Yang, X.; Hu, Z.; Duan, Q.; Wang, C.; Li, H. Signet Ring Cell Detection with a Semi-supervised Learning Framework. In *Information Processing in Medical Imaging, Proceedings of the 26th International Conference IPMI 2019, Hong Kong, China, 2–7 June 2019*; Chung, A.C.S., Gee, J.C., Yushkevich, P.A., Bao, S., Eds.; Springer: Cham, Switzerland, 2019; pp. 842–854. [CrossRef]
7. Wang, S.; Jia, C.; Chen, Z.; Gao, X. Signet Ring Cell Detection with Classification Reinforcement Detection Network. In *Bioinformatics Research and Applications, Proceedings of the 16th International Symposium, ISBRA 2020, Moscow, Russia, 1–4 December 2020*; Cai, Z.; Mandoiu, I., Narasimhan, G., Skums, P., Guo, X., Eds.; Springer: Cham, Switzerland, 2020; pp. 13–25. [CrossRef]
8. Lin, T.; Guo, Y.; Yang, C.; Yang, J.; Xu, Y. Decoupled gradient harmonized detector for partial annotation: Application to signet ring cell detection. *Neurocomputing* **2021**, *453*, 337–346. [CrossRef]
9. Zhang, S.; Yuan, Z.; Wang, Y.; Bai, Y.; Chen, B.; Wang, H. REUR: A unified deep framework for signet ring cell detection in low-resolution pathological images. *Comput. Biol. Med.* **2021**, *136*, 104711. [CrossRef] [PubMed]
10. Chen, Z.; Wang, S.; Jia, C.; Hu, K.; Ye, X.; Li, X.; Gao, X. CRDet: Improving Signet Ring Cell Detection by Reinforcing the Classification Branch. *J. Comput. Biol.* **2021**, *28*, 732–743. [CrossRef] [PubMed]
11. Shelhamer, E.; Long, J.; Darrell, T. Fully Convolutional Networks for Semantic Segmentation. *IEEE Trans. Pattern Anal. Mach. Intell.* **2017**, *39*, 640–651. [CrossRef] [PubMed]

12. Badrinarayanan, V.; Kendall, A.; Cipolla, R. SegNet: A Deep Convolutional Encoder-Decoder Architecture for Image Segmentation. *IEEE Trans. Pattern Anal. Mach. Intell.* **2017**, *39*, 2481–2495. [CrossRef]
13. Shan, P.; Wang, Y.; Fu, C.; Song, W.; Chen, J. Automatic skin lesion segmentation based on FC-DPN. *Comput. Biol. Med.* **2020**, *123*, 103762. [CrossRef]
14. Sharma, A.; Mishra, P.K. DRI-UNet: Dense residual-inception UNet for nuclei identification in microscopy cell images. *Neural Comput. Appl.* **2023**, *35*, 19187–19220. [CrossRef]
15. Zheng, Y.; Song, W.; Du, M.; Chow, S.S.M.; Lou, Q.; Zhao, Y.; Wang, X. Cryptography-Inspired Federated Learning for Generative Adversarial Networks and Meta Learning. In *Advanced Data Mining and Applications, Proceedings of the International Conference on Advanced Data Mining and Applications, Shenyang, China, 27–29 August 2023*; Springer: Cham, Switzerland, 2023; pp. 393–407. [CrossRef]
16. Yuan, J.; Xiao, L.; Wattanachote, K.; Xu, Q.; Luo, X.; Gong, Y. FGNet: Fixation guidance network for salient object detection. *Neural Comput. Appl.* **2023**, 1–16. [CrossRef]
17. Heydarheydari, S.; Birgani, M.J.T.; Rezaeijo, S.M. Auto-segmentation of head and neck tumors in positron emission tomography images using non-local means and morphological frameworks. *Pol. J. Radiol.* **2023**, *88*, e365. [CrossRef] [PubMed]
18. Hosseinzadeh, M.; Gorji, A.; Fathi Jouzdani, A.; Rezaeijo, S.M.; Rahmim, A.; Salmanpour, M.R. Prediction of Cognitive Decline in Parkinson's Disease Using Clinical and DAT SPECT Imaging Features, and Hybrid Machine Learning Systems. *Diagnostics* **2023**, *13*, 1691. [CrossRef] [PubMed]
19. Ronneberger, O.; Fischer, P.; Brox, T. U-Net: Convolutional Networks for Biomedical Image Segmentation. In *Medical Image Computing and Computer-Assisted Intervention—MICCAI 2015, Proceedings of the MICCAI 2015, Munich, Germany, 5–9 October 2015*; Navab, N., Hornegger, J., Wells, W.M., Frangi, A.F., Eds.; Springer: Cham, Switzerland, 2015; pp. 234–241.
20. Lu, Z.; Carneiro, G.; Bradley, A.P. An Improved Joint Optimization of Multiple Level Set Functions for the Segmentation of Overlapping Cervical Cells. *IEEE Trans. Image Process.* **2015**, *24*, 1261–1272. [CrossRef]
21. Chen, H.; Qi, X.; Yu, L.; Heng, P.A. DCAN: Deep Contour-Aware Networks for Accurate Gland Segmentation. In Proceedings of the 2016 IEEE Conference on Computer Vision and Pattern Recognition (CVPR), Las Vegas, NV, USA, 27–30 June 2016; pp. 2487–2496. [CrossRef]
22. Naylor, P.; Laé, M.; Reyal, F.; Walter, T. Segmentation of Nuclei in Histopathology Images by Deep Regression of the Distance Map. *IEEE Trans. Med. Imaging* **2019**, *38*, 448–459. [CrossRef]
23. Graham, S.; Vu, Q.D.; Raza, S.E.A.; Azam, A.; Tsang, Y.W.; Kwak, J.T.; Rajpoot, N. Hover-Net: Simultaneous segmentation and classification of nuclei in multi-tissue histology images. *Med. Image Anal.* **2019**, *58*, 101563. [CrossRef]
24. Zhao, T.; Fu, C.; Tian, Y.; Song, W.; Sham, C.W. GSN-HVNET: A Lightweight, Multi-Task Deep Learning Framework for Nuclei Segmentation and Classification. *Bioengineering* **2023**, *10*, 393. [CrossRef]
25. Zhou, Y.; Onder, O.F.; Dou, Q.; Tsougenis, E.; Chen, H.; Heng, P.A. CIA-Net: Robust Nuclei Instance Segmentation with Contour-Aware Information Aggregation. In *Information Processing in Medical Imaging, Proceedings of the 26th International Conference IPMI 2019, Hong Kong, China, 2–7 June 2019*; Chung, A.C.S., Gee, J.C., Yushkevich, P.A., Bao, S., Eds.; Springer: Cham, Switzerland, 2019; pp. 682–693. [CrossRef]
26. Zhang, J.; Xie, Y.; Zhang, P.; Chen, H.; Xia, Y.; Shen, C. Light-weight hybrid convolutional network for liver tumor segmentation. In Proceedings of the Twenty-Eighth International Joint Conference on Artificial Intelligence (IJCAI-19), Macao, China, 10–16 August 2019; Volume 2019, pp. 4271–4277.
27. Zhao, Y.; Fu, C.; Xu, S.; Cao, L.; Ma, H.F. LFANet: Lightweight feature attention network for abnormal cell segmentation in cervical cytology images. *Comput. Biol. Med.* **2022**, *145*, 105500. [CrossRef]
28. Han, K.; Wang, Y.; Tian, Q.; Guo, J.; Xu, C.; Xu, C. Ghostnet: More features from cheap operations. In Proceedings of the IEEE/CVF Conference on Computer Vision and Pattern Recognition, Seattle, WA, USA, 13–19 June 2020; pp. 1580–1589. [CrossRef]
29. Hou, Q.; Zhou, D.; Feng, J. Coordinate Attention for Efficient Mobile Network Design. In Proceedings of the IEEE/CVF Conference on Computer Vision and Pattern Recognition, Nashville, TN, USA, 20–25 June 2021; IEEE: Piscataway, NJ, USA, 2021; pp. 13708–13717. [CrossRef]
30. Milletari, F.; Navab, N.; Ahmadi, S.A. V-Net: Fully Convolutional Neural Networks for Volumetric Medical Image Segmentation. In Proceedings of the 2016 Fourth International Conference on 3D Vision (3DV), Stanford, CA, USA, 25–28 October 2016; pp. 565–571. [CrossRef]
31. Feng, R.; Liu, X.; Chen, J.; Chen, D.Z.; Gao, H.; Wu, J. A Deep Learning Approach for Colonoscopy Pathology WSI Analysis: Accurate Segmentation and Classification. *IEEE J. Biomed. Health Inform.* **2021**, *25*, 3700–3708. [CrossRef]
32. Iglovikov, V.; Shvets, A. Ternausnet: U-net with vgg11 encoder pre-trained on imagenet for image segmentation. *arXiv* **2018**, arXiv:1801.05746.
33. Pravitasari, A.A.; Iriawan, N.; Almuhayar, M.; Azmi, T.; Irhamah, I.; Fithriasari, K.; Purnami, S.W.; Ferriastuti, W. UNet-VGG16 with transfer learning for MRI-based brain tumor segmentation. *TELKOMNIKA (Telecommun. Comput. Electron. Control.)* **2020**, *18*, 1310–1318. [CrossRef]
34. Diakogiannis, F.I.; Waldner, F.; Caccetta, P.; Wu, C. ResUNet-a: A deep learning framework for semantic segmentation of remotely sensed data. *ISPRS J. Photogramm. Remote Sens.* **2020**, *162*, 94–114. [CrossRef]

35. Chen, L.C.; Papandreou, G.; Kokkinos, I.; Murphy, K.; Yuille, A.L. DeepLab: Semantic Image Segmentation with Deep Convolutional Nets, Atrous Convolution, and Fully Connected CRFs. *IEEE Trans. Pattern Anal. Mach. Intell.* **2018**, *40*, 834–848. [CrossRef]
36. Wang, X.; Yao, L.; Wang, X.; Paik, H.Y.; Wang, S. Global Convolutional Neural Processes. In Proceedings of the 2021 IEEE International Conference on Data Mining (ICDM), Auckland, New Zealand, 7–10 December 2021; IEEE: Piscataway, NJ, USA, 2021; pp. 699–708. [CrossRef]

bioengineering

MDPI

Article

RGSB-UNet: Hybrid Deep Learning Framework for Tumour Segmentation in Digital Pathology Images

Tengfei Zhao [1], Chong Fu [1,2,3,*], Ming Tie [4], Chiu-Wing Sham [5] and Hongfeng Ma [6]

1 School of Computer Science and Engineering, Northeastern University, Shenyang 110819, China
2 Engineering Research Center of Security Technology of Complex Network System, Ministry of Education, Shenyang 110819, China
3 Key Laboratory of Intelligent Computing in Medical Image, Ministry of Education, Northeastern University, Shenyang 110819, China
4 Science and Technology on Space Physics Laboratory, Beijing 100076, China
5 School of Computer Science, The University of Auckland, Auckland 1142, New Zealand
6 Dopamine Group Ltd., Auckland 1542, New Zealand
* Correspondence: fuchong@mail.neu.edu.cn

Abstract: Colorectal cancer (CRC) is a prevalent gastrointestinal tumour with high incidence and mortality rates. Early screening for CRC can improve cure rates and reduce mortality. Recently, deep convolution neural network (CNN)-based pathological image diagnosis has been intensively studied to meet the challenge of time-consuming and labour-intense manual analysis of high-resolution whole slide images (WSIs). Despite the achievements made, deep CNN-based methods still suffer from some limitations, and the fundamental problem is that they cannot capture global features. To address this issue, we propose a hybrid deep learning framework (RGSB-UNet) for automatic tumour segmentation in WSIs. The framework adopts a UNet architecture that consists of the newly-designed residual ghost block with switchable normalization (RGS) and the bottleneck transformer (BoT) for downsampling to extract refined features, and the transposed convolution and 1×1 convolution with ReLU for upsampling to restore the feature map resolution to that of the original image. The proposed framework combines the advantages of the spatial-local correlation of CNNs and the long-distance feature dependencies of BoT, ensuring its capacity of extracting more refined features and robustness to varying batch sizes. Additionally, we consider a class-wise dice loss (CDL) function to train the segmentation network. The proposed network achieves state-of-the-art segmentation performance under small batch sizes. Experimental results on DigestPath2019 and GlaS datasets demonstrate that our proposed model produces superior evaluation scores and state-of-the-art segmentation results.

Keywords: hybrid deep learning framework; tumour segmentation; whole slide image; Residual-Ghost-SN; bottleneck transformer

Citation: Zhao, T.; Fu, C.; Tie, M.; Sham, C.-W.; Ma, H. RGSB-UNet: Hybrid Deep Learning Framework for Tumour Segmentation in Digital Pathology Images. *Bioengineering* **2023**, *10*, 957. https://doi.org/10.3390/bioengineering10080957

Academic Editors: Yan Pei and Jijiang Yang

Received: 31 May 2023
Revised: 6 August 2023
Accepted: 9 August 2023
Published: 12 August 2023

1. Introduction

Colorectal cancer (CRC) is a gastrointestinal tumour that has a higher incidence and mortality rate than common tumours [1,2]. However, early screening with colonoscopy followed by pathological biopsy can significantly reduce the mortality rate [3]. Pathology is considered the gold standard for distinguishing between benign and malignant CRCs. During a diagnosis, physicians analyse the tumour's condition by observing the H&E-stained pathological section, drawing on their clinical expertise [4].

The use of high-resolution, large-scale whole slide images (WSIs) has become a routine diagnostic method with the rapid development of image scanning techniques [5]. WSI technology has great potential for developing and using algorithms for pathological diagnosis [6]. WSIs are widely used for digital pathology analysis, particularly in clinical practice [7]. However, the large size of WSIs can make manual analysis by pathologists time-consuming, and the unavoidable cognitive biases can lead to varying diagnoses.

CRC segmentation in whole slide images presents a unique set of implementation challenges due to the high-resolution and large size of these images, including gigapixel-scale data, computational resources, data handling and preprocessing, and integration with clinical workflow. Addressing these challenges often involves a combination of advanced image processing techniques, deep learning architectures tailored for large images, efficient data handling methods, and collaboration between medical experts and computer scientists. Overcoming these challenges is critical to harness the full potential of whole slide image segmentation in improving the accuracy and efficiency of colon cancer diagnosis and treatment planning.

In recent years, deep learning-based approaches [8] have been widely applied to histopathology image analysis, achieving remarkable results. In [9], Xu et al., proposed a deep learning method based on convolutional neural networks (CNNs) to automatically segment and classify epithelial and stromal regions in histopathology images. In [10], Liu et al., proposed a framework for the automatic detection and localization of breast tumours. In [11], Wang et al. proposed a deep CNN method to automatically identify the tumour in lung cancer images, using the shape feature to predict survival outcomes. In [12], Johnson et al. used Mask-RCNN to segment the nuclei in pathology images. In [13], Fan et al. proposed an improved deep learning method based on a classification pipeline to detect cancer metastases in WSI. In [14], Cho et al. proposed a deep neural network with scribbles for interactive pathology image segmentation. In [15], Zhai et al. proposed deep neural network guided by an attention mechanism for segmentation of liver pathology images. In [16], Deng et al. proposed a interpretable multi-modal image registration network based on disentangled convolutional sparse coding to solve the problem of lack of interpretability. In [17], Jin et al. proposed a two-stage deep learning system named iERM to provide accurate automatic grading of epiretinal membranes for clinical practice. In [18], Xiong et al. proposed DCGNN, a novel single-stage 3D object detection network based on density clustering and graph neural networks. DCGNN utlized density clustering ball query to partition the point cloud space and exploits local and global relationships by graph neural networks.

While histopathological image analysis has shown remarkable results, few studies have investigated deep learning-based methods for CRC tissue segmentation, particularly in WSIs. In [19], Qaiser et al. introduced two versions of our tumour segmentation method: one aimed at achieving faster processing while maintaining accuracy, and the other focused on achieving higher accuracy. The faster version relied on selecting representative image patches from a convolutional neural network (CNN) and classifying the patches by quantifying the difference between the exemplars' persistent homology profiles (PHPs) and the input image patch. In contrast, the more accurate version combined the PHPs with high-level CNN features and utilized a multi-stage ensemble strategy to label image patches. In [20], Zhu et al. proposed an adversarial context-aware and appearance consistency UNet (CAC-UNet) for segmentation and classification tasks, and achieved first place in the DigestPath2019 challenge. In [21], Feng et al. employed a UNet with a VGG backbone for WSI-based colorectal tumour segmentation, and achieved second place in the DigestPath2019 challenge.

Despite the remarkable results achieved by the methods mentioned above, several challenges still persist, including fewer public CRC datasets with expert annotations and difficulty accurately segmenting the refined boundary of the tumour, impeding further research on CRC tissue segmentation. Additionally, most existing deep learning frameworks rely on convolutional stacking, which reduces local redundancy but fails to capture global dependencies owing to the limited receptive field [22]. By contrast, transformers can capture long-distance dependencies through self-attention. However, excessive visual-semantic alignment may lead to redundancy in token representation, making it necessary to balance global dependency and local specificity when designing deep learning models.

This study proposes a hybrid deep learning framework for segmenting the CRC tumour in WSIs with a focus on refining the boundary segmentation and addressing network stability under small batch sizes. The proposed encoder–decoder architecture utilizes a newly designed encoder that includes residual ghost blocks with switchable normalization (RGS) and a bottleneck transformer block (BoT) for downsampling, while the decoder employs transpose convolution for upsampling [23–27]. By leveraging the benefits of CNNs and the transformer, the proposed encoder uses RGS and BoT as downsampling operations to extract more refined features from input images. The operation extracts local information, and the multi-head self-attention (MHSA) in the BoT models global dependency [27]. Experimental results demonstrate that the proposed model can accurately segment the tumour and produce a more refined boundary, leading to improved segmentation accuracy under small batch sizes. The primary contributions of our study are outlined below:

- We propose a deep hybrid network that combines a transformer and CNN for automatic tumour region segmentation in pathology images of the colon.
- A newly-designed feature extraction block RGS is presented. The block can adaptively determine the optimal combination of normalizers for each layer, making our model robust to varying batch sizes.
- Our novel hybrid backbone encoder, which includes RGS and BoT blocks, can extract more refined features.
- Experimental results demonstrate that the proposed RGSB-UNet achieves higher evaluation scores and produces finer segmentation results than state-of-the-art segmentation methods under small batch sizes.

The remainder of this paper is structured as follows. In Section 2, we present the proposed network architecture. Section 3 describes the datasets and evaluation criteria used in our experiments, while Section 4 presents our experimental results. Finally, in Section 5, we summarize the study results and suggest potential avenues for future research.

2. Proposed Method

2.1. Network Architecture

Our proposed deep learning framework for colon pathology WSI analysis is illustrated in Figure 1. As shown in Figure 2, to extract relevant features from original images, we start with $512 \times 512 \times 3$ image patches using dense cropping methods. The encoder includes a novel downsampling operation that combines RGS and BoT blocks as the feature extraction backbone. The details of the design of the encoder and decoder, GBS, RGS, and BoT will be discussed below.

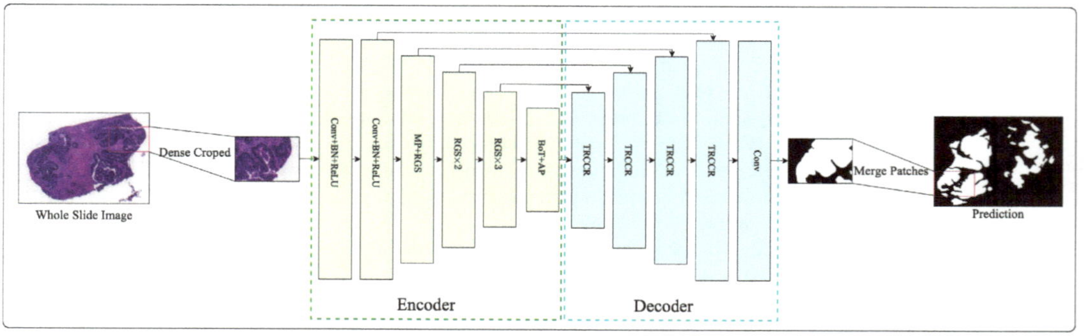

Figure 1. An overview of RGSB-UNet. The TRCCR denotes transposed convolution, ReLU, concatenate, convolution, and ReLU.

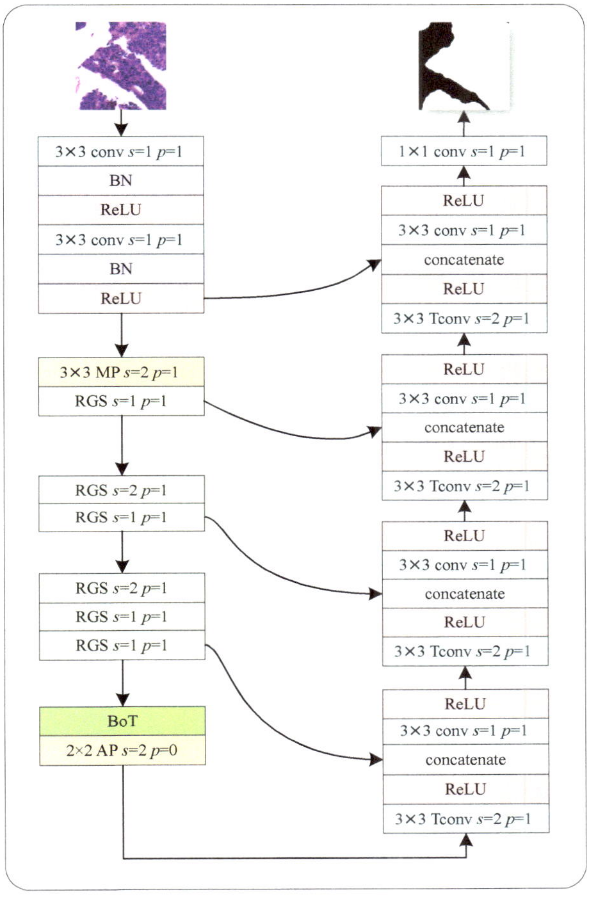

Figure 2. Schematic diagram of RGSB-UNet. RGS denotes the proposed residual ghost block with switchable normalization, and BoT denotes the bottleneck transformer. MP and AP denote the max and average pooling, respectively. Tconv denotes the transposed convolution used for upsampling.

2.1.1. Encoder and Decoder

In order to extract an efficient set of features, we use two 3 × 3 convolutions with batch normalization and ReLU, following a max pooling for downsampling, and devise a new residual ghost network, embedding a BoT at the end of the encoder as part of the encoder in our network architecture. The network employs four downsampling modules, each utilizing a different number of residual ghost blocks. As shown in Figure 2, the first downsampling module uses a 3 × 3 max pooling (MP) and a residual ghost block; the second and third downsampling modules use two and three stacked residual ghost blocks, respectively. By leveraging the ghost convolution technique, our network can generate rich feature maps using significantly fewer input features than traditional convolution methods, which improves the computational efficiency of our encoder. Additionally, the stability of our network is enhanced by the ability to select optimal combinations of different normalizers for each normalization layer, resulting in an accuracy that is not impacted by batch size. The fourth downsampling module incorporates a BoT block and a 2 × 2 average pooling (AP), which significantly boosts the extraction of refined features. Each downsampling module reduces the input spatial resolution by a factor of two.

The decoder is composed of four upsampling modules that utilize a transposed convolution and a 1 × 1 convolution with ReLU [28], increasing the input spatial resolution by a factor of two. The concatenate block concatenates the skip and output features of Tconv-ReLU; this operation attaches more local information extracted from different layers of the encoder directly into their corresponding decoder layers at the same level, which adds detailed information to the general area of the target of judgment. Further elaboration on the RGS and BoT components will be provided in subsequent subsections.

2.1.2. Ghost Block with Switchable Normalization

Our proposed Ghost-Block-SN architecture is presented in Figure 3, which utilizes the Ghost-Block to generate more representative features at a lower computational cost. The Ghost-Block firstly employs traditional convolution to generate intrinsic feature maps and then utilizes cost-effective linear operations to expand the features and channels. The computational cost of linear operations on feature maps is much lower than traditional convolution, making the block more efficient than other existing efficient methods. The size of the primary convolution kernel in Ghost-Block is customizable, and we used a 1 × 1 point-wise convolution in our study. A BN layer is introduced after each Ghost-Block in Residual-Ghost-Block, which provides stability and speeds up the training process.

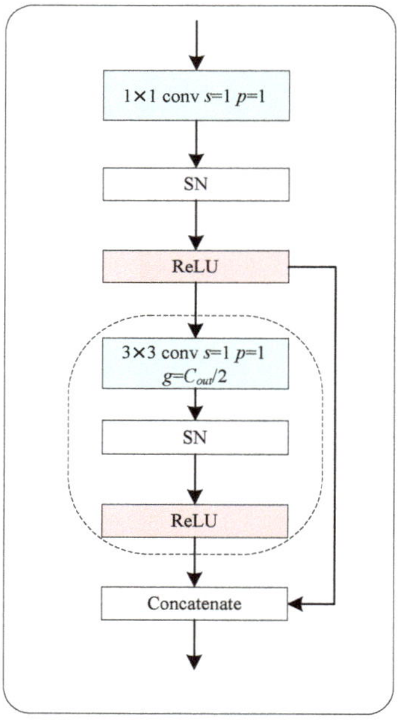

Figure 3. Schematic diagram of Ghost block with switchable normalization. The dash box denotes the cheap operation that uses a 3 × 3 group convolution in the ghost block.

However, the performance of Ghost-Block-BN is restricted by the batch size as BN uses a single normalizer throughout the network, which can be unstable and degrade accuracy under small batch sizes. To overcome this issue, we incorporated switchable normalization (SN) [29], a technique that is robust to a wide range of batch sizes. SN measures channel-wise, layer-wise, and minibatch-wise statistics using BN [30], instance normalization (IN) [31], and layer normalization (LN) [32], respectively, and learns their important weights to find their optimal combination, ensuring network stability and accuracy in the case of small batch sizes.

2.1.3. Residual Ghost Block with Switchable Normalization

As shown in Figure 4a, our RGS is constructed by incorporating the above presented GBS with a residual bottleneck, which is the fundamental building block of a ResNet [23], due to its exceptional performance. The core concept behind a residual block is to reformulate the layers as learning residual functions with respect to the layer inputs, rather than learning unreferenced functions. Compared to ResNet-50, our encoder employs fewer building units, boosting the computational efficiency. Moreover, the proposed RGS is highly robust and can handle a wide range of batch sizes.

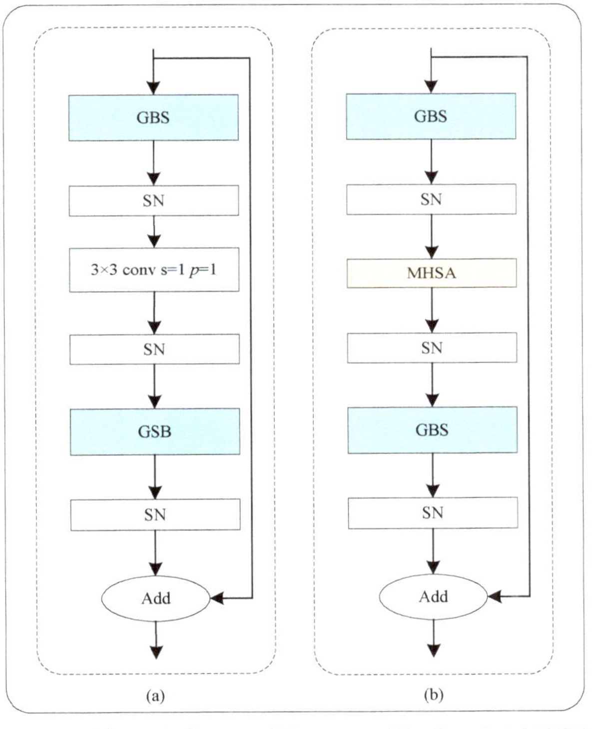

Figure 4. Schematic diagram of the proposed bottleneck. (**a**) RGS Bottleneck. (**b**) Bottleneck transformer. GBS and SN denote the ghost block with switchable normalization and switchable normalization, respectively. MHSA denotes multi-head self-attention.

2.1.4. Bottleneck Transformer

Figure 4b shows the bottleneck transformer (BoT), an important block in the proposed hybrid network, which uses multi-head self-attention (MHSA) to replace the 3×3 convolution compared with RGS. The BoT is embedded in the last layer of the encoder. As is known, the self-attention (Figure 5a) can process and aggregate the information in the feature maps to complement the CNN handle long-distance dependencies. Particularly, the self-attention in MHSA can help the network better understand the relationships between different regions and improve the accuracy of segmentation when working with highly detailed images. In addition, as shown in Figure 5b, the MHSA with sufficient heads is at least as expressive as any convolutional layer [27]. The MHSA produces multiple attention maps and embedding features from an image to encode rich information, enhancing the deep model's robustness towards representation learning. Benefiting from the MHSA, the BoT block can help the network to boost the segmentation performance.

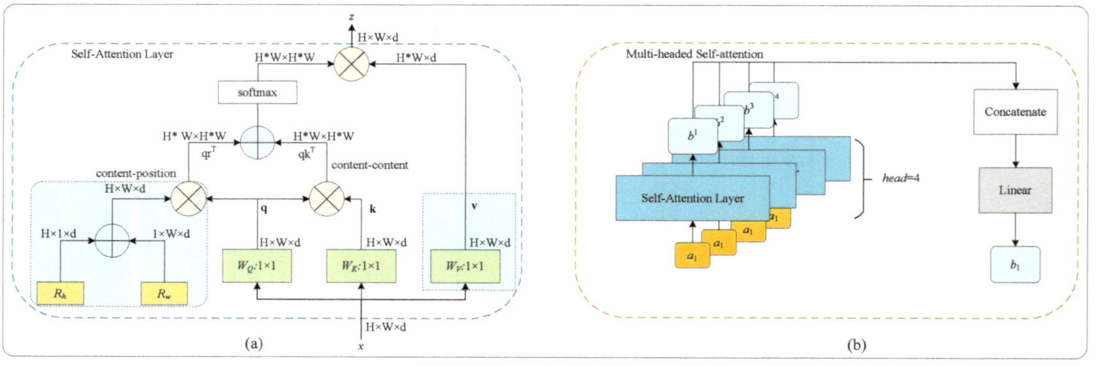

Figure 5. Schematic diagram of (**a**) self-attention [26] and (**b**) multi-head self-attention.

2.2. Loss Function

Dice loss is leveraged as a standard loss function in image segmentation tasks and indicates the difference between the predicted and ground-truth mask [33]. However, there are still some limitations when employing this function. For instance, there is no segmenting target, and the dice loss is 0. Clearly, the dice loss function receives no punishment when predicting a false positive.

To address this issue, the improved class-wise dice loss function is leveraged to compute the background and lesion segmentation dice similarity coefficients (DSCs) for benign and malignant images, respectively [21]. The improved loss function can effectively reduce false positives, including its practicality for clinical applications. The improved class-wise dice loss (CDL) function is described by

$$L_{CDL} = 1 - \sum_{i}^{N}(y_p \frac{y_i \hat{y}_i}{y_i + \hat{y}_i} + \frac{(1-y_p)(1-y_i)(1-\hat{y}_i)+\epsilon}{(1-y_i)+(1-\hat{y}_i)+\epsilon}), \quad (1)$$

where y_i is the binary label of pixel i, $\hat{y}_i$ is the predicted probability, and N is the total number of pixels in a patch. ϵ is a small number to avoid the denominator becoming 0.

The presence of a lesion area determines the patch label (y_p). The CDL function can alleviate pixel-level class imbalance, resulting in an all-zero mask when training negative samples.

3. Evaluation and Datasets

3.1. Evaluation

We use the *DSC*, Jaccard Index (*JI*), and relative volume difference (*RVD*) to measure the segmentation performance of our proposed model [34]. The *DSC* measures the similarity between the network segmentation results when using the proposed method and the gold standard mask in image segmentation. *DSC*, *JI*, and *RVD* are defined as

$$DSC = \frac{2|Y_A \cap Y_P|}{|Y_A| + |Y_P|}, \quad (2)$$

$$JI = \frac{|Y_A \cap Y_P|}{|Y_A| + |Y_P| - |Y_A \cap Y_P|}, \quad (3)$$

and

$$RVD = \frac{|Y_P| - |Y_A|}{|Y_A|}, \quad (4)$$

where Y_A is the set of lesion pixels in the annotation, and Y_P is the corresponding set of lesion pixels in the segmentation result.

We use pixel accuracy (*PA*) and area under the curve (*AUC*) to measure the classification performance of our proposed model. *AUC* is defined as the area of the receiver operating characteristic (ROC) curve, determined by the true positive rate (*TPR*) and false positive rate (*FPR*). *TPR*, *FPR*, and *Precision* are defined as follows:

$$TPR = \frac{TP}{TP + FN}, \tag{5}$$

$$FPR = \frac{FP}{FP + TN}, \tag{6}$$

and

$$Precision = \frac{TP}{TP + FP}, \tag{7}$$

where *TP*, *FP*, *TN*, and *FN* are true positives, false positives, true negatives, and false negatives, respectively.

AUC and *PA* are defined as

$$AUC = \int_{x=0}^{1} TPR(FPR^{-1}(x))dx = P(X_1 > X_0) \tag{8}$$

and

$$PA = \frac{TP + TN}{TP + TN + FP + FN}, \tag{9}$$

where X_0 and X_1 are the scores for the negative and positive instances, respectively.

3.2. Datasets and Implementation

We trained the proposed network on the DigestPath2019 [35] gland segmentation (GlaS) [36] datasets. In these datasets, numerous expert-level annotations on digestive system pathological images are available, which will substantially advance research on automatic segmentation and classification of pathological tissues.

The DigestPath2019 dataset contains positive and negative samples of 872 tissue slices from 476 patients. The average size of a tissue slice is 3000 × 3000. The training set comprises 660 images from 324 patients, from which 250 images from 93 patients are annotated by pathologists. The positive training samples contain 250 tissue images from 93 WSIs, with pixel-level annotation, where 0 indicates the background and 255 indicates the foreground (malignant lesion). Some samples cropped from WSI are shown in Figure 6. The negative training samples contain 410 tissue images from 231 WSIs. These negative images have no annotation because they have no malignant lesions. The entry to DigestPath2019 competition has closed and the official test set is not publicly accessible. To address this issue, we remake a balanced test set by randomly selecting 108 samples with a 54:54 positive to negative ratio from the original training set. We retrained all the compared models on the DigestPath2019 dataset using their original code, and the test set images are not used in training. Defining an objective criteria for distinguishing between benign (negative) and malignant (positive) lesions is difficult. To make it easier for academic research, according to the WHO classification of digestive system tumours, we regarded the following lesions as malignant: high-grade intraepithelial neoplasia and adenocarcinoma, including papillary adenocarcinoma, mucinous adenocarcinoma, poorly cohesive carcinoma, and signet ring cell carcinoma. Low-grade intraepithelial neoplasia and severe inflammation are not included in the dataset because they are generally difficult for pathologists to detect.

The GlaS dataset consists of 165 tissue slices containing both positive and negative samples. The GlaS dataset contains a training set of 85 samples from which we selected 17 samples as the validation data. The dataset offers two different test sets, testA and testB, consisting of 60 and 20 samples, respectively. We used the validation set to select the optimal model and all the performance evaluations are carried out on the joining of testA and testB. Glands are vital histological structures found in various organ systems, serving as the primary mechanism for protein and carbohydrate secretion. Adenocarcinomas, which are malignant tumors originating from glandular epithelium, have been identified as the most prevalent form of cancer. Pathologists routinely rely on gland morphology to assess the malignancy level of several adenocarcinomas, such as those affecting the prostate, breast, lung, and colon. Accurately segmenting glands is often a crucial step in obtaining reliable morphological statistics. However, this task is inherently challenging due to the significant variation in glandular morphology across different histologic grades. Most studies to date have primarily focused on gland segmentation in healthy or benign samples, with limited attention given to intermediate or high-grade cancer. Additionally, these studies often optimize their methods for specific datasets.

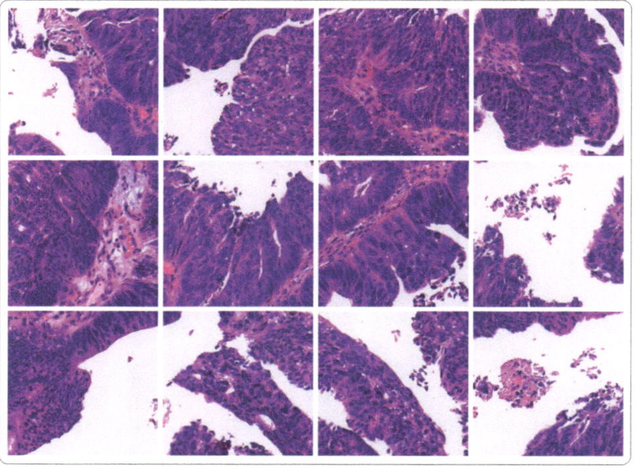

Figure 6. Samples cropped from WSI.

The simulations were run on a station equipped with an NVIDIA GeForce RTX 3090 GPU and Intel(R) Xeon(R) CPU E5-2680v4×2. We augmented the training data during training. Table 1 lists the detailed hyperparameters of the proposed framework. We embarked on an iterative journey of manual tuning, wherein we systematically explored and fine-tuned various hyperparameters within our framework. By meticulously adjusting parameters such as learning rates, batch sizes, and model architecture, we meticulously tracked the impact of each modification on the overall performance metrics. This exhaustive process allowed us to discover the optimal combination of hyperparameters, leading to a highly refined and efficient version of our framework that exhibits superior accuracy and generalization on diverse datasets.

Table 1. Hyperparameters of our framework.

Hyperparameters	Value
Crop Method	Dense Crop
Crop Stride	512
Crop Patch Size	$512 \times 512 \times 3$
Batch Size	2
MHSA Head	4
Optimizer	SGD
Learning Rate	$1.0 \times e^{-2}$
Weight Deacy	$1.0 \times e^{-4}$
Momentum	0.9
Epoch Number	500

4. Experimental Results

Table 2 shows the results of the ablation study, which demonstrate the performance gains when integrating different blocks into UNet, including residual block (RSB), residual ghost block (RGB), RGS, and BoT. Especially, our proposed RGSB-UNet achieves the highest DSC score of 0.8336. We further analyze the performance of different batch sizes and MHSA head numbers based on RGSB-UNet. As is shown in Table 3, the proposed network maintains high performance even with small batch sizes. We tried different small batch sizes in our experiments. We prove that batch size is no longer a strict limitation for the proposed network. In addition, the head numbers of MHSA impact the performance of the proposed network. We have tried different numbers of heads for the MHSA in the proposed network to search for the best results, and our network achieved optimal performance when the heads are four. When integrating RGS and BoT together to the UNet, the segmentation model produces the best performance, which indicates that these blocks can improve the performance of pathology image segmentation.

Table 2. Performance gains by integrating different blocks into UNet on the DigestPath2019 dataset. RSB and RGB denote the residual block and residual ghost block with batch normalization, respectively.

UNet	RSB	RGB	RGS	BoT	DSC
✓					0.8150
✓	✓				0.8197
✓	✓			✓	0.8201
✓		✓			0.8203
✓		✓		✓	0.8261
✓			✓		0.8263
✓			✓	✓	0.8336

Table 3. Effect of batch size and MHSA head on model performance. The best results are marked in bold.

Batch Size	**1**				**2**			
MHSA Head	×	1	2	4	×	1	2	4
DSC	0.8126	0.8241	0.8220	0.8331	0.8263	0.8294	0.8250	**0.8336**

Table 4 compares the performance of the proposed and other popular models in terms of six metrics on the DigestPath2019 dataset; the numbers in bold indicate the best results for each metric. As can be seen from this table, under a small batch of two, our proposed model achieves the highest DSC, PA, JI, and Precision; it also achieves the second best RVD and AUC. Furthermore, although DeepLab with Xception backbone outperforms other models in terms of RVD, and the CAC-UNet (first place) achieves the highest AUC, our model performs significantly better in the other three metrics. In Figure 7, we illustrate the results of tumour segmentation on the sample images and compare them with that of [20,21,37–46]. As shown in this figure, the mask predicted by the proposed network is extremely close to the ground truth. Compared with other leading networks, our proposed network can successfully segment tumour regions with nearly overlapping margins, indicated in the red boxes. Overall, our proposed model can capture more refined features and achieve state-of-the-art accuracy in tumour segmentation tasks.

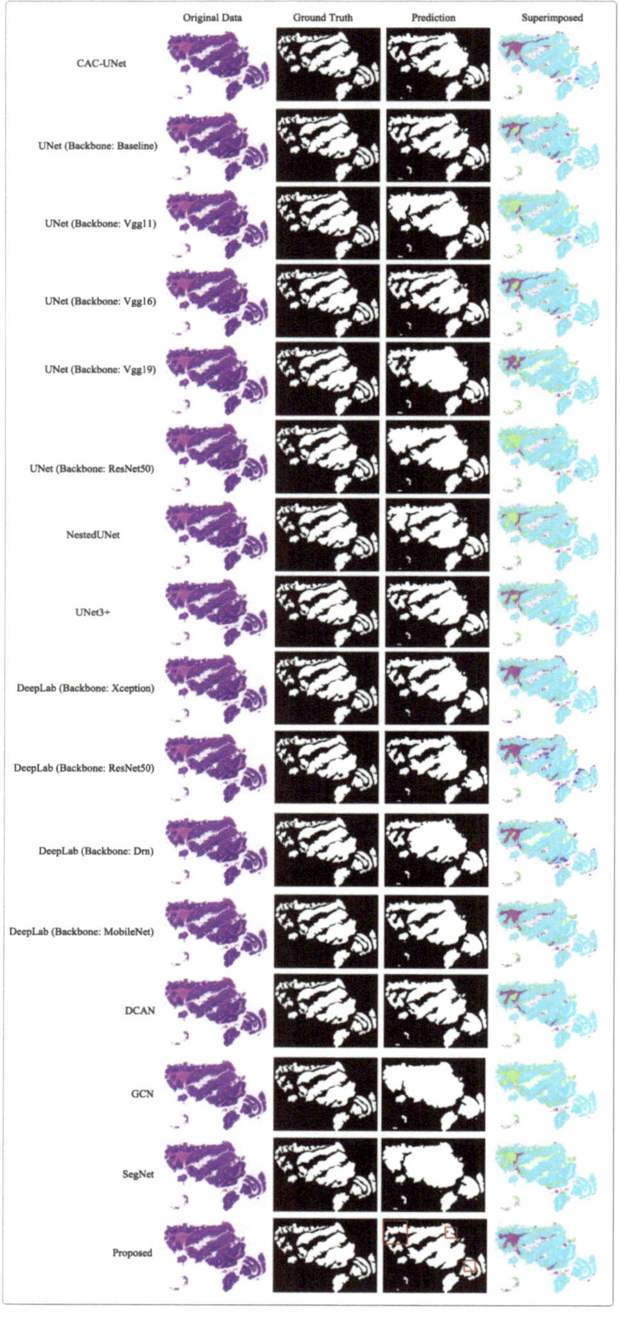

Figure 7. Segmentation results of different networks on the DigestPath2019 dataset. In the superimposed images, the areas marked in green represent the ground truth.

Table 4. Comparative results for tumour segmentation on the DigestPath2019 dataset. The best results are marked in bold.

Methods	DSC	AUC	PA	JI	RVD	Precision
CAC-UNet [20]	0.8292	1.0000	0.8935	0.7082	0.3219	0.9072
UNet (Baseline) [37]	0.8150	0.9060	0.8611	0.6914	0.2852	0.6511
UNet (Backbone: Vgg11) [38]	0.8258	0.9187	0.8796	0.7081	0.2964	0.6829
UNet (Backbone: Vgg16) [39]	0.8323	0.9562	0.9351	0.7177	0.2445	0.8000
UNet (Backbone: Vgg19) [21]	0.7417	0.5875	0.3889	0.5990	0.4803	0.2987
UNet (Backbone: ResNet50) [40]	0.8197	0.9312	0.8981	0.7019	0.3652	0.7179
UNet (Backbone: DenseNet121) [41]	0.2183	0.5758	0.5092	0.1441	0.4825	0.3076
NestedUNet [42]	0.7609	0.7625	0.6481	0.6254	0.5561	0.4242
Unet3+ [43]	0.7467	0.6250	0.4450	0.6127	0.3977	0.3181
DeepLab (Backbone: Xception) [44]	0.6999	0.9500	0.9259	0.5517	**0.1925**	0.7778
DeepLab (Backbone: ResNet50) [44]	0.7964	0.6375	0.4629	0.6684	0.3829	0.3255
DeepLab (Backbone: Drn) [44]	0.7917	0.7125	0.5740	0.6605	0.3214	0.3783
DeepLab (Backbone: MobileNet) [44]	0.7943	0.8250	0.7407	0.6658	0.4206	0.5000
DCAN [45]	0.8322	0.9562	0.9351	0.7169	0.2291	0.8000
GCN [46]	0.6372	0.6625	0.5000	0.4903	0.5051	0.3414
SegNet [47]	0.7564	0.7937	0.6944	0.6174	0.5845	0.4590
Proposed	**0.8336**	**0.9813**	**0.9722**	**0.7190**	0.2122	**0.9032**

To demonstrate our proposed method's generalizability and its performance in different contexts, we use the GlaS dataset to verify the network. As shown in Table 5, our proposed model achieves the highest scores in state-of-the-art accuracy in gland segmentation tasks. Figure 8 shows the results of gland segmentation on the test set and compares them with [21,37–46]. As shown from this figure, compared with other leading works, our proposed network can significantly segment gland boundaries, as indicated in the red box. Our idea can be directly applied to a computer-aided pathological diagnosis system to reduce the workload of pathologists.

Table 5. Comparative results for gland segmentation on the GlaS dataset. The best results are marked in bold.

Methods	DSC	AUC	PA	JI	RVD	Precision
UNet (Baseline) [37]	0.5132	0.4339	0.8125	0.3745	0.4959	0.9285
UNet (Backbone: Vgg11) [38]	0.7486	0.5068	0.9480	0.6195	0.6165	0.9313
UNet (Backbone: Vgg16) [39]	0.7324	0.6328	0.8265	0.6038	0.7378	0.8375
UNet (Backbone: Vgg19) [21]	0.7289	0.5979	0.8975	0.5999	0.7595	0.7928
UNet (Backbone: ResNet50) [40]	0.6511	0.5000	0.9375	0.5065	0.9228	0.9375
UNet (Backbone: DenseNet121) [41]	0.6491	0.5998	0.9263	0.5037	0.9046	0.9261
NestedUNet [42]	0.6003	0.4533	0.8500	0.4651	0.8031	0.9315
Unet3+ [43]	0.6650	0.6725	0.9450	0.5170	0.8459	0.9428
DeepLab (Backbone: Xception) [44]	0.6867	0.4735	0.8875	0.5564	0.4423	0.9342
DeepLab (Backbone: ResNet50) [44]	0.6887	0.4866	0.9125	0.5503	0.5648	0.9358
DeepLab (Backbone: Drn) [44]	0.7367	0.5306	0.9375	0.6039	0.6299	0.9375
DeepLab (Backbone: MobileNet) [44]	0.6839	0.4933	0.9250	0.5410	0.6062	0.9367
DCAN [45]	0.6415	0.6107	0.9177	0.4896	0.9459	0.9370
GCN [46]	0.5696	0.5079	0.6983	0.4220	0.9918	0.7863
SegNet [47]	0.5206	0.5533	0.8625	0.3799	0.3995	0.9445
Proposed	**0.8865**	**0.8920**	**0.9823**	**0.7953**	**0.2128**	**0.9475**

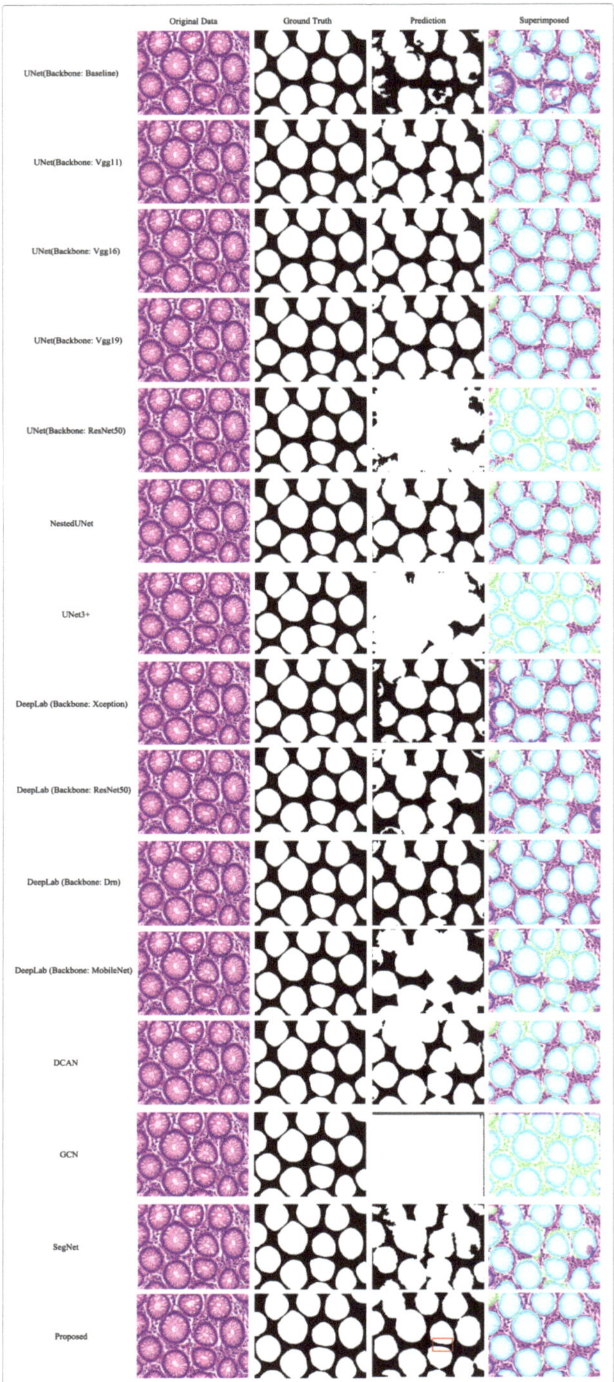

Figure 8. Segmentation results of different networks on the GlaS dataset. In the superimposed images, the areas marked in green represent the ground truth.

5. Conclusions

In this paper, we propose a hybrid deep learning framework for segmenting tumours in WSIs. Our model employs an encoder–decoder architecture, with a newly designed RGS block and a BoT block in the decoder part. These blocks are implemented to capture more refined features and improve network stability, particularly when working with small batch sizes. To evaluate the performance of our approach, we conducted extensive experiments on the DigestPath2019 and GlaS datasets, and the results indicate that our model achieved state-of-the-art segmentation accuracy.

Our proposed framework is generic and can be easily applied to other histopathology image analysis tasks. In addition, the decoder architecture proposed in this study is flexible and can be incorporated into other deep CNNs for histopathology image analysis. However, we are yet to conduct experiments using natural images; therefore the superiority of our approach in this context cannot be guaranteed. We consider this an open problem and plan to conduct further research to provide a theoretical analysis with complete proof.

Author Contributions: Conceptualization, T.Z. and C.F.; methodology, T.Z. and C.F.; validation, T.Z.; formal analysis, T.Z., M.T. and C.-W.S.; investigation, T.Z. and C.F.; resources, T.Z.; data curation, T.Z., H.M. and C.F.; writing—original draft preparation, T.Z.; writing—review and editing, T.Z., C.F. and C.-W.S.; visualization, T.Z.; supervision, T.Z.; project administration, T.Z.; funding acquisition, C.F. All authors have read and agreed to the published version of the manuscript.

Funding: This research is supported by the National Natural Science Foundation of China (No. 62032013), and the Fundamental Research Funds for the Central Universities (No. N2224001-7).

Institutional Review Board Statement: Not applicable.

Informed Consent Statement: Not applicable.

Data Availability Statement: The data that support the findings of this study are available from the corresponding author upon reasonable request.

Conflicts of Interest: The authors declare no conflict of interest.

References

1. Siegel, R.L.; Miller, K.D.; Goding Sauer, A.; Fedewa, S.A.; Butterly, L.F.; Anderson, J.C.; Cercek, A.; Smith, R.A.; Jemal, A. Colorectal cancer statistics, 2020. *CA Cancer J. Clin.* **2020**, *70*, 145–164. [CrossRef] [PubMed]
2. Xia, C.; Dong, X.; Li, H.; Cao, M.; Sun, D.; He, S.; Yang, F.; Yan, X.; Zhang, S.; Li, N.; et al. Cancer statistics in China and United States, 2022: Profiles, trends, and determinants. *Chin. Med. J.* **2022**, *135*, 584–590. [CrossRef]
3. Vega, P.; Valentín, F.; Cubiella, J. Colorectal cancer diagnosis: Pitfalls and opportunities. *World J. Gastrointest. Oncol.* **2015**, *7*, 422. [CrossRef] [PubMed]
4. Song, W.; Fu, C.; Zheng, Y.; Cao, L.; Tie, M. A practical medical image cryptosystem with parallel acceleration. *J. Ambient. Intell. Humaniz. Comput.* **2022**, *14*, 9853–9867. [CrossRef]
5. Wang, S.; Yang, D.M.; Rong, R.; Zhan, X.; Xiao, G. Pathology image analysis using segmentation deep learning algorithms. *Am. J. Pathol.* **2019**, *189*, 1686–1698. [CrossRef]
6. Kumar, N.; Gupta, R.; Gupta, S. Whole slide imaging (WSI) in pathology: Current perspectives and future directions. *J. Digit. Imaging* **2020**, *33*, 1034–1040. [CrossRef]
7. Wright, A.M.; Smith, D.; Dhurandhar, B.; Fairley, T.; Scheiber-Pacht, M.; Chakraborty, S.; Gorman, B.K.; Mody, D.; Coffey, D.M. Digital slide imaging in cervicovaginal cytology: A pilot study. *Arch. Pathol. Lab. Med.* **2013**, *137*, 618–624. [CrossRef]
8. Zhang, W.; Fu, C.; Chang, X.; Zhao, T.; Li, X.; Sham, C.W. A more compact object detector head network with feature enhancement and relational reasoning. *Neurocomputing* **2022**, *499*, 23–34. [CrossRef]
9. Xu, J.; Luo, X.; Wang, G.; Gilmore, H.; Madabhushi, A. A deep convolutional neural network for segmenting and classifying epithelial and stromal regions in histopathological images. *Neurocomputing* **2016**, *191*, 214–223. [CrossRef]
10. Liu, Y.; Gadepalli, K.; Norouzi, M.; Dahl, G.E.; Kohlberger, T.; Boyko, A.; Venugopalan, S.; Timofeev, A.; Nelson, P.Q.; Corrado, G.S.; et al. Detecting Cancer Metastases on Gigapixel Pathology Images. *arXiv* **2017**, arXiv:1703.02442.
11. Wang, S.; Chen, A.; Yang, L.; Cai, L.; Xie, Y.; Fujimoto, J.; Gazdar, A.; Xiao, G. Comprehensive analysis of lung cancer pathology images to discover tumor shape and boundary features that predict survival outcome. *Sci. Rep.* **2018**, *8*, 10393. [CrossRef] [PubMed]
12. Johnson, J.W. Adapting mask-rcnn for automatic nucleus segmentation. *arXiv* **2018**, arXiv:1805.00500.
13. Fan, K.; Wen, S.; Deng, Z. Deep learning for detecting breast cancer metastases on WSI. In *Innovation in Medicine and Healthcare Systems, and Multimedia*; Springer: Berlin/Heidelberg, Germany, 2019; pp. 137–145.

14. Cho, S.; Jang, H.; Tan, J.W.; Jeong, W.K. DeepScribble: Interactive Pathology Image Segmentation Using Deep Neural Networks with Scribbles. In Proceedings of the 2021 IEEE 18th International Symposium on Biomedical Imaging (ISBI), Nice, France, 13–16 April 2021; pp. 761–765.
15. Zhai, Z.; Wang, C.; Sun, Z.; Cheng, S.; Wang, K. Deep Neural Network Guided by Attention Mechanism for Segmentation of Liver Pathology Image. In Proceedings of the 2021 Chinese Intelligent Systems Conference, Fuzhou, China, 16–17 October 2021; Springer: Berlin/Heidelberg, Germany, 2022; pp. 425–435.
16. Deng, X.; Liu, E.; Li, S.; Duan, Y.; Xu, M. Interpretable Multi-Modal Image Registration Network Based on Disentangled Convolutional Sparse Coding. *IEEE Trans. Image Process.* **2023**, *32*, 1078–1091. [CrossRef] [PubMed]
17. Jin, K.; Yan, Y.; Wang, S.; Yang, C.; Chen, M.; Liu, X.; Terasaki, H.; Yeo, T.H.; Singh, N.G.; Wang, Y.; et al. iERM: An Interpretable Deep Learning System to Classify Epiretinal Membrane for Different Optical Coherence Tomography Devices: A Multi-Center Analysis. *J. Clin. Med.* **2023**, *12*, 400. [CrossRef] [PubMed]
18. Xiong, S.; Li, B.; Zhu, S. DCGNN: A single-stage 3D object detection network based on density clustering and graph neural network. *Complex Intell. Syst.* **2022**, *9*, 3399–3408. [CrossRef]
19. Qaiser, T.; Tsang, Y.W.; Taniyama, D.; Sakamoto, N.; Nakane, K.; Epstein, D.; Rajpoot, N. Fast and accurate tumor segmentation of histology images using persistent homology and deep convolutional features. *Med. Image Anal.* **2019**, *55*, 1–14. [CrossRef]
20. Zhu, C.; Mei, K.; Peng, T.; Luo, Y.; Liu, J.; Wang, Y.; Jin, M. Multi-level colonoscopy malignant tissue detection with adversarial CAC-UNet. *Neurocomputing* **2021**, *438*, 165–183. [CrossRef]
21. Feng, R.; Liu, X.; Chen, J.; Chen, D.Z.; Gao, H.; Wu, J. A deep learning approach for colonoscopy pathology WSI analysis: Accurate segmentation and classification. *IEEE J. Biomed. Health Inform.* **2020**, *25*, 3700–3708. [CrossRef]
22. Song, W.; Fu, C.; Zheng, Y.; Cao, L.; Tie, M.; Sham, C.W. Protection of image ROI using chaos-based encryption and DCNN-based object detection. *Neural Comput. Appl.* **2022**, *34*, 5743–5756. [CrossRef]
23. He, K.; Zhang, X.; Ren, S.; Sun, J. Deep residual learning for image recognition. In Proceedings of the IEEE Conference on Computer Vision and Pattern Recognition, Las Vegas, NV, USA, 27–30 June 2016; pp. 770–778.
24. Han, K.; Wang, Y.; Tian, Q.; Guo, J.; Xu, C. GhostNet: More Features From Cheap Operations. In Proceedings of the 2020 IEEE/CVF Conference on Computer Vision and Pattern Recognition (CVPR), Seattle, WA, USA, 13–19 June 2020.
25. Luo, P.; Ren, J.; Peng, Z.; Zhang, R.; Li, J. Differentiable Learning-to-Normalize via Switchable Normalization. In Proceedings of the 7th International Conference on Learning Representations, ICLR, New Orleans, LA, USA, 6–9 May 2019.
26. Srinivas, A.; Lin, T.Y.; Parmar, N.; Shlens, J.; Abbeel, P.; Vaswani, A. Bottleneck transformers for visual recognition. In Proceedings of the IEEE/CVF Conference on Computer Vision and Pattern Recognition, Nashville, TN, USA, 20–25 June 2021; pp. 16514–16524.
27. Cordonnier, J.B.; Loukas, A.; Jaggi, M. On the Relationship between Self-Attention and Convolutional Layers. In Proceedings of the 8th International Conference on Learning Representations, ICLR, Addis Ababa, Ethiopia, 26–30 April 2020.
28. Dumoulin, V.; Visin, F. A guide to convolution arithmetic for deep learning. *Stat* **2018**, *1050*, 11. [CrossRef]
29. Luo, P.; Zhang, R.; Ren, J.; Peng, Z.; Li, J. Switchable Normalization for Learning-to-Normalize Deep Representation. *IEEE Trans. Pattern Anal. Mach. Intell.* **2021**, *43*, 712–728. [CrossRef] [PubMed]
30. Ioffe, S.; Szegedy, C. Batch normalization: Accelerating deep network training by reducing internal covariate shift. *PMLR* **2015**, *37*, 448–456.
31. Huang, X.; Belongie, S. Arbitrary style transfer in real-time with adaptive instance normalization. In Proceedings of the IEEE International Conference on Computer Vision, Venice, Italy, 22–29 October 2017; pp. 1510–1519.
32. Ba, J.L.; Kiros, J.R.; Hinton, G.E. Layer normalization. *arXiv* **2016**, arXiv:1607.06450.
33. Milletari, F.; Navab, N.; Ahmadi, S.A. V-net: Fully convolutional neural networks for volumetric medical image segmentation. In Proceedings of the 2016 Fourth International Conference on 3D Vision (3DV), Stanford, CA, USA, 25–28 October 2016; pp. 565–571.
34. Qadri, S.F.; Lin, H.; Shen, L.; Ahmad, M.; Qadri, S.; Khan, S.; Khan, M.; Zareen, S.S.; Akbar, M.A.; Bin Heyat, M.B.; et al. CT-Based Automatic Spine Segmentation Using Patch-Based Deep Learning. *Int. J. Intell. Syst.* **2023**, *2023*, 2345835. [CrossRef]
35. Da, Q.; Huang, X.; Li, Z.; Zuo, Y.; Zhang, C.; Liu, J.; Chen, W.; Li, J.; Xu, D.; Hu, Z.; et al. DigestPath: A benchmark dataset with challenge review for the pathological detection and segmentation of digestive-system. *Med. Image Anal.* **2022**, *80*, 102485. [CrossRef]
36. Sirinukunwattana, K.; Pluim, J.P.; Chen, H.; Qi, X.; Heng, P.A.; Guo, Y.B.; Wang, L.Y.; Matuszewski, B.J.; Bruni, E.; Sanchez, U.; et al. Gland segmentation in colon histology images: The glas challenge contest. *Med. Image Anal.* **2017**, *35*, 489–502. [CrossRef]
37. Ronneberger, O.; Fischer, P.; Brox, T. U-net: Convolutional networks for biomedical image segmentation. In Proceedings of the International Conference on Medical Image Computing and Computer-Assisted Intervention, Munich, Germany, 5–9 October 2015; Springer: Berlin/Heidelberg, Germany, 2015; pp. 234–241.
38. Sri, S.V.; Kavya, S. Lung Segmentation Using Deep Learning. *Asian J. Appl. Sci. Technol. AJAST* **2021**, *5*, 10–19. [CrossRef]
39. Pravitasari, A.A.; Iriawan, N.; Almuhayar, M.; Azmi, T.; Irhamah, I.; Fithriasari, K.; Purnami, S.W.; Ferriastuti, W. UNet-VGG16 with transfer learning for MRI-based brain tumor segmentation. *TELKOMNIKA Telecommun. Comput. Electron. Control.* **2020**, *18*, 1310–1318. [CrossRef]
40. Diakogiannis, F.I.; Waldner, F.; Caccetta, P.; Wu, C. ResUNet-a: A deep learning framework for semantic segmentation of remotely sensed data. *ISPRS J. Photogramm. Remote Sens.* **2020**, *162*, 94–114. [CrossRef]
41. Li, X.; Chen, H.; Qi, X.; Dou, Q.; Fu, C.W.; Heng, P.A. H-DenseUNet: Hybrid densely connected UNet for liver and tumor segmentation from CT volumes. *IEEE Trans. Med. Imaging* **2018**, *37*, 2663–2674. [CrossRef]

42. Zhou, Z.; Rahman Siddiquee, M.M.; Tajbakhsh, N.; Liang, J. UNet++: A Nested U-Net Architecture for Medical Image Segmentation. In Proceedings of the Deep Learning in Medical Image Analysis and Multimodal Learning for Clinical Decision Support, Granada, Spain, 20 September 2018; Stoyanov, D., Taylor, Z., Carneiro, G., Syeda-Mahmood, T., Martel, A., Maier-Hein, L., Tavares, J.M.R., Bradley, A., Papa, J.P., Belagiannis, V., et al., Eds.; Springer International Publishing: Cham, Switzerland, 2018; pp. 3–11.
43. Huang, H.; Lin, L.; Tong, R.; Hu, H.; Zhang, Q.; Iwamoto, Y.; Han, X.; Chen, Y.W.; Wu, J. UNet 3+: A Full-Scale Connected UNet for Medical Image Segmentation. In Proceedings of the ICASSP 2020—2020 IEEE International Conference on Acoustics, Speech and Signal Processing (ICASSP), Barcelona, Spain, 4–8 May 2020; pp. 1055–1059. [CrossRef]
44. Chen, L.C.; Papandreou, G.; Kokkinos, I.; Murphy, K.; Yuille, A.L. Deeplab: Semantic image segmentation with deep convolutional nets, atrous convolution, and fully connected crfs. *IEEE Trans. Pattern Anal. Mach. Intell.* **2017**, *40*, 834–848. [CrossRef] [PubMed]
45. Chen, H.; Qi, X.; Yu, L.; Heng, P.A. DCAN: Deep contour-aware networks for accurate gland segmentation. In Proceedings of the IEEE Conference on Computer Vision and Pattern Recognition, Las Vegas, NV, USA, 26 June–1 July 2016; pp. 2487–2496.
46. Wang, X.; Yao, L.; Wang, X.; Paik, H.Y.; Wang, S. Global Convolutional Neural Processes. In Proceedings of the 2021 IEEE International Conference on Data Mining (ICDM), Auckland, New Zealand, 7–10 December 2021; pp. 699–708.
47. Badrinarayanan, V.; Kendall, A.; Cipolla, R. SegNet: A Deep Convolutional Encoder-Decoder Architecture for Image Segmentation. *IEEE Trans. Pattern Anal. Mach. Intell.* **2017**, *39*, 2481–2495. [CrossRef] [PubMed]

bioengineering

MDPI

Article

Exploring the Possibility of Measuring Vertebrae Bone Structure Metrics Using MDCT Images: An Unpaired Image-to-Image Translation Method

Dan Jin [1], Han Zheng [2] and Huishu Yuan [1,*]

1 Department of Radiology, Peking University Third Hospital, Beijing 100191, China; jindan1418@bjmu.edu.cn
2 School of Traffic and Transportation, Beijing Jiaotong University, Beijing 100044, China; hanzheng@bjtu.edu.cn
* Correspondence: huishuy@bjmu.edu.cn

Abstract: Bone structure metrics are vital for the evaluation of vertebral bone strength. However, the gold standard for measuring bone structure metrics, micro-Computed Tomography (micro-CT), cannot be used in vivo, which hinders the early diagnosis of fragility fractures. This paper used an unpaired image-to-image translation method to capture the mapping between clinical multidetector computed tomography (MDCT) and micro-CT images and then generated micro-CT-like images to measure bone structure metrics. MDCT and micro-CT images were scanned from 75 human lumbar spine specimens and formed training and testing sets. The generator in the model focused on learning both the structure and detailed pattern of bone trabeculae and generating micro-CT-like images, and the discriminator determined whether the generated images were micro-CT images or not. Based on similarity metrics (i.e., SSIM and FID) and bone structure metrics (i.e., bone volume fraction, trabecular separation and trabecular thickness), a set of comparisons were performed. The results show that the proposed method can perform better in terms of both similarity metrics and bone structure metrics and the improvement is statistically significant. In particular, we compared the proposed method with the paired image-to-image method and analyzed the pros and cons of the method used.

Keywords: micro-CT-like images; unpaired image-to-image translation; vertebrae; bone structure

Citation: Jin, D.; Zheng, H.; Yuan, H. Exploring the Possibility of Measuring Vertebrae Bone Structure Metrics Using MDCT Images: An Unpaired Image-to-Image Translation Method. *Bioengineering* **2023**, *10*, 716. https://doi.org/10.3390/bioengineering10060716

Academic Editors: Yan Pei, Jijiang Yang and Luca Mesin

Received: 19 May 2023
Revised: 5 June 2023
Accepted: 8 June 2023
Published: 12 June 2023

1. Introduction

Bone mineral density (BMD) tests are now internationally recognized as the primary method of diagnosis for vertebral fragility fractures [1,2]. However, even with standardized image quality requirements, diagnostic criteria and operating manuals, the rate of underdiagnosis of fragility fractures remains high [3–9]. A high rate of underdiagnosis means that patients miss out on the timely treatment of vertebral fractures, which can lead to height loss, kyphosis, chronic back pain and back-related dysfunction and can significantly reduce the chance of survival of patients.

Numerous studies [10–13] have found that changes in bone structure decrease bone quality and increase the risk of fragility fractures, suggesting that bone structure also plays a key role in bone strength. For example, Taes Y et al. [14] concluded that fractures in adult men are associated with a smaller cortical bone area and reduced cortical thickness, but not with bone density. Wehrli FW et al. [15] studied the bone structure of the distal radius and tibia in postmenopausal women and found that changes in bone structure explained 96% of the change in bone strength, with trabecular volume alone explaining 37–67% of the change in bone strength. Koester et al. found that increased cortical porosity may lead to a 75% reduction in proximal femur bone strength and that cortical porosity increases with age [11]. When the trabecular structure deteriorates, the trabeculae decrease in number, thin or even disappear; gaps widen; and trabeculae transform from plate-like to rod-like; these changes increase separation and decrease connectivity, which ultimately lead to significant changes in structure metrics [10–12]. Bone structure includes the macrostructure and microstructure of bone [16]. The macrostructure refers to the geometry and topology of bone, and the microstructure refers to the thickness and spatial distribution of cortical and trabecular bone. For convenience of expression, "bone structure" in this paper refers to the microstructure of bone.

Microcomputed tomography (micro-CT) is the gold standard for measuring bone structure metrics and has a resolution of 10 μm or less. However, it cannot be used to measure bone structure at anatomical sites such as the spine and hip due to the small aperture (<10 cm) and high radiation dose. Multidetector computed tomography (MDCT) can be used routinely for measuring the bone density of human medial bones with calibrated body models, which has a wide range of clinical applications. However, due to the low resolution (approximately 200–500 μm), which is much larger than the average thickness of bone trabeculae, MDCT cannot capture the detailed information of bone trabeculae and therefore cannot support the accurate measurement of bone structure metrics. If the relationship between MDCT and micro-CT images can be obtained using deep learning techniques, it will be possible to generate micro-CT-like images on the basis of MDCT, which in turn enables the measurement of bone structure metrics.

The generation of micro-CT-like images from MDCT images themselves has logical self-consistency. Clustering techniques allow us to observe the structural matchings in MDCT and micro-CT images (as shown in Figure 1). The distribution of bone and bone marrow tissues has an obvious spatial mapping. Therefore, it is reasonable to assume that there is also a hidden relationship between low-resolution MDCT and high-resolution micro-CT images in terms of image structure and detail.

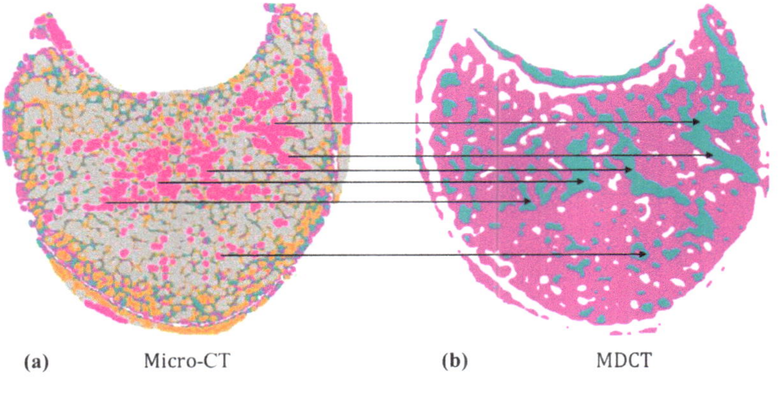

Figure 1. Example of an inherent mapping relationship between micro-CT (**a**) and MDCT (**b**) after the clustering process. Arrows represent the spatial mapping between MDCT and micro-CT images.

Conditional generative adversarial networks (CGANs) [17–21] are currently popular image translation and generation methods. Among these methods, the paired-image-based method has been proven to generate realistic images with sharp details and to have good quantitative performance [22]. Such methods are trained on a paired-image dataset, where an image from the source domain already has a corresponding translated image in the target domain. In the domain of our study, the paired-image-based method requires a large number of paired MDCT and micro-CT images, and finer results can be obtained when a sufficient number of paired samples is obtained. However, this paired dataset requirement imposes a huge practical constraint in the medical field, because micro-CT images can only be obtained from human cadaver specimens. In contrast, the unpaired-image-based method can be trained based on unpaired MDCT and micro-CT images, and the method is less difficult to preprocess than the paired-image-based method.

This paper utilized a method to generate micro-CT-like images from MDCT images using FUNIT [23], a few-shot unpaired-image-based method that enables high-resolution image translation between image domains. This method does not change the clinical scanning technique and measures bone structure metrics that are highly correlated with those of micro-CT images without increasing the cost or radiation dose.

The remainder of the article is organized as follows: in Section 2, we review the history of medical image translations and analyze the need for few-shot unpaired-image-based learning. In Section 3, we systematically present a series of techniques used to measure bone structure metrics. In Section 4, we compare the generation results of the selected method with those of other methods and analyze the properties of unpaired-image-based learning for micro-CT-like image generation.

2. Literature Reviews

For measuring bone structure metrics, image translation methods are used to find associations between MDCT and micro-CT images and generate micro-CT-like images. Such methods have been used and explored in the medical field for numerous applications, such as replenishing missing images [24], cross-scan mode conversion [25], image resolution enhancement [26] and creating labeled datasets [27]. Mathematically, the goal of image translation is to transform the input image x_A from the original domain A to the target domain B, thus acquiring the detailed features of domain B, while preserving features of the source domain. To achieve this goal, a model $G_{A \to B}$ needs to be trained to generate image $x_{AB} \in B$ given the original domain image $x_A \in A$. The generated image cannot be distinguished from the image $x_B \in B$ of the target domain. This process can be expressed as follows:

$$x_{AB} \in B : x_{AB} = G_{A \to B}(x_A) \tag{1}$$

In early studies, translation models $G_{A \to B}(A)$ were implemented via classical image scaling, including four major categories of interpolation methods, frequency domain analysis, instance-based methods and nonlinear learning methods [28–30]. The interpolation methods can be further divided into various specific methods, such as nearest neighbor interpolation [31], bilinear interpolation [31] and bicubic interpolation [32,33]. These methods translate images by filling pixels based on the inter-relationship of pixels after expanding the source image to make the image edges and content clear. Frequency domain analysis methods, such as Fourier sharpening and wavelet denoising, Ref. [34] have also gained wider application in the clinical field [35] and have been applied in low-dose X-ray image resolution enhancement. Example-based methods [36] obtain the relationships between regions to achieve image translation. These methods are good at image translation tasks with regular content, such as the resolution improvement of architectural pictures. In addition, nonlinear learning methods, such as dictionary learning [37] and random forest [38], are used in translating medical images, which are based on features selected by experts. However, manually selected features are limited in their ability to represent complex image information in medical image translations. The aforementioned methods are mainly focused on filling in the pixels of the target image (CT or MRI image), which only ensures a clearer image and makes the boundaries between tissues (i.e., edges or contours) clearer and does not extend and fill in the content or structure details [31,39]. Deep learning [40,41] methods can address this problem by automatically learning features.

Deep learning super-resolution methods [42–45] became popular in medical image translations during 2015 [46]. Convolutional Neural Networks (CNNs) are a dominant class of method [47–50]. CNNs mimic the way the biological visual cortex works [51] and can be simply understood as the extraction of the boundaries between neighboring pixels by using convolutional kernels. Based on these, CNNs were first used for image translation studies within the same scan pattern. Chen et al. [47] proposed a three-layer CNN model to generate relatively high-quality images from low-dose, low-quality CT images of the human body. Chen et al. [48] used a residual CNN model to achieve low-dose CT image resolution enhancement. These studies provided solutions to effectively reduce the radiation dose of CT scans. In the field of MRI, CNN-based image translations have also been used for image resolution enhancement: Zend et al. [49] used ResNet [52] for the resolution enhancement of brain MRI. Chaudhari et al. [50] used a similar approach to study the resolution enhancement of knee MRI images. These studies provided solutions to effectively reduce the scanning time of MRI and lay the technical foundation for the implementation of image post-processing techniques such as 3T to 7T. In addition, more complex CNN models, such as cascaded CNNs [53], have been explored [54] and applied to more complex medical image mapping tasks. For example, Xiang L et al. [55] investigated the conversion method of T1-weighted images to CT images in cranial MRI via a CNN.

However, CNNs tend to use deeper and higher-dimensional models to obtain a larger perceptual field, which makes the model difficult to train and easy to overfit [56]. At the same time, CNN training aims to minimize the loss function, which tends to focus on minimizing the reconstruction error, and the results may have a high peak signal-to-noise ratio and tend to lose high-frequency details [57,58]. This makes CNN-based methods prone to problems such as blurring and noise on edges and detailed textures, and, in general, only able to handle lower-resolution images. The emergence of Generative Adversarial Networks (GANs) [59] and Conditional Generative Adversarial Networks (CGANs) [60] has provided a new solution for this problem, and these networks have achieved promising results [61–63]. These models introduce the concept of adversarial learning based on the powerful feature extraction capability of CNNs and separate the image generation task from the discriminator task to reduce the overall training difficulty.

CGAN-based image translation [18–20,59,64] focuses more on the acquisition of internal mapping relationships between different images [19] and the generation of gold-standard-like images, rather than focusing on simple pixel-based filling or sampling, and tends to be better at content connecting and filling [20]. After years of development, the CGAN and its various derivative models have proven suitable for implementation in image translation and have gained widespread attention [30–34]. These methods have been used [22,50,65,66] in medical imaging. For example, Nie et al. [67] used a cascade GAN technique to implement brain and pelvic MRI to generate corresponding CT images and to accomplish the task of 3T MRI to 7T MRI; Hiasa et al. [68] implemented the process of mapping T1-weighted imaging from the pelvis to the distal femur to CT images via CycleGAN. Dar et al. [69] used the pix2pix (a CGAN-derived model) technique [19] to achieve mapping between T1-weighted images and T2-weighted images.

It is worth noting that CGAN-based image mapping methods can be divided into paired-image-based methods and unpaired-image-based methods. Paired-image-based methods [19,22,53,56] aim to train generators and discriminators based on paired-image training sets to achieve "image-to-image" mapping from the source domain to the target domain, while unpaired-image-based methods [20,23,70] aim to train generators and discriminators to achieve "class-to-class" mapping from the source domain to the target domain based on an unpaired (but containing both the domain and target domain images) training set. Because of this, paired-image-based methods require complex collection and preprocessing for images (images from different image domains need to be collected with the same scan pattern as much as possible, and images need to be paired one by one), while unpaired-image-based methods have relatively simple preprocessing steps and do not require pairing.

Generally, paired-image-based methods can obtain results with high similarity to the gold standard if the training dataset is sufficient [56]. However, the image translation studied in this paper requires in vitro data samples, which are generally collected through cadaver specimens for training and testing. It is difficult to collect large-scale data from various aspects, such as policies, regulations and costs. In addition, a large amount of the CT image pairing work itself is costly, which further hinders the scaling up of paired-image-based methods.

Thus, we need a few-shot unpaired-image-based method that can discover the relationships between MDCT and micro-CT images to capture the overall and local multi-resolution features and achieve the accurate generation of vertebral structure and bone trabecular details in a way that supports the measurement of bone metrics. This is still a challenge for imaging methods with large differences and large image sizes such as MDCT and micro-CT.

3. Methodology

Based on the above discussion, a series of techniques related to image translation were designed in this paper based on a few-shot unpaired-image-based method, and FUNIT [23] was applied as the core. To demonstrate the effectiveness of the chosen method, the unpaired-image-based StarGAN [70] and CycleGAN [20] and paired-image-based pix2pixHD [22,56] methods were selected as the control methods. SSIM and FID metrics and vertebral bone structure metrics, including bone volume fraction (BV/TV), trabecular thickness (Tb.Th) and trabecular spacing (Tb.Sp), were measured to demonstrate the feasibility of measuring vertebrae bone structure metrics using MDCT images. The framework of the methodology is shown in Figure 2.

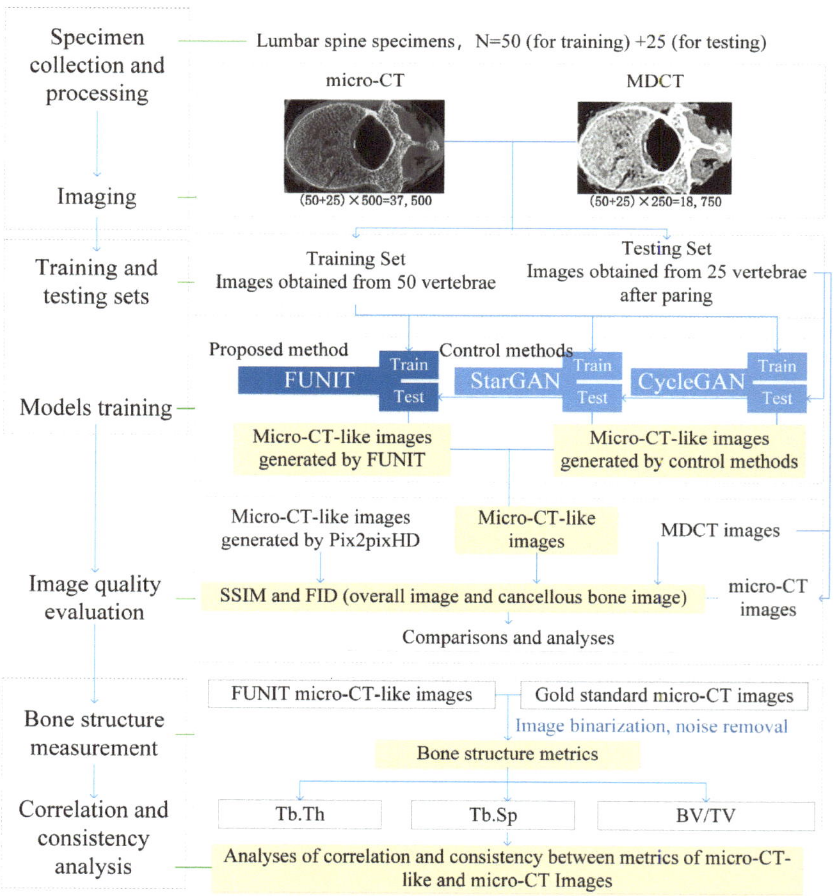

Figure 2. The framework of this study.

This study was an applied basic research study based on scanned images of human cadaveric lumbar spine specimens. The specimens used were from the Department of Anatomy and Research, Faculty of Medicine, Peking University. All donors signed an agreement related to the donation of human remains and agreed that the remains would be used for clinical medical education and research. The study protocol was approved by the Medical Science Research Ethics Committee of Peking University Third Hospital; the ethics number is IRB00006761-M2021179.

3.1. Specimens

In this study, a total of 75 lumbar vertebrae, comprising 15 sets of lumbar spines (L1 to L5), were obtained from 15 formalin-fixed human cadavers (9 males and 6 females; mean age 73 years; age range 62–88 years). These donors had bequeathed their bodies to the local Institute of Anatomy for educational and research purposes, adhering to the relevant institutional and legislative guidelines. Lumbar vertebrae that showed significant compression fractures, bone neoplasms or other substantial bone destruction were excluded from the study. Consequently, all 75 specimens were incorporated into the experiment. The lumbar spine, along with the surrounding muscle, was sectioned into individual segments using a band saw, ensuring the preservation of the pedicle and appendix structures to the greatest extent possible. To minimize trapped gas, the samples were submerged in a phosphate-buffered saline (PBS) solution at 4 °C for a duration of 24 h prior to scanning. The study protocol underwent review and received approval from the local institutional review boards.

3.2. Imaging Techniques

The specimens underwent scanning using both micro-CT (Inveon, Siemens, Erlangen, Germany) and MDCT (SOMATOM Definition Flash, Siemens, Erlangen, Germany) imaging techniques. For micro-CT imaging, the parameters were set at 80 kVp/500 mAs, with a field of view on the $x - y$ plane measuring 80×80 mm^2. A standard matrix size of 1536×1536 pixels was employed, along with 1024 slices at an effective pixel size of 52 μm. The exposure time for each of the 360 rotational steps was 1500 ms. In contrast, the MDCT imaging parameters included 120 kVp/250 mAs, a field of view of 100×100 mm^2, a slice thickness of 0.6 mm, a slice spacing of 0.1 mm, a pitch of 0.8 and a standard matrix size of 512×512 pixels. After excluding images with incomplete, upper and lower endplate views, for all lumbar spine specimens, axial images were captured 1.25 cm above and below the center of the vertebral body. Given that the slice spacing for micro-CT was approximately 0.05 mm and the MDCT slice spacing was approximately 0.1 mm, 500 micro-CT images and 250 MDCT images were captured for each vertebra.

3.3. Few-Shot Unpaired-Image-Based Translation Model for Generating Micro-CT-like Images

The few-shot unpaired-image-based model, FUNIT [23], learns image mapping relationships from unpaired MDCT and micro-CT images. The model simultaneously learns geometric characteristics, internal structures and the distribution of light and dark regions from MDCT images, as well as the detailed texture of bone structures from micro-CT images. After training, the model can generate high-resolution micro-CT-like images with MDCT images as input.

The model mainly consists of two core modules, namely, (1) a structured detail-filled generator G and (2) a multitask adversarial discriminator D. The generator G can extract micro-structure information and generate gold-standard-like images by filling textures, while the discriminator D can discriminate whether the generated image belongs to the target domain. As an unpaired-image-based learning model, the model is designed to translate among multiple types of images. Mathematically, the generator G takes x and K mapping targets $\{y_1, \cdots, y_K\}$ as inputs and outputs generated images $\overline{x}$ with features of K targets.

$$\overline{x} = G(x, \{y_1, \cdots, y_K\}) \tag{2}$$

The low-resolution MDCT is considered to be the input image x. Some high-resolution images such as HR-pQCT [71], micro-CT [72–74], etc., can be treated as the mapping targets $\{y_1, \cdots, y_K\}$. In this paper, we only consider generating micro-CT-like images, so we set $K = 1$, and the micro-CT image is the only y. Thus, Equation (2) can be written as Equation (3).

$$\overline{x} = G(x, y) \tag{3}$$

In Equation (3), the generator G is designed to have the ability to generate micro-CT-like images from MDCT. It consists of three sub-networks, namely, the content encoder E_x, class encoder E_y and decoder F_x, as shown in Figure 3a.

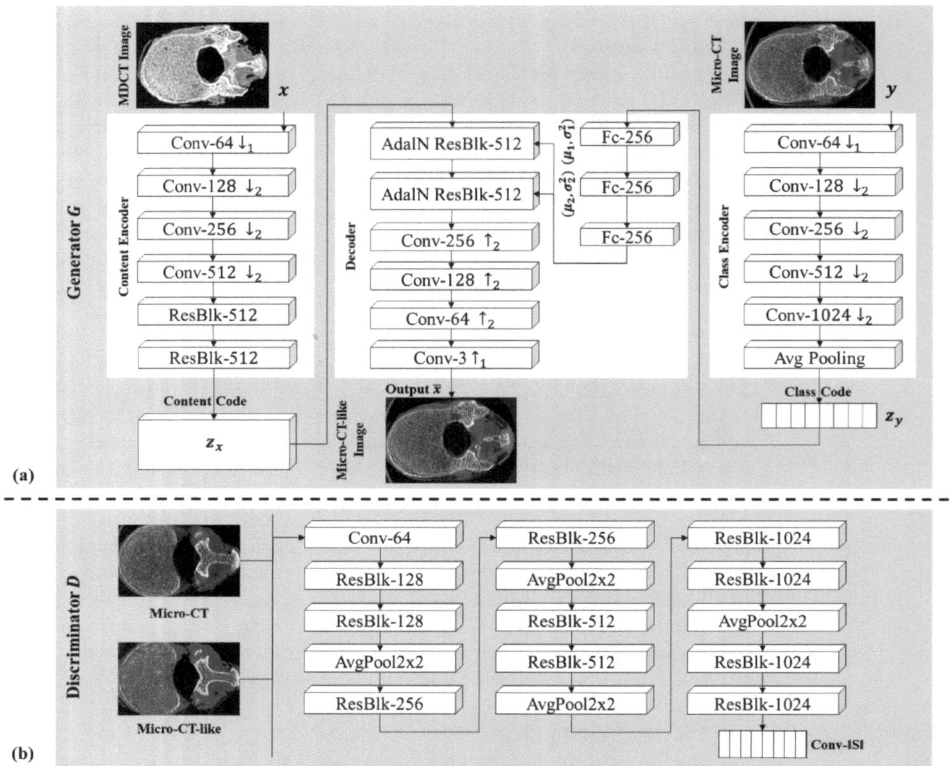

Figure 3. Architecture of (**a**) the generator G and (**b**) the discriminator.

The content encoder E_x is designed to extract texture-independent positional and structural region information, such as the extraction of the vertebral geometry and trabecular layout of the bone. E_x consists of two-dimensional convolutional layers and residual blocks [52,75], and each convolutional layer has normalized functions and ReLU nonlinear functions. The feature maps are scaled by a factor of 2 in each spatial dimension using the nearest-neighbor up-sampling method. The input MDCT image is mapped into a spatial feature map z_x by a 3-stride-2 down-sampling operation.

The class encoder E_y mainly extracts detailed characteristics such as bone trabeculae texture and alignment. It consists of several two-dimensional convolutional layers, which are then averaged along the sampling axis. E_y maps the micro-CT images to a class latent code for describing the texture characteristics of bone trabeculae. This process uses a VGG [57] network to map each input class image to a class latent code z_y. Afterwards, the class latent code is fed to the decoder F_x through the AdaIN layer, where E_y can control detailed characteristics (e.g., texture) and E_x can determine regional characteristics (e.g., the location of regions with different trabecular characteristics). This enables the generation of bone structure details on the basis of reasonable correspondence between MDCT and micro-CT.

The decoder F_x takes latent code z_y as input and obtains a set of mean and variance $(\mu_i, \sigma_i^2) i = 1,2$ through two fully connected networks. These values are then used as affine transformation parameters in the AdaIN residual block, where the σ_i^2s are the scaling factors and the μ_is are the biases [76]. For each residual block, the same affine transformation is applied to each spatial location in the feature map. The affine transformation is spatially invariant and therefore can only be used to obtain global appearance information, which controls how the content is potentially encoded for decoding to generate the output image.

According to the above design, the generator G can map the input MDCT image x to the output micro-CT-like image $\overline{x}$ such that $\overline{x}$ looks like an image belonging to the class c_y of gold-standard micro-CT images, and $\overline{x}$ and x have structural similarity.

The chosen discriminator D is a patch discriminator [19]. This discriminator applies a Leaky ReLU nonlinear activation function and consists of a convolutional layer and 10 activated residual blocks without normalization [77]. The architecture of the discriminator is shown in Figure 3b. It consists of Conv-64 → ResBlk-128 → ResBlk-128 → AvgPool2x2 → ResBlk-256 → ResBlk-256 → AvgPool2x2 → ResBlk-512 → ResBlk-512 → AvgPool2x2 → ResBlk-1024 → ResBlk-1024 → AvgPool2x2 → ResBlk-1024 → ResBlk-1024 → Conv-||S||, where ResBlk-X denotes the residual block of output size X × X [52] and ||S|| is the number of mapped target image classes, which is two in this study, namely, MDCT and micro-CT images.

3.4. Training and Testing

3.4.1. Training Process

The training process of the FUNIT model is a process of solving the minmax optimization problem with the objective function of:

$$\underset{G\quad D}{minmax}\mathscr{L}_{GAN}(D,G) + \lambda_R\mathscr{L}_R(G) + \lambda_F\mathscr{L}_F(G) \tag{4}$$

where $\mathscr{L}_{GAN}$, $\mathscr{L}_R$ and $\mathscr{L}_F$ are the GAN loss function, the loss function of the reconstructed input image with the original input domain and the feature matching loss function, respectively. These functions are defined as follows:

$$\mathscr{L}_{GAN}(D,G) = \mathbb{E}_x[logD^{c_x}(x)] + \mathbb{E}_{x,\{y_1\}}[log(1 - D^{c_y}(G(x,\{y_1\}) = \overline{x}))] \tag{5}$$

where $D(x)$ is a discriminant probability distribution of a discriminator expressing the probability of classifying x as a target gold-standard image, rather than a generated gold-standard-like image, and the superscript indicates the type of target discriminated. That is, $D^{c_x}(x)$ expresses the ability to discriminate the input image as an MDCT image, while $D^{c_y}(\overline{x})$ is the ability to discriminate the generated gold-standard-like image as a micro-CT image, and $1 - D^{c_y}(\overline{x})$ expresses the ability to discriminate the generated gold-standard-like image and not discriminate it as a micro-CT image.

Thus, $\mathscr{L}_{GAN}(D,G)$ expresses the ability of the model to discriminate the input image as an MDCT image and the generated class image as not a micro-CT image. For the discriminator D, the input should be discriminated as an MDCT image and the generated gold-standard-like image should be discriminated as not a micro-CT image as much as possible, so this ability is as large as possible and is taken as max; meanwhile, for the generator G, this ability is as small as possible and is therefore taken as min.

In addition, $\mathscr{L}_R$ can help train the generator G model for image mapping. Specifically, when using the same MDCT image as the input image and the mapped target image (in this case, $K = 1$), this loss function encourages G to produce an output image that is identical to the input MDCT.

$$\mathscr{L}_R(G) = \mathbb{E}_x\left[\|x - G(x,\{x\})\|_1^1\right] \tag{6}$$

The $\mathscr{L}_F$ provides a normalization ability to the training. By removing the last layer of the discriminator D, a feature extractor D_f is obtained. Using D_f to extract features from the class micro-CT image $\overline{x}$ and micro-CT image $\{y_1\}$, respectively, and minimize their differences, we have:

$$\mathscr{L}_F(G) = \mathbb{E}_{x,\{y_1,\cdots,y_K\}}\left[\left\|D_f(\overline{x}) - \textstyle\sum_k \frac{D_f(y_k)}{K}\right\|_1^1\right] \tag{7}$$

The proposed model was trained on a Windows 10 workstation equipped with two Nvidia A6000 GPUs. In the training process, the discriminator G randomly draws two images $c_x, c_y \in S$ and $c_x \neq c_y$ from different classes of source images (MDCT and micro-CT) and performs mapping training to finally obtain the ability to generate micro-CT-like and MDCT-like images. We used the default hyper-parameters of FUNIT for training but changed the image sizes to fit MDCT and micro-CT images.

3.4.2. Image Pairing Method for Testing

In order to test the performance of the model, a ground-truth image pair set was needed. The scheme for preparing the ground-truth image pair set is as follows:

1. Image matching: The scale invariant feature transform (SIFT) algorithm [78] was used to find coupling key points in MDCT and micro-CT images. We calculated the Euclidean distance between key points and set the mean value to be the distance between MDCT and micro-CT images (Figure 4). Based on this, we compared MDCT and micro-CT images one by one and constructed the matrix of distances between all MDCT and micro-CT images. The best matched image pair could be obtained via the dynamic time warping (DTW) algorithm [79].
2. MDCT image amplification and image pair generation: Due to the different layer spacing between the two scanning methods, MDCT images and micro-CT images of the same specimen are not equal in overall number, and approximately two layers of micro-CT images correspond to one layer of MDCT images. Therefore, the MDCT images of each vertebra needed to be replicated (250 × 2) according to the matching relationship to obtain one-to-one paired-image pairs of MDCT and micro-CT images, i.e., 500 image pairs were generated for each vertebral specimen. Applying the above method to all 25 vertebrae in the test set, a total of 25 × 500 = 12,500 image pairs could be obtained.

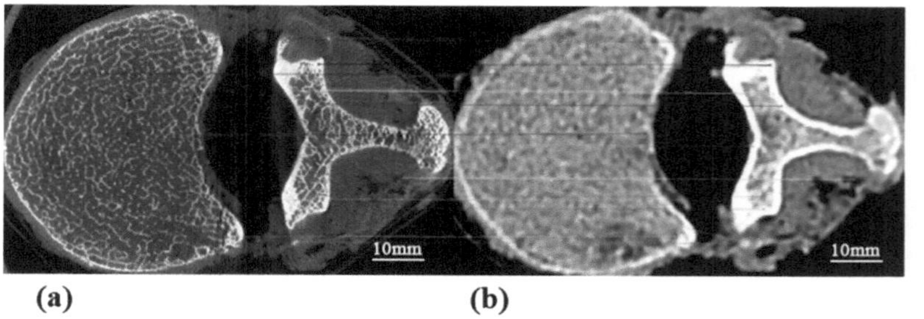

Figure 4. Image matching for micro-CT and MDCT. (**a**) is the micro-CT image, (**b**) is the MDCT image, and the similarity between all micro-CT and MDCT images can be calculated using the average distance of coupling key points. Different colored lines indicate the coupling relationship between key points.

3.5. Assessment Methods

3.5.1. Similarity Metrics

To evaluate the similarity between two images, this study employed the structural similarity (SSIM) [80] and Fréchet inception distance (FID) [81] metrics. The SSIM is designed to evaluate similarity with respect to structure, where a higher SSIM value signifies greater similarity between images [82]. Conversely, the FID metric focuses on evaluating similarity in terms of details, with a lower FID value indicating a higher degree of similarity between images [83]. The definitions of SSIM and FID can be found in the research [22].

3.5.2. Born Structure Metrics

The trabecular microstructure analysis in micro-CT and micro-CT-like images was conducted by employing the BoneJ plug-in [83] within the Fiji (Version 1.53t) software [83]. Utilizing Fiji, which represents a distribution of ImageJ2 developed by the National Institutes of Health [84,85], both micro-CT and micro-CT-like images of vertebrae were processed as 8 bit stack maps. Then, the gray-level images from micro-CT and micro-CT-like sources were binarized into bone and marrow phases by implementing the IsoData algorithm [86], a global thresholding technique. Following this binarization, metrics were computed, including bone volume fraction (BV/TV), trabecular thickness (Tb.Th) and trabecular spacing (Tb.Sp). BV/TV was derived via simple voxel counting, whereby all the foreground voxels were counted and assumed to represent bone and then compared to the total number of voxels in the image. Tb.Th and Tb.Sp were calculated without model assumptions and measured directly by taking foreground voxels as trabeculae and background voxels as spacing [87].

Continuous axial images were required to form a cylindrical volume of interest (VOI) to measure the bone structure metrics. After training the model, all original MDCT images of the 25 vertebrae from the test set were inputted into the model to obtain continuous micro-CT-like images. Subsequently, two cylindrical VOIs (approximately 15 mm in diameter and 5 mm in height) for each vertebra were selected in both the micro-CT and micro-CT-like images. The positioning of the VOI can be found in Figure 5. Identical VOI settings were applied to the MDCT images in order to measure bone structure metrics to serve as a control group.

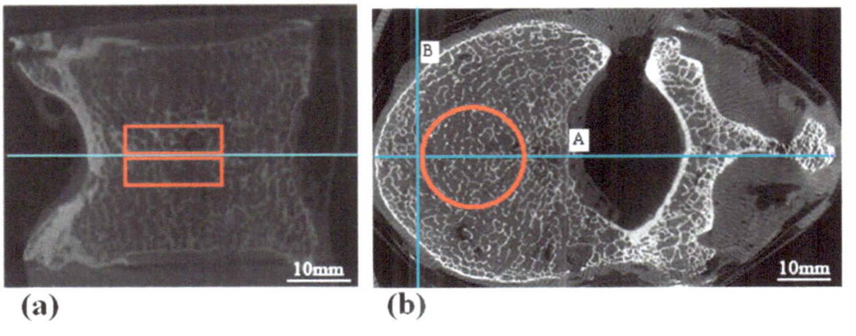

Figure 5. Cylindrical volume of interest (VOI) selection method. (**a**) The sagittal position of the VOI is shown on the vertebral body micro-CT sagittal image via the two areas located 5 mm above and 5 mm below each of the vertebral body sagittal midline. (**b**) The axial position of the VOI is shown on the vertebral body micro-CT axial image. Line A denotes the centerline of the short axis of the vertebral body axial map, line B is perpendicular to line A and the intersection of line A and line B is located 5 mm within the intersection of line A and the anterior edge of the vertebral body. A cylindrical VOI with a diameter of 15 mm was taken with the intersection of line A and line B as the tangent point.

4. Results

4.1. Training Results

Figure 6 shows the process of training by showing metrics of one slice of vertebra in different epochs. After approximately 8000 epochs of learning, the change in the image metrics slowed and stabilized. The figure shows several representative points in the training process, which can be used to observe the learning process of the model for generating micro-CT-like images. The model first learns the contour information of MDCT, starts from the range area, gradually adds the bone cancellous and bone cortical information, and gradually fills in the details of the internal trabecular structure. In the initial stage of the generation process, there are vacant areas, and as the training epoch increases, the vacant areas are gradually reduced and the details of the images are gradually clarified.

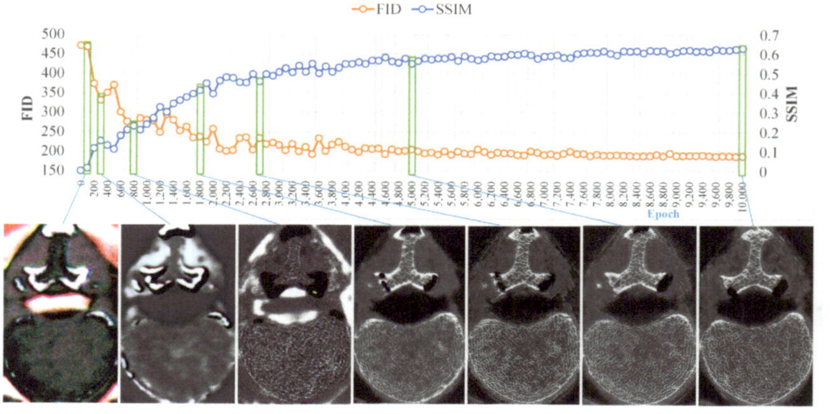

Figure 6. The training effect of the FUNIT model of one slice of vertebra in different epochs. The biaxial line graph at the top of the figure shows the trends of the SSIM value and FID value during training, and the vertebral body images at the bottom are the vertebral micro-CT-like images corresponding to the learning epoch.

After training, MDCT images from the test set were input into the unpaired-image-based model to obtain micro-CT-like images. Figures 7 and 8 show examples of micro-CT-like images.

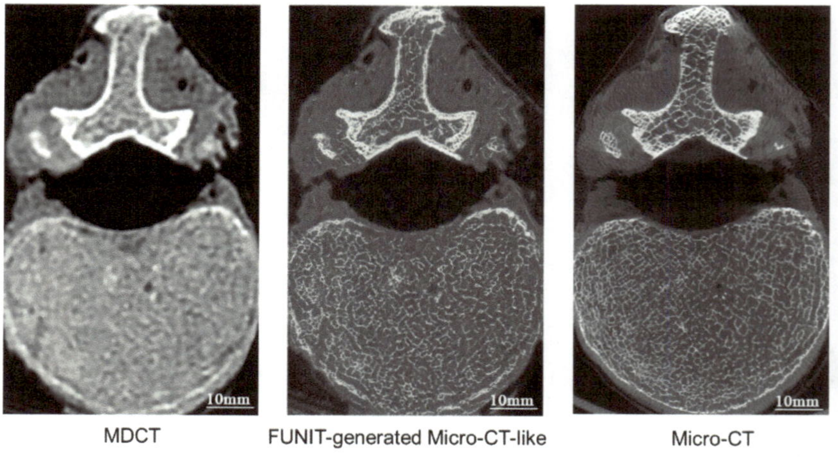

Figure 7. Comparison of FUNIT micro-CT-like image generated from FUNIT, MDCT image and micro-CT image.

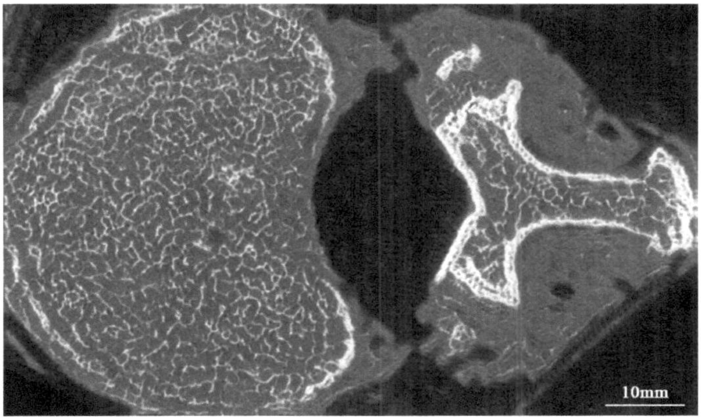

Figure 8. Example of a FUNIT-generated micro-CT-like image.

Although the micro-CT-like images have sufficient similarity with gold-standard images, the micro-CT like images have some shortcomings: (1) there is an obvious "double-border phenomenon" in the bone cortex, i.e., the phenomenon of bone cortex delamination; (2) there is a lack of bone cortex on the surface of the vertebral canal; (3) there is a localized trabecular texture in the peripheral soft tissue of the vertebral body; (4) there is a dense area of bone trabeculae in the cancellous bone, and there is an overfilling of bone trabeculae.

4.2. Comparison of SSIMs and FIDs for Generated Images

Statistical methods were used to more rigorously determine whether the metrics were significantly different. The normality of all continuous variables was verified using the Kolmogorov–Smirnov test, and chi-squaredness was verified using the Levene test. The Friedman test was used to compare the differences in SSIM and FID values between the images generated using FUNIT, StarGAN and CycleGAN, the original MDCT images and the gold-standard micro-CT images. The Mann–Whitney U test was used to compare the SSIM and FID differences between the FUNIT model and the pix2pixHD model for the micro-CT-like images. The differences in bone structure metrics between the FUNIT micro-CT-like and gold-standard micro-CT images were analyzed using paired *t*-test datasets. Linear regression was used to analyze the correlation between bone structure metrics between the FUNIT micro-CT-like and gold-standard micro-CT images. The Z-test was used to compare differences in correlation coefficients between bone structure metrics between FUNIT micro-CT-like, pix2pixHD micro-CT-like images [22] and micro-CT and MDCT images. Intraclass correlation coefficients (ICCs) were used to analyze the consistency between bone structure metrics between FUNIT micro-CT-like and gold-standard micro-CT images. The above statistical analyses were performed using SPSS 26.0 (SPSS Inc., Chicago, IL, USA) and MedCalcv10.002 (Ostend, Belgium) software, and differences were considered statistically significant if the two-sided p value < 0.05. Since the vertebral body consists of cancellous and cortical bone, both of which are of interest for bone strength, we compared the quality of generated images by considering the overall image and local cancellous bone image separately.

4.2.1. Comparing Generated Micro-CT-like Images with MDCT Images

In terms of overall images, using the micro-CT image as the gold standard, the mean values of SSIM between gold-standard images and the micro-CT-like images generated by using three unpaired-image-based models (i.e., FUNIT, StarGAN and CylceGAN) were greater than the SSIM values between the gold-standard and MDCT images, and the differences were statistically significant ($p < 0.001$). Similarly, using micro-CT as the gold standard, the FID values of the generated images were all smaller than the FID values of MDCT. The differences were statistically significant ($p < 0.001$), and these results are shown in Table 1 and Figure 9. Based on these, we found that the micro-CT-like images generated using the three unpaired-image-based models were more similar to the gold-standard images than the original MDCT images in terms of macro-structure and detailed micro-structure. Among the three unpaired-image-based models, the metrics (both SSIM and FID) of the micro-CT-like images generated using FUNIT were better than those of the other two comparison models, and the differences were statistically significant ($p < 0.001$).

Table 1. SSIM and FID values of the four sets of images and the gold-standard micro-CT images.

Scale	Metrics	MDCT	FUNIT	StarGAN	CycleGAN	*p*-Value †
Overall image	SSIM	0.238 ± 0.031	0.519 ± 0.030	0.437 ± 0.025	0.377 ± 0.035	<0.001 ***
	FID	453.425 ± 39.081	201.737 ± 15.031	289.503 ± 18.037	347.311 ± 25.051	<0.001 ***
Localized cancellous bone images	SSIM	0.213 ± 0.052	0.714 ± 0.023	0.589 ± 0.031	0.508 ± 0.037	<0.001 ***
	FID	495.024 ± 54.435	83.696 ± 11.022	175.531 ± 17.035	219.559 ± 16.033	<0.001 ***

Note: † The Friedman test was used to test the differences in metrics among the four sets of images; *** indicates the corresponding image quality evaluation indicators compared between groups have $p < 0.001$.

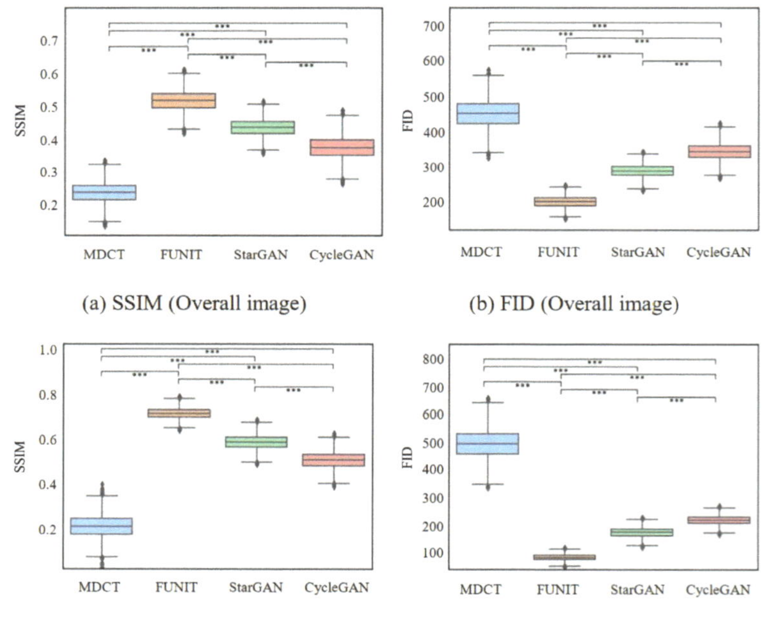

Figure 9. SSIM and FID values of the MDCT and three generated images. The Friedman test was used to test the differences in the metrics between the four groups of images, *** represents $p < 0.001$.

In terms of localized cancellous bone images, the mean values of SSIM and FID of generated micro-CT like images generated by the three unpaired-image-based models improved compared with the values of the overall image. Additionally, FUNIT performed better than the other two methods in SSIM and FID, with statistically significant differences ($p < 0.001$).

4.2.2. Comparison of Micro-CT-like Images Generated Using Unpaired-Image-Based FUNIT Model and Paired-Image-Based pix2pixHD Model

In terms of both the overall image and the local cancellous bone image, the SSIM and FID values of the FUNIT-generated micro-CT-like images were better than the correlation values of the pix2pixHD-generated micro-CT-like images, and the differences were statistically significant ($p < 0.001$). These results are shown in Table 2 and Figure 10.

Table 2. Comparison of the micro-CT-like images generated using the FUNIT model and pix2pixHD model.

Scale	Metrics	FUNIT	pix2pixHD [22]	*p*-Value †
Overall image	SSIM	0.519 ± 0.030	0.804 ± 0.037	<0.001 ***
	FID	201.737 ± 15.031	43.598 ± 9.108	<0.001 ***
Localized cancellous bone images	SSIM	0.714 ± 0.023	0.849 ± 0.021	<0.001 ***
	FID	83.696 ± 11.022	31.724 ± 10.021	<0.001 ***

Note: † The Mann–Whitney U test was used to verify the differences in metrics between micro-CT-like images generated using FUINT and pix2pixHD. *** indicates $p < 0.001$.

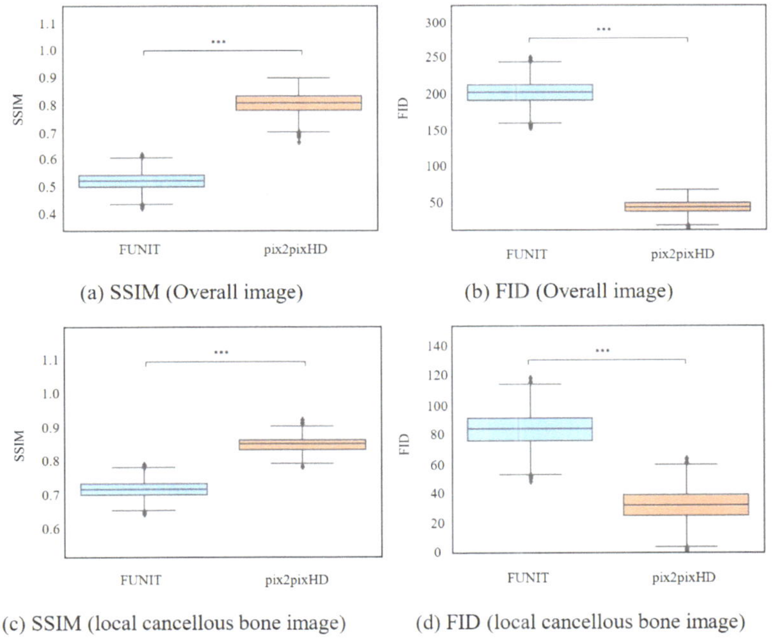

Figure 10. SSIM and FID values of the generated images of paired-image-based pix2pixHD and unpaired-image-based FUNIT. The Mann–Whitney U test was used to test the differences in metrics between the images. *** represents $p < 0.001$.

4.3. Correlation and Consistency of Bone Structure Metrics between Generated Micro-CT-like and Gold-Standard Micro-CT Images

4.3.1. Correlation of Bone Structure between FUNIT-Generated Micro-CT-like and Gold-Standard Micro-CT Images

The bone structure metrics of FUNIT-generated micro-CT-like and gold-standard micro-CT images with their correlations are shown in Table 3. The correlation values of BV/TV and Tb.Th of FUNIT-generated micro-CT-like images were smaller than those of the gold standard, while the Tb.Sp was larger than that of the gold standard, and the difference was statistically significant ($p < 0.001$). Linear regression equations for bone structure metrics of FUNIT-generated micro-CT-like and micro-CT images were: BV/TV: $y = 0.935x - 0.025$; Th.Th: $y = 1.078x - 0.076$ and Tb.Sp: $y = 1.029x + 0.182$, with $R^2_{(FUNIT)}$, and the F values are shown in Table 3. The BV/TV, Tb.Th and Tb.Sp values of FUNIT-generated micro-CT-like images were highly correlated with those of the gold standard, and the correlation was significant ($p < 0.001$).

Table 3. Bone structure metric values and correlation between FUNIT-generated micro-CT-like and micro-CT images.

N=50	FUNIT Micro-CT-like	Micro-CT	*p*-Value †	R^2	F-Value	*p*-Value ‡
BV/TV (%)	0.143 ± 0.018	0.180 ± 0.016	<0.001 ***	0.667	96.102	<0.001 ***
Tb.Th (mm)	0.158 ± 0.021	0.218 ± 0.015	<0.001 ***	0.613	78.69	<0.001 ***
Tb.Sp (mm)	1.144 ± 0.166	0.934 ± 0.126	<0.001 ***	0.603	75.573	<0.001 ***

Note: † Paired *t*-test was used to compare the difference between the two groups of bone structure metrics, *** represents $p < 0.001$. ‡ Linear regression was used to analyze the correlation between the two groups of bone structure metrics, *** represents $p < 0.001$.

4.3.2. Consistency between Bone Structure Metrics of FUNIT Micro-CT-like and Gold-Standard Micro-CT Images

The ICC values of the bone structure metrics of FUNIT-generated micro-CT-like and gold-standard micro-CT images are shown in Table 4. The FUNIT-generated micro-CT-like bone structure metrics are highly consistent with those of the gold standard.

Table 4. ICC values of bone structure metrics of FUNIT-generated micro-CT-like and gold-standard micro-CT images.

	Bone Structure Metrics	ICC	95% CI	*p*-Value
micro-CT-like	BV/TV	0.809	0.887~0.686	<0.001
(FUNIT). vs.	Tb.Th	0.752	0.852~0.601	<0.001
micro-CT	Tb.Sp	0.753	0.852~0.603	<0.001

4.4. Discussion

4.4.1. Characterization of the Proposed Method

From both the overall image and the local cancellous bone image, the SSIM values of the micro-CT-like images generated using the three unpaired-image-based methods were greater than those of MDCT, and the FID values were smaller than those of MDCT ($p < 0.001$). The micro-CT-like images generated using the unpaired-image-based methods had more obvious improvements in structure and details than the original MDCT images, and the generated micro-CT-like images were more similar to the gold-standard images. Comparing the results of three unpaired-image-based models, we found that the FUNIT method had larger SSIM values and smaller FID values than the other two unpaired-image-based models ($p < 0.001$), indicating that the FUNIT method had the best model performance in the image mapping process among the three groups of models.

FUNIT focuses on generating structured images and uses a more systematic generator design, which consists of three main parts: a content encoder and a class encoder and decoder. The content encoder extracts information from MDCT that is not related to detailed texture but highly relevant to the location and regional structure, such as the structure of each region in cancellous bone and the macro layout of bone trabeculae. Then, a content feature code is generated after extraction. The class encoder learns location-independent bone trabeculae detail information from micro-CT, including texture, alignment, etc. The class specific features are generated after extraction [23]. The model simultaneously learns the mapping relationship between MDCT and micro-CT and finally fuses the class features with the content features on the decoder to form micro-CT-like images. Thus, hidden information such as bone material and bone marrow distribution in MDCT is extracted, and bone trabeculae texture is attached to form micro-CT-like images. By the judgment of the discriminator, the formed micro-CT-like image will have the characteristics of the bone trabecular structure in micro-CT. For this reason, FUNIT can perform better in the environment studied in this paper and generate micro-CT-like images that exceed those of other unpaired-image-based methods in quality.

Although the FUNIT model used can generate micro-CT-like images that are more similar to the gold standard than the other two methods, the generated images still have deficiencies. From the results, the SSIM value of the image of the cancellous bone portion of FUNIT-generated micro-CT-like was improved compared to the overall image, indicating that the cancellous bone region was more similar to the gold standard, while there were some problems in the outer contour of the vertebral body, i.e., the bone cortex. Figure 11 shows an example of FUNIT-generated micro-CT-like images, and the problems with the images are shown specifically in the red box in Figure 12. First, there is a clear "double-border phenomenon" in the bone cortex, where the originally compact bone cortex is filled with two or more layers of thin linear bone cortex. The possible reason for this phenomenon is that the model focuses on cancellous bone features when generating the images, and the whole image is filled with the structural pattern of bone trabeculae, so the bone cortex on the MDCT image is replaced by multiple near-parallel bone trabeculae textures.

Additionally, there is a problem of loss of bone cortex in specific areas, especially in the vertebral canal surface where the bone cortex is prone to defects and disruption of continuity, which in turn leads to a situation where the boundary between the bone tissue and the surrounding soft tissue is unclear.

Furthermore, short trabecular texture-like shadows of bone trabeculae appear within the peripheral soft tissues of the vertebral body. This is due to texture within the soft tissues being mistaken for bone trabeculae in MDCT: soft tissues with discrepancies in CT values may be misidentified as bone tissue and then filled. However, this phenomenon is not widespread and does not have an impact on bone structure studies.

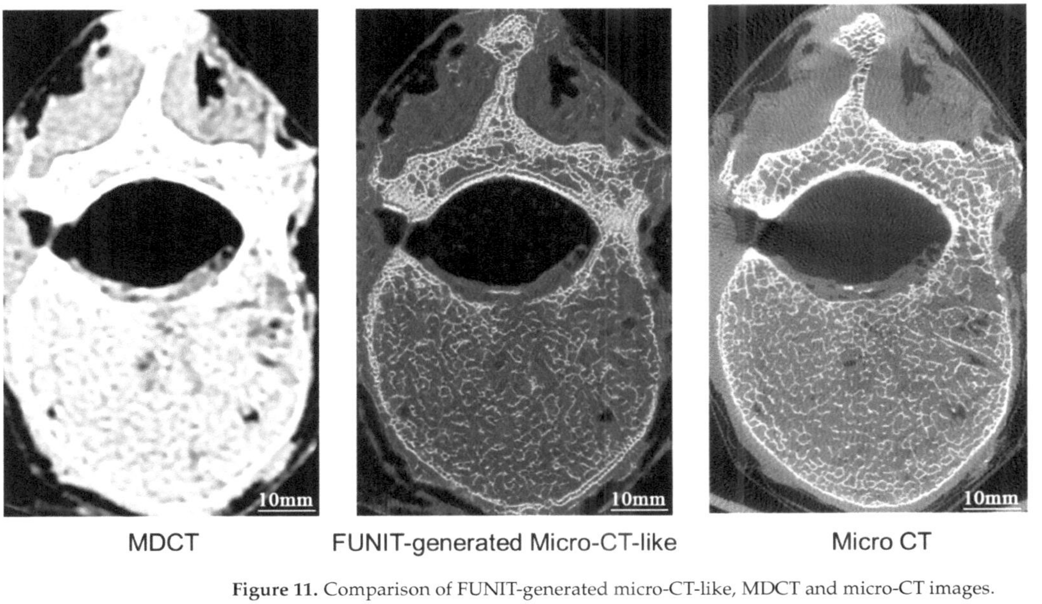

Figure 11. Comparison of FUNIT-generated micro-CT-like, MDCT and micro-CT images.

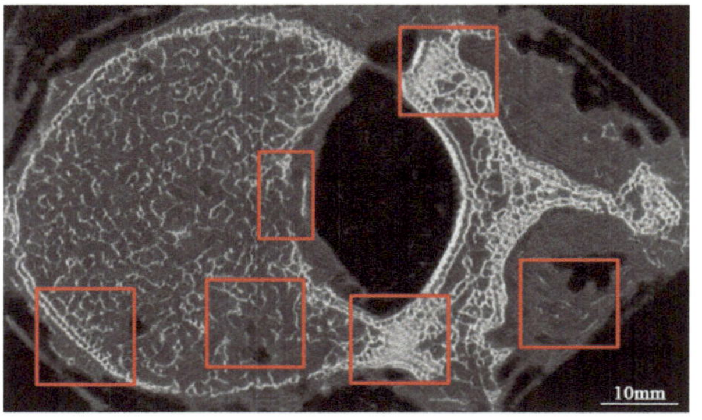

Figure 12. Examples of defects in micro-CT-like images generated via FUNIT. The red boxes are the areas where anomalies exist.

Finally, in the case of vertebral cancellous bone, if there are relatively dense areas in the cancellous bone, FUNIT will overfill the relatively dense areas to a certain extent during the mapping process, as shown by the local thickening of the trabeculae. In contrast, the relatively sparse areas are underfilled, which is reflected by the local absence and thinning of trabeculae.

Although there are some issues in the micro-CT-like images generated via FUNIT, none of them are distributed in core regions of cancellous bone. This is the reason why the localized SSIM and FID values were better than the overall SSIM and FID values. Since cancellous bone is important for the diagnosis of osteoporosis, it can be assumed that the studied FUNIT method meets the requirements of bone structure analysis.

The BV/TV and Tb.Th of FUNIT-generated micro-CT-like images were smaller than those of the gold standard, and the differences were statistically significant ($p < 0.001$). The Tb.Sp of FUNIT-generated micro-CT-like images was greater than that of the gold standard, and the difference was statistically significant ($p < 0.001$). All measured bone structural metrics were moderately correlated with the gold standard (BV/TV: $R^2_{(FUNIT)} = 0.667$, Th.Th: $R^2_{(FUNIT)} = 0.613$, Tb.Sp: $R^2_{(FUNIT)} = 0.603$), the correlation was higher than that of MDCT (BV/TV: $R^2_{(MDCT)} = 0.367$, Th.Th: $R^2_{(MDCT)} = 0.275$, Tb.Sp: $R^2_{(MDCT)} = 0.283$) and the differences were statistically significant. The ICC results showed that acceptable consistency existed between the generated images and the gold standard. However, the smaller BV/TV and Tb.Th and larger Tb.Sp imply that the trabeculae are broken, missing, or unfilled during the mapping process, resulting in wider spacing and a relative decrease in bone volume fraction. This situation may occur because FUNIT is obtained by finding structures in MDCT and later adding details similar to those in micro-CT images to obtain micro-CT-like images. If the structure in MDCT is not very obvious, details are easily missed and the results of its generated images will be biased toward conservatism. On the other hand, the unpaired-image-based method learns the structure in MDCT corresponding to the texture feature in the micro-CT image, and this feature is not learned one-to-one, meaning that unreasonable bone trabeculae orientation, etc., may occur when filling the details. This result may lead to a reduction in predicted bone strength compared to actual bone strength when FUNIT-generated micro-CT-like images are eventually used to predict bone strength, which in turn may lead to an increased false-positive rate in fracture risk prediction. The further optimization of model parameters and increased sample diversity are needed in subsequent studies to remedy this deficiency.

4.4.2. Paired-Image-Based pix2pixHD Model versus Unpaired-Image-Based FUNIT Model

By comparing the SSIM with the FID index, as well as the direct sample shown in Figure 13, we found that the pix2pixHD-micro-CT-like images were more similar to the gold standard than the FUNIT-micro-CT-like images. FUNIT generates less of the bone cortex and is prone to problems such as the "double-border phenomenon" on the bone cortex, missing bone cortex and trabecular texture in the soft tissue. In contrast, the bone cortex of pix2pixHD-generated images is more similar to that of the micro-CT gold standard, with a tighter and more continuous bone cortex and a clear boundary with the soft tissues. As analyzed, this is related to the training mechanism of FUNIT and pix2pixHD, which adopts a "class-to-class" learning model and has a certain tendency to "imagine" in the filling process, i.e., it uses the local information of MDCT for generation. In contrast, the pix2pixHD method adopts an "image-to-image" learning mode, and its "imagination" capability is more convergent; consequently, the mapping results are more realistic, which is one of the advantages of paired-image training. However, pix2pixHD-generated micro-CT-like images also have the problem of overfilling and noise formation in dense and complex bone areas such as attachments. Although there are still some shortcomings in the texture details of both methods, such as reduced local trabecular definition and less natural alignment, which make the measured bone structure metrics not fully consistent with those of the gold standard, there is sufficient correlation between the bone structure metrics and those of the gold standard.

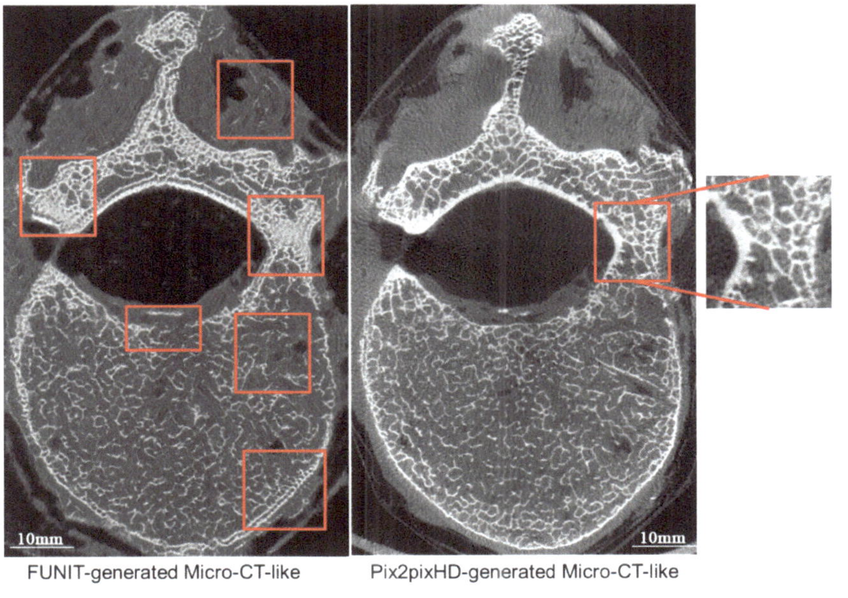

Figure 13. Comparison of micro-CT-like images generated using the paired-image-based pix2pixHD method with those generated using the unpaired-image-based FUNIT method. The red boxes are the areas where anomalies exist.

Unpaired-image-based learning does not require paired images due to its learning mechanism, and it has a greater ability to generalize. The model can find the structural features embedded in MDCT images and find their mapping relationships with micro-CT images to make certain associations and add detailed textures. This property allows the model to transform images to a limited degree even when it encounters MDCT input data of a vertebra type that has not appeared before, making the trained model somewhat robust.

5. Conclusions

As the population ages and life expectancy continues to increase, the incidence of fragility fractures has increased significantly. Therefore, the early identification of fragility fracture risk is critical. In addition, as the age of the population treated with spinal instrumentation increases, clinicians need to pay more attention to bone strength profiles to develop individualized surgical plans and reduce the probability of postoperative complications. BMD cannot fully explain changes in bone strength alone, so it is extremely important to analyze a diversity of bone structure metrics. The primary focus of this study is to investigate the possibility of measuring vertebrae bone structure metrics using MDCT images, of which the core task is establishing a mapping relationship between vertebral MDCT images and micro-CT images using deep learning methods to generate micro-CT-like images based on MDCT images.

From the perspective of computer image science, mapping two images with vastly different resolutions, such as MDCT and micro-CT images, remains an open research challenge. The emergence of CGANs and their derived models has made this feasible [17]. In this study, the above image mapping task was achieved by finding nonlinear feature associations between vertebral MDCT and micro-CT images through the unpaired-image-based FUNIT method.

The bone structure metrics measured using micro-CT-like images are highly correlated with those obtained from the gold standard of micro-CT images. The used method can fully utilize the potential of MDCT images and provides a technical methodological possibility to realize in vivo vertebral bone structure measurement. In terms of image translation, this paper discusses the presence of some phenomena (e.g., the double-border phenomenon), but it mainly focuses on the qualitative discussion. Quantitative description methods of these phenomena should be explored in depth in the future. In terms of model training, although it is currently in the preliminary exploratory stage using a small sample of in vitro vertebral specimens, the deep learning model can be further optimized, and its generalization capability can be improved in the future through measures such as expanding the sample size, increasing sample diversity, and simulating in vivo environments. More detailed and systemic clinical evaluations should be conducted in the future.

Author Contributions: Conceptualization, D.J.; Data Curation, D.J.; Funding Acquisition, H.Y.; Methodology, D.J. and H.Z.; Resources, H.Z. and H.Y.; Supervision, H.Y.; Validation, H.Z.; Writing—Original Draft, D.J.; Writing—Review and Editing, D.J. and H.Y. All authors have read and agreed to the published version of the manuscript.

Funding: This research was supported by the National Natural Science Foundation of China [Grant No. 82171927], the Beijing Natural Science Foundation [Grant No. 7212126] and the Beijing New Health Industry Development Foundation [Grant No. XM2020-02-006].

Institutional Review Board Statement: The study was conducted according to the guidelines of the Declaration of Helsinki and approved by the Institutional Review Board (or Ethics Committee) of Peking University Third Hospital Medical Science Research Ethics Committee (protocol code IRB00006761-M2021179, 9 April 2021).

Informed Consent Statement: Informed consent was waived due to the donors having dedicated their bodies for educational and research purposes to the local Institute of Anatomy prior to death, in compliance with local institutional and legislative requirements.

Data Availability Statement: Data sharing not applicable.

Acknowledgments: We appreciate the support from Beijing Key Laboratory of Spinal Disease Research, Peking University Third Hospital for providing micro-CT scanning and the Department of Anatomy, Peking University Health Science Center for providing spine specimens.

Conflicts of Interest: The authors declare no conflict of interest.

References

1. Cosman, F.; de Beur, S.J.; LeBoff, M.S.; Lewiecki, E.M.; Tanner, B.; Randall, S.; Lindsay, R. Clinician's guide to prevention and treatment of osteoporosis. *Osteoporos. Int.* **2014**, *25*, 2359–2381. [CrossRef]
2. Ammann, P.; Rizzoli, R. Bone strength and its determinants. *Osteoporos. Int.* **2003**, *14* (Suppl. 3), S13–S18. [CrossRef]
3. Delmas, P.D.; van de Langerijt, L.; Watts, N.B.; Eastell, R.; Genant, H.; Grauer, A.; Cahall, D.L. Underdiagnosis of vertebral fractures is a worldwide problem: The impact study. *J. Bone Miner. Res. Off. J. Am. Soc. Bone Miner. Res.* **2005**, *20*, 557–563. [CrossRef] [PubMed]
4. Schuit, S.C.E.; van der Klift, M.; Weel, A.E.A.M.; de Laet, C.E.D.H.; Burger, H.; Seeman, E.; Hofman, A.; Uitterlinden, A.G.; van Leeuwen, J.P.T.M.; Pols, H.A.P. Fracture incidence and association with bone mineral density in elderly men and women: The rotterdam study. *Bone* **2004**, *34*, 195–202. [CrossRef] [PubMed]
5. Cranney, A.; Jamal, S.A.; Tsang, J.F.; Josse, R.G.; Leslie, W.D. Low bone mineral density and fracture burden in postmenopausal women. *CMAJ* **2007**, *177*, 575–580. [CrossRef] [PubMed]
6. Pasco, J.A.; Seeman, E.; Henry, M.J.; Merriman, E.N.; Nicholson, G.C.; Kotowicz, M.A. The population burden of fractures originates in women with osteopenia, not osteoporosis. *Osteoporos. Int.* **2006**, *17*, 1404–1409. [CrossRef]
7. Stone, K.L.; Seeley, D.G.; Lui, L.-Y.; Cauley, J.A.; Ensrud, K.; Browner, W.S.; Nevitt, M.C.; Cummings, S.R. Bmd at multiple sites and risk of fracture of multiple types: Long-term results from the study of osteoporotic fractures. *J. Bone Miner. Res. Off. J. Am. Soc. Bone Miner. Res.* **2003**, *18*, 1947–1954. [CrossRef]
8. Wainwright, S.A.; Marshall, L.M.; Ensrud, K.E.; Cauley, J.A.; Black, D.M.; Hillier, T.A.; Hochberg, M.C.; Vogt, M.T.; Orwoll, E.S. Hip fracture in women without osteoporosis. *J. Clin. Endocrinol. Metab.* **2005**, *90*, 2787–2793. [CrossRef]
9. Wehrli, F.W.; Saha, P.K.; Gomberg, B.R.; Song, H.K.; Snyder, P.J.; Benito, M.; Wright, A.; Weening, R. Role of magnetic resonance for assessing structure and function of trabecular bone. *Top Magn. Reason. Imaging* **2002**, *13*, 335–355. [CrossRef]
10. McCoy, S.; Tundo, F.; Chidambaram, S.; Baaj, A.A. Clinical considerations for spinal surgery in the osteoporotic patient: A comprehensive review. *Clin. Neurol. Neurosurg.* **2019**, *180*, 40–47. [CrossRef]
11. Koester, K.J.; Barth, H.D.; Ritchie, R.O. Effect of aging on the transverse toughness of human cortical bone: Evaluation by r-curves. *J. Mech. Behav. Biomed. Mater.* **2011**, *4*, 1504–1513. [CrossRef] [PubMed]
12. Morgan, E.F.; Bayraktar, H.H.; Keaveny, T.M. Trabecular bone modulus-density relationships depend on anatomic site. *J. Biomech.* **2003**, *36*, 897–904. [CrossRef] [PubMed]
13. Cummings, S.R.; Black, D.M.; Rubin, S.M. Lifetime risks of hip, colles', or vertebral fracture and coronary heart disease among white postmenopausal women. *Arch. Intern. Med.* **1989**, *149*, 2445–2448. [CrossRef] [PubMed]
14. Taes, Y.; Lapauw, B.; Griet, V.; De Bacquer, D.; Goemaere, S.; Zmierczak, H.; Kaufman, J.-M. Prevalent fractures are related to cortical bone geometry in young healthy men at age of peak bone mass. *J. Bone Miner. Res. Off. J. Am. Soc. Bone Miner. Res.* **2010**, *25*, 1433–1440. [CrossRef]
15. Wehrli, F.W.; Rajapakse, C.S.; Magland, J.F.; Snyder, P.J. Mechanical implications of estrogen supplementation in early post-menopausal women. *J. Bone Miner. Res.* **2010**, *25*. [CrossRef]
16. Currey, J.D. Mechanical properties of bone tissues with greatly differing functions. *J. Biomech.* **1979**, *12*, 313–319. [CrossRef]
17. Pang, Y.; Lin, J.; Qin, T.; Chen, Z. Image-to-image translation: Methods and applications. *arXiv* **2021**, arXiv:2101.08629. [CrossRef]
18. Arjovsky, M.; Chintala, S.; Bottou, L. Wasserstein gan. *arXiv* **2017**, arXiv:1701.07875.
19. Isola, P.; Zhu, J.Y.; Zhou, T.; Efros, A.A. Image-to-Image Translation with Conditional Adversarial Networks. In Proceedings of the 30th IEEE Conference on Computer Vision and Pattern Recognition, CVPR 2017, Honolulu, HI, USA, 21–26 July 2016; pp. 1125–1134.
20. Zhu, J.Y.; Park, T.; Isola, P.; Efros, A.A. Unpaired image-to-image translation using cycle-consistent adversarial networks. In Proceedings of the IEEE International Conference on Computer Vision, Venice, Italy, 22–29 October 2017; pp. 2223–2232.
21. Radford, A.; Metz, L.; Chintala, S. Unsupervised representation learning with deep convolutional generative adversarial networks. *arXiv* **2015**, arXiv:1511.06434.
22. Jin, D.; Zheng, H.; Zhao, Q.; Wang, C.; Zhang, M.; Yuan, H. Generation of vertebra micro-ct-like image from mdct: A deep-learning-based image enhancement approach. *Tomography* **2021**, *7*, 767–782. [CrossRef]
23. Liu, M.-Y.; Huang, X.; Mallya, A.; Karras, T.; Aila, T.; Lehtinen, J.; Kautz, J. Few-shot unsupervised image-to-image translation. In Proceedings of the IEEE International Conference on Computer Vision (ICCV), Seoul, Republic of Korea, 27 October—2 November 2019.
24. Van Tulder, G.; De Bruijne, M. Why does synthesized data improve multi-sequence classification? In Proceedings of the Medical Image Computing and Computer-Assisted Intervention—MICCAI, Munich, Germany, 5–9 October 2015; Navab, N., Hornegger, J., Wells, W.M., Frangi, A., Eds.; Springer International Publishing: Cham, Switzerland, 2015; pp. 531–538.
25. Ye, D.H.; Zikic, D.; Glocker, B.; Criminisi, A.; Konukoglu, E. Modality propagation: Coherent synthesis of subject-specific scans with data-driven regularization. *Med. Image Comput. Comput. Assist. Interv.* **2013**, *16*, 606–613.
26. Huang, Y.; Shao, L.; Frangi, A.F. Simultaneous super-resolution and cross-modality synthesis of 3d medical images using weakly-supervised joint convolutional sparse coding. In Proceedings of the 2017 IEEE Conference on Computer Vision and Pattern Recognition (CVPR), Honolulu, HI, USA, 21–26 July 2017; pp. 5787–5796.
27. Costa, P.; Galdran, A.; Meyer, M.I.; Niemeijer, M.; Abramoff, M.; Mendonca, A.M.; Campilho, A. End-to-end adversarial retinal image synthesis. *IEEE Trans. Med. Imaging* **2018**, *37*, 781–791. [CrossRef] [PubMed]

28. Reeth, E.V.; Tham, I.; Tan, C.H.; Poh, C.L. Super-resolution in magnetic resonance imaging: A review. *Concepts Magn. Reson. Part A* **2012**, *40A*, 306–325. [CrossRef]
29. Peleg, I.S. Motion analysis for image enhancement: Resolution, occlusion, and transparency. *J. Vis. Commun. Image Represent.* **1993**, *4*, 324–335.
30. Hayit, G. Super-resolution in medical imaging. *Comput. J.* **2009**, *1*, 43–63.
31. Wang, Z.; Chen, J.; Hoi, S.C.H. Deep learning for image super-resolution: A survey. *IEEE Trans. Pattern Anal. Mach. Intell.* **2021**, *43*, 3365–3387. [CrossRef]
32. Ashikaga, H.; Estner, H.L.; Herzka, D.A.; McVeigh, E.R.; Halperin, H.R. Quantitative assessment of single-image super-resolution in myocardial scar imaging. *IEEE J. Transl. Eng. Health Med.* **2014**, *2*, 1–12. [CrossRef]
33. Bernstein, M.A.; Fain, S.B.; Riederer, S.J. Effect of windowing and zero-filled reconstruction of mri data on spatial resolution and acquisition strategy. *J. Magn. Reason. Imaging* **2001**, *14*, 270–280. [CrossRef]
34. Robinson, M.; Toth, C.; Lo, J.; Farsiu, S. Efficient fourier-wavelet super-resolution. *IEEE Trans. Image Process. A Publ. IEEE Signal Process. Soc.* **2010**, *19*, 2669–2681. [CrossRef]
35. Robinson, M.D.; Farsiu, S.; Lo, J.Y.; Toth, C.A. Efficient restoration and enhancement of super-resolved X-ray images. In Proceedings of the 2008 15th IEEE International Conference on Image Processing, San Diego, CA, USA, 12–15 October 2008; pp. 629–632.
36. Salvador, J. *Example-Based Super Resolution*; Academic Press: Cambridge, MA, USA, 2016; pp. 1–141.
37. Wang, Y.; Ma, G.; An, L.; Shi, F.; Zhang, P.; Lalush, D.S.; Wu, X.; Pu, Y.; Zhou, J.; Shen, D. Semisupervised tripled dictionary learning for standard-dose pet image prediction using low-dose pet and multimodal mri. *IEEE Trans. Biomed. Eng.* **2017**, *64*, 569–579. [CrossRef]
38. Jog, A.; Carass, A.; Roy, S.; Pham, D.L.; Prince, J.L. Random forest regression for magnetic resonance image synthesis. *Med. Image Anal.* **2017**, *35*, 475–488. [CrossRef] [PubMed]
39. Semmlow, J.L.; Griffel, B. *Biosignal and Medical Image Processing*; CRC Press: Boca Raton, FL, USA, 2009.
40. Sahiner, B.; Pezeshk, A.; Hadjiiski, L.M.; Wang, X.; Drukker, K.; Cha, K.H.; Summers, R.M.; Giger, M.L. Deep learning in medical imaging and radiation therapy. *Med. Phys.* **2019**, *46*, e1–e36. [CrossRef]
41. Yi, X.; Walia, E.; Babyn, P. Generative adversarial network in medical imaging: A review. *Med. Image Anal.* **2019**, *58*, 101552. [CrossRef] [PubMed]
42. Fleet, D.; Pajdla, T.; Schiele, B.; Tuytelaars, T. Learning a deep convolutional network for image super-resolution. Lecture Notes in Computer Science (including subseries Lecture Notes in Artificial Intelligence and Lecture Notes in Bioinformatics). In Proceedings of the Computer Vision–ECCV 2014: 13th European Conference, Zurich, Switzerland, 6–12 September 2014; Volume 8692, pp. 184–199.
43. Shi, W.; Caballero, J.; Huszar, F.; Totz, J.; Aitken, A.P.; Bishop, R.; Rueckert, D.; Wang, Z. Real-time single image and video super-resolution using an efficient sub-pixel convolutional neural network. In Proceedings of the IEEE Computer Society Conference on Computer Vision and Pattern Recognition, Las Vegas, NV, USA, 27–30 June 2016; pp. 1874–1883.
44. Kim, J.; Lee, J.K.; Lee, K.M. Deeply-recursive convolutional network for image super-resolution. In Proceedings of the 2016 IEEE Conference on Computer Vision and Pattern Recognition (CVPR), Las Vegas, NV, USA, 27–30 June 2016; pp. 1637–1645.
45. Kim, J.; Lee, J.K.; Lee, K.M. Accurate image super-resolution using very deep convolutional networks. In Proceedings of the 2016 IEEE Conference on Computer Vision and Pattern Recognition (CVPR), Las Vegas, NV, USA, 27–30 June 2016; pp. 1646–1654.
46. Van Nguyen, H.; Zhou, K.; Vemulapalli, R. Cross-domain synthesis of medical images using efficient location-sensitive deep network. In Proceedings of the Medical Image Computing and Computer-Assisted Intervention—MICCAI, Munich, Germany, 5–9 October 2015; Navab, N., Hornegger, J., Wells, W.M., Frangi, A., Eds.; Springer International Publishing: Cham, Switzerland, 2015; pp. 677–684.
47. Chen, H.; Zhang, Y.; Zhang, W.; Liao, P.; Li, K.; Zhou, J.; Wang, G. Low-dose ct via convolutional neural network. *Biomed. Opt. Express* **2017**, *8*, 679–694. [CrossRef]
48. Chen, H.; Zhang, Y.; Kalra, M.K.; Lin, F.; Chen, Y.; Liao, P.; Zhou, J.; Wang, G. Low-dose ct with a residual encoder-decoder convolutional neural network. *IEEE Trans. Med. Imaging* **2017**, *36*, 2524–2535. [CrossRef] [PubMed]
49. Zeng, K.; Zheng, H.; Cai, C.; Yang, Y.; Zhang, K.; Chen, Z. Simultaneous single-and multi-contrast super-resolution for brain mri images based on a convolutional neural network. *Comput. Biol. Med.* **2018**, *99*, 133–141. [CrossRef]
50. Chaudhari, A.S.; Fang, Z.; Kogan, F.; Wood, J.; Stevens, K.J.; Gibbons, E.K.; Lee, J.H.; Gold, G.E.; Hargreaves, B.A. Super-resolution musculoskeletal mri using deep learning. *Magn. Reson. Med.* **2018**, *80*, 2139–2154. [CrossRef]
51. Hubel, D.H.; Wiesel, T.N. Receptive fields, binocular interaction and functional architecture in the cat's visual cortex. *J. Physiol.* **1962**, *160*, 106–154. [CrossRef]
52. He, K.; Zhang, X.; Ren, S.; Sun, J. Deep residual learning for image recognition. In Proceedings of the 2016 IEEE Conference on Computer Vision and Pattern Recognition (CVPR), Las Vegas, NV, USA, 27–30 June 2016.
53. Chen, Q.; Koltun, V. Photographic image synthesis with cascaded refinement networks. In Proceedings of the 2017 IEEE International Conference on Computer Vision (ICCV), Venice, Italy, 22–29 October 2017.
54. Chartsias, A.; Joyce, T.; Giuffrida, M.V.; Tsaftaris, S.A. Multimodal mr synthesis via modality-invariant latent representation. *IEEE Trans. Med. Imaging* **2018**, *37*, 803–814. [CrossRef]

55. Xiang, L.; Wang, Q.; Nie, D.; Zhang, L.; Jin, X.; Qiao, Y.; Shen, D. Deep embedding convolutional neural network for synthesizing ct image from t1-weighted mr image. *Med. Image Anal.* **2018**, *47*, 31–44. [CrossRef]
56. Wang, T.C.; Liu, M.Y.; Zhu, J.Y.; Tao, A.; Kautz, J.; Catanzaro, B. High-resolution image synthesis and semantic manipulation with conditional gans. In Proceedings of the IEEE Computer Society Conference on Computer Vision and Pattern Recognition, Salt Lake City, UT, USA, 18–23 June 2018; pp. 8798–8807.
57. Simonyan, K.; Zisserman, A. Very deep convolutional networks for large-scale image recognition. *Comput. Sci.* **2014**. [CrossRef]
58. Ledig, C.; Theis, L.; Huszár, F.; Caballero, J.; Cunningham, A.; Acosta, A.; Aitken, A.; Tejani, A.; Totz, J.; Wang, Z.; et al. Photo-realistic single image super-resolution using a generative adversarial network. In Proceedings of the 2017 IEEE Conference on Computer Vision and Pattern Recognition (CVPR), Honolulu, HI, USA, 21–26 July 2017; pp. 105–114.
59. Goodfellow, I.; Pouget-Abadie, J.; Mirza, M.; Xu, B.; Warde-Farley, D.; Ozair, S.; Courville, A.; Bengio, Y. Generative adversarial nets. *Adv. Neural Inf. Process. Syst.* **2014**, *27*. [CrossRef]
60. Mirza, M.; Osindero, S. Conditional generative adversarial nets. *arXiv* **2014**, arXiv:1411.1784.
61. Ben-Cohen, A.; Klang, E.; Raskin, S.P.; Amitai, M.M.; Greenspan, H. Virtual pet images from ct data using deep convolutional networks: Initial results. In Proceedings of the International Workshop on Simulation and Synthesis in Medical Imaging, Québec City, QC, Canada, 10 September 2017; Springer: Berlin/Heidelberg, Germany, 2017; pp. 49–57.
62. Bi, L.; Kim, J.; Kumar, A.; Feng, D.; Fulham, M. Synthesis of positron emission tomography (pet) images via multi-channel generative adversarial networks (gans). In *Molecular Imaging, Reconstruction and Analysis of Moving Body Organs, and Stroke Imaging and Treatment*; Springer: Berlin/Heidelberg, Germany, 2017; pp. 43–51.
63. Chartsias, A.; Joyce, T.; Dharmakumar, R.; Tsaftaris, S.A. Adversarial image synthesis for unpaired multi-modal cardiac data. In Proceedings of the International Workshop on Simulation and Synthesis in Medical Imaging, Québec City, QC, Canada, 10 September 2017; Springer: Berlin/Heidelberg, Germany, 2017; pp. 3–13.
64. Mao, X.; Li, Q.; Xie, H.; Lau, R.; Smolley, S.P. Least squares generative adversarial networks. In Proceedings of the 2017 IEEE International Conference on Computer Vision (ICCV), Venice, Italy, 22–29 October 2017.
65. Yu, S.; Dong, H.; Yang, G.; Slabaugh, G.; Dragotti, P.; Ye, X.; Liu, F.; Arridge, S.; Keegan, J.; Firmin, D.; et al. Deep de-aliasing for fast compressive sensing mri. *arXiv* **2017**, arXiv:1705.07137.
66. Gupta, R.; Sharma, A.; Kumar, A. Super-resolution using gans for medical imaging. *Procedia Comput. Sci.* **2020**, *173*, 28–35. [CrossRef]
67. Nie, D.; Trullo, R.; Lian, J.; Wang, L.; Petitjean, C.; Ruan, S.; Wang, Q.; Shen, D. Medical image synthesis with deep convolutional adversarial networks. *IEEE Trans. Biomed. Eng.* **2018**, *65*, 2720–2730. [CrossRef]
68. Hiasa, Y.; Otake, Y.; Takao, M.; Matsuoka, T.; Takashima, K.; Carass, A.; Prince, J.L.; Sugano, N.; Sato, Y. Cross-modality image synthesis from unpaired data using cyclegan. In Proceedings of the International Workshop on Simulation and Synthesis in Medical Imaging, Granada, Spain, 16 September 2018; Springer: Berlin/Heidelberg, Germany; pp. 31–41.
69. Dar, S.U.; Yurt, M.; Karacan, L.; Erdem, A.; Erdem, E.; Çukur, T. Image synthesis in multi-contrast mri with conditional generative adversarial networks. *IEEE Trans. Med. Imaging* **2019**, *38*, 2375–2388. [CrossRef] [PubMed]
70. Choi, Y.; Uh, Y.; Yoo, J.; Ha, J.W. Stargan v2: Diverse image synthesis for multiple domains. In Proceedings of the 2020 IEEE/CVF Conference on Computer Vision and Pattern Recognition (CVPR), Seattle, WA, USA, 13–19 June 2020.
71. Liu, X.S.; Stein, E.M.; Zhou, B.; Zhang, C.A.; Nickolas, T.L.; Cohen, A.; Thomas, V.; McMahon, D.J.; Cosman, F.; Nieves, J.; et al. Individual trabecula segmentation (its)-based morphological analyses and microfinite element analysis of hr-pqct images discriminate postmenopausal fragility fractures independent of dxa measurements. *J. Bone Miner. Res. Off. J. Am. Soc. Bone Miner. Res.* **2012**, *27*, 263–272. [CrossRef] [PubMed]
72. Shuai, B.; Shen, L.; Yang, Y.; Ma, C.; Zhu, R.; Xu, X. Assessment of the impact of zoledronic acid on ovariectomized osteoporosis model using micro-ct scanning. *PLoS ONE* **2015**, *10*, e0132104. [CrossRef]
73. Gomes, C.C.; Freitas, D.Q.; Medeiros Araújo, A.M.; Ramírez-Sotelo, L.R.; Yamamoto-Silva, F.P.; de Freitas Silva, B.S.; de Melo Távora, D.; Almeida, S.M. Effect of alendronate on bone microarchitecture in irradiated rats with osteoporosis: Micro-ct and histomorphometric analysis. *J. Oral Maxillofac. Surg.* **2018**, *76*, 972–981. [CrossRef]
74. Xie, F.; Zhou, B.; Wang, J.; Liu, T.; Wu, X.; Fang, R.; Kang, Y.; Dai, R. Microstructural properties of trabecular bone autografts: Comparison of men and women with and without osteoporosis. *Arch. Osteoporos.* **2018**, *13*, 18. [CrossRef]
75. Johnson, J.; Alahi, A.; Fei-Fei, L. *Perceptual Losses for Real-Time Style Transfer and Super-Resolution*; Springer: Cham, Switzerland, 2016.
76. Huang, X.; Liu, M.Y.; Belongie, S.; Kautz, J. *Multimodal Unsupervised Image-to-Image Translation*; Springer: Cham, Switzerland, 2018.
77. Mescheder, L.; Geiger, A.; Nowozin, S. Which training methods for gans do actually converge? In Proceedings of the International Conference on Machine Learning, Stockholm, Sweden, 10–15 July 2018.
78. Bradski, G. The opencv library. *Dr. Dobb's J. Softw. Tools* **2000**, *25*, 120–123.
79. Keogh, E.; Ratanamahatana, C.A. Exact indexing of dynamic time warping. *Knowl. Inf. Syst.* **2005**, *7*, 358–386. [CrossRef]
80. Zhou, W.; Bovik, A.C.; Sheikh, H.R.; Simoncelli, E.P. Image quality assessment: From error visibility to structural similarity. *IEEE Trans Image Process* **2004**, *13*, 600–612.
81. Heusel, M.; Ramsauer, H.; Unterthiner, T.; Nessler, B.; Hochreiter, S. Gans trained by a two time-scale update rule converge to a local nash equilibrium. *Adv. Neural Inf. Process. Syst.* **2017**, *30*, 6629–6640.
82. Domander, R.; Felder, A.; Doube, M. Bonej2—Refactoring established research software. *Wellcome Open Res.* **2021**, *6*, 37. [CrossRef]

83. Schindelin, J.; Arganda-Carreras, I.; Frise, E.; Kaynig, V.; Longair, M.; Pietzsch, T.; Preibisch, S.; Rueden, C.; Saalfeld, S.; Schmid, B.; et al. Fiji: An open-source platform for biological-image analysis. *Nat. Methods* **2012**, *9*, 676–682. [CrossRef]
84. Rueden, C.T.; Schindelin, J.; Hiner, M.C.; DeZonia, B.E.; Walter, A.E.; Arena, E.T.; Eliceiri, K.W. Imagej2: Imagej for the next generation of scientific image data. *BMC Bioinform.* **2017**, *18*, 1–26. [CrossRef]
85. Schneider, C.A.; Rasband, W.S.; Eliceiri, K.W. Nih image to imagej: 25 years of image analysis. *Nat. Methods* **2012**, *9*, 671–675. [CrossRef]
86. Ridler, T.W.; Calvard, S. Picture thresholding using an iterative selection method. *IEEE Trans. Syst. Man Cybern.* **1978**, *8*, 630–632.
87. Dougherty, R.; Kunzelmann, K.H. Computing local thickness of 3d structures with imagej. *Microsc. Microanal.* **2007**, *13*, 1678–1679. [CrossRef]

bioengineering

MDPI

Article

GSN-HVNET: A Lightweight, Multi-Task Deep Learning Framework for Nuclei Segmentation and Classification

Tengfei Zhao [1], Chong Fu [1,2,3,*], Yunjia Tian [4], Wei Song [1] and Chiu-Wing Sham [5]

1 School of Computer Science and Engineering, Northeastern University, Shenyang 110819, China
2 Engineering Research Center of Security Technology of Complex Network System, Ministry of Education, Shenyang 110819, China
3 Key Laboratory of Intelligent Computing in Medical Image, Ministry of Education, Northeastern University, Shenyang 110819, China
4 State Grid Liaoning Information and Communication Company, Shenyang 110006, China
5 School of Computer Science, The University of Auckland, Auckland 1142, New Zealand
* Correspondence: fuchong@mail.neu.edu.cn

Abstract: Nuclei segmentation and classification are two basic and essential tasks in computer-aided diagnosis of digital pathology images, and those deep-learning-based methods have achieved significant success. Unfortunately, most of the existing studies accomplish the two tasks by splicing two related neural networks directly, resulting in repetitive computation efforts and a redundant-and-large neural network. Thus, this paper proposes a lightweight deep learning framework (GSN-HVNET) with an encoder–decoder structure for simultaneous segmentation and classification of nuclei. The decoder consists of three branches outputting the semantic segmentation of nuclei, the horizontal and vertical (HV) distances of nuclei pixels to their mass centers, and the class of each nucleus, respectively. The instance segmentation results are obtained by combing the outputs of the first and second branches. To reduce the computational cost and improve the network stability under small batch sizes, we propose two newly designed blocks, Residual-Ghost-SN (RGS) and Dense-Ghost-SN (DGS). Furthermore, considering the practical usage in pathological diagnosis, we redefine the classification principle of the CoNSeP dataset. Experimental results demonstrate that the proposed model outperforms other state-of-the-art models in terms of segmentation and classification accuracy by a significant margin while maintaining high computational efficiency.

Keywords: joint nuclei segmentation and classification; lightweight, multi-task deep learning framework; Residual-Ghost-SN; Dense-Ghost-SN

Citation: Zhao, T.; Fu, C.; Tian, Y.; Song, W.; Sham, C.-W. GSN-HVNET: A Lightweight, Multi-Task Deep Learning Framework for Nuclei Segmentation and Classification. *Bioengineering* **2023**, *10*, 393. https://doi.org/10.3390/bioengineering10030393

Academic Editors: Yan Pei, Jijiang Yang and Liang Luo

Received: 12 February 2023
Revised: 13 March 2023
Accepted: 20 March 2023
Published: 22 March 2023

1. Introduction

Over the past several years, deep-learning-based computer vision techniques have been extensively applied to computer-aided diagnosis (CAD). In computational pathology, pathological image analysis based on the deep learning method has proven powerful in improving efficiency and accuracy in cancer detection [1]. The morphology of the nuclei is the essential feature used by pathologists in cancer diagnosis and further cancer prognoses, such as predicting survival [2] and pathological grading of tumors [3]. Accurate nuclei segmentation and classification can advance the quality of tissue segmentation [4,5]. Nuclei segmentation is the crucial first step to obtaining the morphological features used in the downstream analysis. However, the morphological heterogeneity of nuclei makes studies challenging. The karyomorphism shows variability, while different diseases may cause chromatin abnormalities to exhibit variable size and shape patterns. Another severe problem is that the cells in a cancerous tumor are usually densely packed and even have more than one nucleus, causing overlapping nuclei. This overlapping brings difficulty for further research on separating neighboring instances via automatic segmentation.

Extracting each nucleus and distinguishing its type can promote the diagnostic potential in present-day digital pathology pipelines. For instance, precisely distinguishing each nucleus from tumors or lymphocytes can significantly facilitate downstream analysis of tumor-infiltrating lymphocytes (TIL), which has been proven effective in predicting cancer recurrence [6]. The nucleus-by-nucleus classification has become another problem researchers have been interested in recently due to the high variability and diversity of nuclei appearance in a whole slide image.

The current deep models for histopathology image diagnosis are mainly based on single-task learning. Single-task learning is designing a model for a specific task and then optimizing iteratively. In this case, the nuclei segmentation and classification tasks require two independent models, one for detecting the location of each nucleus and the other for classifying the type of nuclei [7,8]. For more complicated tasks, we are accustomed to modeling each part of the task by disassembling. However, there exists an obvious problem in this way. When modeling each sub-task, it is easy to ignore the relationships, conflicts, and constraints between different sub-tasks, resulting in the downgrading of the overall performance of the entire task.

To address the above issue, multi-task models have drawn much attention [9–12]. The multi-task models have the following advantages: (1) multiple tasks share the same model, reducing the amount of memory; (2) multiple tasks obtain results through a forward calculation at one time, and the inference speed increases; (3) associated tasks share information and complement each other, improving each tasks' performance.

Recently, several multi-task deep models for histopathology image diagnosis have been suggested and achieved promising results [13–15]. Unfortunately, these approaches still suffer from efficiency issues, such as dealing with a cumbersome model with a huge amount of parameters. In addition, the classification on the CoNSeP dataset [13] seems hard to meet the needs of practical pathological diagnosis.

The present paper proposes a lightweight, multi-task deep learning framework for segmenting and classifying nuclei simultaneously. To address the problem of network stability encountered by batch normalization (BN) when dealing with small batch sizes, we introduce two newly designed blocks. We devise an efficient encoder–decoder architecture, where the encoder adopts our proposed RGS for down-sampling, while the decoder uses Dense-Ghost-Module (DGM) and convolution for up-sampling. By encoding the HV distance of nuclei pixels, we can obtain more representative features on the instance with fewer layers. Here, HV distance can be used to segment overlapping nuclei instances accurately. Later, the decoder using the output features of the encoder predicts nuclei types. According to the above characteristics, we call the proposed network GSN-HVNET. Our experimental results show that the proposed model can retain shallow features on nuclei to improve segmentation and classification accuracy. Our main contributions are outlined below:

- We propose a novel, lightweight, multi-task deep learning framework containing a unified model for segmentation and classification of nuclei instances simultaneously with superior efficiency and accuracy.
- We propose the newly designed RGS and DGS to improve accuracy and compress the training model.
- We redefine the classification principle of the CoNSeP dataset so that the auxiliary diagnostic results have practical significance in pathological diagnosis.
- Our experiments on the CoNSeP, Kumar, and CPM-17 datasets confirm the improvements to existing works [13,14]. Compared with the state-of-the-art HoVer-Net [13], the number of parameters is reduced by 64%. In addition, we try different batch sizes in our experiments and prove that batch size is no longer a strict limitation on the proposed network; even when a small batch is presented, the proposed network can maintain a high performance.

The remainder of this paper is organized as follows: Section 2 introduces the current research on applying learning algorithms in histopathology image analysis. Our new network architecture is presented in Section 3. We conduct experiments and show desirable results in Section 4. Finally, Section 5 concludes our work and gives a brief discussion of future work.

2. Related Work

2.1. Nuclei Segmentation

Nuclei segmentation is the crucial first step in computer-aided systems for cancer detection. Low level information analysis of histopathology images, such as histograms analysis[16–20], were often used for early nuclei segmentation algorithms. There was an obvious shortcoming that occurred to those algorithms. A certain threshold was hard to be determined to adapt to all scenarios. In [21], the authors proposed a fast and flexible segmentation algorithm based on computing the watersheds in digital grayscale images. Unfortunately, related experiments reported several false-positive segmentation cases. In [22], the authors proposed a novel, marker-controlled watershed based on mathematical morphology to segment clustered cells with less oversegmentation, designing a tracking method based on modified mean shift algorithm to segment undersegmented cells or merge oversegmented cells. In [23], the authors proposed a method combining region growing and machine learning to segment touching nuclei and classify them. In [24], the authors proposed an improved method, which used a joint optimization of a multiple-level set function to segment the cytoplasm and nuclei from clumps of overlapping cervical cells. In [25], the authors proposed using the graph theory technique to segment glands and computed a gland-score for estimating how similar a segmented region is to a gland. In [26], the authors proposed a superpixel-based segmentation technique with different morphological and clustering algorithms. Unfortunately, these existing segmentation algorithms cannot provide utterly reliable results because they need to manually extract nucleus features, which, thus, are inflexible and laborious to extend to a complex scenario.

Rather than manual feature extraction in traditional algorithms, deep learning methods can automatically extract a distinct set of features, and have been widely applied to nuclei segmentation [27]. For instance, U-Net has presented an outstanding performance in biomedical image segmentation [28]. In [29], the authors proposed a deep multi-scale neural network for accurately segmenting nuclei by improving sensitivity to hematoxylin intensity. In [30], to meet the challenge of segmenting overlapping or touching nuclei, the authors formulated the segmentation problem as a regression task of the distance map, and the nuclei boundary information was used as prior knowledge for a segmentation network. In [31], the authors proposed a contour-aware informative aggregation network with a multi-level information aggregation module between two task decoders: one of these segments the nuclei, and the other segments the contours.

2.2. Nuclei Classification

Nuclei classification is a vital step in histopathology image analysis, promoting downstream analysis such as evaluating cancer progression. Early studies utilize manually extracted features to classify the nuclei automatically. Typically, an SVM-based method [32] applied iterative feedback to obtain subtle and complex features of cellular morphology. Albeit showing good performance in high-penetrant phenotypes, it can hardly achieve a satisfying performance in lower-penetrant phenotypes. In [33], Ada-boost was used as the classifier to classify the nuclei after segmentation. The classifier was constructed based on intensity, texture, and morphology features. However, these machine learning methods manually extract features, and their representation ability and stability can still be affected by subjective factors to some extent.

Generally, a deep-learning-based nuclei classification model consists of two main phases. Firstly, each nucleus is segmented or detected using a deep model; then, those features are fed into a classifier to confirm nuclei types. For instance, in [34], the nuclei in colon cancer histology images were firstly detected using a spatially constrained CNN.

Then, each nucleus with associated patches was fed into the convolution network to predict its type, i.e., inflammatory, healthy, or malignant epithelium.

3. Proposed Method

Figure 1 shows an overview of the GSN-HVNET for simultaneous nuclei instance segmentation and classification. The network input starts with 80 × 80 × 3 images, which are center patches cropped out from the sample images of size 270 × 270 × 3. The model can simultaneously segment the nuclei and predict nuclei types and HV-Maps (horizontal and vertical maps). After a post-processing procedure, the nuclei instance can be obtained using HV-Map and nuclei pixel predictions. The final output results can be obtained by combining the segmentation results with the nuclei-type predictions. In other words, the network can complete the segmentation and classification of nuclei instances at the last step.

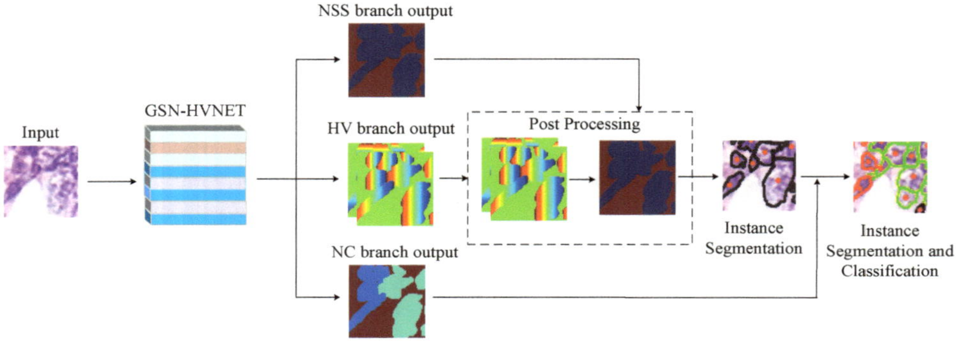

Figure 1. An overview of the GSN-HVNET for simultaneous nuclei instance segmentation and classification. The NSS branch achieves nuclei semantic segmentation, and the HV branch predicts the HV distances of nuclei pixels to their mass centers. Nuclei types are predicted in the NC branch. The nuclei instance segmentation can be accomplished by combining the output of the NSS and HV branches.

3.1. Network Architecture

Figure 2 illustrates the detailed structure of the proposed GSN-HVNET. The proposed network consists of an encoder and a decoder for automatic segmentation and classification of nuclei instances. The encoder can extract an effective set of features; then, the output result of the encoder is used as the decoder input. The decoder contains three branches. Branch I (NSS) is used in nuclei semantic segmentation, and branch II (HV) predicts the HV distances of nuclei pixels to their mass centers. Nuclei types are predicted in branch III (NC). We combine the output of branch I and branch II to accomplish the instance segmentation. Then, the instance segmentation result combines the branch III output to accomplish automatic segmentation and classification of the nuclei instance. The encoder employs the proposed RGS, as discussed in Section 3.1.1. The details of GBS and RGS will be introduced in Sections 3.1.2 and 3.1.3, respectively. In Section 3.1.4, the decoder designed with DGS will be described. The details of DGS will be presented in Section 3.1.5.

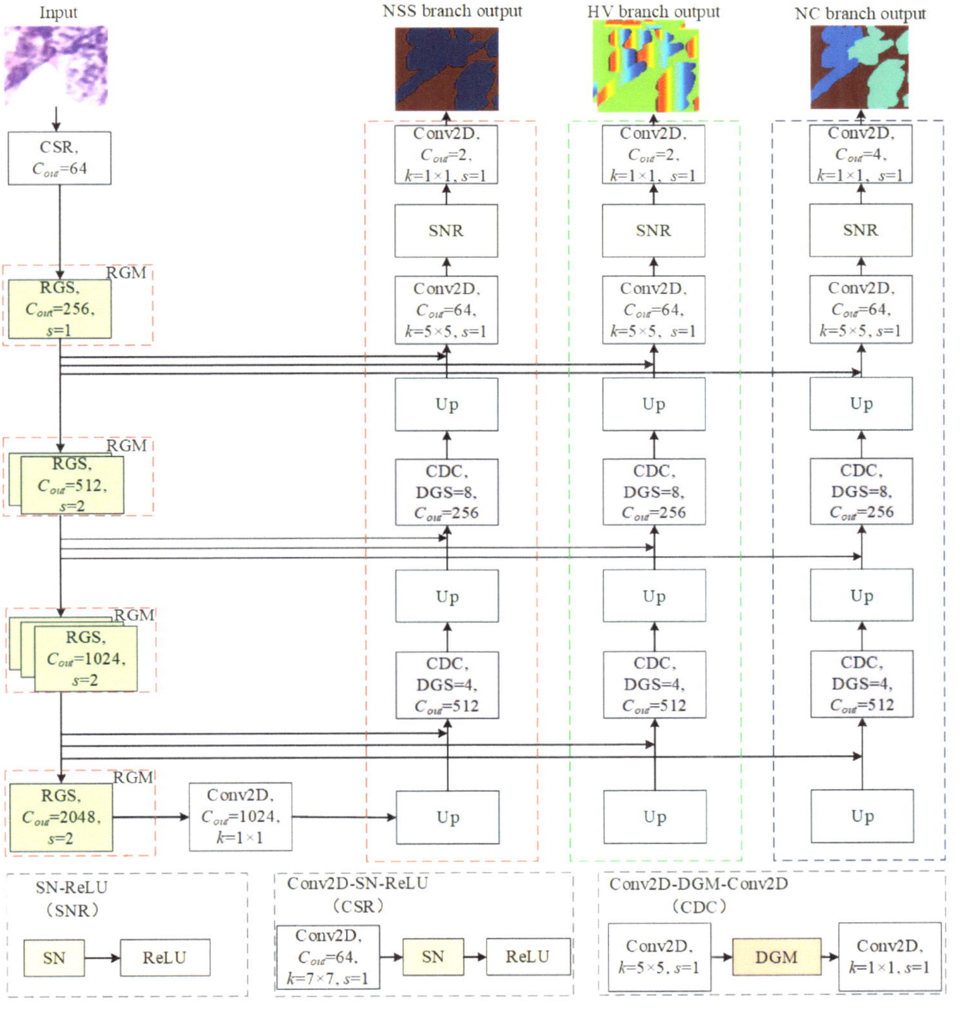

Figure 2. The structure of GSN-HVNET. Our proposed network contains an encoder and a decoder. The encoder, which extracts an effective set of features, is composed of a CSR block, four RGMs, and a Conv2D. The decoder is composed of three branches to achieve accurate nuclei instance segmentation and classification simultaneously.

3.1.1. Encoder

To extract a practical set of features, we design a novel residual ghost network as part of the encoder in the overall network. The network employs a Conv2D-SN-ReLU (CSR) and a series of 4 Residual-Ghost-Modules (RGMs) for down-sampling. Here, the CSR block is composed of a Conv2D, SN, and ReLU. An RGM consists of multiple instances of our improved Ghost-Block—Residual-Ghost-Block with switchable normalization (RGS) [35]. Benefiting from ghost convolution, our network requires much fewer parameters to generate abundant feature maps compared with using ordinary convolution, resulting in an improved computational efficiency of our encoder. Moreover, the SN can select an optimal combination of different normalizers for different normalization layers, improving the network stability, i.e., the accuracy is not affected by the batch size. Each RGM is used as a down-sampling level of 2, which means that the spatial resolution of the input is reduced by a factor of 2. We will give a detailed discussion on RGS in the two subsequent subsections.

3.1.2. Ghost Block with Switchable Normalization

Figure 3 compares the structure of Ghost-Block-BN (GBB) [36] and our suggested Ghost-Block-SN (GBS). As is known, Ghost-Block can help a convolutional neural network to generate more features at a much lower cost. To do that, a Ghost-Block first generates several intrinsic feature maps using ordinary convolution operation and then uses cheap linear operations to expand the features and increase the channels. The computational cost of the linear operations on feature maps is much lower than traditional convolution and transcends other existing efficient works. We can customize the kernel size of the primary convolution in a Ghost-Block, and 1×1 point-wise convolution is employed in this paper. In the Residual-Ghost-Block (RGB), each Ghost-Block is followed by a BN layer, which offers several advantages, including stabilizing and speeding up the training procedure. However, the performance of GBB is severely restricted by the batch size. This is because BN only utilizes a single normalizer in the entire network, which can be unstable and hurt the accuracy in the case of a small batch size.

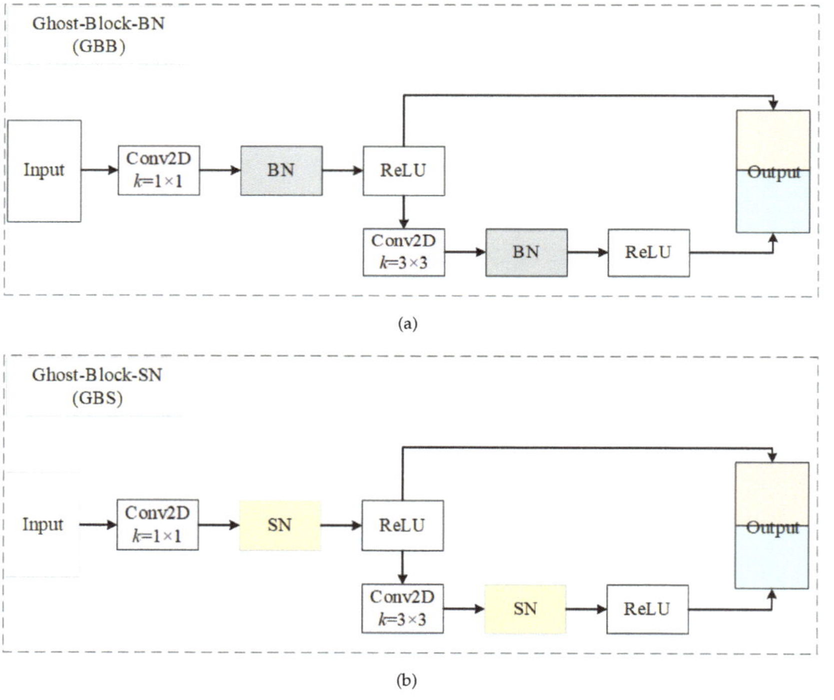

Figure 3. An illustration of the ghost block and the improved ghost block with switchable normalization. (**a**) Ghost block with batch normalization. (**b**) Ghost block with switchable normalization.

To solve the above problem, we apply switchable normalization (SN), which is robust to a wide range of batch sizes, whether a small batch size or not. As shown in Figure 4, SN measures channel-wise, layer-wise, and minibatch-wise statistics by using instance normalization (IN) [37], layer normalization (LN) [38], and batch normalization (BN) [39], respectively, and tries to find an optimal combination by learning their important weights, ensuring the stability and accuracy of the network in the case of small batch size.

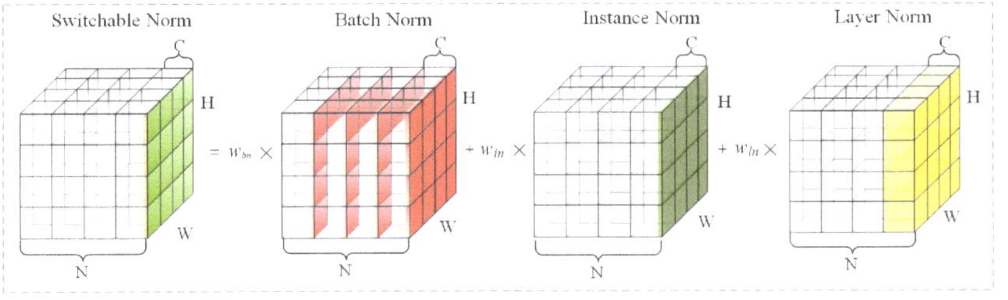

Figure 4. Switchable normalization. It learns to select different normalizers for different normalization layers of a deep neural network.

3.1.3. Residual Ghost Block with Switchable Normalization

Our RGS adopts the structure of residual block—the essential building unit of residual neural network (ResNet) [40]—owing to its outstanding performance. The key idea behind residual block is to reformulate the layers as learning residual functions with reference to the layer inputs, instead of learning unreferenced functions. As shown in Figure 5, we embed the proposed GBS in a residual block as RGS. Later, several RGSs are stacked to form the RGM. Our network contains of four stacked RGMs with 1, 2, 3, and 1 RGS, respectively. Compared with original ResNet-50, our network employs fewer building units to extract feature maps and reduce redundant features, leading to a reduction in model size. In addition, our proposed RGS is generic and can be used in the construction of other lightweight deep learning architectures.

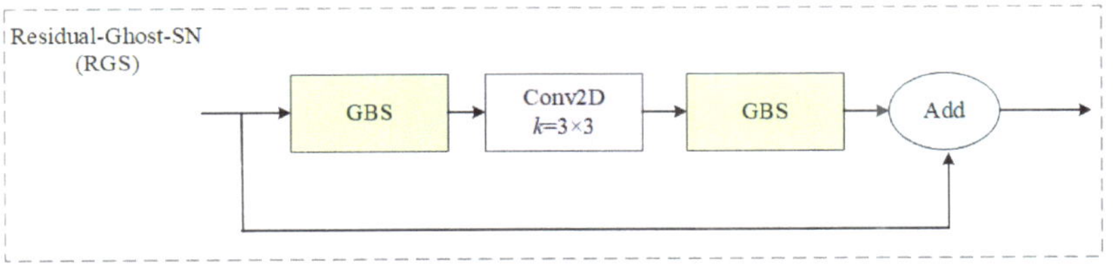

Figure 5. Residual ghost block with switchable normalization. The GBS denotes the ghost block with switchable normalization.

3.1.4. Decoder

As aforementioned, the decoder contains three branches to obtain accurate nuclei instance segmentation and classification simultaneously. These three branches adopt the same architecture consisting of a series of up-sampling operations and two Dense-Ghost-Modules. A DGM contains a series of cascading DGSs. Through stacking multiple DGSs, we can enrich the receptive field with relatively fewer parameters compared with the most popular Dense-Block, resulting in increased computational efficiency. As is known, low-level information is critical in segmentation tasks because it precisely helps to determine object boundaries. To make use of it, we adopt the skip connections to merge feature from each RGS in the encoder via the concatenation operation. The DGM follows the first and second up-sampling operations. There are eight and four DGSs in the first and second DGM, respectively. Each of the three branches contains three up-sampling steps, making the output feature the same dimension as the input image, i.e., $80 \times 80 \times 3$. By combining the results of the two up-sampling branches, NSS and HV, we can obtain accurate boundaries of each individual cell nucleus, and thereby accomplish the nuclei instance segmentation. Compared with independent networks for different tasks, the proposed network is a unified model to simultaneously accomplish nuclei segmentation and classification, thus reducing the total training time.

3.1.5. Dense Ghost Module with Switchable Normalization

In this part, we propose a novel module applied to the decoder of GSN-HVNET. An example of the proposed module is shown in Figure 6, in which $n = 4$. Each DGS connects to other DGSs with forwarding feedback and employs GBS to extract feature maps. The feature maps from all preceding layers are utilized as current inputs, and the feature maps output by a DGS are used as inputs for all subsequent layers.

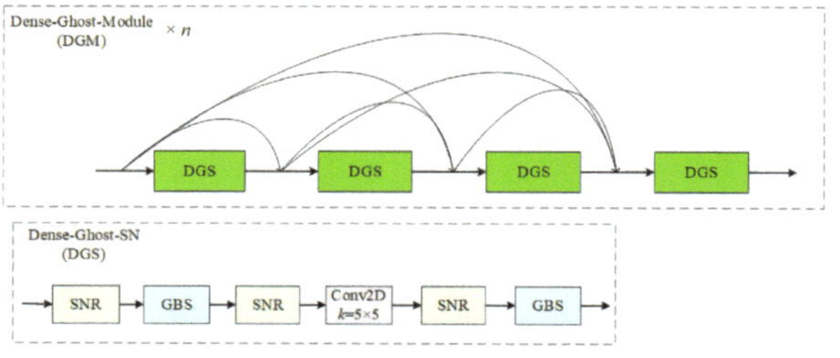

Figure 6. Dense ghost module with switchable normalization. The GBS and SNR denote the ghost block with switchable normalization and switchable normalization with ReLU, respectively.

Thus, the proposed module can retain more abundant features as inputs of subsequent layers.

Similarly, benefiting from the lightweight nature of GBS, our proposed DGS utilizes fewer parameters to generate abundant feature maps and valid features compared with Dense-Block [41]. Moreover, it helps to avoid unnecessary calculations by reducing redundant feature maps. Particularly, the DGM can maintain its performance under a small mini-batch size.

3.1.6. Joint Loss Function of GSN-HVNET

We design different loss functions for each different task. In Table 1, we define the notations for our works. The joint loss function L_{Join} is defined by

$$L_{Join} = L_{NSS} + L_{HV} + L_{NC}. \tag{1}$$

The NSS branch corresponds to a semantic segmentation task, and its loss function is designed using BCE loss and dice loss. It is defined by

$$L_{NSS} = \lambda_a L_{BCE} + \lambda_b L_{DICE}, \tag{2}$$

where L_{BCE} and L_{DICE} represent the binary cross-entropy loss function and dice loss function for the output of the NSS branch, respectively. The λ_a and λ_b are scalars that give weights to their associated loss function. The above two functions are defined by

$$L_{BCE} = -[\frac{1}{n}\sum_{i=1}^{N}\sum_{k=1}^{K} X_{i,k}(I)\log Y_{i,k}(I) + \sum_{i=1}^{N}\sum_{k=1}^{K}(1 - X_{i,k}(I))\log(1 - Y_{i,k}(I))] \tag{3}$$

and

$$L_{DICE} = 1 - \frac{2 \times \sum_{i=1}^{N}(Y_i(I) \times X_i(I)) + \epsilon}{\sum_{i=1}^{N} Y_i(I) + \sum_{i=1}^{N} X_i(I) + \epsilon}, \tag{4}$$

where X represents the ground truth, Y denotes the prediction, and K represents the number of categories. In order to avoid zero denominators, we set ϵ to $1.0 \times e^{-4}$ in the numerator and denominator.

The loss function for the HV branch is defined by

$$L_{HV} = \lambda_c L_{MSE} + \lambda_d L_{MSGE}, \tag{5}$$

where L_{MSE} represents the mean squared error measuring the difference between the HV distances prediction and the ground truth, λ_c and λ_d are the weights of their associated loss function. The loss function L_{MSGE} is used to calculate the gradients of the mean squared error between HV maps and ground truth. L_{MSE} and L_{MSGE} are defined by

$$L_{MSE} = \frac{1}{n}\sum_{n=1}^{n}(p_i(I) - \Gamma_i(I))^2 \tag{6}$$

and

$$L_{MSGE} = \frac{1}{m}\sum_{i \in M}(\nabla_x(p_{i,x}(I) - \Gamma_{i,x}(I)))^2 + \frac{1}{m}\sum_{i \in M}(\nabla_y(p_{i,y}(I) - \Gamma_{i,y}(I)))^2, \tag{7}$$

where I represents the input image and $p_i(I)$ is defined as the regression output of HV branch. The pixel-wise softmax predictions of NSS and NC branches are represented by $q_i(I)$ and $r_i(I)$, respectively. $\Gamma_i(I)$ denotes the ground truth of the HV distance of nuclei pixels to their mass centers.

The loss function of L_{NC} is defined by

$$L_{NC} = \lambda_e L_{BCE} + \lambda_f L_{DICE}. \tag{8}$$

Similarly, λ_e and λ_f are used to balance the two loss functions L_{BCE} and L_{DICE}.

Table 1. The definition of notations.

Notation	Definition
h_{ncij}	The value of a pixel before normalization.
$\hat{h}_{ncij}$,	The value of a pixel after normalization.
γ, β	Scale and shift parameter
$I_k, \|I_k\|$	A set of pixels, and the number of pixels in I_k.
L, λ	L denotes the loss function and λ represents its parameters.
I	The input image.
$\Gamma_i(I)$	The HV distance of nuclei pixels to their mass centers.
$p_i(I)$	The regression output of HV branch.
$q_i(I)$	The pixel-wise and softmax predictions of NSS branch.
$r_i(I)$	The pixel-wise and softmax predictions of NC branch.
E	The energy landspace.
F_c^t	The whole measurement for nuclei type classification and nuclei instance segmentation.
FP, FN	False-positive, false-negative.
TP, TN	True-positive, true-negative.

3.2. Post-Processing

The proposed network produces three outputs. To obtain the nuclei location and separate overlapping or clustered nuclei, we need to post-process the output of NSS and HV. Within each HV map, there are significant differences between pixels in adjacent instances. Using this property, we can calculate the gradient so as to separate the clustered nuclei. To do that, we have

$$S_m = max(Hor(p_{hor}), Ver(p_{ver})), \tag{9}$$

where p_{hor} and p_{ver} represent the horizontal and vertical predictions produced by the HV branch, and Hor and Ver refer to the horizontal and vertical components, respectively, of the Sobel operator, which calculates the horizontal and vertical derivative approximations.

In Figure 1, S_m highlights the regions where pixels in adjacent regions of two instances differ significantly in the horizontal and vertical maps.

We compute the marker M according to

$$M = \sigma(\tau(q,h) - \tau(S_m,k)), \tag{10}$$

where q is the output probability map of the NSS branch and $\tau(q,h)$ is a threshold function acting on q and sets values above h to 1 or 0; otherwise, σ is a rectifier setting all negative value to 0 and M is the output marker. We can obtain desired segmentation results by choosing appropriate h and k.

Next, we compute the energy landscape E according to

$$E = [1 - \tau(S_m,k)] * \tau(q,h). \tag{11}$$

Finally, given the energy landscape E, a marker-controlled watershed is carried out using M as the marker to determine how to split $\tau(q,h)$, given the energy landscape E. The task of joint segmentation and classification of nuclei requires converting per-pixel nuclei type prediction in the NSS branch to the prediction of the type of nuclei instances. To do that, we combine the post-processing result with NC branch output.

4. Experiment

4.1. Datasets and Implementation

In our experiment, we adopt three authoritative nuclei datasets: CoNSeP [13], Kumar [42], and CPM-17 [43]. Table 2 describes these datasets used in our experiment. The CoNSeP dataset, extracted from 16 colorectal adenocarcinoma (CRA) WSIs, consists of 41 hematoxylineosin (H&E) staining images, each of size 1000 × 1000 at 40× objective magnification. In CoNSeP dataset, tumor regions, stroma, muscular, fat, glandular, and collagen can be observed. In addition to containing different tissue components, seven nuclei types are provided, including malignant/dysplastic epithelial nuclei, normal epithelium, inflammatory, fibroblast, muscle, endothelial, and miscellaneous. In [13], the authors combined the original seven categories into four categories, of which malignant/dysplastic epithelial and normal epithelial were combined into a single type corresponding to the epithelial class, and fibroblast, muscle, and endothelial were combined into a single type corresponding to the spindle-shaped class. However, in practical clinical diagnosis, a CAD system should mainly focus on the identification of lesion area. To address this issue, we reclassified this dataset in our experiment. Specially, the normal epithelium, fibroblast, muscle, endothelial, and miscellaneous were combined into a single type corresponding to normal region, and the malignant/dysplastic epithelial and inflammatory are considered as two separate types—that is, the reclassified contain three nuclei categories as well as the background category. With this classification rule, our model can directly report the types of nuclei in lesion areas.

Kumar is an annotated dataset containing over 13,000 segmented nuclei from four different organs—breast, kidney, liver, and prostate—of 16 patients. The CPM-17 dataset provides the tissue image with labels for nuclei segmentation and classification. It is obtained from patients with head and neck squamous cell (HNSCC), glioblastoma multiforme (GBM), non-small cell lung cancer (NSCLC), and lower-grade glioma tumors (LGG). Some examples taken from these datasets are shown in Figure 7.

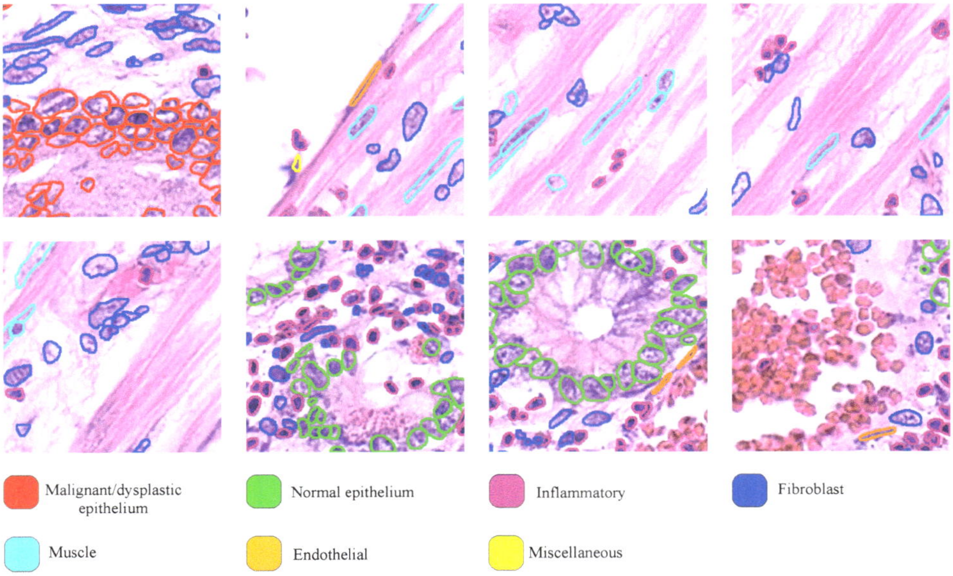

Figure 7. The sample clipping region is extracted from the CoNSeP dataset, where the color of each nuclear boundary indicates its category.

Table 2. Description of the dataset used in our experiment. The Seg denotes the dataset with segmentation labels and the Class denotes the dataset with classification labels.

	CoNSeP	CPM-17	Kumar
Total numbers of nuclei	24,319	7570	21,623
Labeled nuclei	24,319	0	0
Number of images	41	32	30
Origin	UHCW	TCGA	TCGA
Magnification	40×	40× & 20 ×	40×
Size of images	1000×1000	500×500 to 600×600	1000×1000
Seg/Class	Seg&Class	Seg	Seg
Number of cancer types	1	4	8

We run our code on a server equipped with an NVIDIA Geforce RTX 3090 GPU and Intel(R) Xeon(R) Gold 5118 CPU. During the training phase, we performed data augmentation to augment the training data. We randomly combined zooming, channel shifting, shearing, rotating, and horizontal/vertical flipping, which cropped the original image into 270 × 270 sub-images. We used Kaiming normalization [44] to initialize weights and set initial bias as false. We used Adam [45] as the optimizer, with a trainable batch size of 4. We set an initial learning rate as $1.0 \times e^{-4}$ and weight decay as 0.1. The six hyper-parameters λ_a, λ_b, λ_c, λ_d, λ_e, and λ_f used for balancing the joint loss function are tuned to be $\{1, 1, 1, 1, 2, 1\}$ on the validation set.

4.2. Evaluation Metrics

4.2.1. Nuclei Instance Segmentation Evaluation

The segmentation of the nuclei instances can be divided into three sub-tasks; these three sub-tasks are the separation of the nuclei from the background, the detection of individual nuclei instances, and the segmentation of each detected instance. The Ensemble

Dice [43] and Aggregated Jaccard Index [42] are two popular metrics used to measure the performance of nuclei instance segmentation. To better investigate the proposed method, we need to measure the performance of each sub-task. The dice coefficient (F_1 score) is defined by

$$Dice\_coef = \frac{|TP|}{|TP| + \frac{1}{2}|FP| + \frac{1}{2}|FN|} = \frac{2 \times (X \cap Y)}{(|X| + |Y|)}, \tag{12}$$

where TP represents the true-positive rate, FP represents the false-positive rate, and FN represents the false-negative rate. X and Y represent the ground truth and prediction, respectively.

The AJI calculates the ratio of an aggregated intersection cardinality to an aggregated union cardinality between the ground truth and prediction. It is defined by

$$AJI = \frac{\sum_{i=1}^{N} |G_i \cap P_M^i|}{\sum_{i=1}^{N} |G_i \cup P_M^i| + \sum_{F \in U} |P_F|}, \tag{13}$$

where G_i is the ith nucleus from the ground truth with N nuclei. P_M^i represents the Mth connected component in prediction, which has the largest Jaccard Index with G_i, and where each M cannot be utilized more than once. U is a set representing the connected component in the prediction without the corresponding ground truth.

Unfortunately, F_1 score and AJI only calculate an overall score for the instance segmentation quality. In addition, the two metrics suffer from a limitation that they will produce excessive penalization and result in an abnormal score for overlapping regions.

To take a measurement of each sub-task, we take advantage of panoptic quality [46] with accurate quantification and interpretability to measure the performance of nuclei instance segmentation. The panoptic quality for nuclei instance segmentation is defined by

$$PQ = DQ \times SQ = Dice\_coef \times \frac{\sum_{(x,y) \in TP} IoU(x,y)}{|TP|}, \tag{14}$$

where x and y denote a ground truth component and a prediction component, respectively. The IoU represents the intersection over union. Each (x, y) must be unique over the whole set of prediction and ground truth segments, if their $IoU(x, y) > 0.5$. DQ and SQ help to give a direct insight into detecting individual nuclear instances and segmenting each detected instance. Therefore, PQ can serve as the objective evaluation criteria for measuring the performance of the nuclei instance segmentation task.

To demonstrate the effectiveness of the proposed method, we use the following three metrics. Dice coefficient and PQ are used to measure the separation of all nuclei from the background and serve as a unified score for comparison, respectively. The AJI is used for the comparison with other methods. In this study, these three metrics serve as objective evaluation criteria. As the most reliable assessment of the segmentation quality, the subjective evaluation can also be carried out in practical applications.

4.2.2. Nuclei Classification Evaluation

Nuclei classification is influenced by nuclei instance segmentation. The whole measurement for nuclei type classification should include nuclei instance segmentation. HoVer-Net [13] defines an efficient evaluation, which is defined by

$$F_c^t = \frac{2(TP_c + TNc)}{2(TP_c + TN_c) + 2(FP_c + FN_c) + (FP_d + FN_d)}, \tag{15}$$

where FP_d and FN_d are false-positive and false-negative in detecting ground truth instances, respectively. TP_c, TN_c, FP_c, and FN_c denote true-positive, true-negative, false-positive, and false-negative, respectively.

4.3. Experimental Results

Table 3 compares the number of trainable parameters of the proposed and other popular models. As can be seen from this table, our model gives the smallest size among all others in terms of the nuclei segmentation task and the second smallest size in term of the joint segmentation and classification task. Consequently, our model offers a high degree of computational efficiency. Table 4 compares the Dice scores of the proposed and two state-of-the-art models working with small mini-batch sizes. The results indicate that the proposed model appears more stable, i.e., our model can work well on small-memory-capacity GPUs, such as the NVIDIA 1080ti or 2080ti, thus reducing the hardware cost. The model size is significantly smaller than other compared networks.

Table 3. Comparative results for the number of trainable parameters of different networks for nuclei segmentation and classification. The Seg denotes the single-task network for segmentation. The Seg&Class denotes the multi-tasking network for simultaneous segmentation and classification.

Method	Seg/Class	Parameters
HoVer-Net [13]	Seg	42.94M
HoVer-Net [13]	Seg&Class	52.20M
Micro-Net [14]	Seg&Class	183.67M
DIST [30]	Seg&Class	8.81M
DCAN [47]	Seg	39.54M
SegNet [48]	Seg	28.07M
FCN8 [49]	Seg	128.05M
U-Net [28]	Seg&Class	35.23M
Mask-RCNN [15]	Seg&Class	44.17M
Our proposed	Seg	15.03M
Our proposed	Seg&Class	32.52M

Table 4. Comparative results for different mini-batch sizes presenting in three multi-tasking networks. The Dice coefficient is used to evaluate the segmentation performance on the CoNSeP, Kumar, and CPM-17 datasets.

Batch Size	Our Proposed			HoVer-Net			Micro-Net		
		Dice			Dice			Dice	
	CoNSeP	Kumar	CPM-17	CoNSeP	Kumar	CPM-17	CoNSeP	Kumar	CPM-17
1	0.821	0.851	0.865	0.816	0.794	0.843	0.752	0.759	0.828
2	0.830	0.844	0.870	0.806	0.804	0.875	0.764	0.785	0.857
3	0.839	0.842	0.870	0.835	0.819	0.879	0.758	0.794	0.859

The proposed network is measured by the three kinds of metrics discussed above, compared with baselines and other state-of-the-art networks, and the results are reported in Table 5. The results indicate that our proposed network achieves the highest accuracy among all the others. Moreover, even though the DIST model has fewer parameters than ours on joint nuclei segmentation classification task, its segmentation performance is worse than ours by a large margin on all three datasets. Therefore, our network offers an optimal trade-off between accuracy and efficiency.

As aforementioned, the 4-class nuclei classification carried out in HoVer-Net is impractical for use in practical pathological diagnosis. Accordingly, we have reclassified the data. Table 6 lists the comparative results for 3-class nuclei classification on the CoNSeP dataset. Here, F_d denotes the F_1 score for nuclei detection. F_c^1, F_c^2, and F_c^3 denote the classification score for healthy, inflammatory, and malignant/dysplastic epithelium classes, respectively. The results show that the proposed network outperforms all the others in terms of F_c^1, F_c^2, and F_c^3 scores. In Figure 8, we illustrate the results of nuclei segmentation and classification on the sample images and compare them with those of [13–15,30].

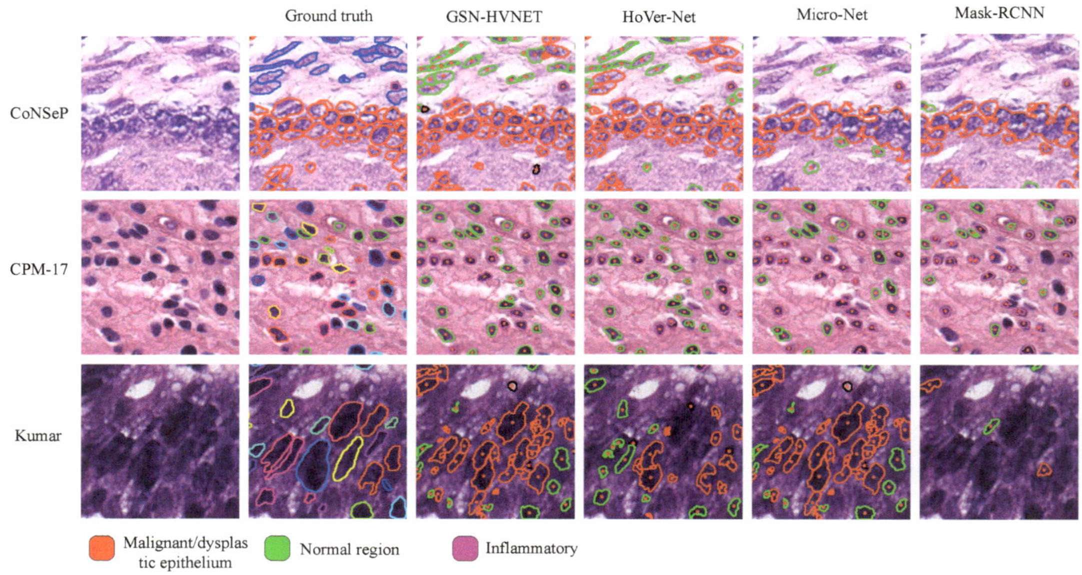

Figure 8. Comparative results for nuclei classification and segmentation. The normal epithelium, fibroblast, muscle, endothelial, and miscellaneous are combined into a single type corresponding to the normal region, and the malignant/dysplastic epithelial and inflammatory are considered as two separate types.

As can be seen from this figure, our lightweight method is successful in segmenting overlapping and clustered nuclei. It is also excellent to complete the task of nuclei classification at the same time. Overall, our proposed model achieves state-of-the-art accuracy on nuclei segmentation and classification tasks while maintaining low computation cost. Our idea can be directly deployed in the cell pathology diagnosis system to reduce the workload of pathologists.

Table 5. Comparative results for nuclei segmentation. The Dice coefficient, AJI, and PQ are used to evaluate the instance segmentation performance of ten networks on the CoNSeP, Kumar, and CPM-17 datasets.

Method	CoNSeP			Kumar			CPM-17		
	Dice	AJI	PQ	Dice	AJI	PQ	Dice	AJI	PQ
HoVer-Net [13]	0.838	0.525	0.494	0.826	0.618	0.597	0.869	0.705	0.697
SegNet [48]	0.796	0.194	0.270	0.811	0.377	0.407	0.857	0.491	0.531
FCN8 [49]	0.756	0.123	0.163	0.797	0.281	0.312	0.840	0.397	0.435
U-Net [28]	0.724	0.482	0.328	0.758	0.556	0.478	0.813	0.643	0.578
DIST [30]	0.798	0.495	0.386	0.789	0.559	0.443	0.826	0.616	0.504
DCAN [47]	0.733	0.289	0.256	0.792	0.525	0.492	0.828	0.561	0.545
Micro-Net [14]	0.784	0.518	0.421	0.797	0.560	0.519	0.857	0.668	0.661
Mask-RCNN [15]	0.740	0.474	0.460	0.760	0.546	0.509	0.850	0.684	0.674
CIA-Net [31]	-	-	-	0.818	0.620	0.577	-	-	-
Our proposed	0.861	0.602	0.566	0.879	0.635	0.644	0.899	0.701	0.683

Table 6. Comparative results for 3-class nuclei classification on the CoNSeP dataset. F_d denotes the F_1 score for nuclei detection. F_c^1, F_c^2, and F_c^3 denote the classification score for healthy, inflammatory, and malignant/dysplastic epithelium classes, respectively.

Method	F_d	F_c^1	F_c^2	F_c^3
HoVer-Net [13]	0.784	0.488	0.525	0.517
Micro-Net [14]	0.812	0.487	0.549	0.546
DIST [30]	0.782	0.489	0.569	0.526
Mask-RCNN [15]	0.701	0.413	0.568	0.514
Our proposed	0.820	0.514	0.572	0.519

5. Conclusions

In this paper, we designed a lightweight, multi-task deep learning framework for nuclei segmentation and classification. Our model follows an encoder–decoder architecture, and the decoder consists of three branches, each outputting a prediction for a sub-task. To sufficiently use the correlation among the three branches, we employ NSS and HV branches to complete the nuclei instance segmentation and use NC branch to predict the classes of each nucleus in a learning process. Two newly designed blocks, Residual-Ghost-SN and Dense-Ghost-SN, are employed in the encoder and decoder parts, respectively, to reduce the computational cost and improve the network stability under small batch sizes. Extensive experiments have been carried out on the CoNSeP, Kumar, and CPM-17 datasets, and the results demonstrate that our model offers a state-of-the-art trade-off between computational efficiency and both segmentation and classification accuracy.

Ultimately, our idea is generic, and can be easily deployed to other histopathology images analysis works. Moreover, the blocks proposed in this paper, including Residual-Ghost-SN and Dense-Ghost-SN, are also generic and can be flexibly embedded into other deep CNNs for histopathology image diagnostic tasks. However, regarding their application in the field of natural images, we have not conducted experiments, and the effects cannot be guaranteed. Thus, we pose this as an open problem and expect to provide a theoretical analysis with complete proof in further research.

Author Contributions: Conceptualization, T.Z. and C.F.; methodology, T.Z. and C.F.; validation, T.Z. and Y.T.; formal analysis, T.Z., W.S. and C.-W.S.; investigation, T.Z. and C.F.; resources, T.Z. and W.S.; data curation, T.Z. and C.F.; writing—original draft preparation, T.Z. and Y.T.; writing—review and editing, T.Z., C.F., Y.T., W.S. and C.-W.S.; visualization, T.Z. and Y.T.; supervision, T.Z.; project administration, T.Z.; funding acquisition, C.F. All authors have read and agreed to the published version of the manuscript.

Funding: This research is supported by the National Natural Science Foundation of China (No. 62032013), the Fundamental Research Funds for the Central Universities (Nos. N2224001-7 and N2116020), and the Natural Science Foundation of Liaoning Province (No. 2021-YGJC-24).

Institutional Review Board Statement: Not applicable.

Informed Consent Statement: Not applicable.

Data Availability Statement: The data that support the findings of this study are available from the corresponding author upon reasonable request.

Conflicts of Interest: The authors declare no conflict of interest.

References

1. Pantanowitz, L. Digital images and the future of digital pathology. *J. Pathol. Inform.* **2010**, *1*, 15. [CrossRef]
2. Alsubaie, N.; Sirinukunwattana, K.; Raza, S.E.A.; Snead, D.; Rajpoot, N. A bottom-up approach for tumour differentiation in whole slide images of lung adenocarcinoma. In *Medical Imaging 2018: Digital Pathology*; Tomaszewski, J.E., Gurcan, M.N., Eds.; International Society for Optics and Photonics, SPIE: Houston, TX, USA, 2018; Volume 10581, p. 105810E. [CrossRef]

3. Lu, C.; Romo-Bucheli, D.; Wang, X.; Janowczyk, A.; Ganesan, S.; Gilmore, H.; Rimm, D.; Madabhushi, A. Nuclear shape and orientation features from H&E images predict survival in early-stage estrogen receptor-positive breast cancers. *Lab. Investig.* **2018**, *98*, 1438–1448. [CrossRef]
4. Javed, S.; Fraz, M.M.; Epstein, D.; Snead, D.; Rajpoot, N.M. Cellular community detection for tissue phenotyping in histology images. In *Computational Pathology and Ophthalmic Medical Image Analysis*; Springer: Berlin/Heidelberg, Germany, 2018; pp. 120–129. [CrossRef]
5. Sirinukunwattana, K.; Snead, D.; Epstein, D.; Aftab, Z.; Mujeeb, I.; Tsang, Y.W.; Cree, I.; Rajpoot, N. Novel digital signatures of tissue phenotypes for predicting distant metastasis in colorectal cancer. *Sci. Rep.* **2018**, *8*, 13692. [CrossRef] [PubMed]
6. Corredor, G.; Wang, X.; Zhou, Y.; Lu, C.; Fu, P.; Syrigos, K.; Rimm, D.L.; Yang, M.; Romero, E.; Schalper, K.A.; et al. Spatial architecture and arrangement of tumor-infiltrating lymphocytes for predicting likelihood of recurrence in early-stage non–small cell lung cancer. *Clin. Cancer Res.* **2019**, *25*, 1526–1534. [CrossRef]
7. Sharma, H.; Zerbe, N.; Heim, D.; Wienert, S.; Behrens, H.M.; Hellwich, O.; Hufnagl, P. A multi-resolution approach for combining visual information using nuclei segmentation and classification in histopathological images. In Proceedings of the VISAPP, Berlin, Germany, 11–14 March 2015; Volume 3, pp. 37–46.
8. Wang, P.; Hu, X.; Li, Y.; Liu, Q.; Zhu, X. Automatic cell nuclei segmentation and classification of breast cancer histopathology images. *Signal Process.* **2016**, *122*, 1–13. [CrossRef]
9. Girshick, R. Fast r-cnn. In Proceedings of the IEEE International Conference on Computer Vision, Santiago, Chile, 7–13 December 2015; pp. 1440–1448. [CrossRef]
10. Kendall, A.; Gal, Y.; Cipolla, R. Multi-task learning using uncertainty to weigh losses for scene geometry and semantics. In Proceedings of the IEEE Conference on Computer Vision and Pattern Recognition, Salt Lake City, UT, USA, 18–23 June 2018; pp. 7482–7491. [CrossRef]
11. Zhou, Y.; Chen, H.; Li, Y.; Liu, Q.; Xu, X.; Wang, S.; Yap, P.T.; Shen, D. Multi-task learning for segmentation and classification of tumors in 3D automated breast ultrasound images. *Med. Image Anal.* **2021**, *70*, 101918. [CrossRef] [PubMed]
12. Shan, P.; Wang, Y.; Fu, C.; Song, W.; Chen, J. Automatic skin lesion segmentation based on FC-DPN. *Comput. Biol. Med.* **2020**, *123*, 103762. [CrossRef]
13. Graham, S.; Vu, Q.D.; Raza, S.E.A.; Azam, A.; Tsang, Y.W.; Kwak, J.T.; Rajpoot, N. Hover-net: Simultaneous segmentation and classification of nuclei in multi-tissue histology images. *Med. Image Anal.* **2019**, *58*, 101563. [CrossRef]
14. Raza, S.E.A.; Cheung, L.; Shaban, M.; Graham, S.; Epstein, D.; Pelengaris, S.; Khan, M.; Rajpoot, N.M. Micro-Net: A unified model for segmentation of various objects in microscopy images. *Med. Image Anal.* **2019**, *52*, 160–173. [CrossRef]
15. Abdulla, W. Mask R-CNN for Object Detection and Instance Segmentation on Keras and TensorFlow. 2017. Available online: https://github.com/matterport/Mask_RCNN (accessed on 31 January 2023).
16. Song, W.; Zheng, Y.; Fu, C.; Shan, P. A novel batch image encryption algorithm using parallel computing. *Inf. Sci.* **2020**, *518*, 211–224. [CrossRef]
17. Song, W.; Fu, C.; Zheng, Y.; Cao, L.; Tie, M. A practical medical image cryptosystem with parallel acceleration. *J. Ambient. Intell. Humaniz. Comput.* **2022**, 1–15.
18. Otsu, N. A threshold selection method from gray-level histograms. *IEEE Trans. Syst. Man Cybern.* **1979**, *9*, 62–66. [CrossRef]
19. Song, W.; Fu, C.; Zheng, Y.; Cao, L.; Tie, M.; Sham, C.W. Protection of image ROI using chaos-based encryption and DCNN-based object detection. *Neural Comput. Appl.* **2022**, *34*, 5743–5756.
20. Song, W.; Fu, C.; Zheng, Y.; Tie, M.; Liu, J.; Chen, J. A parallel image encryption algorithm using intra bitplane scrambling. *Math. Comput. Simul.* **2023**, *204*, 71–88. [CrossRef]
21. Vincent, L.; Soille, P. Watersheds in digital spaces: An efficient algorithm based on immersion simulations. *IEEE Comput. Archit. Lett.* **1991**, *13*, 583–598. [CrossRef]
22. Yang, X.; Li, H.; Zhou, X. Nuclei segmentation using marker-controlled watershed, tracking using mean-shift, and Kalman filter in time-lapse microscopy. *IEEE Trans. Circuits Syst. Regul. Pap.* **2006**, *53*, 2405–2414. [CrossRef]
23. Mohammed, J.G.; Boudier, T. Classified region growing for 3D segmentation of packed nuclei. In Proceedings of the 2014 IEEE 11th International Symposium on Biomedical Imaging (ISBI), Beijing, China, 29 April–2 May 2014; pp. 842–845.
24. Lu, Z.; Carneiro, G.; Bradley, A.P. An improved joint optimization of multiple level set functions for the segmentation of overlapping cervical cells. *IEEE Trans. Image Process.* **2015**, *24*, 1261–1272. [CrossRef]
25. Nguyen, K.; Sarkar, A.; Jain, A.K. Prostate cancer grading: Use of graph cut and spatial arrangement of nuclei. *IEEE Trans. Med. Imaging* **2014**, *33*, 2254–2270. [CrossRef]
26. Mitra, S.; Dey, S.; Das, N.; Chakrabarty, S.; Nasipuri, M.; Naskar, M.K. Identification of Benign and Malignant Cells from cytological images using superpixel based segmentation approach. In Proceedings of the Annual Convention of the Computer Society of India, Kolkata, India, 19–21 January 2018; Springer: Berlin/Heidelberg, Germany, 2018; pp. 257–269. [CrossRef]
27. LeCun, Y.; Bengio, Y.; Hinton, G. Deep learning. *Nature* **2015**, *521*, 436–444. [CrossRef]
28. Ronneberger, O.; Fischer, P.; Brox, T. U-net: Convolutional networks for biomedical image segmentation. In Proceedings of the International Conference on Medical Image Computing and Computer-Assisted Intervention, Munich, Germany, 5–9 October 2015; Springer: Berlin/Heidelberg, Germany, 2015; pp. 234–241. [CrossRef]

29. Graham, S.; Rajpoot, N.M. SAMS-NET: Stain-aware multi-scale network for instance-based nuclei segmentation in histology images. In Proceedings of the 2018 IEEE 15th International Symposium on Biomedical Imaging (ISBI 2018), Washington, DC, USA, 4–7 April 2018; pp. 590–594. [CrossRef]
30. Naylor, P.; Laé, M.; Reyal, F.; Walter, T. Segmentation of nuclei in histopathology images by deep regression of the distance map. *IEEE Trans. Med. Imaging* **2018**, *38*, 448–459. [CrossRef]
31. Zhou, Y.; Onder, O.F.; Dou, Q.; Tsougenis, E.; Chen, H.; Heng, P.A. Cia-net: Robust nuclei instance segmentation with contour-aware information aggregation. In Proceedings of the International Conference on Information Processing in Medical Imaging, Hong Kong, China, 2–7 June 2019; Springer: Berlin/Heidelberg, Germany, 2019; pp. 682–693. [CrossRef]
32. Jones, T.R.; Carpenter, A.E.; Lamprecht, M.R.; Moffat, J.; Silver, S.J.; Grenier, J.K.; Castoreno, A.B.; Eggert, U.S.; Root, D.E.; Golland, P.; et al. Scoring diverse cellular morphologies in image-based screens with iterative feedback and machine learning. *Proc. Natl. Acad. Sci. USA* **2009**, *106*, 1826–1831. [CrossRef]
33. Yuan, Y.; Failmezger, H.; Rueda, O.M.; Ali, H.R.; Gräf, S.; Chin, S.F.; Schwarz, R.F.; Curtis, C.; Dunning, M.J.; Bardwell, H.; et al. Quantitative image analysis of cellular heterogeneity in breast tumors complements genomic profiling. *Sci. Transl. Med.* **2012**, *4*, 157ra143. [CrossRef] [PubMed]
34. Sirinukunwattana, K.; Raza, S.E.A.; Tsang, Y.W.; Snead, D.R.; Cree, I.A.; Rajpoot, N.M. Locality sensitive deep learning for detection and classification of nuclei in routine colon cancer histology images. *IEEE Trans. Med. Imaging* **2016**, *35*, 1196–1206. [CrossRef]
35. Luo, P.; Ren, J.; Peng, Z.; Zhang, R.; Li, J. Differentiable Learning-to-Normalize via Switchable Normalization. In Proceedings of the International Conference on Learning Representation (ICLR), Vancouver, BC, Canada, 30 April–3 May 2019.
36. Han, K.; Wang, Y.; Tian, Q.; Guo, J.; Xu, C.; Xu, C. Ghostnet: More features from cheap operations. In Proceedings of the IEEE/CVF Conference on Computer Vision and Pattern Recognition, Seattle, WA, USA, 13–19 June 2020; pp. 1580–1589. [CrossRef]
37. Huang, X.; Belongie, S. Arbitrary style transfer in real-time with adaptive instance normalization. In Proceedings of the IEEE International Conference on Computer Vision, Venice, Italy, 22–29 October 2017; pp. 1501–1510. [CrossRef]
38. Ba, J.L.; Kiros, J.R.; Hinton, G.E. Layer normalization. *arXiv* **2016**, arXiv:1607.06450.
39. Ioffe, S.; Szegedy, C. Batch normalization: Accelerating deep network training by reducing internal covariate shift. In Proceedings of the International Conference on Machine Learning, Lille, France, 7–9 July 2015; pp. 448–456.
40. He, K.; Zhang, X.; Ren, S.; Sun, J. Deep residual learning for image recognition. In Proceedings of the IEEE Conference on Computer Vision and Pattern Recognition, Las Vegas, NV, USA, 27–30 June 2016; pp. 770–778. [CrossRef]
41. Huang, G.; Liu, Z.; Van Der Maaten, L.; Weinberger, K.Q. Densely connected convolutional networks. In Proceedings of the IEEE Conference on Computer Vision and Pattern Recognition, Honolulu, HI, USA, 21–26 July 2017; pp. 4700–4708. [CrossRef]
42. Kumar, N.; Verma, R.; Sharma, S.; Bhargava, S.; Vahadane, A.; Sethi, A. A dataset and a technique for generalized nuclear segmentation for computational pathology. *IEEE Trans. Med. Imaging* **2017**, *36*, 1550–1560. [CrossRef]
43. Vu, Q.D.; Graham, S.; Kurc, T.; To, M.N.N.; Shaban, M.; Qaiser, T.; Koohbanani, N.A.; Khurram, S.A.; Kalpathy-Cramer, J.; Zhao, T.; et al. Methods for segmentation and classification of digital microscopy tissue images. *Front. Bioeng. Biotechnol.* **2019**, *7*, 53. [CrossRef] [PubMed]
44. He, K.; Zhang, X.; Ren, S.; Sun, J. Delving deep into rectifiers: Surpassing human-level performance on imagenet classification. In Proceedings of the IEEE International Conference on Computer Vision, Santiago, Chile, 7–13 December 2015; pp. 1026–1034. [CrossRef]
45. Kingma, D.P.; Ba, J. Adam: A method for stochastic optimization. *arXiv* **2014**, arXiv:1412.6980.
46. Kirillov, A.; He, K.; Girshick, R.; Rother, C.; Dollár, P. Panoptic segmentation. In Proceedings of the IEEE/CVF Conference on Computer Vision and Pattern Recognition, Long Beach, CA, USA, 15–20 June 2019; pp. 9404–9413. [CrossRef]
47. Chen, H.; Qi, X.; Yu, L.; Heng, P.A. DCAN: Deep contour-aware networks for accurate gland segmentation. In Proceedings of the IEEE Conference on Computer Vision and Pattern Recognition, Las Vegas, NV, USA, 27–30 June 2016; pp. 2487–2496. [CrossRef]
48. Badrinarayanan, V.; Kendall, A.; Cipolla, R. Segnet: A deep convolutional encoder-decoder architecture for image segmentation. *IEEE Trans. Pattern Anal. Mach. Intell.* **2017**, *39*, 2481–2495. [CrossRef]
49. Long, J.; Shelhamer, E.; Darrell, T. Fully convolutional networks for semantic segmentation. In Proceedings of the IEEE Conference on Computer Vision and Pattern Recognition, Boston, MA, USA, 7–12 June 2015; pp. 3431–3440. [CrossRef]

bioengineering

MDPI

Article

Physiological Noise Filtering in Functional Near-Infrared Spectroscopy Signals Using Wavelet Transform and Long-Short Term Memory Networks

So-Hyeon Yoo [1], Guanghao Huang [2] and Keum-Shik Hong [1,2,*]

1 School of Mechanical Engineering, Pusan National University, 2 Busandaehak-ro, Geumjeong-gu, Busan 46241, Republic of Korea; bsh00156@pusan.ac.kr
2 Institute for Future, School of Automation, Qingdao University, Qingdao 266071, China; hgh@qdu.edu.cn
* Correspondence: kshong@pusan.ac.kr

Abstract: Activated channels of functional near-infrared spectroscopy are typically identified using the desired hemodynamic response function (dHRF) generated by a trial period. However, this approach is not possible for an unknown trial period. In this paper, an innovative method not using the dHRF is proposed, which extracts fluctuating signals during the resting state using maximal overlap discrete wavelet transform, identifies low-frequency wavelets corresponding to physiological noise, trains them using long-short term memory networks, and predicts/subtracts them during the task session. The motivation for prediction is to maintain the phase information of physiological noise at the start time of a task, which is possible because the signal is extended from the resting state to the task session. This technique decomposes the resting state data into nine wavelets and uses the fifth to ninth wavelets for learning and prediction. In the eighth wavelet, the prediction error difference between the with and without dHRF from the 15-s prediction window appeared to be the largest. Considering the difficulty in removing physiological noise when the activation period is near the physiological noise, the proposed method can be an alternative solution when the conventional method is not applicable. In passive brain-computer interfaces, estimating the brain signal starting time is necessary.

Keywords: functional near-infrared spectroscopy; filtering; physiological noise; maximal overlap discrete wavelet transform; long-short term memory

Citation: Yoo, S.-H.; Huang, G.; Hong, K.-S. Physiological Noise Filtering in Functional Near-Infrared Spectroscopy Signals Using Wavelet Transform and Long-Short Term Memory Networks. *Bioengineering* **2023**, *10*, 685. https://doi.org/10.3390/bioengineering10060685

Academic Editors: Yan Pei and Jijiang Yang

Received: 3 May 2023
Revised: 26 May 2023
Accepted: 2 June 2023
Published: 4 June 2023

1. Introduction

In processing functional near-infrared spectroscopy (fNIRS) signals, a task-related hemodynamic signal cannot be identified if a physiological noise period is overlapped with the designed task period. This study proposes a novel method to identify physiological noises from the resting state and remove those noises during the task period using wavelet techniques and neural networks-based prediction. FNIRS is a brain-imaging technique that uses two or more wavelengths of light in near-infrared bands to measure changes in oxidized and deoxidized hemoglobin concentration in the cerebral cortex [1]. When a person moves, thinks, or receives an external stimulus, the nerve cells in cerebral cortical layers become excited. As the cells require more energy, the oxidized hemoglobin concentration around the nerve cells increases, and the deoxygenated hemoglobin concentration decreases [1]. Based on this principle, fNIRS can measure brain activity in real time. Because fNIRS is inexpensive, easy to use, and harmless to the human body, it has been used in brain disease diagnosis [2,3], brain-computer interface (BCI) [4], decoding sensory signals [5,6], child development [7], and psychology research [8].

An fNIRS channel consists of one source and one detector. When the light is emitted from a light source, photons pass through several layers, including the scalp, skull, cerebrospinal fluid, capillaries, and cerebral cortex, before returning to a detector. Through this process, the detected light contains various noises that make it challenging to know the hemodynamic responses. These noises include heartbeat, breathing, and motion artifacts [9]; more problematically, very low-frequency noise around 0.01 Hz has been reported [10–12].

In improving the accuracy of the measured signal, noise removal/reduction techniques are indispensable. Various techniques can remove physiological noises such as heartbeat, breathing, and Mayer waves. For instance, the superficial noise in the scalp can be removed using short separated channels [13], additional external devices, or applying denoising techniques such as adaptive filtering [14] and correlation analysis methods [15]. In addition, since the frequency bands of physiological noise are roughly known, a band-pass filter has become one of the most easily applied noise reduction techniques [16].

A general linear model (GLM) method has been widely used to find the task-related hemodynamic response in the fNIRS signal after preprocessing [17]. The desired hemodynamic response function (dHRF), which should be used for the GLM method, is designed considering the experimental paradigm. However, in the case that the essential frequency of the dHRF overlaps with a specific frequency of physiological noise, the conventional GLM method will not work and may result in mistaking noise for the hemodynamic response. Therefore, a new different denoising technique must be pursued.

A discrete wavelet transform (DWT) is a mathematical tool used to analyze signals in the time-frequency domain [18]. In fNIRS research, DWT has been used for denoising [19,20] and connectivity analysis [21,22]. The maximal overlap discrete wavelet transform (MODWT) is a type of DWT often used in signal processing and time series analysis [23]. It decomposes a signal into a series of wavelet coefficients at different widths and time locations. Unlike the usual DWTs, which use non-overlapping sub-signal windows to perform the wavelet decomposition, the MODWT uses overlapping sub-signal windows. This nested-window approach allows the MODWT to improve time-frequency localization and reduce the boundary effects that can occur in DWTs [24]. Due to this advantage, MODWT has been applied to a wide range of signals, including audio signals [25], weather information [26,27], and biomedical signals [28]. MODWT is powerful when the signals are abnormal or have complex frequency components.

Deep learning, a subfield of artificial intelligence, is based on artificial neural networks. In recent years, brain research has increasingly used it to analyze large, complex data sets, such as those generated by biomedical devices [29]. Research has been conducted to analyze health data such as magnetic resonance imaging (MRI) [30], electrocardiograms (ECG) [31] and electroencephalograms (EEG) [32,33], or to decode brain waves to control BCI [34,35]. Furthermore, analyzing brain neuroimaging data and identifying patterns associated with specific diseases can help with early diagnosis and personalized treatment.

Long short-term memory (LSTM) is a type of recurrent neural network (RNN) architecture designed to overcome the limitations of traditional RNNs in handling long-term dependencies in sequential data [36]. It has been used in a wide range of applications for time-series data classification and forecasting [37–40]. LSTMs are particularly useful in tasks that require modeling long-term dependencies in sequential data. LSTMs' ability to selectively remember and forget information over time is vital for accurate forecasting.

MODWT-LSTM-based prediction research has shown excellent results in predicting periodic data such as water level [41], ammonia nitrogen [42], weather [43], etc. In brain research, MODWT has been applied as a preprocessing method for EEG-based seizure detection [28,44], Alzheimer's diagnosis [45], and resting state network analysis of fMRI [46]. Since brain signals are measured in time series, active research on brain signal classification [47,48] uses LSTM. However, to our knowledge, this is the first study to predict the noise in fNIRS signals despite many of the noise components being periodic.

In this study, one thousand synthetic data are generated, assuming 600-s rest and 40-s task. Each data is decomposed into eight levels by the MODWT. Five wavelets containing low-frequency components from the 600-s data are used to train an LSTM network. The trained LSTM networks are used to predict the next 40 s, presumably the predicted signals of the low-frequency oscillations. The predicted signals are then subtracted from the task period data. For validation purposes, the predicted signal and original data are compared by calculating mean absolute errors (MAEs), and root mean square errors (RMSEs). Finally, the proposed method is demonstrated by analyzing the actual fNIRS data from humans.

This paper is organized as follows: Section 2 describes the proposed method on the synthetic data, Section 3 demonstrates the proposed method with actual fNIRS data, Section 4 discusses the results of this study and its applications, and Section 5 presents conclusions.

2. Method Development

This section describes the development of the proposed method with the following four subsections. In the first subsection, the method of synthetic fNIRS data generation is described. The second and third subsections explain the operation of MODWT and LSTM, respectively. The fourth subsection describes the validation of the proposed method. The last subsection presents the results of the data analysis.

2.1. Synthetic fNIRS Data Generation

One thousand synthetic data are generated according to the method of Germignani et al. [49] with a sampling frequency of 8.138 Hz. For each data, thirty orders of autoregressive noise are added to the baseline noise [50]. The synthetic physiological noises include frequency ranges of 1 ± 0.1 Hz, 0.25 ± 0.01 Hz, and 0.1 ± 0.01 Hz for cardiac, respiratory, and Mayer waves, respectively. In addition, a sine wave with a frequency of 0.01 ± 0.001 Hz was generated for the very low-frequency component [11]. The amplitudes of five signals in a synthetic fNIRS signal were set randomly in the range of 0.01 to 0.03. In this paper, the resting period is set to 10 min, considering that the concerned low-frequency noise is near 0.01 Hz.

For five hundred data samples only, the desired hemodynamic function (dHRF) based on a 2-gamma function with 20 s of task and 20 s of resting state after the 10 min resting state were added. The amplitude of this signal was randomized between 0.1 and 0.35 and added to the previously generated noise. All data were set to zero at the starting point before processing the signals. Figure 1 depicts synthetic signals for various noises and the resultant HbO signal assumed.

2.2. Maximal Overlap Discrete Wavelet Transform

The discrete wavelet transform (DWT) is a signal processing technique that decomposes a signal into different frequency components at multiple levels of resolution. The DWT works by convolving the signal with a set of filters, called wavelet filters, which capture different frequency bands. The signal is decomposed into approximation and detail coefficients [19], which represent low-frequency components and high-frequency components, respectively. This decomposition is applied recursively to the approximation coefficients to obtain a multi-resolution representation. However, the DWT has several drawbacks, including the introduction of boundary artifacts due to the filtering process, the lack of shift invariance in the decomposition, and the potential loss of fine detail at higher decomposition levels.

Zhang et al. (2018) [51] utilized the DWT in forecasting vehicle emissions and specifically compared four cases: The autoregressive integrated moving average (ARIMA) model, LSTM, DWT-ARIMA, and DWT-LSTM. They reported that adopting DWT improved the performance overall. Individually, between ARIMA and LSTM, LSTM performed better; between ARIMA and DWT-ARIMA, DWT-ARIMA generated improved results; between LSTM and DWT-LSTM, DWT-LSTM was superior; and between DWT-LSTM and DWT-ARIMA, DWT-LSTM demonstrated the best forecasting.

MODWT is a mathematical technique that transforms a signal into a multilevel wavelet and scaling factor. MODWT has several advantages over DWT. For example, the MODWT can be adequately defined for signals of arbitrary length, whereas the DWT is only for signals of integer length to the power of two.

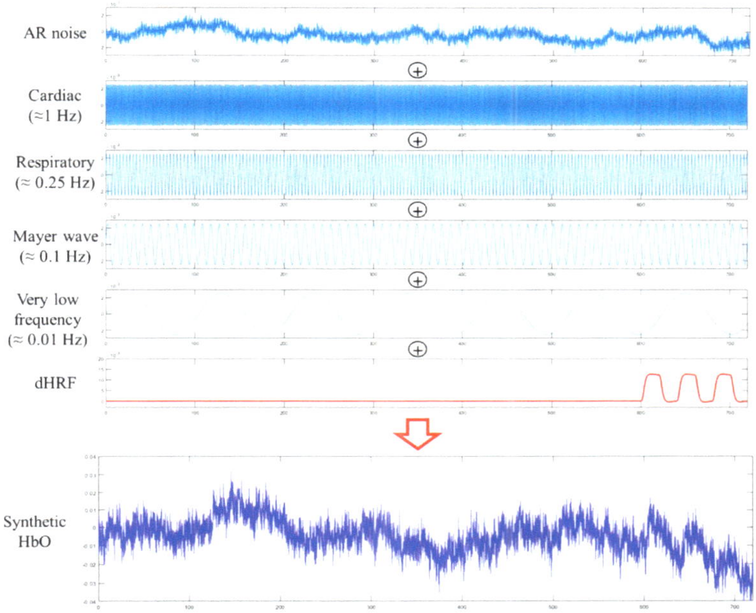

Figure 1. A synthetic HbO signal is made of six components.

For discrete signal $X = \{X_t, t = 0, 1, \cdots, n-1\}$, the jth element W_j and scaling factor V_j of the MODWT are defined as follows.

$$W_{j,t} = \sum_{l=0}^{n-1} \overset{\sim\circ}{h}_{j,l} X_{t-l\mathrm{mod}n}, \; j = 1, 2, \cdots, L, \tag{1}$$

$$V_{j,t} = \sum_{l=0}^{n-1} \overset{\sim\circ}{g}_{j,l} X_{t-l\mathrm{mod}n}, \tag{2}$$

where $W_{j,t}$ is the wavelet coefficient of the tth element of the jth level of the MODWT; $V_{j,t}$ is the scaling factor of the tth element of the jth level; $\overset{\sim\circ}{h}_{j,l}$ and $\overset{\sim\circ}{g}_{j,l}$ are the jth level's high- and low-pass filters (wavelet and scaling filters) of MODWT generated by periodizing $\tilde{h}_{j,l}$ and $\tilde{g}_{j,l}$, respectively, with n lengths; $\tilde{h}_{j,l}$ and $\tilde{g}_{j,l}$ are the jth level MODWT high $(\tilde{h}_{j,l} \equiv h_{j,l}/2^{\frac{j}{2}})$ and low $(\tilde{g}_{j,l} \equiv g_{j,l}/2^{\frac{j}{2}})$ pass filters; $h_{j,l}$ and $g_{j,l}$ are the jth level DWT high-pass and low-pass filters, where L is the maximum decomposition level. The filters are determined by the mother wavelet as in the DWT [52]. The MODWT based multiresolution analysis is expressed as follows.

$$X = \sum_{j=1}^{L} D_j + A_{J0}, \tag{3}$$

$$D_{j,t} = \sum_{l=0}^{n-1} \overset{\sim\circ}{h}_{j,l} W_{j,t+l\mathrm{mod}n}, \tag{4}$$

$$A_{j,t} = \sum_{l=0}^{n-1} \overset{\sim\circ}{g}_{j,l} V_{j,t+l\mathrm{mod}n}, \tag{5}$$

where A_L is the approximation component and D_j is the detail components ($j = 1, 2, \cdots, L$). Figure 2 shows a scheme of MODWT-based multiresolution analysis.

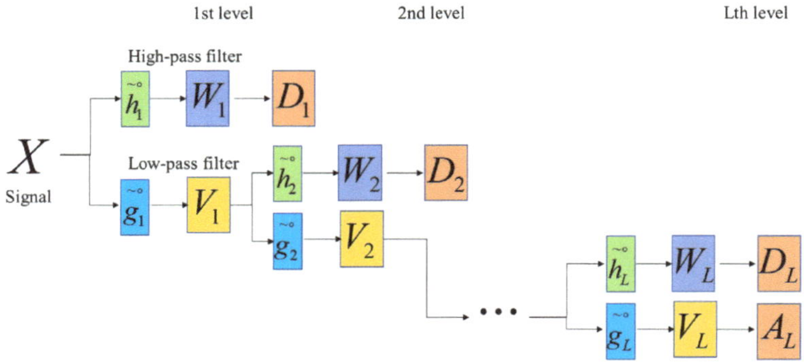

Figure 2. Schematic of the MODWT decomposition.

In this study, Sym4 was selected as the mother wavelet because it resembles the canonical hemodynamic response function. Let the number of data be N. Then, the maximum decomposition level becomes less than $\log_2(N)$. Considering our case's shortest resting state of 60 s, the data size is 60 s $\times$ 8.13 Hz = 487.8. Therefore, the decomposition level in our work was selected by 8, which is the largest integer less than $\log_2(487.8)$. The eight decompositions result in nine signals, of which only five signals belonging to low frequencies will be predicted.

2.3. *Long Short-Term Memory*

LSTM is a type of RNN architecture that addresses the vanishing gradient problem and allows for capturing long-term dependencies in sequential data. LSTM consists of memory cells that store and update information over time. The primary function of an LSTM is to use memory cells that can hold information for long periods. Memory cells can selectively forget or remember information based on input data and past states. This allows the network to learn and remember important information while ignoring irrelevant or redundant information. An LSTM network has three gates (input gate, forget gate, and output gate) that control the flow of information into and out of the memory cells. The input gate $i(t)$ determines which information is stored in the memory cell $c(t)$, the forget gate $f(t)$ determines which information is discarded, and the output gate $o(t)$ controls the output of the memory cell (Figure 3) [53].

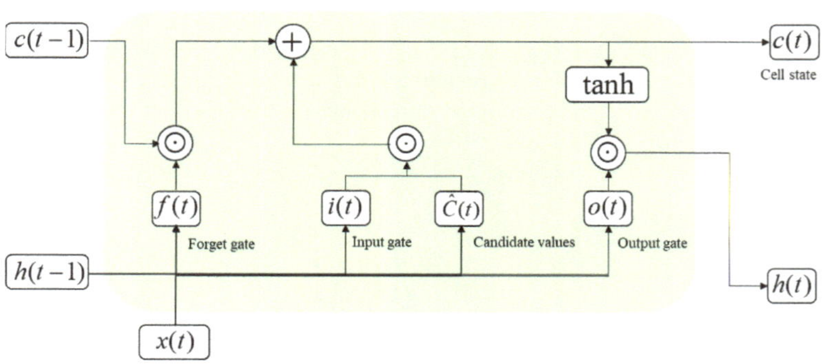

Figure 3. A structure of LSTM layers.

The LSTM model is represented by the following equations:

$$a(t) = \sigma(W_i x(t) + U_i h(t-1) + b_i), \quad (6)$$

$$f(t) = \sigma(W_f x(t) + U_f h(t-1) + b_f), \quad (7)$$

$$\tilde{c}(t) = \tanh(W_c x(t) + U_c h(t-1) + b_c), \quad (8)$$

$$c(t) = f_t \times c(t-1) + i_t \times \tilde{c}(t), \quad (9)$$

$$o(t) = \sigma(W_o x(t) + U_o h(t-1) + b_o), \quad (10)$$

$$h(t) = o(t) \times \tanh(c(t)), \quad (11)$$

where $c(t-1)$ and $c(t)$ are the cell states at $t-1$ and t, and at each gate, b_i, b_f, b_c, b_0 are the bias vectors, W_i, W_f, W_c, W_0 are the weight matrices, and U_i, U_f, U_c, U_0 are the recurrent weights. σ is a sigmoid function, tanh is a hyperbolic tangent activation function, and $\times$ denotes the cross product of two vectors.

In this study, three LSTM layers were utilized, with the number of hidden units set to [128, 64, 32], and a dropout layer was employed between the LSTM layers with a probability of 0.2 to prevent overfitting (Figure 4). To train the LSTM network, the Adam optimizer was used with a maximum epoch of 100 and a minibatch size of 128. All data were normalized before training.

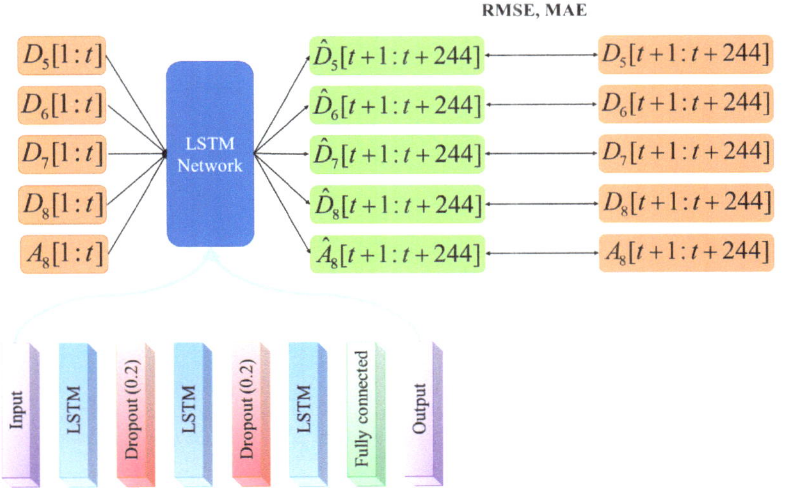

Figure 4. Diagram of the proposed time-series prediction based on MODWT-LSTM.

For the synthetic data, nine hundred data were randomly selected from the thousand data to train the network, and then one hundred data were tested. The number of data points trained was divided into five conditions ([600 s, 300 s, 150 s, 90 s, 60 s] $\times$ sampling rate (8.13 Hz) = [4883, 2441, 1221, 732, 488]), and then 244 data points (30 s $\times$ 8.13 Hz) were predicted.

For actual fNIRS data, a leave-one-out method was used to avoid splitting data from the same person for training and testing. For example, to train an LSTM network to predict the 48 channels of a subject, a total of 432 channels (nine subjects $\times$ 48 channels) were used. Since the data was only 600 s long, 570 s of data were used for training, and the trained LSTM network predicted the next 30 s.

2.4. Validation

To determine the accuracy of the signal predicted by the LSTM, the mean absolute error (MAE) and root mean squared error (RMSE) were calculated compared to the original signal, divided by the signal with and without dHRF. The data was segmented, analyzed, and predicted to find the required resting-state length to achieve optimal prediction accuracy, as shown in Figure 5. MAE and RMSE can be calculated using the following equations.

$$\mathrm{MAE} = \frac{1}{n}\sum_{i=1}^{n}|y_i - \hat{y}_i|, \tag{12}$$

$$\mathrm{RMSE} = \sqrt{\frac{\sum_{i=1}^{n}(y_i - \hat{y}_i)^2}{n}}, \tag{13}$$

where y_i is the original signal, $\hat{y}_i$ is the predicted signal, i is the timestep, and n is the number of data. The calculated MAEs and RMSEs of the signal with and without the dHRF were compared using a two-sample t-test.

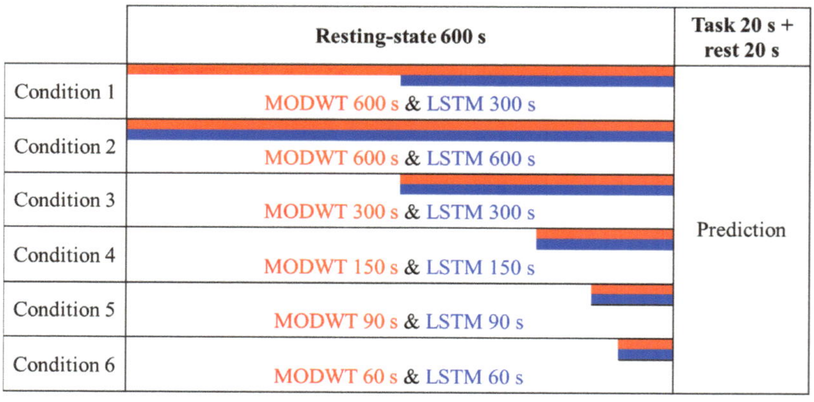

Figure 5. Data segmentation for validation (red: MODWT data length, blue: LSTM data length for training).

2.5. Synthetic Data Analysis

The synthetic data were decomposed into nine components using MODWT, and the components used for prediction were the fifth through ninth. The frequency of the fifth wavelet was between 0.13 and 0.26 Hz, the sixth between 0.067 and 0.13 Hz, the seventh between 0.035 and 0.067 Hz, the eighth between 0.017 and 0.035 Hz, and the ninth consisted of signals below 0.017 Hz. Figure 6 shows the prediction results of the signal with and without dHRF. The signal with dHRF showed a significant fluctuation during the task period in the low-frequency signals of Wavelets 6–9, and the predicted signal did not follow this fluctuation.

Figure 7 and Table 1 show the calculated MAEs and RMSEs. In all conditions, the MAEs and RMSEs of the signal with dHRF corresponding to Wavelets 6–9 and the signal without dHRF were statistically significantly different. The only statistically significant difference between with and without dHRF was found in the RMSE of Wavelet 5 when the MODWT-LSTM analysis was performed with 300 s of data (Figure 7c). To compare the prediction results for each condition, MAEs and RMSEs for all conditions are shown in Figure 8. The error of the dHRF signal was the largest in Condition 2 (MODWT-LSTM at 600 s) and the smallest in Condition 6 (MODWT-LSTM with 60 s of data). In particular, the difference in prediction accuracy between with and without dHRF signals of Wavelet 8 was the largest in all conditions.

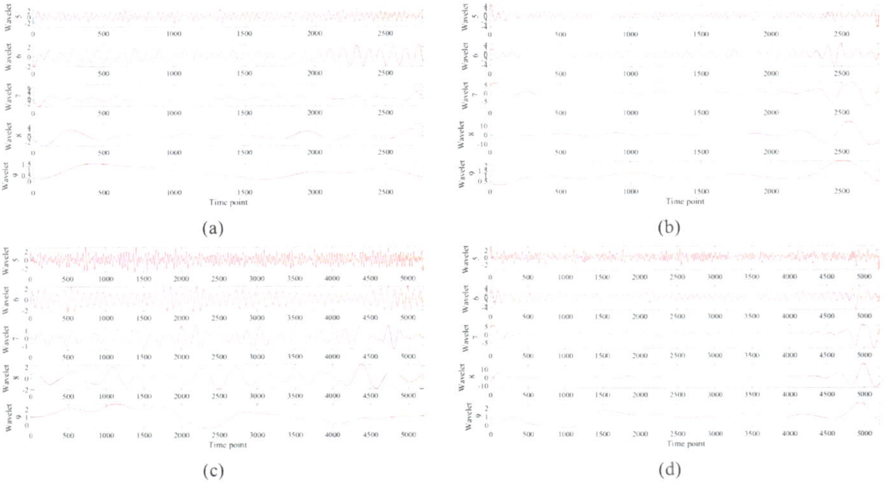

Figure 6. Prediction results from synthetic data: (**a**) Without dHRF (MODWT 300 s, LSTM 300 s), (**b**) with dHRF (MODWF 300 s, LSTM 300 s), (**c**) without dHRF (MODWT 600 s, LSTM 600 s), and (**d**) with dHRF (MODWT 600 s, LSTM 600 s) (blue line: training data, red line: MODWT results including test time, orange dotted line: predicted result).

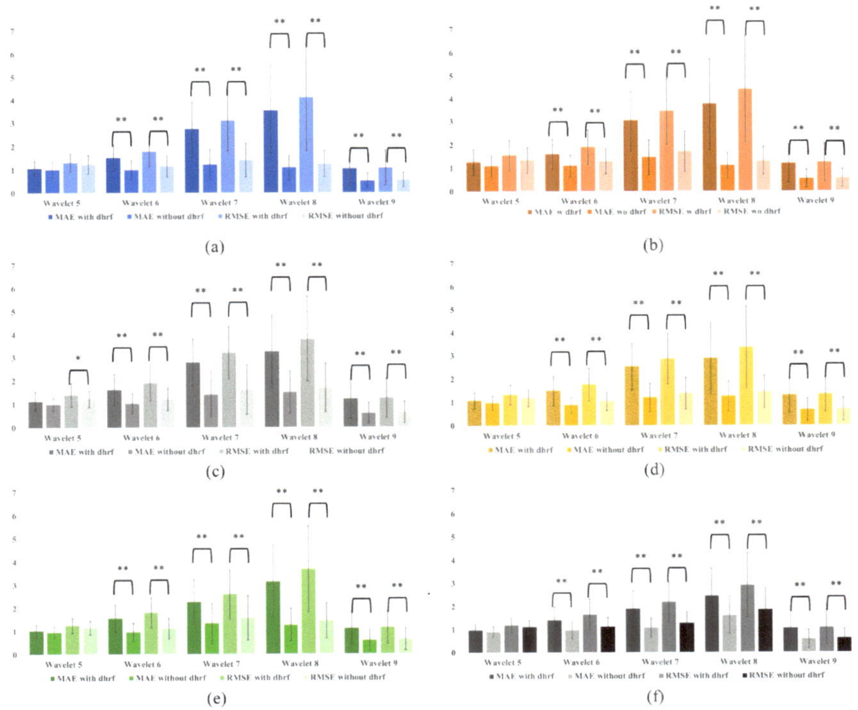

Figure 7. MAEs and RMSEs for wavelets 5–9: (**a**) MODWT 600 s and LSTM training 300 s, (**b**) MODWT 600 s and LSTM training 600 s, (**c**) MODWT 300 s and LSTM training 300 s, (**d**) MODWT 150 s and LSTM training 150 s, (**e**) MODWT 90 s and LSTM training 90 s, and (**f**) MODWT 60 s and LSTM training 60 s (* $p < 0.05$, ** $p < 0.01$).

Table 1. Mean and standard deviation of MAEs and RMSEs for synthetic data (* $p < 0.05$, ** $p < 0.01$).

		Wavelet 5				Wavelet 6				Wavelet 7				Wavelet 8				Wavelet 9			
		MAE with dHRF	MAE w/o dHRF	RMSE with dHRF	RMSE w/o dHRF	MAE with dHRF	MAE w/o dHRF	RMSE with dHRF	RMSE w/o dHRF	MAE with dHRF	MAE w/o dHRF	RMSE with dHRF	RMSE w/o dHRF	MAE with dHRF	MAE w/o dHRF	RMSE with dHRF	RMSE w/o dHRF	MAE with dHRF	MAE w/o dHRF	RMSE with dHRF	RMSE w/o dHRF
1	Mean	1.07	1.02	1.32	1.24	1.53	1.01	1.79	1.17	2.74	1.23	3.10	1.42	3.54	1.10	4.12	1.26	1.06	0.53	1.09	0.56
	Std	0.32	0.33	0.38	0.39	0.57	0.41	0.64	0.46	1.18	0.64	1.29	0.72	1.97	0.52	2.30	0.57	0.76	0.32	0.76	0.32
	p	0.40		0.35		0.00 **		0.00 **		0.00 **		0.00 **		0.00 **		0.00 **		0.00 **		0.00 **	
2	Mean	1.27	1.10	1.57	1.35	1.62	1.12	1.91	1.31	3.04	1.48	3.45	1.72	3.78	1.13	4.42	1.32	1.22	0.55	1.25	0.58
	Std	0.54	0.44	0.64	0.54	0.64	0.46	0.73	0.53	1.30	0.73	1.42	0.85	1.96	0.54	2.30	0.60	0.81	0.38	0.81	0.39
	p	0.09		0.08		0.00 **		0.00 **		0.00 **		0.00 **		0.00 **		0.00 **		0.00 **		0.00 **	
3	Mean	1.13	0.99	1.40	1.22	1.62	1.03	1.91	1.21	2.79	1.41	3.22	1.62	3.27	1.50	3.81	1.70	1.21	0.60	1.25	0.64
	Std	0.41	0.27	0.50	0.34	0.67	0.43	0.75	0.48	1.02	0.96	1.13	1.05	1.58	0.91	1.84	1.04	0.83	0.44	0.83	0.46
	p	0.05		0.04 *		0.00 **		0.00 **		0.00 **		0.00 **		0.00 **		0.00 **		0.00 **		0.00 **	
4	Mean	1.08	0.97	1.33	1.19	1.50	0.88	1.76	1.04	2.54	1.21	2.87	1.39	2.90	1.26	3.37	1.44	1.30	0.67	1.34	0.70
	Std	0.35	0.30	0.43	0.37	0.63	0.32	0.71	0.39	1.00	0.61	1.09	0.69	1.53	0.64	1.77	0.70	0.74	0.48	0.74	0.49
	p	0.11		0.09		0.00 **		0.00 **		0.00 **		0.00 **		0.00 **		0.00 **		0.00 **		0.00 **	
5	Mean	1.02	0.95	1.26	1.17	1.57	0.97	1.82	1.13	2.28	1.35	2.61	1.58	3.15	1.28	3.69	1.46	1.13	0.61	1.16	0.64
	Std	0.27	0.25	0.32	0.30	0.59	0.39	0.66	0.45	0.97	0.86	1.06	0.96	1.58	0.71	1.85	0.76	0.71	0.47	0.71	0.48
	p	0.17		0.13		0.00 **		0.00 **		0.00 **		0.00 **		0.00 **		0.00 **		0.00 **		0.00 **	
6	Mean	0.98	0.91	1.20	1.12	1.41	0.98	1.66	1.13	1.90	1.08	2.19	1.28	2.46	1.63	2.92	1.86	1.05	0.57	1.09	0.62
	Std	0.26	0.24	0.32	0.29	0.57	0.36	0.66	0.40	0.77	0.41	0.86	0.46	1.18	0.81	1.38	0.91	0.77	0.39	0.76	0.40
	p	0.19		0.23		0.00 **		0.00 **		0.00 **		0.00 **		0.00 **		0.00 **		0.00 **		0.00 **	

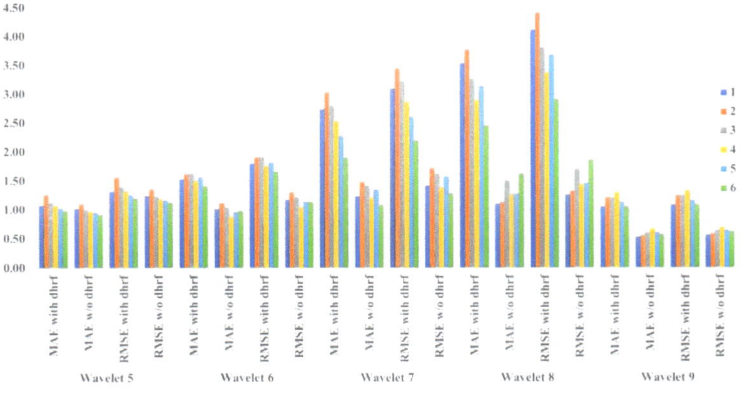

Figure 8. Comparison of MAE and RMSE for all data segmentations: (1) MODWT 600 s and LSTM training 300 s, (2) MODWT 600 s and LSTM training 600 s, (3) MODWT 300 s and LSTM training 300 s, (4) MODWT 150 s and LSTM training 150 s, (5) MODWT 90 s and LSTM training 90 s, and (6) MODWT 60 s and LSTM training 60 s.

For the 600 s data prediction results, MAEs and RMSEs were calculated for 1 s, 3 s, 5 s, 10 s, 15 s, and 30 s (Figure 9). In all cases, there were statistically significant differences in Wavelets 6–9 between with and without dHRF. Especially for Wavelet 7, with the most significant difference at 1 s and a decrease after that, but for Wavelet 8, the difference started at 10 s and was most extensive at 15 s.

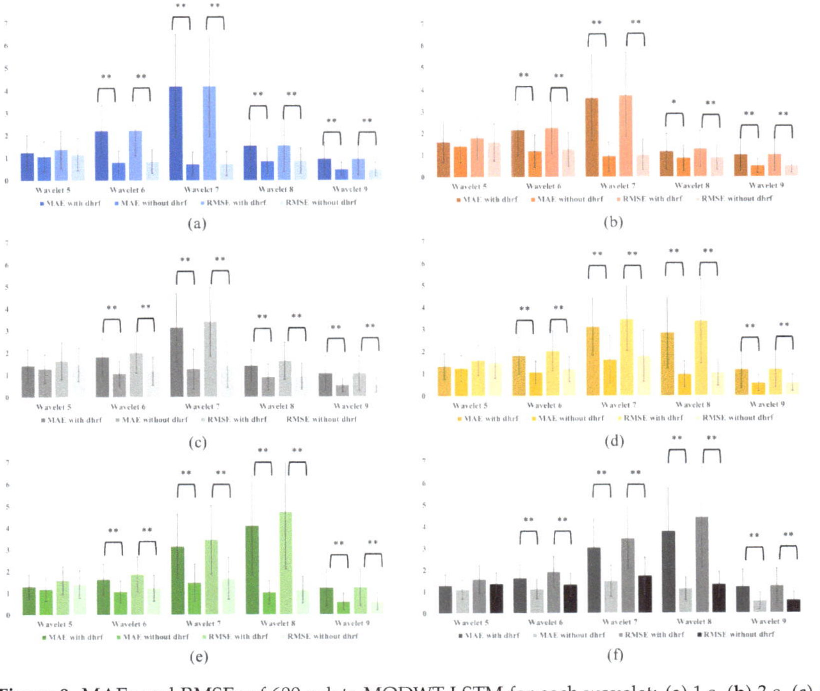

Figure 9. MAEs and RMSEs of 600 s data MODWT-LSTM for each wavelet: (**a**) 1 s, (**b**) 3 s, (**c**) 5 s, (**d**) 10 s, (**e**) 15 s, and (**f**) 30 s (* $p < 0.05$, ** $p < 0.01$).

3. Human Data Application

In this section, actual fNIRS data from human subjects were used to validate the proposed method. The actual fNIRS data were obtained in the authors' previous study, but only resting state data were used [2]. In the first subsection, the fNIRS data acquisition is briefly described. The second subsection describes the results of the application of the proposed method.

3.1. fNIRS Data Acquisition

Resting state data with a data length of 10 min were selected from ten healthy subjects. The selected subjects are five males and five females (age: 68 ± 5.95 years). Prior to the experiment, each subject was fully informed about the purpose of the study. Written informed consent was obtained from each subject. The entire experiment was approved by the ethics committee of Pusan National University Yangsan Hospital (Institutional Review Board approval number: PNUYH-03-2018-003).

Hemodynamic responses in PFC were measured with a portable fNIRS device (NIRSIT; OBELAB, Seoul, Republic of Korea) equipped with 24 sources (laser diode) and 32 detectors (a total of 204 channels, including short channel separation) at a sampling rate of 8.138 Hz. NIRSIT uses two wavelengths of near-infrared light (780 nm and 850 nm) to measure concentration changes of HbO and HbR. Only 48 channels with 3 cm of channel distance out of 204 channels were used for this study.

3.2. Human Data Analysis

The prediction results for the actual HbO data are shown in Figure 10. Unlike the synthetic data, the amplitude of the ninth wavelet was significantly lower than the other wavelets. A spike appeared in all the wavelets at a particular time, presumably a motion artifact. The MODWT results differed at both ends of the wavelets for the 570 s data and the 600 s data.

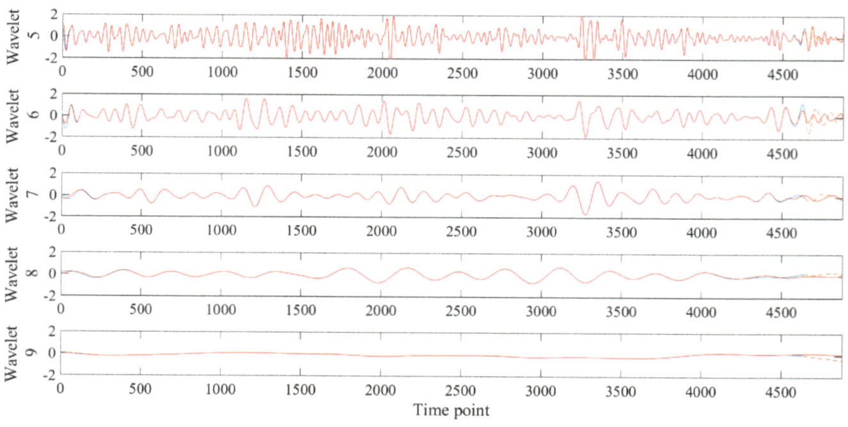

Figure 10. Prediction results of MODWT-LSTM for actual HbO data (blue line: training data, red line: MODWT results including test time points, orange dotted line: predicted result).

Table 2 shows the results of calculating the mean and standard deviation of the MAEs and RMSEs of the predictions on the real HbO data. Among them, the average value is plotted for easy comparison (Figure 11). The ninth wavelet had the slightest error but the most significant standard deviation across all cases. The fifth and sixth wavelets showed increasingly significant errors until 3 s and 5 s, respectively, then decreased. The seventh and eighth wavelets had more significant errors as the time window increased.

Table 2. Mean and standard deviation of MAEs and RMSEs for real data.

		MAE						RMSE					
		1 s	**3 s**	**5 s**	**10 s**	**15 s**	**30 s**	**1 s**	**3 s**	**5 s**	**10 s**	**15 s**	**30 s**
Wavelet 5	Mean	0.538	0.775	0.677	0.556	0.494	0.469	0.594	0.865	0.786	0.677	0.617	0.619
	Std	0.835	1.056	1.009	0.856	0.796	0.744	0.873	1.138	1.141	0.995	0.932	1.063
Wavelet 6	Mean	0.359	0.612	0.698	0.635	0.580	0.598	0.385	0.682	0.779	0.731	0.682	0.737
	Std	0.643	0.714	0.988	0.905	0.814	0.953	0.669	0.804	1.120	1.056	0.950	1.356
Wavelet 7	Mean	0.309	0.429	0.527	0.529	0.552	0.645	0.316	0.456	0.566	0.585	0.622	0.759
	Std	0.828	1.063	1.297	1.090	1.247	1.520	0.835	1.105	1.356	1.204	1.385	1.802
Wavelet 8	Mean	0.335	0.322	0.334	0.428	0.525	0.683	0.338	0.336	0.358	0.481	0.595	0.805
	Std	1.295	1.172	1.144	1.377	1.680	1.880	1.298	1.208	1.226	1.541	1.852	2.156
Wavelet 9	Mean	0.363	0.372	0.375	0.379	0.378	0.362	0.364	0.374	0.378	0.385	0.388	0.387
	Std	2.157	2.188	2.214	2.225	2.176	1.838	2.157	2.189	2.215	2.227	2.180	1.886

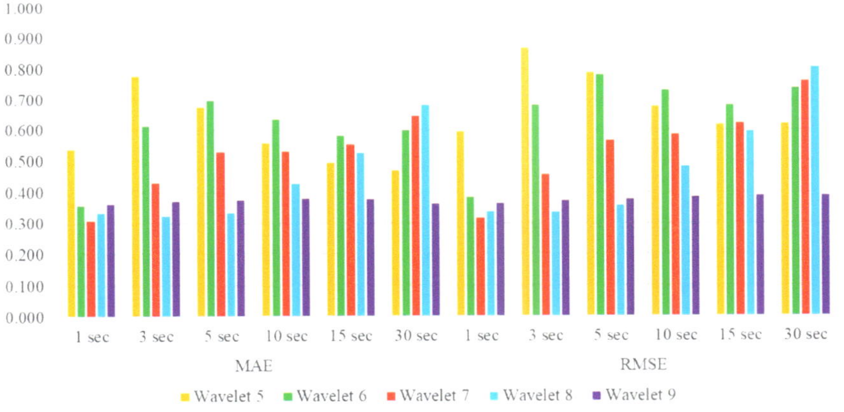

Figure 11. Averaged MAEs and RMSEs of real HbO data MODWT-LSTM for each wavelet by predicted time windows.

4. Discussion

In fNIRS studies, cognitive tasks are used to evaluate cognitive abilities such as working memory, conflict processing, language processing, emotional processing, and memory encoding and retrieval [54]. For example, *N*-back, Stroop, and verbal fluency tasks evaluate working memory, conflict processing, language processing, etc. Such cognitive tasks are also often used to detect brain diseases such as schizophrenia, depression, cognitive impairment, attention-deficit hyperactivity disorder, etc. [55].

Cortical activations caused by cognitive tasks are investigated by a *t*-map, a connectivity map, or extracted features from HbO signals [2,3]. The *t*-map is reconstructed with *t*-values from the GLM method, indicating the dHRF's weight at each channel. The connectivity map is an image map of correlation coefficients between two channels, which reflects how those two channels are interrelated. Hemodynamic features such as the mean, slope, and peak value have also been used to diagnose brain diseases. Cognitive task analysis can identify activated/deactivated regions and differences between healthy and non-healthy people.

The proposed method was validated in two ways: (i) By comparing synthetic data with and without dHRF, and (ii) by predicting the resting state data. In the synthetic data, the proposed method showed statistically significant differences in the prediction errors between with and w/o dHRF. The prediction errors in human resting state data also showed concordance with the results of synthetic data without dHRF. The agreement between the synthetic data without dHRF and the human resting state data demonstrates that the task-related response can also be differentiated from the proposed method.

Since the hemodynamic signal in this study consisted of 20 s of task and 20 s of rest and had a frequency of 0.025 Hz, it was expected that the eighth wavelet would show a significant difference with and without dHRF. As shown in Figure 6, the wavelet decomposition of the signal with dHRF was different from the signal without in the sixth through ninth wavelets. As expected, a statistically significant difference was found in the eighth wavelet, but the sixth, seventh, and ninth wavelets also showed significant differences. This is likely due to the decomposition of the dHRF into multiple levels when performing the MODWT.

The LSTM results show that the difference between with and without dHRF is more pronounced when the number of training data points increases. (Figure 7a,b). In addition, the smaller the number of training data points, the smaller the prediction error of the signal with dHRF and the larger the prediction error of the signal without dHRF. This is not surprising, since sufficient data is required for practical training of the LSTM.

To investigate whether the occurrence of hemodynamic signals can be predicted early, MAEs and RMSEs were estimated by dividing the predicted data into 1 s, 3 s, 5 s, 10 s, 15 s, and 30 s, and the difference in error between the seventh wavelet with and without dHRF was significant early. The difference between the eighth wavelet with and without dHRF was significant at 15 s because it took more than 10 s for the dHRF to rise to the maximum, since it takes time for the dHRF to rise.

When the proposed method was applied to real data, the error was similar to that of the synthetic data without dHRF. The lowest error occurred in the ninth wavelet, which seems to be due to the lowest signal strength of the ninth wavelet. Initially, wavelets with higher frequencies produced relatively higher errors, but the opposite was true as the prediction time increased. This suggests that as the data length varies, the results of the MODWT change as well, as this is more pronounced at both ends of the data.

Methods to estimate the hemodynamic response and remove noise from fNIRS signals include Kalman filtering [56], Bayesian filtering [57], block averaging [58], general linear models [59], and adaptive filtering [14,60,61]. In addition, initial-dip detection has also been studied for early detection of hemodynamic responses [62,63]. However, these methods rely heavily on the desired hemodynamic function as a reference signal (Table 3). The hemodynamic signal is designed by gamma functions [64], the balloon model [65], the finite element method [66], the state-space method [67,68], etc. These hemodynamic signals are not suitable for use in unknown areas because they depend on the brain region or task being measured. However, the proposed method is differentiated from existing methods in that it does not require a reference signal and can be applied without external devices.

Table 3. Comparison with the existing methods (adaptive filtering and general linear model) and the proposed method.

Category \ Method	Adaptive Filtering [14]	Bandpass Filter [5]	The Proposed Method
Low-freq. noise removal capacity	Middle	Low	High
Experiment paradigm (dHRF)	Required	Required	Not required
Processing type	Online	On/offline	Offline
Unknown task period	Cannot handle	Cannot handle	Can handle
Dataset size	Small	Small	Large

5. Conclusions

The following three implications are made:

(i) Alleviating the dHRF's trap: In the conventional methods (i.e., general linear model [59], recursive estimation method [60], etc.), the brain signal is identified by comparing HbO signals with a dHRF. If the correlation coefficient between two signals is high, the measured HbO is attributed to the task. The dHRF computed by convolving a gamma function with the task period contains multiple frequencies, not a single frequency. For example, for a 20 s task followed by a 20 s rest, the dHRF has 0.025 Hz (=1/40 s), and all other components are considered noises. Such multiple frequencies are also seen from the synthetic data analysis, showing that the added 0.025 Hz dHRF affected neighboring frequency bands, see Figure 6. Therefore, if the brain signal is identified with only the dHRF, the neighboring signals are unwillingly included (which could be noises). Hence, the proposed method can alleviate the dHRF's trap.

(ii) Can handle an unknown task period: In neuroscience, fNIRS has been used to identify brain regions associated with specific tasks and to understand how neural networks function. In particular, regular examinations in daily life are essential for the early detection of cognitive decline due to brain disease or aging. Research on the classification of cognitive decline and brain disease diagnosis using fNIRS is being actively conducted. However, it is challenging to establish classification criteria because hemodynamic signals vary depending on various factors such as age and gender. In particular, it is necessary to compare behavioral data and fNIRS signals for classification, and the duration of cognitive function tests belonging to neuropsychological tests should be pre-designed. Thus, the proposed method can be used when the task period to be observed is unknown or very long.

(iii) Starting time estimation for passive BCI: Recently, passive BCI has become essential for fault-free automotive cars, pilots, etc. In this case, the brain signal's starting time has to be identified. To estimate the starting time, a moving-window approach can be adopted. If the prediction error becomes large while moving the window, the instance of a significant error can be considered as the starting time of a passive brain signal, and we can generate a BCI command.

The proposed method can overcome the variability in the resting state, which varies from person to person, by predicting the subsequent signal. The predicted signal ought to be removed from the measured signal, and the remaining signal should be analyzed for brain activity. Although the proposed method has some limitations, e.g., large volumes of training data and computation time to train the model for the first time, it is expected to play a significant role in improving the temporal resolution of fNIRS in the future.

Author Contributions: Conceptualization, S.-H.Y., K.-S.H.; methodology, S.-H.Y.; software, S.-H.Y.; formal analysis, S.-H.Y., K.-S.H.; resources, G.H.; writing—original draft preparation, S.-H.Y.; writing—review and editing, K.-S.H.; visualization, S.-H.Y.; supervision, K.-S.H. All authors have read and agreed to the published version of the manuscript.

Funding: This research was funded by the National Research Foundation of Korea under the Ministry of Science and ICT, Korea (grant no. RS-2023-00207954).

Institutional Review Board Statement: The study was conducted in accordance with the Declaration of Helsinki and approved by the Pusan National University Yangsan Hospital ethics committee (Institutional Review Board approval no: PNUYH-03-2018-003).

Informed Consent Statement: Informed consent was obtained from all subjects involved in the study.

Data Availability Statement: The data and code that support the findings of this study are openly available in https://github.com/sohyeonyoo/MODWT-LSTM.

Conflicts of Interest: The authors declare no conflict of interest.

References

1. Ayaz, H.; Baker, W.B.; Blaney, G.; Boas, D.A.; Bortfeld, H.; Brady, K.; Brake, J.; Brigadoi, S.; Buckley, E.M.; Carp, S.A.; et al. Optical imaging and spectroscopy for the study of the human brain: Status report. *Neurophotonics* **2022**, *9*, S24001. [CrossRef] [PubMed]
2. Yoo, S.-H.; Woo, S.-W.; Shin, M.-J.; Yoon, J.A.; Shin, Y.-I.; Hong, K.-S. Diagnosis of mild cognitive impairment using cognitive tasks: A functional near-infrared spectroscopy study. *Curr. Alzheimer Res.* **2020**, *17*, 1145–1160. [CrossRef] [PubMed]
3. Yang, D.L.; Hong, K.-S.; Yoo, S.-H.; Kim, C.-S. Evaluation of neural degeneration biomarkers in the prefrontal cortex for early identification of patients with mild cognitive impairment: An fNIRS study. *Front. Hum. Neurosci.* **2019**, *13*, 317. [CrossRef] [PubMed]
4. Naseer, N.; Hong, K.-S. fNIRS-based brain-computer interfaces: A review. *Front. Hum. Neurosci.* **2015**, *9*, 3. [CrossRef] [PubMed]
5. Yoo, S.-H.; Santosa, H.; Kim, C.-S.; Hong, K.-S. Decoding multiple sound-categories in the auditory cortex by neural networks: An fNIRS study. *Front. Hum. Neurosci.* **2021**, *15*, 63619. [CrossRef]
6. Hong, K.-S.; Bhutta, M.R.; Liu, X.; Shin, Y.-I. Classification of somatosensory cortex activities using fNIRS. *Behav. Brain Res.* **2017**, *333*, 225–234. [CrossRef]
7. Zhang, D.D.; Chen, Y.; Hou, X.L.; Wu, Y.J. Near-infrared spectroscopy reveals neural perception of vocal emotions in human neonates. *Hum. Brain Mapp.* **2019**, *40*, 2434–2448. [CrossRef]
8. Flynn, M.; Effraimidis, D.; Angelopoulou, A.; Kapetanios, E.; Williams, D.; Hemanth, J.; Towell, T. Assessing the effectiveness of automated emotion recognition in adults and children for clinical investigation. *Front. Hum. Neurosci.* **2020**, *14*, 70. [CrossRef]
9. Huang, R.; Qing, K.; Yang, D.; Hong, K.-S. Real-time motion artifact removal using a dual-stage median filter. *Biomed. Signal Proces.* **2022**, *72*, 103301. [CrossRef]
10. Bicciato, G.; Keller, E.; Wolf, M.; Brandi, G.; Schulthess, S.; Friedl, S.G.; Willms, J.F.; Narula, G. Increase in low-frequency oscillations in fNIRS as cerebral response to auditory stimulation with familiar music. *Brain Sci.* **2022**, *12*, 42. [CrossRef]
11. Rojas, R.F.; Huang, X.; Hernandez-Juarez, J.; Ou, K.L. Physiological fluctuations show frequency-specific networks in fNIRS signals during resting state. In Proceedings of the 39th Annual International Conference of the IEEE Engineering in Medicine and Biology Society (EMBC), Jeju, Republic of Korea, 11–15 July 2017.
12. Vermeij, A.; Meel-van den Abeelen, A.S.S.; Kessels, R.P.C.; van Beek, A.H.E.A.; Claassen, J.A.H.R. Very-low-frequency oscillations of cerebral hemodynamics and blood pressure are affected by aging and cognitive load. *Neuroimage* **2014**, *85*, 608–615. [CrossRef] [PubMed]
13. Noah, J.A.; Zhang, X.; Dravida, S.; DiCocco, C.; Suzuki, T.; Aslin, R.N.; Tachtsidis, I.; Hirsch, J. Comparison of short-channel separation and spatial domain filtering for removal of non-neural components in functional near-infrared spectroscopy signals. *Neurophotonics* **2021**, *8*, 015004. [CrossRef] [PubMed]
14. Nguyen, H.D.; Yoo, S.-H.; Bhutta, M.R.; Hong, K.-S. Adaptive filtering of physiological noises in fNIRS data. *Biomed. Eng. Online* **2018**, *17*, 180. [CrossRef]
15. Sutoko, S.; Chan, Y.L.; Obata, A.; Sato, H.; Maki, A.; Numata, T.; Funane, T.; Atsumori, H.; Kiguchi, M.; Tang, T.B.; et al. Denoising of neuronal signal from mixed systemic low-frequency oscillation using peripheral measurement as noise regressor in near-infrared imaging. *Neurophotonics* **2019**, *6*, 015001. [CrossRef] [PubMed]
16. Hong, K.-S.; Khan, M.J.; Hong, M.J. Feature extraction and classification methods for hybrid fNIRS-EEG brain-computer interfaces. *Front. Hum. Neurosci.* **2018**, *12*, 246. [CrossRef]
17. von Luhmann, A.; Ortega-Martinez, A.; Boas, D.A.; Yucel, M.A. Using the general linear model to improve performance in fNIRS single trial analysis and classification: A perspective. *Front. Hum. Neurosci.* **2020**, *14*, 30. [CrossRef]
18. Shensa, M.J. The discrete wavelet transform—Wedding the a trous and Mallat algorithms. *IEEE Trans. Signal Process.* **1992**, *40*, 2464–2482. [CrossRef]
19. Molavi, B.; Dumont, G.A. Wavelet-based motion artifact removal for functional near-infrared spectroscopy. *Physiol. Meas.* **2012**, *33*, 259–270. [CrossRef]
20. Duan, L.; Zhao, Z.P.; Lin, Y.L.; Wu, X.Y.; Luo, Y.J.; Xu, P.F. Wavelet-based method for removing global physiological noise in functional near-infrared spectroscopy. *Biomed. Opt. Express* **2018**, *9*, 330656. [CrossRef]
21. Dommer, L.; Jager, N.; Scholkmann, F.; Wolf, M.; Holper, L. Between-brain coherence during joint n-back task performance: A two-person functional near-infrared spectroscopy study. *Behav. Brain Res.* **2012**, *234*, 212–222. [CrossRef]
22. Holper, L.; Scholkmann, F.; Wolf, M. Between-brain connectivity during imitation measured by fNIRS. *Neuroimage* **2012**, *63*, 212–222. [CrossRef]
23. Zhu, L.; Wang, Y.X.; Fan, Q.B. MODWT-ARMA model for time series prediction. *Appl. Math. Model.* **2014**, *38*, 1859–1865. [CrossRef]
24. Percival, D.B.; Walden, A.T. *Wavelet Methods for Time Series Analysis*; Cambridge University Press: Cambridge, UK, 2000.
25. Abdelhakim, S.; El Amine, D.S.M.; Fadia, M. Maximal overlap discrete wavelet transform-based abrupt changes detection for heart sounds segmentation. *J. Mech. Med. Biol.* **2023**, *23*, 2350017. [CrossRef]
26. Adib, A.; Zaerpour, A.; Lotfirad, M. On the reliability of a novel MODWT-based hybrid ARIMA-artificial intelligence approach to forecast daily Snow Depth (Case study: The western part of the Rocky Mountains in the USA). *Cold Reg. Sci. Technol.* **2021**, *189*, 103342. [CrossRef]
27. Yousuf, M.U.; Al-Bahadly, I.; Avci, E. Short-Term Wind Speed Forecasting based on hybrid MODWT-ARIMA-Markov model. *IEEE Access* **2021**, *9*, 79695–79711. [CrossRef]

28. Li, M.Y.; Chen, W.Z.; Zhang, T. Application of MODWT and log-normal distribution model for automatic epilepsy identification. *Biocybern. Biomed. Eng.* **2017**, *37*, 679–689. [CrossRef]
29. Bizzego, A.; Gabrieli, G.; Esposito, G. Deep neural networks and transfer learning on a multivariate physiological signal dataset. *Bioengineering* **2021**, *8*, 35. [CrossRef]
30. ElNakieb, Y.; Ali, M.T.; Elnakib, A.; Shalaby, A.; Mahmoud, A.; Soliman, A.; Barnes, G.N.; El-Baz, A. Understanding the role of connectivity dynamics of resting-state functional MRI in the diagnosis of autism spectrum disorder: A comprehensive study. *Bioengineering* **2023**, *10*, 56. [CrossRef]
31. Madan, P.; Singh, V.; Singh, D.P.; Diwakar, M.; Pant, B.; Kishor, A. A hybrid deep learning approach for ECG-based arrhythmia classification. *Bioengineering* **2022**, *9*, 152. [CrossRef]
32. Lee, P.L.; Chen, S.H.; Chang, T.C.; Lee, W.K.; Hsu, H.T.; Chang, H.H. Continual learning of a transformer-based deep learning classifier using an initial model from action observation EEG data to online motor imagery classification. *Bioengineering* **2023**, *10*, 186. [CrossRef]
33. Nafea, M.S.; Ismail, Z.H. Supervised machine learning and deep learning techniques for epileptic seizure recognition using EEG signals-A systematic literature review. *Bioengineering* **2022**, *9*, 781. [CrossRef] [PubMed]
34. Ghonchi, H.; Fateh, M.; Abolghasemi, V.; Ferdowsi, S.; Rezvani, M. Deep recurrent-convolutional neural network for classification of simultaneous EEG-fNIRS signals. *IET Signal Process.* **2020**, *14*, 142–153. [CrossRef]
35. Hekmatmanesh, A.; Azni, H.M.; Wu, H.; Afsharchi, M.; Li, M.; Handroos, H. Imaginary control of a mobile vehicle using deep learning algorithm: A brain computer interface study. *IEEE Access* **2021**, *10*, 20043–20052. [CrossRef]
36. Hochreiter, S.; Schmidhuber, J. Long short-term memory. *Neural Comput.* **1997**, *9*, 1735–1780. [CrossRef] [PubMed]
37. Cao, J.; Li, Z.; Li, J. Financial time series forecasting model based on CEEMDAN and LSTM. *Phys. A* **2019**, *519*, 127–139. [CrossRef]
38. Esmaeili, F.; Cassie, E.; Nguyen, H.P.T.; Plank, N.O.V.; Unsworth, C.P.; Wang, A.L. Predicting analyte concentrations from electrochemical aptasensor signals using LSTM recurrent networks. *Bioengineering* **2022**, *9*, 529. [CrossRef]
39. Shao, E.Z.; Mei, Q.C.; Ye, J.Y.; Ugbolue, U.C.; Chen, C.Y.; Gu, Y.D. Predicting coordination variability of selected lower extremity couplings during a cutting movement: An investigation of deep neural networks with the LSTM structure. *Bioengineering* **2022**, *9*, 411. [CrossRef]
40. Phutela, N.; Relan, D.; Gabrani, G.; Kumaraguru, P.; Samuel, M. Stress classification using brain signals based on LSTM network. *Comput. Intell. Neurosci.* **2022**, *2022*, 767592. [CrossRef]
41. Barzegar, R.; Aalami, M.T.; Adamowski, J. Coupling a hybrid CNN-LSTM deep learning model with a boundary corrected maximal overlap discrete wavelet transform for multiscale lake water level forecasting. *J. Hydrol.* **2021**, *598*, 126196. [CrossRef]
42. Li, Y.T.; Li, R.Y. Predicting ammonia nitrogen in surface water by a new attention-based deep learning hybrid model. *Environ. Res.* **2023**, *216*, 114723. [CrossRef]
43. Li, Y.M.; Peng, T.; Zhang, C.; Sun, W.; Hua, L.; Ji, C.L.; Shahzad, N.M. Multi-step ahead wind speed forecasting approach coupling maximal overlap discrete wavelet transform, improved grey wolf optimization algorithm and long short-term memory. *Renew. Energy* **2022**, *196*, 1115–1126. [CrossRef]
44. Raghu, S.; Sriraam, N.; Temel, Y.; Rao, S.V.; Hegde, A.S.; Kubben, P.L. Complexity analysis and dynamic characteristics of EEG using MODWT based entropies for identification of seizure onset. *J. Biomed. Res.* **2020**, *34*, 213–227. [CrossRef] [PubMed]
45. Ge, Q.; Lin, Z.C.; Gao, Y.X.; Zhang, J.X. A robust discriminant framework based on functional biomarkers of eeg and its potential for diagnosis of Alzheimer's disease. *Healthcare* **2020**, *8*, 476. [CrossRef] [PubMed]
46. Cordes, D.; Kaleem, M.F.; Yang, Z.S.; Zhuang, X.W.; Curran, T.; Sreenivasan, K.R.; Mishra, V.R.; Nandy, R.; Walsh, R.R. Energy-period profiles of brain networks in group fMRI resting-state data: A comparison of empirical mode decomposition with the short-time Fourier transform and the discrete wavelet transform. *Front. Neurosci.* **2021**, *15*, 663403. [CrossRef]
47. Martin-Chinea, K.; Ortega, J.; Gomez-Gonzalez, J.F.; Pereda, E.; Toledo, J.; Acosta, L. Effect of time windows in LSTM networks for EEG-based BCIs. *Cogn. Neurodyn.* **2023**, *17*, 385–398. [CrossRef]
48. Van Houdt, G.; Mosquera, C.; Napoles, G. A review on the long short-term memory model. *Artif. Intell. Rev.* **2020**, *53*, 5929–5955. [CrossRef]
49. Gemignani, J.; Gervain, J.; IEEE. A practical guide for synthetic fNIRS data generation. In Proceedings of the 43rd Annual International Conference of the IEEE Engineering in Medicine and Biology Society (EMBC), Electronic Network, Virtual, 1–5 November 2021; pp. 828–831.
50. Gemignani, J.; Middell, E.; Barbour, R.L.; Graber, H.L.; Blankertz, B. Improving the analysis of near-spectroscopy data with multivariate classification of hemodynamic patterns: A theoretical formulation and validation. *J. Neural Eng.* **2018**, *15*, 045001. [CrossRef]
51. Zhang, Q.; Li, F.; Long, F.; Ling, Q. Vehicle emission forecasting based on wavelet transform and long short-term memory network. *IEEE Access* **2018**, *6*, 56984–56994. [CrossRef]
52. Seo, Y.; Choi, Y.; Choi, J. River stage modeling by combining maximal overlap discrete wavelet transform, support vector machines and genetic algorithm. *Water* **2017**, *9*, 525. [CrossRef]
53. Abbasimehr, H.; Paki, R. Improving time series forecasting using LSTM and attention models. *J. Ambient Intell. Humaniz. Comput.* **2022**, *13*, 673–691. [CrossRef]
54. Cutini, S.; Basso Moro, S.; Bisconti, S. Functional near infrared optical imaging in cognitive neuroscience: An introductory review. *J. Near Infrared Spectrosc.* **2012**, *20*, 75–92. [CrossRef]

55. Ernst, L.S.; Schneider, S.; Ehlis, A.-C.; Fallgatter, A.J. Functional near infrared spectroscopy in psychiatry: A critical review. *J. Near Infrared Spectrosc.* **2012**, *20*, 93–105. [CrossRef]
56. Hu, X.S.; Hong, K.-S.; Ge, S.Z.S.; Jeong, M.-Y. Kalman estimator- and general linear model-based on-line brain activation mapping by near-infrared spectroscopy. *Biomed. Eng. Online* **2010**, *9*, 82. [CrossRef] [PubMed]
57. Scarpa, F.; Brigadoi, S.; Cutini, S.; Scatturin, P.; Zorzi, M.; Dell'Acqua, R.; Sparacino, G. A reference-channel based methodology to improve estimation of event-related hemodynamic response from fNIRS measurements. *Neuroimage* **2013**, *72*, 106–119. [CrossRef]
58. von Luhmann, A.; Li, X.E.; Muller, K.R.; Boas, D.A.; Yucel, M.A. Improved physiological noise regression in fNIRS: A multimodal extension of the general linear model using temporally embedded canonical correlation analysis. *Neuroimage* **2020**, *208*, 116472. [CrossRef]
59. Barker, J.W.; Aarabi, A.; Huppert, T.J. Autoregressive model based algorithm for correcting motion and serially correlated errors in fNIRS. *Biomed. Opt. Express* **2013**, *4*, 1366–1379. [CrossRef]
60. Zhang, Q.; Strangman, G.E.; Ganis, G. Adaptive filtering to reduce global interference in non-invasive NIRS measures of brain activation: How well and when does it work? *Neuroimage* **2009**, *45*, 788–794. [CrossRef]
61. Aqil, M.; Hong, K.-S.; Jeong, M.-Y.; Ge, S.S. Cortical brain imaging by adaptive filtering of NIRS signals. *Neurosci. Lett.* **2012**, *514*, 35–41. [CrossRef]
62. Zafar, A.; Hong, K.-S. Neuronal activation detection using vector phase analysis with dual threshold circles: A functional near-infrared spectroscopy study. *Int. J. Neural Syst.* **2018**, *28*, 1850031. [CrossRef]
63. Hong, K.-S.; Zafar, A. Existence of initial dip for BCI: An illusion or reality. *Front. Neurorobot.* **2018**, *12*, 69. [CrossRef]
64. Shan, Z.Y.; Wright, M.J.; Thompson, P.M.; McMahon, K.L.; Blokland, G.; de Zubicaray, G.I.; Martin, N.G.; Vinkhuyzen, A.A.E.; Reutens, D.C. Modeling of the hemodynamic responses in block design fMRI studies. *J. Cereb. Blood Flow Metab.* **2014**, *34*, 316–324. [CrossRef] [PubMed]
65. Buxton, R.B.; Uludag, K.; Dubowitz, D.J.; Liu, T.T. Modeling the hemodynamic response to brain activation. *Neuroimage* **2004**, *23*, S220–S233. [CrossRef] [PubMed]
66. Huang, R.S.; Hong, K.-S. Multi-channel-based differential pathlength factor estimation for continuous-wave fNIRS. *IEEE Access* **2021**, *9*, 37386–37396. [CrossRef]
67. Hong, K.-S.; Nguyen, H.D. State-space models of impulse hemodynamic responses over motor, somatosensory, and visual cortices. *Biomed. Opt. Express* **2014**, *5*, 1778–1798. [CrossRef]
68. Aqil, M.; Hong, K.-S.; Jeong, M.-Y.; Ge, S.S. Detection of event-related hemodynamic response to neuroactivation by dynamic modeling of brain activity. *Neuroimage* **2012**, *63*, 553–568. [CrossRef] [PubMed]

bioengineering

MDPI

Article

Named Entity Recognition of Diabetes Online Health Community Data Using Multiple Machine Learning Models

Qian Xu [1,2,3,4,5,†], Yue Zhou [1,2,4,5,†], Bolin Liao [3], Zirui Xin [1,4,5], Wenzhao Xie [4,5], Chao Hu [6] and Aijing Luo [1,4,5,*]

1. Second Xiangya Hospital, Central South University, Changsha 410011, China; qian_xu2000@csu.edu.com (Q.X.); zhouyue8@csu.edu.cn (Y.Z.); xinzirui@csu.edu.cn (Z.X.)
2. School of Life Sciences, Central South University, Changsha 410013, China
3. College of Computer Science and Engineering, Jishou University, Jishou 416000, China; mulinliao8184@163.com
4. Key Laboratory of Medical Information Research, Central South University, College of Hunan Province, Changsha 410013, China; xie_wenzhao@126.com
5. Clinical Research Center for Cardiovascular Intelligent Healthcare in Hunan Province, Changsha 410011, China
6. Big Data Institute, Central South University, Changsha 410011, China; huchao@csu.edu.cn

* Correspondence: luoaj@csu.edu.cn

† These authors contributed equally to this work.

Abstract: The rising prevalence of diabetes and the increasing awareness of self-health management have resulted in a surge in diabetes patients seeking health information and emotional support in online health communities. Consequently, there is a vast database of patient consultation information in these online health communities. However, due to the heterogeneity and incompleteness of the content, mining medical information and patient health data from these communities can be a challenge. To address this issue, we built the RoBERTa-BiLSTM-CRF (RBC) model for identifying entities in the online health community of diabetes. We selected 1889 question–answer texts from the most active online health community in China, Good Doctor Online, and used these public data to identify five types of entities. In addition, we conducted a comparative evaluation with three other commonly used models to validate the performance of our proposed model, including RoBERTa-CRF (RC), BilSTM-CRF (BC), and RoBERTa-Softmax (RS). The results showed that the RBC model achieved excellent performance on the test set, with an accuracy of 81.2% and an F1 score of 80.7%, outperforming the performance of traditional entity recognition models in named entity recognition in online medical communities for doctors and diabetes patients. The high performance of entity recognition in online health communities will provide a crucial knowledge source for constructing medical knowledge graphs. This integration would help alleviate the growing demand for medical consultations and the strain on healthcare resources, while assisting healthcare professionals in making informed decisions and providing personalized services to patients.

Keywords: diabetes; online healthcare data; named entity recognition; RoBERTa-BiLSTM-CRF; online health community

Citation: Xu, Q.; Zhou, Y.; Liao, B.; Xin, Z.; Xie, W.; Hu, C.; Luo, A. Named Entity Recognition of Diabetes Online Health Community Data Using Multiple Machine Learning Models. *Bioengineering* **2023**, *10*, 659. https://doi.org/10.3390/bioengineering10060659

Academic Editor: Luca Mesin

Received: 21 April 2023
Revised: 19 May 2023
Accepted: 25 May 2023
Published: 29 May 2023

1. Introduction

In 2030, it is expected that 11.3% of adults will have diabetes, which would affect roughly 643 million people. Diabetes is one of the most rapidly expanding global crises of the 21st century [1]. Relevant studies have indicated that roughly half of web-based health information users with chronic health issues may benefit from accessing online health information [2]. The Q&A structure of online health communities (OHCs) is becoming more and more popular, with diabetic patients seeking medical knowledge and diabetic self-management assistance [3–5].Online health communities store a significant amount of case information, medical knowledge, and prescription data, which serves as the hotspots for medical big data applications.

In the doctor–patient Q&A texts from the online health community, entities can be identified as linked to diseases, medications, tests, treatments, and symptoms for diabetes patients, and used to provide various intelligent services to diabetic patients. We can also gain a deeper understanding of patients' needs and interests in health-related information through entity recognition [6].With this knowledge mined from online health communities, we can then offer patients individualized medical care, health information, decision-making participation, emotional support services, and improvements in online medical services.

Historically, vast sets of rules or lexicons had to be manually created by professionals for both rule-based and lexicon approaches to medical entity recognition [7–9]. Using benchmark data from the i2b2 2009 drug challenge and a hybrid lexicon-based and rule-based model, [10] achieved an F1 score of 66.97% for the named entity recognition of pharmaceuticals. Statistics-based machine learning algorithms leveraging manually annotated corpora for supervised training have exhibited a significant increase in accuracy over rule-based and lexicon-based entity recognition approaches [11,12]. With the advent of deep learning, numerous neural-network-based models have effectively been used for the textual entity recognition of biological documents [13,14], electronic medical records [15–17], and online health communities [18–20] Dreyfus Dreyfus. Based on the entity recognition infrastructure deep learning model LSTM-CRF, Guillaume Lample et al. [21] proposed a neural network model that combines bidirectional long short-term memory (BiLSTM) and conditional random fields (CRFs). This bidirectional structure enables the capture of sequential information in context, leading to widespread applications in entity recognition. Wang, Z. et al. [22] retrieved input patient fundamental information and illness information, annotated entities on medical community Q&A texts, and trained a BiLSTM-CRF to recognize and extract entities linked to diabetes in the medical domain. However, the BiLSTM-CRF model focuses on extracting features between words and characters from the text while disregarding the contextual meaning of context. To address this issue, Jacob Devlin et al. from Google introduced a BERT pre-training model [23]. This model improved the quality of embedding words and reduced the workload of downstream classification tasks, resulting in better recognition performance. In recent years, named entity identification in electronic medical records [24–26] and biomedical literature [27] has been successfully implemented with BERT, a pre-trained model with enhanced contextual long-range semantic learning capability based on word vectors.

Due to their lack of medical knowledge, users of the online healthcare communities for diabetes produce texts that contain inaccurate or slang expressions. Entity recognition of Q&A text in online health communities is challenged with semantic ambiguity, content heterogeneity, high complexity, and imperfect recognition; hence, it is difficult to achieve the desired outcome. Some studies have shown that applying the RoBERTa model to named entity recognition tasks improves the entity recognition performance (F1 score) [28]. To mitigate the impact of Chinese online health data on the performance of entity recognition, this paper utilizes a combined model of RoBERTa-BiLSTM-CRF to accomplish medical entity recognition tasks related to diabetes. This method primarily addressed the following tasks: (1) We standardized the diabetes annotation corpus of the online health community using the diabetes entity classification standards of Ruijin Hospital; (2) The pre-trained model RoBERTa-BiLSTM-CRF was used to identify named entities in Q&A text from the Good Doctor Online health community, and evaluated by comparing it with the other three models; (3) The entity recognition performance of the Q&A texts from the perspective of the patient was compared with that of electronic medical records from the clinician's perspective.

2. Method

2.1. Data Collection and Preprocessing

We chose the top Chinese online doctor–patient Q&A platform, "Good Doctor Online" (https://www.haodf.com/, accessed on 5 December 2021), searched the Q&A section of the diabetes-specific disease section, and collected 9446 questions from November 2020 to November 2021. When consulting doctors, patients submitted content using a specified information description framework, as shown in Figure 1.

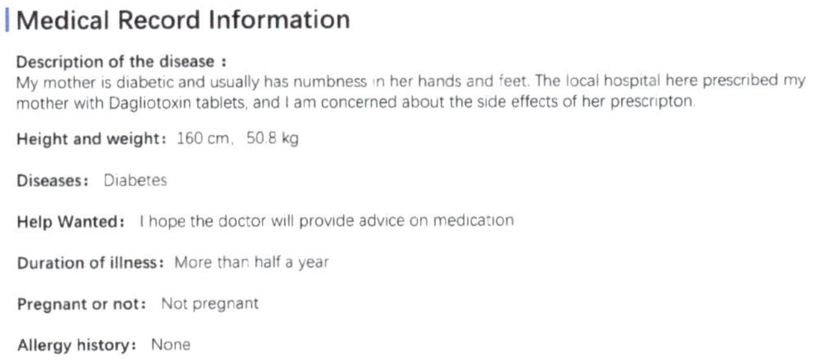

Figure 1. Online doctor-patient Q&A text structure (from the website's original screenshot).

For this study, the following preprocessing processes were carried out: (1) Removed all non-textual content (replacing emoji icons with emoji-related codes); (2) Filtered 2000 values at random from the acquired dataset of 9445 values, deleted duplicate and nonsensical data to obtain 1889 values, and converted the data to JSON format; (3) Annotated the questions of health community Q&A text into eight categories (check, disease, drug, mood, life, social, symptom, and treat) using the Doccano annotation tool. Figure 2 depicts the annotation interface; (4) To process the exported text, it was divided into 6669 values. The dataset was then further split into a training set consisting of 6019 data slices and a test set consisting of 650 data slices. The ratio of this split was approximately 9:1. Within the training set, the data were divided into a training subset and a validation subset, at a ratio of 5:1. Next, we converted the JSON format files into a data format for generic named entity recognition tasks using BIO tagging; (5) Utilized the RoBERTa word vector model made available by the Harbin Institute of Technology as an open source. Figure 3 depicts the specific data preprocessing procedure.

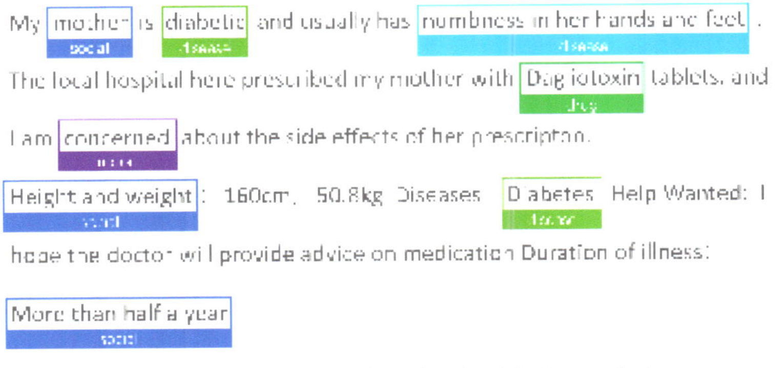

Figure 2. Annotation tool interface (from the website's original screenshot).

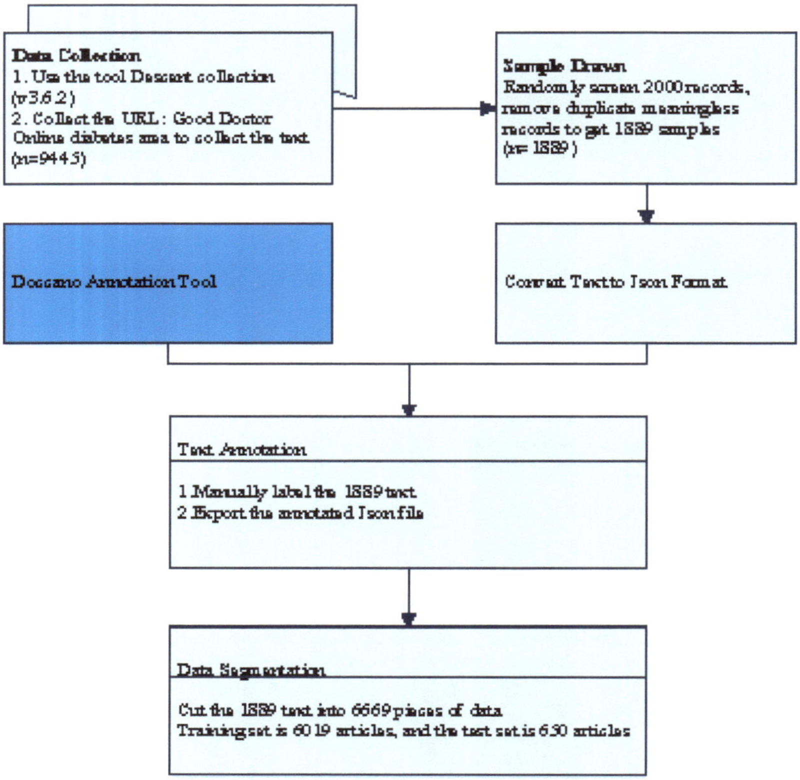

Figure 3. Data preprocessing flow chart.

2.2. RoBERTa-BiLSTM-CRF Model Construction

This article employed the RoBERTa-BiLSTM-CRF model, which is composed of three layers: the RoBERTa word vector layer, the BiLSTM layer, and the CRF layer. In the word vector layer, word embedding and model construction were carried out by applying the Chinese pre-training model from the HUST Xunfei Lab in order to obtain word-level vector information and a semantic representation suitable for the Chinese language. The BiLSTM layer is utilized for semantic encoding, and forward and backward LSTM networks are used for each training sequence; the forward and backward networks were connected to the same output layer. The CRF layer, which effectively evaluated the labeling information before and after the sequence, filtered out entities that did not conform to the labeling rules and outputs a sequence with the best likelihood of being correctly categorized. Figure 4 depicts the general structure of the RoBERTa-BiLSTM-CRF model.

2.2.1. RoBERTa Pre-Training Layer to Construct Word Vectors

Each input word of the encoder generated three vectors, denoted by vectors, accordingly. After calculating the inner product between and producing the similarity weights, the similarity was calculated. Then, the weights were normalized to a value between 0 and 1, and the similarity vector was processed using the function shown in Equation (1).

$$\alpha_i = softmax(f(Q, K_i)) = \frac{exp(f(Q, K_i))}{\sum_i exp(f(Q, K_i))} \tag{1}$$

Scaling was accomplished by multiplying $\frac{1}{\sqrt{d_k}}$ with the result of the inner product of Q and K. The attentional mechanism is presented in Equation (2).

$$Attention(Q, K, V) = softmax\left(\frac{QK^T}{\sqrt{d_k}}\right)V \quad (2)$$

Combining the outcomes of attention processes yielded the multi-headed attention module, as determined using Equation (3).

$$MultiHead(Q, K, V) = Concat(head_1, head_2, head_3, \ldots, head_h)W^o \quad (3)$$

The output of the multi-headed attention layer was then passed to the feed-forward neural network, the module described in Equation (4).

$$FFN(Z) = max(0, ZW_1 + b_1)W_2 + b_2 \quad (4)$$

The output layer employed a self-supervised approach to estimate the probability that the masked target word and the two phrases shared a contextual link. After multiple training iterations, the likelihood and the weight parameter with the largest value for the two tasks are determined.

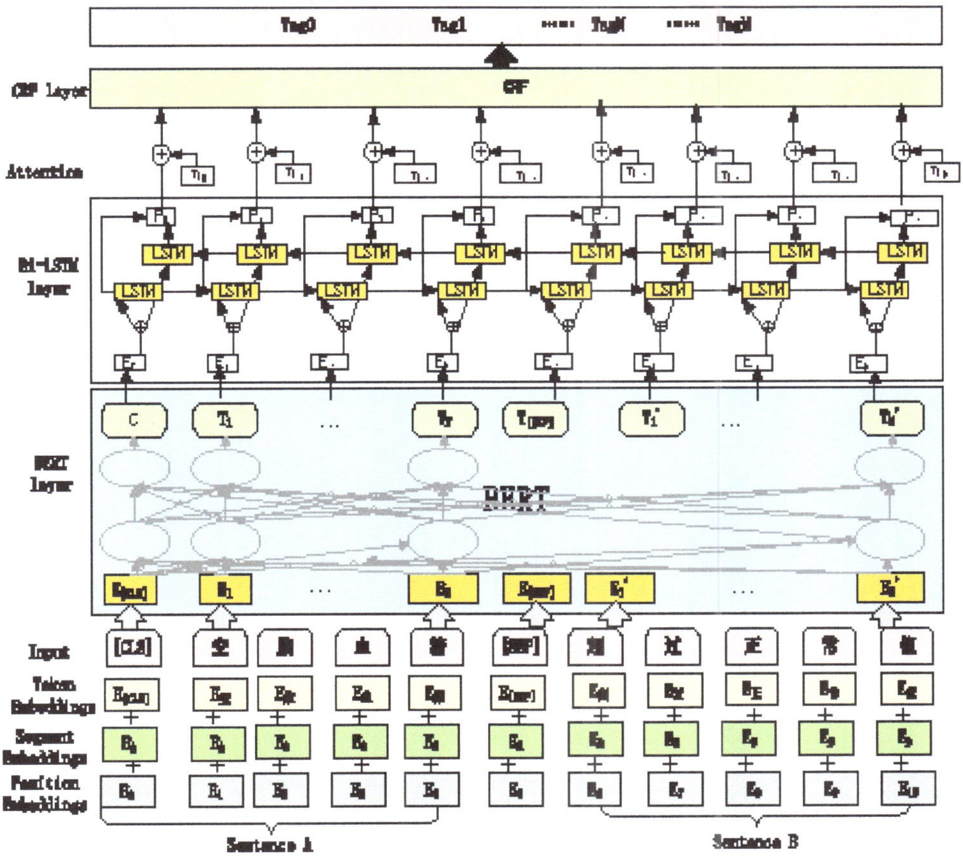

Figure 4. Structural diagram of the Bert-BilSTM-CRF model.

2.2.2. Layer of BiLSTM for Semantic Encoding

Long short-term memory networks incorporate memory units in the hidden layer, which can better solve the problem of gradient disappearance caused by excessively long sequences in the training of conventional recurrent neural networks, enabling them to be more effectively used in the named entity recognition task. Its structure consists of the following equations:

$$i_t = \sigma\left(x_t \cdot w^i_{xh} + h_{t-1} \cdot w^i_{hh'} + b^i_h\right) \tag{5}$$

$$f_t = \sigma\left(x_t \cdot w^f_{xh} + h_{t-1} \cdot w^f_{hh'} + b^f_h\right) \tag{6}$$

$$o_t = \sigma(x_t \cdot w^o_{xh} + h_{t-1} \cdot w^o_{hh'} + b^o_h) \tag{7}$$

$$c\prime_t = \tanh(x_t \cdot w^c_{xh} + h_{t-1} \cdot w^c_{hh'} + b^c_h) \tag{8}$$

$$c_t = i_t \otimes c'_t + f_t \otimes c'_{t-1} \tag{9}$$

$$h_t = o_t \otimes \tanh(c_t) \tag{10}$$

The σ denotes the $Sigmoid$ activation function, $\otimes$ is the dot product operation, and x_t is used as the unit input; i_t,f_t,o_t denotes the input gate, forgetting gate, and output gate at a specific moment, respectively; $tanh$ denotes the hyperbolic tangent activation function; w,b represent the weight matrix and bias vector of the input gate, forgetting gate, and output gate, respectively; c'_t represents the state at time, which is the intermediate state obtained only from the current input and is used to update the state at time t; h_t represents the output at time t.

The BiLSTM bi-directional long and short-term memory network with forward and reverse $LSTM$ for each word sequence was used to decode the text sentences in the input layer, and data conversion and transfer through forward $LSTM$ and backward $LSTM$ were used to acquire contextual feature vectors in both directions. First, the output calculated the error existing in the output layer at each moment, followed by the derivatives of parameters of the forward $LSTM$ from moment t to moment 1. For the network portion of the backward $LSTM$, loss needs to be calculated from moment 1 to moment t, and reverse differentiation be conducted. The formula for the output is provided in the following equations:

$$\overrightarrow{h}_t = LSTM_L\left(\overrightarrow{x}_t, \overrightarrow{h}_{t-1}\right) \tag{11}$$

$$\overleftarrow{h}_t = LSTM_R\left(\overleftarrow{x}_t, \overleftarrow{h}_{t-1}\right) \tag{12}$$

$$h_t = \left[\overrightarrow{h}_t, \overleftarrow{h}_t\right] \tag{13}$$

2.2.3. CRF Optimized Tag Sequence

CRFs can compensate for the shortcomings of BiLSTM by providing an ideal sequence of predictions based on the relationship between surrounding labels. The output score matrix of BiLSTM is supposed to be P for any arbitrary sequence $X = (x_1, x_2, \ldots, x_n)$. The size of P is $n \times k$, where n represents the number of words, k represents the number of tags, and P_{ij} represents the score of the jth tag of the word. Equation (14) describes the score function for the sequence of predictions $Y = (y_1, y_2, \ldots, y_n)$.

$$s(X, Y) = \sum_{i=0}^{n} Ay_i, y_{i+1} + \sum_{i=0}^{n} P_i, y_i \quad (14)$$

A denotes the matrix of transferred scores, A_{ij} represents the scores transferred from label i to label j, and the size of A is $k + 2$. Equation (15) describes the probability of generating the predicted sequence Y.

$$p(Y|X) = \frac{e^{s(X,Y)}}{\sum_{\widetilde{Y} \in Y_X} s\left(X, \widetilde{Y}\right)} \quad (15)$$

The probability function of the expected sequence could be obtained by taking the logarithm at both ends.

$$\ln(p(Y|X)) = s(X, Y) - \ln\left(\sum_{\widetilde{Y} \in Y_X} s\left(X, \widetilde{Y}\right)\right) \quad (16)$$

In Equation (17), $\widetilde{Y}$ denotes the true labeled sequence, whereas Y_X denotes all conceivable labeled sequences. Decoding yielded the output sequence corresponding to the maximum score.

$$Y^* = \arg \underset{\widetilde{Y} \epsilon Y_X}{maxs}\left(X, \widetilde{Y}\right) \quad (17)$$

3. Result

3.1. Text Annotation

Health Community Q&A texts are self-reported by patients to their physicians; therefore, the language of the text differed from that of the medical literature and electronic medical records. When annotating, it is important to note the frequent abbreviations and misspellings. The original words were precisely aligned with the common words. Under the supervision of two medical informatics professionals and one medical expert, we coded each record in terms of the classification criteria for diabetes mellitus at Ruijin Hospital. This labeling was divided into eight categories (check, disease, drug, lifestyle, mood, social context, symptom, and treatment). Table 1 summarizes the classification criteria.

Table 1. Labeling classification standards.

Classification	Description	Labeling Case
Check	Test and examination items, physical examination, review, etc.	A review at the hospital the previous day; a check-up at the hospital.
Disease	Disease names, such as hypertension, diabetes, etc.	No diabetes in the family either.
Drug	The name of the drug, such as nifedipine, metformin, nifedipine, etc.	The medications being taken are Metformin Hydrochloride and Vildagliptin.
Lifestyle	Patient's lifestyle, e.g., smoking, alcohol consumption, sleep, etc.	Smoking; drinking; staying up late.
Mood	Irritable, anxious, worried	So now it is confusing.
Social context	Dad (my dad), wife (my wife), medical history, occupation, height, weight, age, gender (pregnancy and gestation), wanting children.	Height and weight: 171 cm, 70 kg. Pregnancy: not pregnant.
Symptom	Patient's subjective description of feelings and signs (skin jaundice), such as dizziness, non-dizziness, nocturia, puffy eyelids, and frequent need to urinate.	Feeling of vertigo when standing suddenly; I urinate frequently and often, but each time the amount of urine is not much, nausea, vomiting, weakness, stomach pain, and breast swelling.
Treatment	Chinese medicine treatment, immunotherapy, ventilator, and stent release.	Immunotherapy.

3.2. Experimental Setup

This study was based on the Python + PyTorch + GPU deep neural network learning framework. The cross-entropy loss function was used as the loss function, and the AdamW method was employed for model training optimization. A five-fold cross-validation procedure was utilized to run our proposed model. During the training process, we performed fine-tuning on Roberta. The input dimension, sequence_length, was set to 128. The initial learning rate of the model was set to 3×10^{-5}, while the learning rates of BiLSTM and CRF were set to five times greater than that of Roberta, namely, 1.5×10^{-4}. A cosine schedule with a warmup was used to adjust the learning rate. We set the warmup steps to one-tenth of the total training epochs, and the learning rate decay rate was set to 0.01 (weight_decay). The word embedding dimension (pooler_fc_size) was set to 768, and the batch size was set to 16. Dropout was applied with rates of 0.1 at the input layer and hidden layers. The total number of training epochs was set to 50, and the F1 score was calculated on the validation set after each epoch. The best model was saved accordingly. The patient number was set to 10, which means that if the model did not show improvement on the validation set over 10 consecutive epochs, the training would be terminated early. The weights, biases, and other parameters were continuously optimized during the training process. To prevent issues such as gradient explosion or vanishing gradients during code execution, the gradient clipping technique was employed. The performance of the best model was tested on the final test set, which was not used during model training. The F1 score was calculated for each category, and the average score was taken as the F1 score on the test set. The average F1 score from the five rounds of cross-validation was calculated as the final F1 score. The experimental parameters for model training are summarized in Table 2.

Table 2. Experimental parameters.

Experimental Parameters	Value
Sequence_length	128
Batch_size	Train set 16, test set 16
Pooler_fc_size	768
Epoch	50
Learning rate	3×10^{-5}
Optimizer	Adam
Input layer dropout	0.1
Hidden layers dropout	0.1

3.3. Evaluation

This study examined the performance of the model by calculating its precision, recall, accuracy, and F1 scores. TP, TN, FP, and FN are the number of positive samples correctly predicted for the positive class, the number of samples correctly predicted for the negative class, the number of samples incorrectly predicted to be in the positive class, and the number of samples incorrectly predicted to be in the negative category, respectively. In this study, the entity array obtained through manual annotation was referred to as the truth entity set, while the array of entities predicted by the machine learning model after training was called the predicted entity set. Taking the intersection of these two arrays, the number of entities that appear in both arrays was defined as true positives (TPs), indicating that the machine successfully predicted the true entities. The number of entities in the truth entity set that were not correctly predicted was defined as false negatives (FNs), while the portion of entities in the predicted entity set that were not correctly predicted was defined as false positives (FPs). Figure 5 presents the confusion matrices for the four models.

$$Pre = \frac{TP}{TP + FP} \tag{18}$$

$$Re = \frac{TP}{TP + FN} \tag{19}$$

$$F1 = \frac{2 \times Pre \times Re}{Pre + Re} \quad (20)$$

$$Accuracy = \frac{TP + TN}{TP + TN + FP + FN} \quad (21)$$

Confusion Matrix(**RBC**)		Predicted entities: Positive	Predicted entities: Negative
true entities	Positive	1365	282
	Negative	372.8	1455.8

Confusion Matrix(**RC**)		Predicted entities: Positive	Predicted entities: Negative
true entities	Positive	1362.6	284.4
	Negative	417.4	1495.6

Confusion Matrix(**RS**)		Predicted entities: Positive	Predicted entities: Negative
true entities	Positive	1363.8	283.2
	Negative	443	1523.6

Confusion Matrix(**BC**)		Predicted entities: Positive	Predicted entities: Negative
true entities	Positive	1250.8	396.2
	Negative	435	1289.6

Figure 5. Confusion matrix for the four models.

Precision refers to the ratio of actual positive samples to expected positive samples. Recall, also known as sensitivity, is the percentage of predicted true-positive samples to the total number of true-positive samples. The F1 value is a combined precision and recall rating. Accuracy reflects a model's ability to correctly classify the overall samples, i.e., the proportion of samples that are correctly predicted among all samples.

3.4. Model Performance

To verify the validity and feasibility of the model, the total experimental results of our RBC model and other excellent models are shown in Table 3; the baseline model was BC. Other models are RC and RS. The experimental results demonstrate that our suggested RBC model enhanced precision by 4.3%, recall by 7%, F1 values by 5.6%, and Acc values by 5.8%, and had a better overall performance when compared with the BiLST-CRF baseline model. We used five-fold cross-validation, training, and testing on the corpus; the final results are shown in Table 3.

Table 3. Comparative experimental results of four models on the test set.

Models	F1	P	R	Acc
RBC	0.807	0.786	0.829	0.812
RC	0.795	0.755	0.827	0.803
RS	0.790	0.755	0.828	0.799
BC	0.751	0.743	0.759	0.754

Table 4 presents a statistical evaluation of the effectiveness of eight distinct entity recognition categories. We observed that two entity types, emotional and social attributes, achieved superior results with significantly higher precision, recall, and F1 values than other entity types, whereas two entity types, symptoms and therapies, were significantly less effective. For the identical CRF model based on words, the RBC and RC impacts were extremely similar.

Table 4. Evaluation of the effect of different entity recognition of four models.

Model	Index	Check	Disease	Drug	Lifestyle	Mood	Social	Symptoms	Treat
RBC	P	0.739	0.787	0.730	0.754	0.865	0.941	0.609	0.571
	R	0.774	0.863	0.823	0.723	0.922	0.926	0.709	0.585
	F1	0.756	0.823	0.774	0.738	0.892	0.933	0.655	0.578
RC	P	0.719	0.761	0.730	0.696	0.903	0.918	0.596	0.542
	R	0.771	0.850	0.821	0.732	0.933	0.924	0.710	0.639
	F1	0.744	0.803	0.773	0.713	0.917	0.921	0.647	0.586
RS	P	0.717	0.742	0.704	0.697	0.878	0.915	0.582	0.516
	R	0.772	0.866	0.824	0.726	0.956	0.923	0.690	0.624
	F1	0.743	0.799	0.759	0.710	0.915	0.919	0.631	0.564
BC	P	0.682	0.722	0.697	0.702	0.857	0.918	0.533	0.559
	R	0.731	0.766	0.684	0.668	0.756	0.909	0.521	0.610
	F1	0.704	0.748	0.687	0.681	0.795	0.913	0.526	0.582

We identified eight entity types from Q&A texts: check, disease, drug, lifestyle, emotion, social attribute, symptom, and treatment. Table 5 displays the distribution of the eight types of entities in the 1890 records. The entities with the highest frequency were social properties, diseases, and tests, which accounted for 89.68, 80.05, and 80.00%, respectively; followed by drugs, symptoms, lifestyle, and treatment, which accounted for 56.4, 36.40, and 25.93%, respectively. The less frequent entities were symptoms and emotions, which accounted for 7.61 and 7.59%, respectively. In addition, we counted the top 10 highest-frequency words of each entity type. For example, among 4259 check entity types, fasting blood glucose, postprandial blood glucose, glycated hemoglobin, and glucose tolerance tests were the most common tests for diabetes; these high-frequency words accounted for 70.86% of the examination entity categories. Among the disease entity types, hypertension, fatty liver, coronary heart disease, cerebral infarction, and stroke were the most frequently occurring diseases; these high-frequency words accounted for 73.15% of the disease entity categories, indicating that diabetic patients are often afflicted by other types of cardiovascular diseases and complications. Table 5 describes the details of the top 10 entities.

Table 5. Related statistics of entity frequency.

Entity Type	Entity Frequency	Rate	Top 10 Entities	Top 10 Number of Entities	Top 10 Rate
Check	1512/1890	80%	Blood glucose, fasting blood glucose, fasting, postprandial, glycated hemoglobin, physical examination, high blood glucose, glucose tolerance, review, and postprandial blood glucose.	3019/4259	70.86%
Disease	1513/1890	80.05%	Diabetes, hypertension, hyperglycemia, type 2 diabetes, fatty liver, coronary heart disease, cerebral infarction, obesity, hyperlipidemia, and complications of diabetes.	1790/2447	73.15%
Drug	1066/1890	56.4%	Insulin, Metformin, Acarbose, Glucose, Dapagliflozin, Glucagon, Glimepiride, Bystolic, Gleevec, and Chinese medicine.	1306/2504	52.16%
Life	490/1890	25.92%	Blood sugar control, exercise, diet control, poor sleep, stopping the medication, exercise, not taking medication, losing weight, watching what you eat, and staying up late.	509/787	64.68%
Mood	144/1890	7.61%	Worry, doubt, fear, anxiety, hurry, tension, tiredness, anger, uneasiness, and fear.	132/185	71.35%

Table 5. *Cont.*

Entity Type	Entity Frequency	Rate	Top 10 Entities	Top 10 Number of Entities	Top 10 Rate
Social	1695/1890	89.68%	Height and weight, greater than six months, pregnant, not pregnant, within six months, within one month, within one week, self, allergy, and father.	3864/4690	82.39%
Symptom	688/1890	16.4%	Thirst, bitterness and dryness, dizziness and lightheadedness, excessive urination, weakness, weight loss, nausea, sweating, panic attacks, and frequent urination.	465/1564	29.73%
Treat	421/1890	22.08%	Surgery, chemotherapy, radiotherapy, drug therapy, inpatient treatment, weight loss, Chinese medicine, stents, minimally invasive, and immunization.	429/655	65.50%

4. Discussion

Combining the diabetes entity classification criteria of Shanghai Ruijin Hospital, our model demonstrated that the RoBERTa-BiLSTM-CRF-based deep learning model could perform the online Q&A text-based diabetes entity recognition task with an F1 value of 81.51%, outperforming previously published online healthcare entity recognition results using the BiLSTM-CRF model (68.43%) [29]. This is comparable to the recently reported BERT-BiLSTM-CRF-model-based named entity recognition system for the diabetes literature (79.89%) [30]. The benefit of the RoBERTa-BiLSTM-CRF model (F1 value of 81.51%) over the benchmark model, BiLSTM-CRF (F1 value of 75.28%), is that BERT produces better word-level vectors than the phrase vectors acquired using Word2vec. Pre-training in the biomedical corpus improves BERT's ability to comprehend difficult biomedical literature.

The semi-structured doctor–patient health community requires patients to fill in socio-demographic data and provides optional fixed-word input, which may indicate that socio-demographic information descriptions are relatively standardized and fixed, and the accuracy and sensitivity of entity recognition were improved, with F1 values exceeding 90% for all four models. In addition, the patients' inputs in the text boxes of "chief complaint" and "help wanted" were relatively free text, and the majority of patients used colloquial language to describe their symptoms and treatments due to a lack of professional knowledge. The "Help" text box contained a highly free-form description written primarily in colloquial language, with a certain number of misspellings and ambiguities regarding the concept of professional terms, which are significantly different from the electronic medical records portrayed from the physician's perspective. The language style of the doctor–patient Q&A community is information-oriented language expression, which is characterized by specific, certain, and objective vocabularies, while the language style of the patient–patient community is social-support-oriented language expression, characterized by ambiguous and empathic features. In the Chinese electronic medical record dataset, CCKS, based on the BERT model, published studies demonstrating that the F1 values for the symptom-sign category all exceeded 95% and the F1 values for the treatment entity category all exceeded 82% [31,32]. Additionally, the entity recognition was superior to the entity recognition in online health communities [33,34]. It has been demonstrated that biomedical experts and the general public differently perceive medical entities in diabetes.

In addition, we analyzed named entities extracted from online health communities to investigate the key topics discussed and emphasized in patients' online health Q&As for the purpose of studying the health information needs of patients. Table 5 shows the frequency of entity occurrences in each category and the proportion of TOP10 entities in the respective entity type. The frequency indicates the number of times a category of entities is mentioned in relevant posts. In 1890 relevant posts, for example, the test and examination category entities were mentioned 1512 times. The experimental data suggest that they focus on diseases (possibly assessment screening for diabetes and complications of diabetes), tests and examinations (on diabetes screening and concerns about glycemic control management), and medications (possibly counseling on medication involving diabetes), as confirmed by previously published studies on entity identification in online health communities for diabetes [35,36].

The limitations of this study include the following: (1) The data sources only comprised single online health community doctor–patient Q&A texts, without considering differences in recognition performance of the BERT model on datasets from other online health communities with different language styles. Furthermore, the study lacked a comprehensive investigation into the connection between language expression features of different chronic diseases and the applicability of the chosen model. Therefore, further research must be conducted on the applicability of our model to online health community texts. (2) The diabetes data used in the study were cross-sectional static data of patients, and longitudinal cohorts of different stages of chronic disease progression were not collected without patient tracking [37]. Future research must continue to standardize the annotated corpus, expand its coverage, and optimize the outcomes of the model.

5. Conclusions

For the named entity recognition of the online medical community of diabetes, the RoBERTa-LiSTM-CRF model outperforms the other three models: RoBERTa-CRF (RC), BilSTM-CRF (BC), and RoBERTa-Softmax (RS). The proposed model, consisting of a pre-trained model with enhanced contextual long-range semantic learning ability based on word vectors, can effectively address entity recognition challenges within the health community. In addition, we found that patients with different disease stages have distinct focused topics and that the extracted entity type and attribute values will also vary. The high-performance entity recognition in online health communities represents a crucial knowledge source for constructing medical knowledge graphs. It can be applied to intelligent question-answering systems, clinical decision support systems, and other applications. This integration helps alleviate the growing demand for medical consultations and the strain on healthcare resources while assisting healthcare professionals in making informed decisions and providing personalized services to patients. In our future research, we will implement the BERT model for pre-training on additional websites of online healthcare communities.

Author Contributions: Y.Z. collected the data, coded the program, and plotted the results; Q.X. wrote the manuscript and analyzed the data. Q.X., A.L., B.L. and Z.X. revised these drafts for significant scientific content. W.X. and C.H. contributed to the study's concepts. A.L. is the guarantor of the funding acquisition. All authors have read and agreed to the published version of the manuscript.

Funding: This study was supported by both the Key Laboratory of Medical Information Research of Central South University in China within the project "Clinical Research Center for Cardiovascular Intelligent Healthcare in Hunan Province" agreement no. 2021SK4005, and Science and Technology Plan Project of Changsha (grant no. kq1901133).

Institutional Review Board Statement: Not applicable.

Informed Consent Statement: Not applicable.

Data Availability Statement: The datasets used and/or analyzed during the current study are available from the corresponding author upon approved written requests.

Acknowledgments: We appreciate the participation of all investigators, study teams, and subjects for participating in these studies. We appreciate the assistance of the doctors at the Second Xiangya Hospital for their participation in this study.

Conflicts of Interest: The authors declare that the research was conducted in the absence of any commercial or financial relationships that could be construed as potential conflict of interest.

References

1. Sun, H.; Saeedi, P.; Karuranga, S.; Pinkepank, M.; Ogurtsova, K.; Duncan, B.B.; Stein, C.; Basit, A.; Chan, J.C.N.; Mbanya, J.C.; et al. IDF Diabetes Atlas: Global, regional and country-level diabetes prevalence estimates for 2021 and projections for 2045. *Diabetes Res. Clin. Pract.* **2022**, *183*, 13. [CrossRef] [PubMed]
2. Lee, K.; Hoti, K.; Hughes, J.D.; Emmerton, L.M. Consumer Use of "Dr Google": A Survey on Health Information-Seeking Behaviors and Navigational Needs. *J. Med. Internet Res.* **2015**, *17*, 16. [CrossRef] [PubMed]
3. Nakikj, D.; Mamykina, L. A park or a highway: Overcoming tensions in designing for socio-emotional and informational needs in online health communities. In Proceedings of the 2017 ACM Conference on Computer Supported Cooperative Work and Social Computing, Portland, OR, USA, 25 February–1 March 2017; pp. 1304–1319.
4. Nakikj, D.; Mamykina, L. Lost in migration: Information management and community building in an online health community. In Proceedings of the 2018 CHI Conference on Human Factors in Computing Systems, Montreal, QC, Canada, 21–26 April 2018; pp. 1–14.
5. Chen, Z.H.; Zhang, C.C.; Fan, G.H. Interrelationship between Interpersonal Interaction Intensity and Health Self-Efficacy in People with Diabetes or Prediabetes on Online Diabetes Social Platforms: An In-Depth Survey in China. *Int. J. Environ. Res. Public Health* **2020**, *17*, 21. [CrossRef] [PubMed]
6. Chen, D.H.; Zhang, R.T.; Liu, K.C.; Hou, L. Knowledge Discovery from Posts in Online Health Communities Using Unified Medical Language System. *Int. J. Environ. Res. Public Health* **2018**, *15*, 16. [CrossRef] [PubMed]
7. Coden, A.; Savova, G.; Sominsky, I.; Tanenblatt, M.; Masanz, J.; Schuler, K.; Cooper, J.; Guan, W.; de Groen, P.C. Automatically extracting cancer disease characteristics from pathology reports into a Disease Knowledge Representation Model. *J. Biomed. Inform.* **2009**, *42*, 937–949. [CrossRef] [PubMed]
8. Song, M.; Yu, H.; Han, W.S. Developing a hybrid dictionary-based bio-entity recognition technique. *BMC Med. Inform. Decis. Mak.* **2015**, *15*, 8. [CrossRef]
9. Khan, A.T.; Cao, X.W.; Liao, B.L.; Francis, A. Bio-inspired Machine Learning for Distributed Confidential Multi-Portfolio Selection Problem. *Biomimetics* **2022**, *7*, 20. [CrossRef]
10. Pletscher-Frankild, S.; Pallejà, A.; Tsafou, K.; Binder, J.X.; Jensen, L.J. DISEASES: Text mining and data integration of disease–gene associations. *Methods* **2015**, *74*, 83–89. [CrossRef]
11. Rais, M.; Lachkar, A.; Lachkar, A.; Ouatik, S.E.A. A comparative study of biomedical named entity recognition methods based machine learning approach. In Proceedings of the 2014 Third IEEE International Colloquium in Information Science and Technology (CIST), Tetouan, Morocco, 20–22 October 2014; pp. 329–334.
12. Saha, S.K.; Sarkar, S.; Mitra, P. Feature selection techniques for maximum entropy based biomedical named entity recognition. *J. Biomed. Inform.* **2009**, *42*, 905–911. [CrossRef]
13. Francis, A.; Nazeer, K.A. Literature search tool for the extraction of disease-associated genes using frequent itemset mining. In Proceedings of the 2016 International Conference on Computing, Analytics and Security Trends (CAST), Pune, India, 19–21 December 2016; pp. 306–311.
14. Liang, T.; Shih, P.-K. Empirical textual mining to protein entities recognition from PubMed corpus. In Proceedings of the Natural Language Processing and Information Systems: 10th International Conference on Applications of Natural Language to Information Systems, NLDB 2005, Alicante, Spain, 15–17 June 2005; pp. 56–66.
15. Cai, X.L.; Dong, S.B.; Hu, J.L. A deep learning model incorporating part of speech and self-matching attention for named entity recognition of Chinese electronic medical records. *BMC Med. Inform. Decis. Mak.* **2019**, *19*, 9. [CrossRef] [PubMed]
16. Dong, X.S.; Chowdhury, S.; Qian, L.J.; Li, X.F.; Guan, Y.; Yang, J.F.; Yu, Q.B. Deep learning for named entity recognition on Chinese electronic medical records: Combining deep transfer learning with multitask bi-directional LSTM RNN. *PLoS ONE* **2019**, *14*, e0216046. [CrossRef] [PubMed]
17. Dong, X.; Qian, L.; Guan, Y.; Huang, L.; Yu, Q.; Yang, J. A multiclass classification method based on deep learning for named entity recognition in electronic medical records. In Proceedings of the 2016 New York Scientific Data Summit (NYSDS), New York, NY, USA, 14–17 August 2016; pp. 1–10.
18. Dreyfus, B.; Chaudhary, A.; Bhardwaj, P.; Shree, V.K. Application of natural language processing techniques to identify off-label drug usage from various online health communities. *J. Am. Med. Inf. Assoc.* **2021**, *28*, 2147–2154. [CrossRef] [PubMed]
19. Zhang, Y.L.; Li, X.M.; Zhang, Z. Disease-Pertinent Knowledge Extraction in Online Health Communities Using GRU Based on a Double Attention Mechanism. *IEEE Access* **2020**, *8*, 95947–95955. [CrossRef]
20. Whitfield, C.; Liu, Y.; Anwar, M. Surveillance of COVID-19 Pandemic using Social Media: A Reddit Study in North Carolina. In Proceedings of the 12th ACM Conference on Bioinformatics, Computational Biology, and Health Informatics, Gainesville, FL, USA, 1–4 August 2021; Association for Computing Machinery: New York, NY, USA; pp. 1–8.
21. Lample, G.; Ballesteros, M.; Subramanian, S.; Kawakami, K.; Dyer, C. Neural Architectures for Named Entity Recognition. *arXiv* **2016**, arXiv:1603.01360.
22. Wang, Z.; Guan, H. Research on named entity recognition of doctor-patient question answering community based on bilstm-crf model. In Proceedings of the 2020 IEEE International Conference on Bioinformatics and Biomedicine (BIBM), Seoul, Republic of Korea, 16–19 December 2020; pp. 1641–1644.
23. Devlin, J.; Chang, M.-W.; Lee, K.; Toutanova, K. BERT: Pre-training of Deep Bidirectional Transformers for Language Understanding. *arXiv* **2019**, arXiv:1810.04805.

24. Zhang, Q.; Wu, M.; Lv, P.; Zhang, M.; Yang, H. Research on named entity recognition of chinese electronic medical records based on multi-head attention mechanism and character-word information fusion. *J. Intell. Fuzzy Syst.* **2022**, *42*, 4105–4116. [CrossRef]
25. Zhu, Z.; Li, J.; Zhao, Q.; Wei, Y.-C.; Jia, Y. Medical named entity recognition of Chinese electronic medical records based on stacked Bidirectional Long Short-Term Memory. In Proceedings of the 2021 IEEE 45th Annual Computers, Software, and Applications Conference (COMPSAC), Madrid, Spain, 12–16 July 2021; pp. 1930–1935.
26. Fang, A.; Hu, J.H.; Zhao, W.Q.; Feng, M.; Fu, J.; Feng, S.S.; Lou, P.; Ren, H.L.; Chen, X.L. Extracting clinical named entity for pituitary adenomas from Chinese electronic medical records. *BMC Med. Inform. Decis. Mak.* **2022**, *22*, 14. [CrossRef]
27. Han, P.; Li, X.; Wang, X.; Wang, S.; Gao, C.; Chen, W. Exploring the effects of drug, disease, and protein dependencies on biomedical named entity recognition: A comparative analysis. *Front. Pharmacol.* **2022**, *13*. [CrossRef]
28. Yang, X.; Bian, J.; Hogan, W.R.; Wu, Y. Clinical concept extraction using transformers. *J. Am. Med. Inf. Assoc.* **2020**, *27*, 1935–1942. [CrossRef]
29. Yang, H.; Gao, H. Toward Sustainable Virtualized Healthcare: Extracting Medical Entities from Chinese Online Health Consultations Using Deep Neural Networks. *Sustainability* **2018**, *10*, 3292. [CrossRef]
30. Yang, L.; Fu, Y.F.; Dai, Y. BIBC: A Chinese Named Entity Recognition Model for Diabetes Research. *Appl. Sci.* **2021**, *11*, 14. [CrossRef]
31. Li, X.; Zhang, H.; Zhou, X.-H. Chinese clinical named entity recognition with variant neural structures based on BERT methods. *J. Biomed. Inform.* **2020**, *107*, 103422. [CrossRef] [PubMed]
32. Ji, B.; Liu, R.; Li, S.; Tang, J.; Yu, J.; Li, Q.; Xu, W. A BILSTM-CRF method to Chinese electronic medical record named entity recognition. In Proceedings of the 2018 International Conference on Algorithms, Computing and Artificial Intelligence, Sanya, China, 21–23 December 2018; pp. 1–6.
33. Zhang, Y.; Zhang, Y.; Yin, Y.; Xu, J.; Xing, C.; Chen, H. Chronic disease related entity extraction in online chinese question and answer services. In Proceedings of the Smart Health: International Conference, ICSH 2015, Phoenix, AZ, USA, 17–18 November 2015; pp. 55–67.
34. Zhang, Q.; Sun, Y.; Zhang, L.; Jiao, Y.; Tian, Y. Named entity recognition method in health preserving field based on BERT. *Procedia Comput. Sci.* **2021**, *183*, 212–220. [CrossRef]
35. Lu, Y.; Zhang, P.; Liu, J.; Li, J.; Deng, S. Health-Related Hot Topic Detection in Online Communities Using Text Clustering. *PLoS ONE* **2013**, *8*, e56221. [CrossRef] [PubMed]
36. Lu, Y.; Wu, Y.; Liu, J.; Li, J.; Zhang, P. Understanding Health Care Social Media Use From Different Stakeholder Perspectives: A Content Analysis of an Online Health Community. *J. Med. Internet Res.* **2017**, *19*, e109. [CrossRef]
37. Zhang, S.; Grave, E.; Sklar, E.; Elhadad, N. Longitudinal analysis of discussion topics in an online breast cancer community using convolutional neural networks. *J. Biomed. Inform.* **2017**, *69*, 1–9. [CrossRef]

bioengineering

MDPI

Review

A Comprehensive Review on Synergy of Multi-Modal Data and AI Technologies in Medical Diagnosis

Xi Xu [1], Jianqiang Li [1], Zhichao Zhu [1], Linna Zhao [1], Huina Wang [1], Changwei Song [1], Yining Chen [1], Qing Zhao [1,*], Jijiang Yang [2] and Yan Pei [3]

1 Faculty of Information Technology, Beijing University of Technology, Beijing 100124, China; xuxi@bjut.edu.cn (X.X.); lijianqiang@bjut.edu.cn (J.L.); zhuzc@emails.bjut.edu.cn (Z.Z.); zhao_linna@emails.bjut.edu.cn (L.Z.); huinawang@emails.bjut.edu.cn (H.W.); songchangwei1996@emails.bjut.edu.cn (C.S.); chenyn0317@emails.bjut.edu.cn (Y.C.)

2 Tsinghua National Laboratory for Information Science and Technology, Tsinghua University, Beijing 100084, China; yangjijiang@tsinghua.edu.cn

3 School of Computer Science and Engineering, The University of Aizu, Aizuwakamatsu 965-8580, Japan; peiyan@u-aizu.ac.jp

* Correspondence: zhaoqing@bjut.edu.cn

Abstract: Disease diagnosis represents a critical and arduous endeavor within the medical field. Artificial intelligence (AI) techniques, spanning from machine learning and deep learning to large model paradigms, stand poised to significantly augment physicians in rendering more evidence-based decisions, thus presenting a pioneering solution for clinical practice. Traditionally, the amalgamation of diverse medical data modalities (e.g., image, text, speech, genetic data, physiological signals) is imperative to facilitate a comprehensive disease analysis, a topic of burgeoning interest among both researchers and clinicians in recent times. Hence, there exists a pressing need to synthesize the latest strides in multi-modal data and AI technologies in the realm of medical diagnosis. In this paper, we narrow our focus to five specific disorders (Alzheimer's disease, breast cancer, depression, heart disease, epilepsy), elucidating advanced endeavors in their diagnosis and treatment through the lens of artificial intelligence. Our survey not only delineates detailed diagnostic methodologies across varying modalities but also underscores commonly utilized public datasets, the intricacies of feature engineering, prevalent classification models, and envisaged challenges for future endeavors. In essence, our research endeavors to contribute to the advancement of diagnostic methodologies, furnishing invaluable insights for clinical decision making.

Keywords: multi-modal data; artificial intelligence; disease diagnosis; machine learning; deep learning; large model

Citation: Xu, X.; Li, J.; Zhu, Z.; Zhao, L.; Wang, H.; Song, C.; Chen, Y.; Zhao, Q.; Yang, J.; Pei, Y. A Comprehensive Review on Synergy of Multi-Modal Data and AI Technologies in Medical Diagnosis. *Bioengineering* **2024**, *11*, 219. https://doi.org/10.3390/bioengineering11030219

Academic Editor: Larbi Boubchir

Received: 29 December 2023
Revised: 15 February 2024
Accepted: 21 February 2024
Published: 25 February 2024

1. Introduction

The task of disease diagnosis holds significant importance within the medical domain. Timely diagnosis not only facilitates the prompt implementation of therapeutic interventions but also mitigates the risks associated with disease progression and complications, particularly concerning global health challenges such as Alzheimer's disease, breast cancer, depression, heart disease, and epilepsy. Nonetheless, achieving this objective remains challenging, particularly in developing areas and regions with limited medical resources. The high incidence and growth rates of the aforementioned diseases further compound the challenges confronting the healthcare system in terms of diagnosis. This challenge primarily stems from two key factors: firstly, the low specialist-to-patient ratio, and secondly, the time-consuming and labor-intensive nature of the manual diagnosis, which heavily relies on specialized expertise. These issues often result in delayed treatment, exacerbating illness severity, and escalating medical costs. Consequently, there exists an urgent need for automated diagnostic approaches to address these pressing concerns.

AI-driven healthcare, emerging as a transformative force in the medical landscape, seeks to revolutionize clinical practices leveraging the capabilities of information technology. It represents a promising avenue for addressing critical disease diagnosis challenges in regions characterized by disparities in medical resources, garnering significant attention from both scholars and practitioners [1]. AI-driven healthcare entails the integration of medical data with intelligent technologies to enhance healthcare quality and productivity.

The clinical diagnostic process is inherently intricate, involving the generation and analysis of diverse data types encompassing images, speech, text, and genetic information (as depicted in Figure 1). This complexity stems from the synergistic interaction of multiple data sources, including images capturing anatomical structures, speech elucidating patient symptoms, textual descriptions of medical history, genetic information delineating inherent susceptibility, and physiological signals acquired through electrocardiograms (ECGs) and electroencephalograms (EEGs). Each modality furnishes unique and valuable insights that collectively contribute to a holistic understanding of patients' physiological states.

- **Image.** Medical imaging tools such as computed tomography (CT), X-rays, magnetic resonance imaging (MRI), and digital pathology offer visual representations of internal structures and anomalies. These images serve as foundational components of a diagnosis, unveiling intricate details crucial for identifying and characterizing various medical conditions.
- **Text.** Textual data encompassing electronic health records, clinical notes, and medical literature constitute a narrative thread weaving through the patient's medical journey, history, and contextual information vital for precise diagnosis.
- **Speech.** Speech recordings provide a unique avenue for understanding patients' experiences and symptoms. This modality captures nuances such as tone, pace, and articulation, thereby adding a qualitative dimension to the diagnostic process.
- **Genetic data.** Genetic data introduce a molecular layer to elucidate inherent predispositions, susceptibilities, and genetic markers potentially influencing disease manifestation.
- **Physiological signals.** Signal data offer real-time snapshots of cardiac and neural activities. This dynamic modality effectively captures temporal variations, offering critical insights into abnormalities and patterns associated with cardiac or neurological diseases.

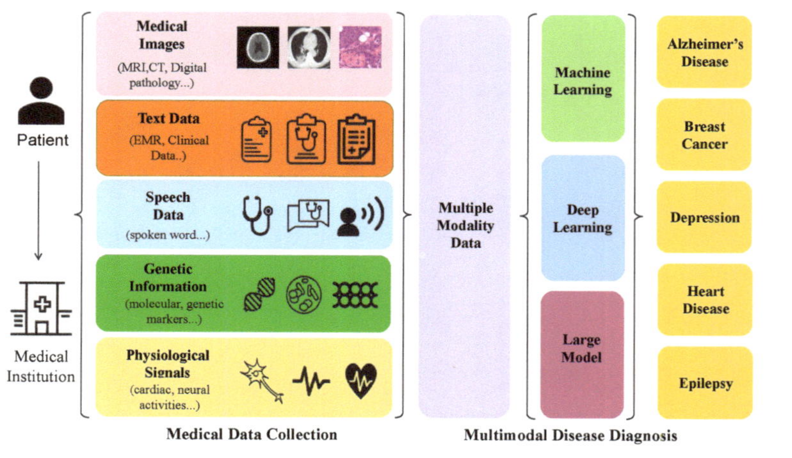

Figure 1. The diverse data types including images, speech, text, and genetic information can be produced in the clinical diagnostic process.

Numerous experts and scholars have actively participated in the collection and integration of medical data for diagnostic tasks, as evidenced by their contributions to various datasets [2–6]. Remarkably, these individuals not only curated and refined these datasets but also advocated for their accessibility and openness. For instance, the ADNI dataset, cited in references [7,8], has emerged as a cornerstone in neuroimaging and dementia research. This dataset incorporates diverse modalities such as structural and functional MRI, positron emission tomography (PET), and cerebrospinal fluid biomarkers, thereby offering a comprehensive perspective on disease progression. The availability of such datasets establishes a standardized framework for the development and evaluation of advanced diagnostic algorithms, particularly those leveraging machine learning and deep learning techniques. These methodologies play a pivotal role in extracting discernible features from multi-modal medical data and have witnessed significant advancements in recent years.

- **Machine learning approaches.** Machine learning methodologies have emerged as pivotal tools for medical diagnosis tasks, exemplified by techniques like Support Vector Machines (SVMs) [9] and Random Forests (RFs). SVMs excel in establishing optimal decision boundaries for classification, and are particularly adept at discerning intricate patterns within multidimensional data. On the other hand, RFs harness the strength of ensemble learning by amalgamating predictions from numerous decision trees, thereby enhancing model performance. The deployment of such machine learning techniques constitutes a substantial advancement in automated disease diagnosis, particularly in handling structured and well-defined datasets.
- **Deep learning models.** Deep learning models, as referenced in the literature [10–12], employ hierarchical neural networks to extract inherent patterns from medical data. For instance, Convolutional Neural Networks (CNNs) specialize in spatial feature extraction and prove beneficial in medical imaging applications, such as tumor detection in radiological scans. Conversely, Recurrent Neural Networks (RNNs) are well suited for sequence data analysis, enabling proficient performance in tasks like time series analysis or monitoring disease progression over time.
- **Large models.** Large models are designed to learn intricate feature representations from vast datasets [13–18]. In the field of medical data, large model approaches are expected to further improve the ability to capture and generalize complex features [19–25].

Existing reviews have offered insightful perspectives on research about automated disease diagnosis utilizing either machine learning or deep learning methodologies. However, these reviews predominantly concentrate on a singular modality or a single disease, whether focusing on a specific disease within multi-modal contexts, various disorders within a specific modality, or a single disease with exclusive reliance on a particular data type. In contrast, our review endeavors to explore the diverse modalities employed in the automatic diagnosis of distinct diseases. Although medical datasets generated by different disease diagnosis processes exhibit commonalities, distinct preferences for specific modalities prevail across different diseases. Consequently, this paper emphasizes general AI techniques applicable to different modalities and diseases, rather than solely focusing on a single disease or modality. Additionally, the latest advancements in large model-based specific disease diagnosis are introduced herein. To elucidate, we initially delineate available public datasets and the AI framework in automatic disease diagnosis, encompassing data pre-processing, feature engineering, model selection, and performance evaluation metrics. Subsequently, we expound upon reported works associated with various diseases. Lastly, a comprehensive discussion and outline of future avenues of exploration are presented to guide innovative solutions in this domain.

The remainder of this paper is structured as follows. In Section 2, we delve into the utilization of multi-modal data and AI in disease diagnosis, encompassing an exploration of public datasets and an overview of the overall processing framework. Section 3 provides a detailed exposition of the reported work, elucidating the methodologies, findings, and insights gleaned from recent research endeavors. In Section 4, we delineate the intricate challenges encountered in this field and outline potential avenues for future research and development. Finally, we encapsulate our findings and insights in the conclusion of this review in Section 5.

2. Multi-Modal and AI Used in Disease Diagnosis

Most diseases are typically only recognized by patients themselves after they manifest, and continuous data collection and monitoring can assist patients in achieving effective disease prevention. The advent of artificial intelligence has rendered the process of data accumulation more intelligent and efficient, thereby holding significant implications for disease prevention and control. This section elaborates on the comprehensive framework of artificial intelligence technology in medical diagnosis applications, encompassing data collection, model architecture construction, and model evaluation.

2.1. Datasets in AI-Based Disease Diagnosis Studies

Data collection plays a pivotal role in the development of machine learning models for disease diagnosis, serving as the bedrock upon which these models are constructed and trained. Many studies on AI-based disease diagnosis choose to utilize established open datasets to augment the research's credibility and scope. In this section, we concentrate on the datasets employed in the research process across various diseases. For more detailed information on the data, please consult Table A1 in the Appendix A.

Alzheimer's disease. The Alzheimer's Disease Neuroimaging Initiative (ADNI) database [7,8], established in 2003, is widely recognized as one of the most prominent datasets for predicting AD. It encompasses various types of data, including brain imaging data such as MRI and PET scans, clinical data, biospecimen information, and genetic data. The patients in the ADNI database are categorized into different stages such as AD, MCI (Mild Cognitive Impairment), and NC (Normal Cognition). Another typical database is the longitudinal dataset called OASIS-3, which integrates multiple modalities [2], including neuroimaging, clinical biomarkers, and cognitive assessment. This dataset primarily investigates the progression of AD in 1378 individuals. Available at: http://www.oasis-database.org (accessed on 29 November 2023). Additionally, since 2006, the UK Biobank (UKB) [3–5] has amassed a substantial amount of data from participants, encompassing various fields such as environmental factors, lifestyle choices, sociodemographic information, overall health and well-being, as well as cognitive and physical assessments [6].

Breast cancer. The Cancer Genome Atlas (TCGA) [26] is a widely utilized dataset for predicting breast cancer. It involves MRI and CT scans, clinical records and genetic information. In the TCGA dataset, breast cancer is categorized into different subtypes, including Luminal A, Luminal B, HER2+, Basel, etc. The SAFHS [27] is a large-scale population-based natural language processing dataset developed by Harvard Medical School. Available at: http://www.ncbi.nlm.nih.gov/ (accessed on 29 November 2023). The Breast Ultrasound Images (BUSI) [28] was created in 2018 and contains normal, benign and malignant breast ultrasound images. Available at: https://scholar.cu.edu.eg/ (accessed on 29 November 2023). In the gene domain, Gene Expression Omnibus (GEO) [29] collects high-throughput functional genomics data for researchers, including microarrays, next-generation sequencing, and other forms. Available at: https://www.ncbi.nlm.nih.gov/geo/(accessed on 29 November 2023).

Heart disease. TLGS [30] is a long-term epidemiological research project for assessing the risk factors for cardiovascular diseases among residents of Tehran, Iran. Available at: https://endocrine.ac.ir/page/Tehran-Lipid-and-Glucose-Study-TLGS (accessed on 29 November 2023). In the text domain, the Acute Myocardial Infarction Dataset of the World Health Organization (WHO) collects from medical institutions and public health departments across various countries. Available at: http://www.who.int/ (accessed on 29 November 2023). It mainly studies the epidemiology, clinical characteristics, treatment methods, and prognosis of acute myocardial infarction and includes patient clinical information, diagnostic results, treatment measures, and other data. In the image domain, the Sunnybrook Cardiac Data (SCD) [31] dataset consists of 45 cine MRI images from different patients with various pathological conditions, including healthy individuals, hypertrophy, ischemic heart failure, and non-ischemic heart failure. Available at: https://www.cardiacatlas.org/sunnybrook-cardiac-data/ (accessed on 29 November 2023). In addition, the Automated Cardiac Diagnosis Challenge (ACDC) [32] database includes medical image data of normal subjects, ischaemic heart failure, dilated cardiomyopathy, hypertrophic cardiomyopathy, and right ventricular abnormalities. Available at: https://www.creatis.insa-lyon.fr/Challenge/acdc/ (accessed on 29 November 2023).

Depression. The Distress Analysis Interview Corpus-Wizard of OZ (DAIC-WOZ) [33] stands as one of the most popular speech datasets utilized for depression prediction. Available at: https://dcapswoz.ict.usc.edu/ (accessed on 29 November 2023). Its objective is to capture individuals' verbal expressions of psychological distress and emotional stress through simulated interactions with AI. The corpus encompasses a broad spectrum of psychological disorders, including depression, anxiety, and post-traumatic stress disorder. Each entry within the dataset includes emotional annotations to furnish quantitative insights into the patient's emotional state. The Multi-modal Open Dataset for Mental Disorder Analysis (MODMA) [34] is a multi-modal dataset tailored for mental disorders, featuring both clinically depressed patients and individuals from the normal population. Available at: http://modma.lzu.edu.cn/data/index/ (accessed on 29 November 2023). It comprises speech data and ECG data. Moreover, the Bipolar Disorder Corpus compiles textual data pertinent to bipolar disorder, aimed at facilitating researchers' comprehension of the disorder's characteristics, diagnosis, and treatment. The textual content within this repository encompasses diaries, medical records, clinical assessment reports, and other pertinent literature from individuals with bipolar disorder.

Epilepsy. The CHB-MIT [35] Database comprises EEG recordings collected from 22 pediatric subjects with intractable seizures and was established in 2010. Available at: http://physionet.org/ (accessed on 29 November 2023). The Bonn EEG time series database [36] involves EEG data obtained from a 128-channel acquisition system, featuring recordings from 5 patients identified as A, B, C, D, and E. Sets C and D encompass intracranial EEG recordings taken during seizure-free intervals, with set C recorded from within the seizure-generating area and set D from outside the seizure-generating area of epileptic patients. Available at: http://www.ukbonn.de/epileptologie/ag-lehnertz-downloads/ (accessed on 29 November 2023). Set E contains intracranial EEG data captured during epileptic seizures. Each set consists of 100 text files, each containing a single EEG time series represented in ASCII code and comprising 4097 samples. This database is devoid of artifacts, obviating the necessity for preprocessing prior to classifying the signals as healthy (non-epileptic) or unhealthy (epileptic). The Temple University EEG corpus database [37] represents an extensive collection of EEG data acquired between 2000 and 2013. Available at: http://isip.piconepress.com/projects/tuh$_$eeg/ (accessed on 29 November 2023). This repository encompasses diverse EEG clinical settings from approximately 10,874 patients. By incorporating a large cohort of patients and spanning a significant timeframe, the Temple University EEG corpus database affords opportunities for multifaceted analyses in EEG research. Researchers can exploit this invaluable repository to explore various facets of EEG data and advance the understanding of neurological conditions.

2.2. Framework for AI in Disease Diagnosis Modeling

Up to now, AI models have been developed for a wide range of disease diagnoses. These models have undergone architecture designing and fine-tuning by leveraging diverse modalities of data such as medical images, medical texts, genetics, medical speeches, EEG, and ECG. Their applications span diagnostic classification, phenotype discovery, and other disease diagnosis tasks. In this section, we will focus on introducing well-known AI models and their intricate framework designs, including data preprocessing, feature engineering, and model selection (as shown in Figure 2).

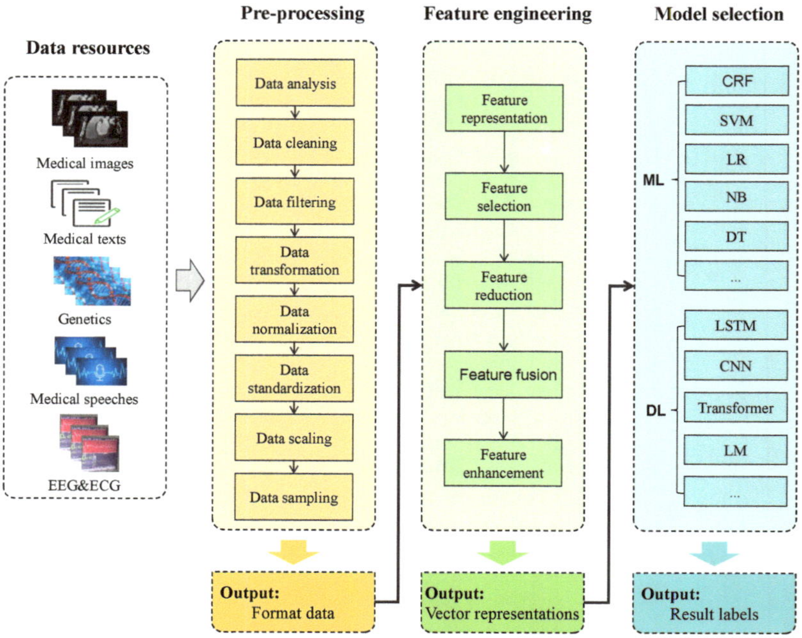

Figure 2. The framework for AI in disease diagnosis modeling (ML and DL denote machine learning and deep learning, respectively).

2.2.1. Pre-Processing

Pre-processing using machine learning and deep learning technologies is a crucial step for disease diagnosis. By preprocessing raw data, inaccurate or irrelevant information can be removed and key features relevant to disease diagnosis are extracted. Common preprocessing operations include data research and analysis, data cleaning, data filtering, data transformation, data normalization, data standardization, data scaling, data sampling, etc. Specifically:

Data exploration. It involves analyzing the number of samples, features, and their distributions of the dataset, which not only reveals the intrinsic properties of the dataset but also provides a solid foundation for the subsequent selection of preprocessing techniques.

Data cleaning. It aims to handle noisy or erroneous data, including removing duplicate entries, handling missing values, and correcting data errors or inconsistencies.

Data filtering. It is used to remove noise from a dataset, including low-pass filtering and high-pass filtering.

Data transformation. It involves converting raw data into different representations or forms.

Data normalization. It scales the data to a standard range or distribution, including min–max normalization, clipping normalization, standard deviation normalization, and z-score normalization.

Data standardization. Its primary function is to convert data from varying ranges and scales into a uniform standard format, such as FHIR HL7 [38], SNOMED CT [39] and DICOM [40], thus making data more suitable for machine learning and statistical analysis.

Data scaling. Data scaling enables data to map to specific ranges or intervals, ensuring comparability at different scales and effectively mitigating biases caused by scale differences.

Data sampling. The purpose of data sampling is to choose a subset of data from the primary dataset, thus forming a representative sample for analysis. In the case of imbalanced datasets, various sampling strategies can be utilized, including random sampling, stratified sampling, or oversampling/undersampling. These strategies can effectively address the issue of disparate class distributions in the dataset, ensuring accurate predictions for each class.

The above preprocessing operations aim to address issues such as noise, missing values, inconsistency, or specific data challenges. Facing different types of data (such as medical imaging, medical texts, genetic data, audio, and electrocardiogram signals), different preprocessing methods are usually required. Specifically:

Medical imaging data. Medical imaging data have a rich and complex spatial structure, consisting of a multidimensional matrix of pixels, each containing information about color and brightness. The preprocessing of medical imaging data mainly focuses on image resolution (number of pixels), color depth (color details in each pixel), and format (encoding methods such as portable network graphics (PNG)). For example, in the imaging process of medical images (such as X-rays, CT scans, and MRI), metal objects in the patient's body (such as implants, dental restorations, surgical screws, etc.) and natural movements (appearing blurry or deformed in the image) can cause artifacts that affect the visualization of surrounding tissues. Metal artifact correction and motion correction are designed to handle such artifact situations. The imaging process is often susceptible to factors such as long or insufficient exposure time, scanning speed, radiation dose, and environmental interference, which can introduce random noise into the image. This requires the use of denoising methods such as wavelet denoising and median filtering. The lesions in medical images are often local abnormal changes, with some lesions having unclear boundaries and no clear boundaries with surrounding tissues. Data filtering operations such as smoothing filters and high-pass filters are needed to enhance the density, texture, and edge features of the image. In addition, images typically have various spatial resolutions, coordinate systems, and storage formats, so resampling techniques are needed to convert them to standard formats, such as from Medical Digital Imaging and Communications (DICOM) to PNG.

Medical text data. The first step in preprocessing medical text data is usually to decompose them into smaller units based on tokenization. During this process, special characters, punctuation, stopwords, and even spelling and morphological corrections will be removed to reduce data noise and redundancy. Additionally, because text data typically contain a large amount of vocabulary and semantic information, preprocessing typically considers factors such as word frequency, text length, and semantic association to reduce data dimensionality.

Genetic data. Genetic expression data usually include the expression levels of thousands of genes under different conditions or at different time points, complex and multidimensional. Also, gene expression data typically have a right-skewed distribution: most genes are concentrated at lower expression levels and a few genes have very high expression. Therefore, in preprocessing, apart from basic steps like data cleaning and normalization, logarithmic transformations (log), log base 10 (log10), square root transformations, etc., is required to convert the raw gene expression data into a form closer to a normal distribution.

Medical speech data. Original Speech data involve the target speaker's voice and the other interference (e.g., background noise, voices of non-target speakers, reverberation, silence). Endpoint detection, pre-emphasis, framing, windowing, and other techniques are typically used to effectively suppress these interferences. Endpoint detection can detect silent segments in audio signals and segment audio sentences by threshold and short-term energy methods. Pre-emphasis technology is used to increase the importance of the high-frequency part for uniform information since important information in audio signals is often concentrated in the low-frequency part. Framing aims to slice the data to obtain short-term stable audio signals. Moreover, windowing effectively improves the issue of information leakage, with common window functions including the Hamming window, Hanning window, and rectangular window.

EEG and ECG data. Electroencephalogram (EEG) and electrocardiogram (ECG) signals are often interfered with by factors like blinking, movement of the body or electrodes, environmental noise, heartbeat fluctuations, power interference, or baseline drift. The preprocessing process is mainly to ensure signal purity. The Independent Component Analysis (ICA) technique is used to eliminate the interference from blinking and eye movement. Artifacts from cardiovascular and musculoskeletal system electrical activity can be removed using band-pass filters or the Discrete Wavelet Transform (DWT). Noise from power sources, harmonics, and movement of electrodes and wiring can be eliminated using filters of different frequencies.

2.2.2. Feature Engineering

Feature engineering plays a crucial role in disease diagnosis using artificial intelligence technologies. It involves extracting, selecting, and transforming important information from original medical data to construct meaningful features for models. Specifically, feature engineering typically encompasses feature representation, feature selection, feature reduction, feature fusion, and feature enhancement.

Feature representation. Feature representation can transform raw input data into numerical representations that can be utilized by the model.

Feature selection. The redundant features can confuse machine learning models, while few features might not effectively and correctly classify data. Therefore, many researchers adopt feature selection techniques to choose appropriate features from extracted features. Common feature selection techniques include Information Gain, Chi-square Test, Mutual Information, Recursive Feature Elimination (RFE), Regularization, etc.

Feature reduction. When the number of extracted features is huge or they have not been properly normalized or scaled, feature reduction techniques are used to alleviate this problem. The most commonly used feature reduction technique is Principal Component Analysis (PCA), followed by other techniques such as Linear Discriminant Analysis (LDA), Sparse Encoding, and Factor Analysis.

Feature fusion. Feature fusion can enhance the efficiency of classifiers in detection tasks. It involves combining features extracted, selected, or reduced through different methods into a single set of parameters. This integration of features from various perspectives and methodologies offers a more comprehensive and in-depth understanding of the data. Typical feature fusion techniques include Topic Models, Multi-view Learning, and Knowledge Graph Fusion, among others.

Feature enhancement. Feature enhancement can enhance the representation of important features in data while weakening or eliminating the influence of irrelevant or noisy features. In disease diagnosis tasks, feature enhancement helps to more accurately distinguish different disease categories, thereby improving the accuracy and robustness of the model.

2.2.3. Model Selection

According to the diagnostic methods of various diseases, artificial intelligence models are divided into two categories: traditional machine learning methods and deep learning methods.

In the era of rapid advancements in deep learning algorithms, traditional machine learning algorithms continue to be favored in the development of AI diagnostic models due to their unique advantages. They require fewer data points and offer better interpretability. However, traditional machine learning algorithms have clear drawbacks. They often require domain experts to pre-define the features to be learned before model training, resulting in additional manual costs and increased resource expenses. In the following sections, we will introduce commonly used machine learning methods in building AI diagnostic models.

Conditional random fields (CRF). CRF [41] has found numerous applications in disease diagnosis. It is a probabilistic graphical model that predicts labels by capturing contextual information of input sequences and considering the dependencies between adjacent labels in the sequence. In the context of disease diagnosis, the CRF model utilizes patient-specific input sequences (such as images, text, or genetic features) to model the conditional probability of the output sequence, representing different disease classifications or subtypes. This is achieved by defining feature functions and weights that represent the relationship between input and output sequences. Feature functions can include observation features (relating the current input to the output label) and transition features (relating the current output label to the previous output label).

Support vector machine (SVM). The SVM [9] is another commonly employed algorithm in disease diagnosis [42–44]. The SVM, introduced by Vapnik in 1990, operates on labeled data. It begins with extracting meaningful features from the input data (e.g., shape features, texture features, or local features for medical images; or disease-related features like biomarkers or keywords for biological signals or clinical text data). Then, leveraging the extracted features to train the SVM. The SVM seeks an optimal hyperplane that distinguishes different classes based on the position of input samples relative to the hyperplane in the feature space. Finally, disease diagnosis is derived from the predicted labels.

Logistic regression (LR). LR [45] maps the results of linear regression to the range (0, 1) using a logistic function, enabling the estimation of the probability of a sample belonging to a particular class. LR has been widely applied in disease diagnosis. It adjusts model parameters to maximize the likelihood function of the training data by learning the relationship between patient features (such as images, text, signals, or genes) and disease labels. Optimization algorithms like gradient descent are used to minimize the loss function and find the optimal model parameters.

Naive Bayes (NB). NB [46] is a probabilistic algorithm that does not rely on networks and performs well with high-dimensional features. In disease diagnosis tasks, the NB classifier learns the relationship between patient data features (such as medical images, clinical text, or biological signals) and disease labels, classifying patients into specific disease categories [47]. Furthermore, NB simplifies learning by independently classifying features within each class.

Decision tree (DT). The DT [48] is a commonly used data analysis algorithm [49]. It consists of terminal and non-terminal nodes, with each non-terminal node describing a condition or test for a data item. This technique is often employed in disease classification and is beneficial for association and regression tasks. Decision trees facilitate easy visualization and identification of various data aspects [1]. Numerous studies have utilized decision trees for disease diagnosis [50].

In addition to the aforementioned methods, many other typical traditional machine learning methods (e.g., K-means, RF, etc.) have been successfully applied to disease diagnosis tasks.

Unlike traditional machine learning approaches, deep learning methods can leverage all the information present in the data as features for training models, eliminating the need for predefined features. This significantly reduces the resource requirements associated with traditional machine-learning methods. Particularly in tasks such as AI diagnosis and prediction, deep learning methods demonstrate a compelling advantage over traditional machine learning methods, especially when abundant data are available. In the medical domain, where high precision is paramount, traditional machine learning methods are progressively being substituted by deep learning methods. The subsequent sections will highlight several widely used deep learning methods.

Long short-term memory (LSTM). LSTM [12], an improved version of the recurrent neural network (RNN), is composed of a series of fundamental units designed to address the issues of gradient vanishing and exploding in RNN through the use of gates and controlled features. Each unit includes an input gate, a cell state, a forget gate and an output gate. The input gate decides which feature information to update, while the forget gate is used to decide the amount of original feature information to discard. The cell state serves as a storage unit for feature information, and the output gate determines which feature information to output. Notably, LSTM excels in capturing contextual relationships and predicting subsequent data based on the preceding sequence. In the realm of disease diagnosis, LSTM finds utility in processing and modeling sequential data, including clinical texts and speech. Furthermore, LSTM has several variants, such as Bidirectional Long Short-Term Memory (Bi-LSTM) and Bidirectional Gated Recurrent Unit (BiGRU), which simultaneously predict the current state based on both the previous states and the future states.

Convolutional neural networks (CNNs). A CNN [10] possesses parallelism characteristics that LSTM does not have. Recently, the CNN has been widely applied in various medical imaging, laboratory reports, pathology reports, etc., and has achieved remarkable success in the field of AI-based diagnosis [51–58]. The concept of the "receptive field" in a CNN is essential as it decides the time frame for the CNN to make predictions based on contextual relationships. The window size and stride used in convolutions are parameters used to control the receptive field. In a CNN, a larger window size generates a larger receptive field, thus capturing more contextual relationships. However, this diminishes the influence of words closest to the prediction target in terms of their positional importance. Setting a larger stride in the CNN ignores certain contextual relationships while significantly increasing the overall computational speed.

Transformer. A transformer [11] is a deep learning model widely used for sequence-to-sequence tasks, having garnered significant acclaim in the field of natural language processing, particularly for machine translation, and subsequently finding broad research applications in other domains, including image processing. In the realm of medical diagnosis, A transformer proves valuable for processing and modeling diverse modalities of medical data, encompassing clinical texts, medical images, and time series data [59–61]. Primarily, leveraging the self-attention mechanism, the transformer computes relevance scores between each position in the input sequence and other positions. These scores facilitate weighted aggregation of input features, empowering each position to capture both global and local contextual information.

Moreover, to bolster modeling capabilities, the transformer introduces a multi-head attention mechanism, employing multiple self-attention sub-layers that focus on distinct facets of relevant information, effectively extracting features at varying levels and perspectives. Simultaneously, to retain positional information within the sequence, Transformer incorporates positional encoding, embedding positional details into the input representation, enabling the model to discern between different positions. Lastly, employing an encoder-decoder architecture, Transformer initially encodes the input sequence into high-dimensional representations, adeptly capturing the input data's features, and subsequently, the decoder generates disease prediction outcomes based on the encoder's output and target labels.

Large model (LM). With the emergence of foundational models [62,63], researchers have introduced a new paradigm that leverages deep learning methods, primarily relying on the emerging capabilities of large models (LMs) to handle more complex tasks through scale expansion. Unlike traditional specialized models trained for specific problems, a large universal foundational model only requires one training session to acquire a wide range of general knowledge and can subsequently adapt to various downstream tasks through prompts. This approach was initially introduced by language models as few-shot learners [64] and has gained widespread recognition with the introduction of groundbreaking models such as GPT-3.5 [13], GPT-4 [14], the LLaMA series (including LLaMA [15] and Llama2 [16]), PaLM [17], FLAN-T5 [65], and Alpaca [18].

Alongside technological advancements, large models targeting different data types, such as images (SAM [66]) and time series (TimeGPT-1 [67]), have also been developed, demonstrating their powerful performance. While these LMs have proven effective in various general domain tasks, they have yet to reach their full potential in specific medical domain tasks. In comparison to specialized models, LMs still exhibit certain gaps because specialized models are not only meticulously designed for specific tasks in terms of architecture but also guided by medical knowledge to better understand and capture subtle differences and semantic features in the data. In contrast, LMs currently fall short in this aspect. Consequently, there has been extensive research on LMs tailored for specific medical domains to better fulfill the requirements. XrayGPT [68] and XrayGLM serve as notable examples of large models applied in medical imaging. XrayGPT is an innovative conversational medical visual language model capable of analyzing and answering open-ended questions regarding chest X-rays. XrayGLM aims to become the first Chinese multi-modal medical LM proficient in interpreting chest X-ray images, showcasing remarkable potential in medical image diagnosis and multi-turn interactive dialogues. Available at: http://github.com/WangRongsheng/XrayGLM (accessed on 29 November 2023). Several LMs focused on medical text and speech have also emerged, including the Med-PaLM series (Med-PaLM [19] and PaLM 2 [20]), HuaTuo Algorithm [21], ChatDoctor [22], DoctorGLM [23], BianQue [24], and BioGPT [25], which have demonstrated significant potential in providing valuable assistance across various healthcare-related domains. In the realm of genetic data, Yang et al. [49] introduced GeneCompass, the first knowledge-based cross-species milestone foundational model, surpassing competitive state-of-the-art models in multiple tasks within a single species.

2.3. Performance Evaluation Metrics

In disease diagnosis tasks using artificial intelligence technology, performance evaluation metrics are commonly calculated based on the confusion matrix for binary classification tasks [69], which include four types of classifications: True Positive (TP), False Positive (FP), True Negative (TN), and False Negative (FN). As shown in Table 1, TP represents the correctly identified positive instances, i.e., the positive class correctly classified as positive. TN represents the correctly identified negative instances, i.e., the negative class correctly classified as negative. FP represents the falsely identified positive instances, i.e., instances of the negative class mistakenly classified as positive. FN represents the falsely identified negative instances, i.e., instances of the positive class mistakenly classified as negative. Total Positive refers to the sum of TP and FN, while Total Negative refers to the sum of TN and FP. True Classification is the sum of TP and FP, and False Classification is the sum of FN and TN. The definition of performance evaluation metrics is shown in Table 2.

Table 1. Definition of the confusion matrix in binary classification.

		Actual Outcome	
		Positive	**Negative**
Predicted Outcome	Positive	TP	FP
	Negative	FN	TN

Table 2. The definition of performance evaluation metrics (note that the N, p_i and y_i in equation Brier score represent the number of samples, the predicted result for sample i, and the observed result (true label) of sample i, respectively).

Metric	Definition
Accuracy (ACC)	ACC = (TP + TN)/(TP + TN + FP + FN)
Precision (P)	P = TP/(TP + FP)
Recall (R)	R = TP/(TP + FN)
F1-score (F1)	F1 = 2 × P × R/(P + R)
Specificity (Sp)	Sp = TN/(TN + FP)
Brier score	Brier score = $(1/N) \times \sum[(p_i - y_i)]^2$

In addition, other classification metrics such as the Area Under the ROC Curve (AUC-ROC) are also commonly adopted. The ROC curve plots the True Positive Rate (TPR) on the y-axis against the False Positive Rate (FPR) on the x-axis, where TPR = Recall(R) = TP/(TP + FN), FPR = FP/(FP + TN). The ROC curve illustrates the relationship among TPR and FPR at different classification thresholds. The AUC measures the area under the ROC curve, ranging from 0 to 1. An AUC of 1 indicates a model with perfect classification ability, while an AUC equals to 0.5 denotes that a model's predictive performance is no better than random guessing.

3. Reported Works

3.1. Diagnosis of Alzheimer's Disease

Alzheimer's disease constitutes a progressive neurodegenerative disorder, characterized by cognitive decline, memory impairment, and compromised communicative abilities. In the realm of AI-driven diagnostic investigations for Alzheimer's disease, medical imaging modalities such as MRI and PET are universally recognized as indispensable tools. They offer profound insights into the alterations of brain structure and functionality, thus furnishing critical information for diagnosis. Concurrently, the analysis of speech patterns has also surfaced as a promising domain. Changes in language and communication frequently serve as precursors to cognitive deterioration, making them significant markers for early detection. This section delves into and evaluates the pertinent literature on automated Alzheimer's disease diagnosis, leveraging MRI, PET, speech, and other multi-modal strategies. A consolidated synopsis of the model and its attributes is presented herein, with detailed elaborations provided in Table 3.

Magnetic resonance imaging (MRI). MRI is pivotal in Alzheimer's disease (AD) diagnostics, offering a non-invasive modality that provides intricate images capturing the brain's structural and tissue details. There has been a substantial focus on harnessing morphological attributes from MRI scans as the central criterion for facilitating automated AD diagnosis. To illustrate, Li et al. [52] initiate the process by pinpointing the hippocampal regions in structural MRI (sMRI) images that are productive for diagnosis, drawing on prior knowledge. Subsequently, they deploy a deep learning architecture to distill distinctive patterns pertinent to AD diagnosis. Building upon this, Lian et al. [70] amalgamate a discriminative localization phase for brain atrophy with the subsequent stages of feature extraction and classification framework development. They introduce a Hierarchical Fully Convolutional Network (H-FCN) designed to autonomously and systematically discern patch-level and region-level indicative sites within the entire brain MRI scan. This model embraces a data-driven strategy that concurrently learns and amalgamates feature representations spanning multiple scales—from patch to region to subject level—to formulate a comprehensive AD diagnostic model. Addressing the nuances of brain atrophy, which pose significant diagnostic challenges in MRI imaging, Zhu et al. [59] unveil DA-MIDL, a novel deep learning framework endowed with a dual attention mechanism. This mechanism is adept at singling out the most salient pathological locales for AD diagnosis. DA-MIDL is composed of a patch network replete with spatial attention blocks, an attention Multiple Instance Learning (MIL) pooling module, and an attention-aware global classifier. The patch network is engineered to extract salient structural features from myriad local sMRI patches disseminated throughout the brain. The attention MIL pooling phase is adept at assigning variable weights to patch-level features, orchestrating them into a holistic representation of the entire brain's architecture. This global representation forms the foundation for the subsequent AD diagnostic classifier.

Furthermore, the quantification of hippocampal volume attrition has been recognized as a seminal marker for AD diagnosis. Uysal et al. leverage semi-automatic segmentation software ITK-SNAP to calculate hippocampal volume metrics. They construct a dataset incorporating parameters such as age, gender, diagnostic status, and volumetric data for left and right hippocampal regions. Utilizing this dataset, they apply machine learning algorithms to effectively differentiate between Alzheimer's disease (AD), Mild Cognitive Impairment (MCI), and cognitively normal (CN) cohorts.

Positron emission tomography (PET). While MRI images primarily yield extensive data on brain structure, they fall short of providing insights at the molecular level. This is where Positron Emission Tomography (PET) imaging gains its prominence. As a molecular imaging technique, PET scrutinizes specific biological processes such as protein aggregation, metabolic rates, or receptor concentrations using radiolabeled tracers. PET imaging thus offers an intricate depiction of biological and metabolic dynamics within the brain and is routinely employed in diagnosing and monitoring Alzheimer's disease (AD). In the study by Chen et al. [60], a novel contrastive learning paradigm is introduced, utilizing brain 18F-FDG PET images to surmount the challenges associated with the paucity of data and the low signal-to-noise ratio, which are typical in PET images pertinent to AD prediction. They implement a data augmentation strategy to amplify the volume of training data, and they apply the adversarial loss to expand the distances between features of different classes while consolidating the similarities within the same class.

Furthermore, they develop a dual convolutional mixed attention module, fine-tuning the network's proficiency in discerning diverse perceptual fields. By aligning the predictive outcomes of individual PET slices with clinical neuropsychological evaluations, they advance a diagnostic methodology conducive to refining AD diagnoses. Baydargil et al. [71] deliver an unsupervised adversarial parallel model tailored for the anomaly analysis in AD, sharply delineating AD, mild cognitive impairment (MCI), and normal control groups. The model exhibits robust classification with rates and area under the curve (AUC) scores reaching 96.03% and 75.21%, respectively, underscoring its effective discriminative performance. Lu et al. lay the groundwork for a cutting-edge deep learning infrastructure, utilizing FDG-PET metabolic imaging to pinpoint subjects with symptomatic pre-AD in the MCI phase, setting them apart from other MCI cohorts (non-AD/non-progressive). They pioneer a multi-scale deep neural network that reports a classification precision of 82.51%, relying solely on a single-modal metric (FDG-PET metabolic data). Cheng et al. [53] present an innovative classification scheme that amalgamates a two-dimensional Convolutional Neural Network (CNN) with a Recurrent Neural Network (RNN). Their strategy is oriented towards deconstructing 3D images into a succession of 2D slices to capture the features inherent to 3D PET imagery. Within this framework, they architect a hierarchical 2D cellular neural network tasked with the extraction of intra-slice features, while the Gated Recurrent Unit (GRU) within the RNN is deployed to elucidate inter-slice features that contribute to the final classification outcome.

Speech. The manifestation of Alzheimer's disease (AD) in speech signals offers a distinctive avenue for diagnosis, as individuals with AD exhibit notable speech pattern alterations compared to those without the condition. Employing speech recognition technology for AD diagnostics is not only non-invasive and safe but also cost efficient, making it an appealing methodology for widespread application. Before the infusion of deep learning into the field, traditional approaches to speech analysis for AD diagnosis relied heavily on manual feature extraction. Techniques such as analysis of static features, utilization of feature sets like ComParE 2016 and eGeMAPS, as well as Mel-Frequency Cepstral Coefficients (MFCC), were common practices. These extracted features were then analyzed using machine learning classifiers, including logistic regression, random forests, and support vector machines, to distinguish between affected and healthy individuals. Studies by Hason et al. [72], Hernández et al. [73], and Yu et al. [74] are examples of such research efforts.

With the advent of deep learning, there has been a paradigm shift in research methodologies for AD diagnosis. Deep learning techniques have taken precedence, given their ability to automatically extract complex patterns from raw data without the need for manual feature selection. In this context, Lopez et al. [55] have made strides in early AD detection by implementing classical Multilayer Perceptrons (MLPs) and Convolutional Neural Networks (CNNs), illustrating the potential of deep learning in enhancing diagnostic accuracy. Further advancing the field, Liu et al. [75] leveraged an Automatic Speech Recognition (ASR) model to derive speaker-independent bottleneck features, which are highly discriminative and robust. They coupled this with a CNN for modeling local context and an RNN for capturing the global context within speech. An attention mechanism was integrated to selectively focus on the most salient features for AD detection, improving the model's interpretability and effectiveness. Additionally, Bertini et al. [76] introduced an end-to-end model for AD detection, innovatively applying SpecAugment [77] for data augmentation to enhance the robustness and generalizability of the model against variability in speech data. They then utilized the auDeep [78] autoencoder, followed by fully connected layers for feature learning and classification, streamlining the process from raw speech input to the diagnostic output. This end-to-end approach simplifies the pipeline and potentially improves the model's accuracy and applicability in clinical settings.

MRI-PET image fusion. The integration of MRI and PET imaging modalities has yielded a synergistic approach in medical diagnostics, particularly for disorders such as Alzheimer's disease (AD). This technique of image fusion leverages the unique strengths of each imaging method to offer a more holistic representation of the brain's structure and function. The pioneering work of Shi et al. [79] introduced the multi-modal Stacked Denoising Predictive Network (MM-SDPN). This algorithm is structured in two phases specifically tailored to merge and learn from the feature representations of multi-modal neuroimaging data. This integration enhances the diagnostic process for Alzheimer's disease, offering a deepened insight into the complex interactions between different types of brain changes associated with the disease. Sharma et al. [80] took a different approach, utilizing wavelet packet transform as their method of fusing MRI and PET images. Their methodology involves an eight-layer Convolutional Neural Network (CNN) that meticulously extracts features across multiple layers. The extracted features are then processed through an ensemble of non-iterative Random Vector Functional Link (RVFL) networks. This ensemble strategy aims to robustly capture the intricate patterns from the fused data for accurate AD diagnosis.

Further advancing the field, Zhou et al. [81] proposed a unique method for latent representation learning that encompasses data from various modalities, including MRI, PET, and genetic information. Their approach focuses on deducing latent representations and then projects these representations into the label space for diagnostic purposes. This technique underscores the potential of combining structural, functional, and biological data to enhance the accuracy of Alzheimer's disease diagnostics. Addressing the potential issue of overfitting when dealing with the fusion of high-dimensional data, Ning et al. [72] developed a relation-induced multi-modal shared representation learning approach. Their model is an integrative framework that combines the processes of representation learning, dimensionality reduction, and classifier design. It operates by learning bidirectional mappings between the original feature space and a shared representation space, thereby distilling the essence of multi-modal inputs into a cohesive, shared format that is conducive to diagnostic analysis. These studies illustrate a growing trend in leveraging sophisticated computational models and algorithms to enhance the accuracy and reliability of Alzheimer's disease diagnostics by capitalizing on complementary information from multiple imaging modalities.

Speech–Text fusion. The nuanced extraction of acoustic features from speech datasets, coupled with the semantic analysis of textual data, fosters an enriched comprehension of Alzheimer's disease (AD). By amalgamating speech and text data, a more extensive spectrum of AD-related features is captured, bolstering the diagnostic accuracy for this condition. Historically, the nascent stages of AD research leveraged machine learning techniques for analytical purposes. Shah et al. [42] focused on the extraction of word-level duration features, datasets on pause rates, and measures of speech clarity. They explored a variety of models, such as logistic regression, random forest, support vector machine (SVM), extreme gradient boosting, and neural networks in isolation and in combination, targeting both classification and regression tasks. Martinc et al. [43] commenced with spectrum subtraction for noise abatement, progressing to the use of a bag-of-n-grams approach for textual feature extraction. Concurrently, they extracted eGeMAPS features from speech data. A suite of classifiers, including XGBoost, SVM, random forest, logistic regression, and linear discriminant classifiers, was then deployed for classification tasks.

In the landscape of recent advancements, deep learning techniques have increasingly been harnessed for the automated diagnosis of Alzheimer's disease. Cai et al. [82] applied

Graph Neural Networks (GNNs) for the extraction of textual features and introduced audio data by utilizing the WavLM model to extract salient audio features. They then integrated these features with text features via various methodologies. Mei et al. [83] extracted a plethora of features comprising static acoustic features, the ComParE 2016 feature set, the eGeMAPS feature set, along with feature vectors from the wav2vec2 pre-trained model, and the Hubert pre-trained model for AD detection. They meticulously fine-tuned the wav2vec2.0 model on speech from assorted frequency bands, culminating in a remarkable accuracy of 87% and an RSME of 3.727. Agbavor et al. [84] procured deep representation features through data2vec and wav2vec2, subsequently refining an end-to-end model with fully connected layers for enhanced AD detection efficacy.

Other models. A diverse array of molecular and multi-omics approaches, including RNA-seq, single nucleotide polymorphisms (SNPs), protein sequences, and integrated omics data, have been employed to unravel the complexities of Alzheimer's disease diagnosis. For instance, groundbreaking work by Li et al. [84], Taeho et al. [85], Xu et al. [86], Javier et al. [87], and Park et al. [88] has significantly contributed to the field by leveraging these techniques. Further, Park et al. [88] have pioneered a deep learning approach tailored for AD prediction that synergistically utilizes multiple heterogeneous omics data. In a similar vein, Golovanevsky et al. [89] have devised a multi-modal Alzheimer's Disease Diagnostic framework (MADDi), ingeniously combining neural networks with attention mechanisms to harness the power of imaging, genetic, and clinical data for enhanced AD diagnostic precision. In addition to these genomic and proteomic strategies, electrophysiological methods such as EEG have been instrumental in AD diagnosis. Notable research by Djemili et al. [90], Pandya et al. [91], Kim et al. [92], along with studies cited as [93], have demonstrated the utility of EEG in capturing the neurophysiological hallmarks of Alzheimer's disease, adding a valuable dimension to the diagnostic toolkit.

Table 3. Summary of different medical features for Alzheimer's disease diagnosis.

Literature	Feature Name	Modality	Dateset	Results
Li et al. [52]	Hippocampal morphology feature	MRI	ADNI	0.939 (AUC)
Lian et al. [70]	Original MRI scan feature	MRI	ADNI	0.9 (ACC); 0.95 (AUC:AD vs. NC)
Zhu et al. [59]	Patch proposals selected from the MRI scans	MRI	ADNI, AIBL	0.9193 (ACC: AD vs. NC vs. MCI) 0.9287 (AUC)
Chen et al. [60]	optimized anchor data from brain 18F-FDG PET slices	PET	ADNI	0.9193 (ACC: AD vs. NC vs. MCI) 0.9287 (AUC)
Baydargil et al. [71]	Original PET slices	PET	ADNI	0.9603 (ACC: AD vs. NC vs. MCI) 0.7521 (AUC)
Cheng et al. [53]	a sequence of 2D slice groups from 3D PET	PET	ADNI	0.9528 (AUC: AD vs. NC)
Shi et al. [79]	high-level features of MRI and PET	MRI, PET	ADNI	0.9713 ± 0.0444 (ACC: AD vs. NC)
Sharma et al. [80]	Fused image by wavelet packet transform (WPT)	MRI, PET	ADNI	0.9603 (ACC: AD vs. NC vs. MCI) 0.7521 (AUC)
Zhou et al. [81]	magnetic resonance imaging (MRI), positron emission tomography (PET), and genetic data	MRI, PET, Gene	ADNI	-
Ning et al. [72]	magnetic resonance imaging (MRI) and positron emission tomography (PET)	MRI, PET	ADNI	0.976 (AUC: AD vs. NC) 0.969 (ACC: AD vs. NC)
Li et al. [84]	RNA-seq	Gene-based	GEO	0.859 (AUC), 0.781 (ACC)
Taeho et al. [85]	SNP	Gene-based	ADNI	0.82 (AAUC)

Table 3. *Cont.*

Literature	Feature Name	Modality	Dateset	Results
Xu et al. [86]	protein	sequence Gene-based	UniProt	0.857 (ACC)
Javier et al. [87]	genetic variation data	Gene-based	ADNI	0.719 (ACC)
Park et al. [88]	Multi-omics data	Gene-based	GEO	0.823 (ACC)
Golovanevsky et al. [89]	imaging, genetic, and clinical data	Gene-based	GEO	0.9688 (ACC)
Djemili et al. [90]	statistical characteristics (1. Maximum value in each IMF. 2. Minimum value in each IMF. Mean of the absolute values in each IMF. 4. Standard deviation in each IMF.)	EEG	Bonn dataset	The classification accuracy for normal and abrupt cessation electroencephalogram (EEG) signals is 1, while the classification accuracy for intermittent and abrupt cessation EEG signals reaches 0.977
Pandya et al. [91]	Amplitude, period and waveform offset of K-Complex	EEG	Private dataset	-
Kim et al. [92]	EEG segment with respect to RP(Absolute power of EEG signals in three different frequency bands)	EEG	Private dataset	0.75 (ACC)
Deepthi et al. [93]	Frequency domain features extracted by Fast Fourier Transform (FFT)	EEG	ADNI	-
Hason et al. [72]	MFCC	speech	ADReSS	Accuracy: 0.822
Hernández et al. [73]	Speech duration, descriptive statistical variables	specch	private dataset	Accuracy: 0.8
Yu et al. [74]	Based on phoneme characteristics, pronunciation coordination characteristics, and pitch variance	speech	private dataset	Accuracy: 0.93
Lopez et al. [55]	Linear features include spectral domain features and time domain features, such as harmonicity, spectrum centroid, formants, etc. Nonlinear characteristics include fractal dimension, permutation entropy, multiscale permutation entropy, etc.	speech	private dataset	Accuracy: 0.89
Liu et al. [75]	Bottleneck feature vector (depth representation feature)	speech	DementiaBank Pitt	F1: 0.7802
Bertini et al. [76]	spectrogram	specch	DementiaBank Pitt	Accuracy is 0.933, F1 score is 0.885
Shah et al. [42]	Word-level duration feature set, pause rate data set, speech intelligibility feature set	speech, text	ADReSS-M	Accuracy: 0.696, RMSE: 4.8
Martinc et al. [43]	bag-of-n-grams features (text) eGeMAPS feature set (voice)	speech, text	DementiaBank Pit	Accuracy: 0.9167
Cai et al. [82]	GNN (text features) WavLM (voice features)	Speech, text	DementiaBank Pit	Accuracy: 0.8484 ± 0.0544
Mei et al. [83]	Silent characteristics ComParE 2016 feature set, eGeMAPS feature set wav2vec2 pre-trained model feature vector Hubert pre-trained model feature vector	Speech, text	AADReSS-M	Accuracy: 0.87, RMSE: 3.727
Agbavor et al. [84]	data2vec, wav2vec2	Speech, text	ADReSSo	F1: 0.728, RMSE: 3.493

3.2. Diagnosis of Breast Cancer

Breast cancer, originating in the breast cell tissue, stands as a pivotal health challenge for individuals across the globe. The key to enhancing survival and ensuring a better quality of life for those impacted by this disease lies in early detection and an integrated approach to treatment, involving a diverse team of medical professionals. The conventional diagnostic toolkit for breast cancer includes mammography, which is instrumental in visualizing breast tissue and identifying any irregularities that may indicate the presence of cancerous cells. Clinical breast exams conducted by healthcare professionals also play a significant role in early detection, as they involve a thorough palpation of the breast tissue to detect lumps or other changes. Additionally, gene screening is becoming increasingly important in breast cancer diagnosis, particularly for women with a family history of the disease, as it can identify inherited genetic mutations that may elevate the risk of breast cancer, such as mutations in the BRCA1 and BRCA2 genes. In this section, the diagnostic methodologies driven by the aforementioned modalities are rigorously explored and demonstrated. To provide a clear and concise representation of the various models and their attributes, reference is made to the details encapsulated in the accompanying tables, labeled as Table 4. These tables present a summarized outlook of the models, delineating their features, performance metrics, and other pertinent details that contribute to the overarching domain of breast cancer diagnosis.

X-ray mammography. Breast Lesion Classification is a critical facet of breast cancer diagnosis, as it aims to accurately differentiate between benign and malignant lesions discovered during screenings. X-ray mammography remains the cornerstone of early breast cancer detection, enabling physicians to spot minuscule masses or calcifications that could indicate the presence of cancer cells within the breast tissue. To augment the diagnostic efficiency for breast lesions, Al-antari et al. [94] have presented a comprehensive Computer-Aided Diagnosis (CAD) system that harnesses the power of deep learning, leveraging data from the DDSM and INbreast databases, which are prominent digital mammography datasets. The innovation began with the utilization of a You Only Look Once (YOLO) [95] deep learning detector specifically calibrated for the identification of breast lesions across whole mammograms. Subsequently, Al-antari et al. assessed and fine-tuned three deep learning classifiers—the standard feedforward CNN, ResNet-50, and InceptionResNet-V2—for the nuanced task of breast lesion classification.

Furthering the advancement in this domain, Yeman et al. [96] introduced an inventive approach employing a parallel deep Convolutional Neural Network (CNN) designed to analyze and learn from the symmetrical deep features extracted from the bilateral views of breast X-ray images. They innovatively computed the probability of pixels being part of a lesion by examining the local line and gradient direction features distribution, which then pinpointed the centers of suspected lesions. A global threshold was applied to these likelihood images to discern potential lesion-bearing regions. Ensuring symmetry, right and left breast X-ray images were horizontally flipped for congruent orientation, and the analysis proceeded with patched images fed into two mirrored deep CNN structures. The concatenated deep features from this twin-CNN setup were introduced into a Neural Network (NN) classifier, which achieved a remarkable prediction accuracy rate of 93.33%. In another groundbreaking work, Riyadh et al. [97] conceived a novel mixed deep learning Computer-Aided Diagnosis system for breast lesions, which combined a backbone residual deep learning network to generate profound features with a transformer that incorporates self-attention mechanisms for the classification of cancer. This innovative model achieved a perfect 100% accuracy rate for binary classification and an impressive 95.80% for multi-class prediction tasks, a testament to the potential of mixed AI models in discerning between benign and malignant breast tissues with high precision.

Magnetic resonance imaging. Breast MRI is a powerful diagnostic tool that excels in providing detailed insights into breast cancer lesions, surpassing other imaging modalities in delivering precise evaluations of lesion size, location, and type. The robust magnetic field and non-ionizing radiation technique of MRI make it a choice modality for comprehensive breast cancer assessment. Abunasser et al. [98] have made significant strides in the realm of breast MRI by training six advanced deep learning models, each with the capability to classify eight specific types of breast cancer, encompassing both benign and malignant forms. Their study incorporated a diverse set of models including their own proposed Breast Cancer Neural Network (BCNN), as well as Xception, InceptionV3, VGG16, MobileNet, and ResNet50, all fine-tuned to analyze MRI images for this purpose. These models demonstrated remarkable accuracy in their classification tasks, with rates of 97.54%, 95.33%, 98.14%, 97.67%, 93.98%, and 98.28% respectively, showcasing their potential to serve as reliable diagnostic aides. Complementing these efforts, Huang et al. [99] embarked on a comprehensive study involving the extraction of an extensive array of 4198 radiomic features from pre-biopsy multiparametric MRI datasets, which included dynamic contrast-enhanced T1-weighted images, fat-suppressed T2-weighted images, and apparent diffusion coefficient maps. In their pursuit of optimal feature selection, they employed a suite of methodologies such as the Least Absolute Shrinkage and Selection Operator (LASSO), Recursive Feature Elimination (RFE), Maximum Relevance Minimum Redundancy (mRMR), Boruta, and Pearson correlation analysis. Leveraging these strategically chosen features, Huang et al. proceeded to construct 120 diagnostic models that varied by classification algorithms, MRI sequence-segmented feature sets, and the employed selection strategies. These models were adeptly designed to not just categorize breast cancer lesions but also to predict cancer molecular subtypes and androgen receptor expression, potentially offering a nuanced approach to personalized cancer care.

Ultrasound images. The field of medical imaging for breast cancer diagnosis has been greatly enhanced by the incorporation of artificial intelligence, with ultrasound imaging being a key focus due to its safety and non-invasive nature. Jabeen et al. [100] introduced a cutting-edge classification framework specifically designed for ultrasound images, which effectively combines the prowess of deep learning with optimal feature selection techniques. This framework is composed of a structured five-step process: (i) Data augmentation is applied to expand the dataset, thereby providing a more robust foundation for training Convolutional Neural Network (CNN) models. (ii) The pre-trained DarkNet-53 model is adapted by modifying its output layer to align with the categories of the augmented dataset. (iii) Transfer learning is employed to train this modified model, with feature extraction carried out from the global average pooling layer. (iv) Two enhanced optimization algorithms, the Improved Differential Evaluation (RDE) and Improved Grey Wolf (RGW), are utilized for the selection of the most discriminative features. (v) A novel, probability-based sequential method is used to combine these optimally selected features, followed by the application of machine learning algorithms for the final classification task. The implementation of this framework on the Augmented Breast Ultrasound Images (BUSI) dataset resulted in an impressive highest accuracy of 99.1%, demonstrating its potential to significantly improve diagnostic processes.

Building on the momentum of innovation in the field, Ragab et al. [101] spearheaded the development of an Integrated Deep Learning Clinical Decision Support System for Breast Cancer Diagnosis and Classification (EDLCDS-BCDC). This innovative technology is engineered to detect the presence of cancer through the analysis of ultrasound images. The process involves an initial preprocessing stage using Wiener filtering and contrast enhancement to prepare the images. Image segmentation is then carried out using the Chaos Krill Herd Algorithm (CKHA) and Kapur Entropy (KE). The feature extraction is performed through an ensemble of three sophisticated deep-learning models, namely VGG-16, VGG-19, and SqueezeNet. The final stage of the classification process employs the Cat Swarm Optimization (CSO) algorithm to optimize a Multi-Layer Perceptron (MLP) model, ensuring precise categorization of the cancer images. Both these studies showcase the innovative intersection of deep learning and optimization algorithms in improving the accuracy and efficiency of breast cancer classification using ultrasound imaging.

Medical text data. The use of advanced natural language processing (NLP) techniques to analyze and classify medical data, including patient self-reports and medical records, has become increasingly prevalent in breast cancer research. Leveraging the power of these techniques can provide valuable insights and assist in the early detection and treatment of breast cancer. Kumar et al. [102] tailored a BERT-based model to specifically address the classification of breast cancer-related posts on Twitter, as described in Shared Task 8 of SMM4H-2021. Their approach was to employ BlueBERT [103], which is pre-trained on a comprehensive biomedical corpus acquired from PubMed, enhancing the model's understanding of medical terminology and context. To bolster the model's resilience against adversarial inputs, they incorporated gradient-based adversarial training, which ultimately resulted in the model achieving F1 scores of 0.8625 on the development set and 0.8501 on the test set, reflecting high accuracy in the automatic classification of breast cancer mentions in social media posts.

Further innovations in NLP, as seen in the works of Chen et al. [104] and Zhou et al. [105], push the boundaries of model interpretability and domain-specific accuracy. Chen et al. [104] took the capabilities of BERT further by integrating semantic trees into the model, thus constructing an interpretable neural network. They harnessed a capsule network with multiple attention heads to refine the semantic representations, while backpropagation and dynamic routing algorithms were implemented to provide local interpretability. This level of interpretability is particularly important in medical applications where understanding the reasoning behind a model's prediction is as crucial as the prediction itself. Zhou et al. [105] explored the benefits of pre-training BERT on a cancer-specific dataset, which aimed to enhance the model's ability to extract breast cancer phenotypes from pathology reports and clinical records. Their findings underscore the significance of domain-specific pre-training, as it substantially improved the performance of the model, making it more attuned to the nuances of cancer-related data. Addtionally, Deng et al. [106] investigated the potential assistance provided by advanced language models like GPT-4 in the context of breast cancer diagnosis. The authors emphasized GPT-4's capability to rapidly mine crucial information from extensive medical records, which could potentially influence the diagnosis of breast cancer. By automating the extraction of key data points, GPT-4 could enhance the accuracy and efficiency of diagnostic procedures, supporting healthcare professionals in making informed decisions. These studies collectively highlight the transformative impact that state-of-the-art NLP models can have on the medical field, particularly in the realm of breast cancer diagnosis and classification.

Genetic data. Human cancer is a heterogeneous disease caused by stochastic cellular mutations and driven by various genomic alterations [107,108]. Currently, numerous research efforts are focused on utilizing genetic data and artificial intelligence algorithms to develop diagnostic models to enhance the clinical efficiency and accuracy of breast cancer diagnosis [109–111]. Presently, artificial intelligence techniques in breast cancer diagnosis research based on genomics primarily focus on RNA-seq data, single nucleotide polymorphisms (SNPs), protein sequences, and the integration of multi-omics data. (1) RNA-seq. Xu et al. [112] proposed a multi-granularity cascade forest (gcForest) for predicting four subtypes of breast cancer (Basal, Her2, Luminal A, and Luminal B). They compared the gcForest classifier with three different machine learning methods (KNN, SVM, and MLP). The results showed that gcForest showed a higher accuracy score of 92%. (2) MicroRNA. Sherafatian et al. [50] employed three tree-based algorithms (Random Forest, Rpart, and tree bag) to classify breast cancer subtypes (Luminal, HER2-enriched, basal) using miRNA data from TCGA. The results showed that Rpart achieved the best classification performance. For the Luminal subtype, the accuracy, sensitivity, and specificity were 88.9%, 82.4%, and 95.4%, respectively. For the HER2-enriched subtype, the accuracy, sensitivity, and specificity were 90.2%, 93.9%, and 86.4%, respectively. For the basal subtype, the accuracy, sensitivity, and specificity were 84.5%, 75%, and 94%, respectively. (3) Multi-omics data. Mohaiminul et al. [58] proposed a comprehensive deep-learning framework for classifying molecular subtypes of breast cancer. The framework utilized copy number alteration and gene expression data from the METABRIC. The results achieved an accuracy of 76.7% and an AUC of 83.8%.

Table 4. Summary of different medical features for breast cancer diagnosis.

Literature	Feature Name	Modality	Dateset	Results
Al-Antari et al. [94]	Original X-ray mammographic data	X-ray	CBIS-DDSM and DDSM	0.985 (ACC)
Yeman et al. [96]	Breast lesion detection from entire mammograms by object detection model	X-ray	DDSM and INbreast	ACC of three models: 94.50%, 95.83%, and 97.50%
Riyadh et al. [97]	Extracted patches centered on the points from the original X-ray	X-ray	General Electric, Siemens, and Hologic	0.933 (AUC)
Abunasser et al. [98]	Original MRI data	MRI	Kaggle depository	98.28 (F1-score)
Huang et al. [99]	multi-parametric MRI	MRI	Private dataset	Multilayer Perceptron (MLP): 0.907 (AUC) and 85.8% (ACC)
Jabeen et al. [100]	Original ultrasound images data	Ultrasound Images	BUSI dataset	99.1% (ACC)
Ragab et al. [101]	Segmented regions from original	ultrasound images Ultrasound Images	-	96.92% (ACC)
Kumar et al. [102], Peng et al. [103]	Word embedding	Text	witter self-report	F1: 0.8501
Chen et al. [104]	Word embedding, syntactic structure	Text	Shanghai Ruijin Hospital Molybdenum Mammography X-ray Report	Mi-P(%) = 91.58 Mi-R(%) = 91.58 Mi-F1(%) = 91.58 Ma-P(%) = 75.95 Ma-R(%) = 79.73 Ma-F1(%) = 77.14
Zhou et al. [105]	mutil feature	Text	private dataset	exact match and lenient match, macro-F1: 0.876, 0.904
Xu et al. [112]	RNA-seq	Gene-based	Medical Records	-
Sherafatian et al. [50]	miRNA	Gene-based	TCGA	92% (ACC)
Mohaiminul Islam M et al. [58]	Copy number alteration (CNA), RNA-seq	Gene-based	METABRIC	76.7% (ACC), 83.8% (AUC)
Sun et al. [108]	Clinical, CNV, RNA-seq	Gene-based	METABRIC	82% (AUC)

3.3. Diagnosis of Depression

Depression is a common mental health disorder characterized by persistent feelings of sadness, hopelessness, and a lack of interest or pleasure in daily activities. It can affect a person's thoughts, emotions, and physical well-being, often leading to challenges in daily functioning. Depression varies in severity, and its impact on individuals can range from mild to severe. In the realm of diagnosis, text, speech, and EEG analysis have emerged as crucial tools for assessing and understanding depression. These modalities offer valuable insights into an individual's mental state, providing a nuanced understanding of their emotional well-being. This section aims to delve into various approaches and methodologies related to the diagnosis of depression using these modalities. This section provides a summarized overview of the model and its features, as detailed in the accompanying Table 5.

Medical text data. Aragon et al. [58] introduced a sophisticated deep emotional attention model tailored for the detection of anorexia and depression. This model integrates nuanced sub-emotion embeddings with the advanced architectures of Convolutional Neural Networks (CNNs), Gated Recurrent Units (GRUs), and attention mechanisms to attain high predictive accuracy. Verma et al. [113] explored depression detection through the analysis of tweet data, utilizing four established machine learning models: Naive Bayes, Support Vector Machines (SVMs), K-Nearest Neighbors (KNNs), and Random Forest. Of these, the Random Forest model demonstrated superior performance, achieving an impressive accuracy peak of 78%.

Furthering the field, Ghosh et al. [114] adopted a novel deep multi-task learning strategy that simultaneously addresses emotion recognition and depression detection. Their findings suggest that the multi-tasking framework significantly boosts the efficacy of both tasks when learned concurrently. Xu et al. [115] ventured into the domain of psychological health with the introduction of their Linguistic Landscape Model (LLM). This model was rigorously tested across a spectrum of tasks, including psychological stress classification, depression severity assessment, suicide ideation detection, and suicide risk evaluation. The empirical results underscored the LLM's robust performance, placing it on par with the leading task-specific models in the field. Lastly, Qi et al. [116] presented an all-encompassing benchmark that capitalizes on supervised learning techniques alongside the LLM framework, with a specific emphasis on the capabilities of the GPT series. Their research offers an in-depth analysis of these advanced LLMs, particularly in their application to cognitive distortion diagnosis and suicide risk stratification. This study not only highlights the models' proficiency in capturing and interpreting complex emotional states but also provides a critical examination of their inherent potential and current limitations within the psychological domain.

Speech. From the initial forays into the realm of machine learning for depression diagnosis, a vast array of approaches has emerged. Liu et al. [117] introduced a multi-task ensemble learning technique that utilizes speaker embeddings to facilitate depression classification. Long et al. [118] devised an innovative multi-classifier system dedicated to depression recognition, distinguished by its synthesis of various speech types and emotional nuances. Jiang et al. [119] developed the Ensemble Logistic Regression Model for Depression Detection (ELRDD), representing a significant stride in predictive modeling. Complementing this, Liu et al. [120] proposed an inventive decision tree-based method for the fusion of speech segments, aimed at bolstering the accuracy of depression recognition.

As deep learning forges ahead, its methodologies are increasingly being adopted for diagnosing depression. Yin et al. [121] presented a deep learning model that harnesses the strengths of parallel Convolutional Neural Networks (CNNs) and Transformers, balancing effective information extraction with computational tractability for depression detection. Adding to this body of work, Tasnim et al. [122] examined the predictive utility of two acoustic feature sets—conventional handcrafted features and those derived from deep representations—in assessing depression severity through speech analysis. He et al. [123] proposed a hybrid approach combining handcrafted elements with deep learning features to precisely gauge depression severity from speech. Dubagunta et al. [124] conducted an exploration into methods for modeling speech source-related information in the context of depression, mindful of the potential neural physiological changes impacting vocal cord function. Zhao et al. [125] sought to advance depression detection by tapping into inherent speech information, advocating for a Long Short-Term Memory (LSTM) model augmented with multi-head temporal attention. In a similar vein, Dong et al. [126] recommended the application of pre-trained models for the extraction of deep Speaker Recognition (SR) and Speech Emotion Recognition (SER) features. Their approach synergizes these two profound speech features to capture the complementary data embedded within speaker voice characteristics and emotional variances.

EEG. The field of depression diagnosis has witnessed the burgeoning integration of electroencephalogram (EEG) and machine learning techniques, marking a pivotal research trajectory. In the reported literature [127], a novel deep learning method named the Asymmetry Matrix Image (AMI) is introduced, which constructs spatial distribution maps from EEG signals by assessing the asymmetry between cerebral hemispheres. AMI has been shown to outperform traditional methods, delivering superior classification accuracy and enhancing the distinction between depression patients and healthy controls. Additional research [128] delves into the utilization of nonlinear EEG signal features, such as Higuchi's fractal dimension (HFD) and sample entropy (SampEn), which serve as indicators of signal complexity and irregularity. These nonlinear metrics have proven efficacious in segregating depression patients from healthy individuals, with high accuracy figures reported across a range of machine learning classifiers. In a different approach, literature [129] focuses on power spectral features and asymmetry measures within the alpha, beta, delta, and theta frequency bands. Notably, findings suggest that asymmetries in the alpha2 and theta bands, particularly when analyzed with a Support Vector Machine (SVM), lead to higher diagnostic precision, with an accuracy rate of 88.33%. Explorations into the use of EEG data for depression diagnosis have also extended to single-channel and multi-channel formats [130]. By refining feature selection and classification models via genetic algorithms, it has been discovered that single-channel analysis can effectively differentiate depression patients, underscoring the potential for employing portable EEG devices in preliminary depression screening despite a noted limitation in clinical generalizability due to small sample sizes. The literature [131] investigates four feature selection techniques and five classification algorithms for processing EEG data. Through rigorous data preprocessing and feature extraction—identifying noise types and harnessing both linear and nonlinear features—the critical role of the data preparation phase is emphasized for achieving optimal classification accuracy.

A novel article [47] presents a multi-modal feature fusion method that integrates EEG with eye movement (EM) signals, aiming to refine the identification of mild depression. The application of deep learning to fuse these multi-modal data sets enables real-time monitoring and detection of mild depression, with the fusion approach in the hidden layers yielding improved recognition accuracy over single-feature methods, and showcasing the benefits of combining diverse physiological signals. The melding of EEG and machine learning has advanced the diagnostic and treatment prediction capabilities for depression. Although challenges such as limited sample sizes and variability in feature extraction persist, forthcoming research endeavors are expected to tackle these issues, thereby enhancing the precision and utility of predictive models. Importantly, these advancements lay the groundwork for tailored treatment modalities, contributing to the delivery of more accurate and efficacious interventions for those suffering from depression.

Multi-modal. The landscape of depression diagnosis is rapidly evolving with the advent of multi-modal approaches, harnessing the rich data from speech, text, and video to create more nuanced and comprehensive diagnostic tools. Ehghaghi et al. [132] embarked on an interpretable analysis to discern the distinct characteristics between dementia and depression. They pinpointed a spectrum of differentiators such as auditory anomalies, repetitive speech patterns, word retrieval struggles, coherence degradation, and variance in lexical density and richness—all of which are pivotal in distinguishing these disorders. Diep et al. [133] ventured further by proposing a model that synthesizes deep learning features from both audio and text modalities, enriched with manually curated attributes deriving from domain expertise. Mao et al. [134] introduced a novel approach using an attention-based multi-modal framework to generate a joint speech and text representation, specifically for the prediction of depression. Exploring the intersection of speech and video modalities, Jan et al. [135] investigated the capability of cognitive machines and robots to autonomously recognize psychological states. By analyzing gestures and facial expressions, these intelligent systems aim to play a role in monitoring depressive states. Uddin et al. [136] optimized the data processing workflow by segmenting audio and video into fixed-length units for input into a spatiotemporal network. This network is tailored to extract both spatial and temporal characteristics, with the introduction of dynamic feature descriptors like the Volume Local Directional Structure Pattern (VLDSP) to capture the nuances of facial dynamics.

Not content with dual-modal analyses, some studies have ambitiously integrated all three modalities—speech, text, and video—to push the boundaries of depression detection. Yang et al. [137] contributed to this growing body of work by discussing a multi-modal depression analysis framework comprising deep convolutional neural networks (DCNNs) and deep neural networks (DNNs). This composite approach leverages the strengths of each modality, offering a more robust and potentially accurate detection system. The convergence of such diverse modalities represents a significant step forward in the field of mental health diagnostics. By combining distinct but complementary data sources, these integrated approaches aim to mirror the complex nature of depression more closely, offering promising directions for future research and potential clinical applications. The ultimate goal is to refine these tools for enhancing early detection and personalizing treatment strategies, thus providing a beacon of hope for individuals grappling with depression.

Table 5. Summary of different medical features for depression disease diagnosis.

Literature	Feature Name	Modality	Dataset	Results
Aragon et al. [58]	Word embedding, hashtag	Text	eRisk 2018 and 2019	0.79 (F1) for Anorexia, 0.58 (F1) for Depression
Verma et al. [113], Ghosh et al. [114]	Word embedding	Text	Twitter data collected by Twitter API	78% (ACC)
Xu et al. [115], Qi et al. [116]	Multiple characteristics	Text	Dreaddit, DepSeverity, SDCNL, CSSRS-Suicide	0.816 (ACC) for Dreaddit, 0.775 (ACC) and 0.756 (ACC) for DepSeverity, 0.724 (ACC) for SDCNL, 0.868 (ACC) and 0.481 (ACC) for CSSRS-Suicide
Liu et al. [117]	MFCC, PLP, FBANK, TDNN × vector, Resnet × vector, I-vector	Speech	CN-Celeb, Depression speech database-20	accuracy: 74.72%
Liu et al. [118]	Short-term energy (power), intensity, loudness, zero crossing rate (ZCR), F0, jitter, flicker, formants and mel frequency cepstral coefficients (MFCC)), linear prediction coefficient (LPC), line spectrum pair (LSP)), perceptual linear prediction coefficient (PLP), etc.	Speech	private dataset	78.02% Accuracy
Jiang et al. [119]	Prosodic, spectral, and glottal features	Speech	private dataset	The accuracy was 75.00% in women and 81.82% in men, and the sensitivity/specificity ratio was 79.25%/70.59% in women and 78.13%/85.29% in men
Liu et al. [120]	MFCC, LPC, Jitter, Fundamental Frequency, etc.	Speech	private dataset	The recognition accuracy for males and females was 75.8% and 68.5% respectively
Yin et al. [121]	MFCC	Speech	DAIC-WOZ, MODM	F1: 92.7, Recall: 92.7, Precision: 92.8
Tasnim et al. [122]	Spectral features, depth representation features	Speech	DAIC-WOZ	F1: 69%
He et al. [123]	eGeMAPS, MRELBP, raw waveform, spectrogram	Speech	AVEC2013, AVEC2014	AVEC2013: RMSE 9.0000, MAE7.4210; AVEC2014: RMSE10.0012, MAE 8.201

Table 5. *Cont.*

Literature	Feature Name	Modality	Dataset	Results
Dubagunta et al. [124]	original speech signal, Low profile filtered signal (LPF), Linear Prediction Residual Signal (LPR), Homomorphically filtered speech source signal (HFVS), Zero frequency filtered signal (ZFF)	Speech	AVEC2013, AVEC2014	RMSE: 8.549, MAE: 6.650, F1: 0.824
Zhao et al. [125]	ComParE, some frame-level features	Speech	DAIC-WOZ, MODM	This model improves 2.3% and 10.3% compared to the LSTM model in public databases
Dong et al. [126]	Depth representation features	Speech	AVEC2013, AVEC2014	MSE: 8.549, MAE: 6.650, F1: 0.82
Kang et al. [127]	Matrix image of asymmetric feature transformation of EEG	EEG	Public dataset HUSM	Accuracy 98.85%
Čukić et al. [128]	HFD and SampEn of EEG signals	EEG	Private dataset (23 patents)	average accuracy 90.24% 97.56%
Mahato et al. [129]	Combined characteristics of alpha, alpha1, alpha2, beta, delta and theta power and theta asymmetry (delta, theta, alpha, beta, alpha1, alpha2) and theta asymmetry (average theta asymmetry and paired theta asymmetry)	EEG	Public dataset	average accuracy 88.33%
Wan et al. [130]	The feature extraction methods of time domain, frequency domain, wavelet, and nonlinear analysis are used to extract features from the sub-band components corresponding to the EEG samples.	EEG	Private (Beijing Anding Hospital, 12 normal people, 23 patients)	accuracy 86.67%
Cai et al. [131]	The linear characteristics are as follows: peak, variance, dip, kurtosis, and Hjorth parameters. Nonlinear characteristics include C0 complexity, correlation dimension, Shannon entropy, Kolmogorov entropy, and power spectral entropy.	EEG	Private dataset: 152 depressed patients and 113 healthy subjects	accuracy 71.32%
Zhu et al. [47]	1760 features (22 EEG features $\times$ 5 frequency bands $\times$ 16 electrodes)	EEG	Public dataset Ad-hoc	accuracy 83.42%
Ehghaghi et al. [132]	The acoustic features comprise spectral and sound-related characteristics, such as statistical functions of Mel-frequency cepstral coefficients (MFCC), fundamental frequency (F0), and zero-crossing rate (ZCR). Text features include syntactic complexity, semantic complexity, and discourse coherence, among others.	Speech, text	Dementia- Bank, Healthy Aging, ADReSS, DEPAC+, AD Clinical Trial	F1: 0.89 $\pm$ 0.03
Diep et al. [133]	Handcrafted features provided by domain experts include acoustic features, semantic features, and lexical-syntactic features.	Speech, text	DEPAC	F1: 63.0%

Table 5. *Cont.*

Literature	Feature Name	Modality	Dataset	Results
Mao et al. [134]	For speech, the features encompass prosodic features (NAQ, QOQ, H1–H2, PSP, MDQ, Peaklope, Rd), voice quality features (F0, VUV), and spectral features (MCEP, HMPDM, HMPDD). In the realm of text, GloVe word vectors are utilized.	Speech, text	DAIC-WOZg	accuracy 95.80%
Jan et al. [135]	Visual feature extraction includes Local Binary Pattern (LBP), Edge Orientation Histogram (EOH), Local Phase Quantization (LPQ), and deep feature extraction using pre-trained models like VGG-face and AlexNet. For audio feature extraction, Mel-frequency cepstral coefficients (MFCC) are employed. Additionally, the feature dynamic historical histogram involves MHH.	Speech, video	AVEC2013, AVEC2014	MAE: 6.14 RMSE: 7.43
Uddin et al. [136]	raw wav, image	Speech, video	AVEC2013, AVEC2014	AVEC2013: MAE 6.92, RMSE 8.54; AVEC2014: MAE 6.75, RMSE 8.45
Yang et al. [137]	For speech, statistical features are extracted. In the domain of text, paragraph vectors are utilized. For video, the feature extraction involves Displacement Range Histogram (DRH).	Speech, text, video	DAIC-WOZ	RMSE: 5.974, MAE: 5.163

3.4. Diagnosis of Heart Disease.

Heart diseases, particularly Cardiovascular Diseases (CVD), stand as the leading cause of death worldwide. Hypertrophic Cardiomyopathy (HCM) poses significant challenges due to the thickening of the left ventricular walls of the heart. The modern era has seen a paradigm shift in heart disease diagnosis, leveraging advanced technologies across various modalities. This chapter will diagnostic methods for heart disease using hypertrophic cardiomyopathy (HCM) as an example. We will gain a deeper understanding of HCM-assisted diagnostic techniques based on echocardiography, medical text data, and electrocardiograms (ECG) and explore other heart disease diagnostic methods based on genetic data. The comprehensive application of these diagnostic tools provides support for the early identification and treatment of heart disease and is of great significance for improving patient prognosis and quality of life. This section provides a summarized overview of the model and its features, as detailed in the accompanying Table 6.

Echocardiography. Deep learning frameworks have shown remarkable promise in enhancing the accuracy and efficiency of heart disease detection and classification. Among these advancements, the work of Almadani et al. [138] stands out with the introduction of the HCM Dynamic Echo, an end-to-end deep learning framework designed for the binary classification of echocardiography videos into hypertrophic cardiomyopathy (HCM) or normal categories. This system includes two analytical components: Branch 1, dubbed the Slow Path, which focuses on extracting spatial features, and Branch 2, known as the Fast Path, which is dedicated to capturing temporal structure information, thereby improving the accuracy of video recognition. They applied transfer learning and pre-trained HCM Dynamic Echo on the large Stanford EchoNet Dynamic Echocardiography dataset, enabling HCM detection in smaller echocardiography video datasets. In rigorous evaluations, HCM

Dynamic Echo outperformed state-of-the-art baselines, with an accuracy of 93.13%, an F1 score of 92.98%, a Positive Predictive Value (PPV) of 94.64%, a specificity of 94.87%, and an Area Under the Curve (AUC) of 93.13%.

Parallel to these developments, other researchers have also made significant contributions to the field. For instance, Madani et al. [139] developed a high-efficiency deep learning classifier for binary Left Ventricular Hypertrophy (LVH) diagnosis using echocardiography images. The core framework of their model included a U-Net for eliminating auxiliary information from image and a series of convolutional neural networks, resulting in an accuracy of 91.2%. To counter data scarcity, they proposed data augmentation using semi-supervised Generative Adversarial Networks (GANs). GANs demonstrated superior performance than traditional CNNs with limited data, attaining a test accuracy of 92.3%. Nasimova et al. [140] introduced a deep convolutional neural network for classifying echocardiography videos as Dilated Cardiomyopathy or Hypertrophic Cardiomyopathy. Their study initially generated an Echo dataset from internet-sourced Echo videos and EchoNet database videos. The team trimmed the collected videos to 2–5 s to remove unnecessary echo information and redundant frames before segmenting them into $112 \times 112 \times 3$ images for manual feature extraction. These images and extracted features were input into a six-layer CNN for classification, achieving a test accuracy of 98.2%.

Moreover, some studies have contributed to the field by applying deep learning models to diagnose various cardiac conditions from echocardiography. Zhang et al. [141] utilized the VGG-16 model to automatically detect three diseases from echocardiography: Hypertrophic Cardiomyopathy, Pulmonary Arterial Hypertension, and Cardiac Amyloidosis. They trained separate networks for each disease, using three random images per video. The images were processed through the VGG-16 model with a fully connected layer featuring two output units, achieving an AUC of 93% and p-value of 0.23 for HCM detection. Ghorbani et al. [142] analyzed 3312 consecutive comprehensive non-stress echocardiography studies collected from June to December 2018. The process started with the first frame of each video, sampling 20 frames at intervals of 100 milliseconds. The Inception-Resnet-v1 network processed each frame individually, and the final prediction was determined by averaging the predictions from all individual frames. This method achieved an AUC-ROC of 0.75 and an F1 score of 0.57.

Medical text data. Sundaram et al. [143] developed a Random Forest (RF) model to automatically identify patients with Hypertrophic Cardiomyopathy (HCM) using features extracted from Cardiac Magnetic Resonance (CMR) imaging reports. The Random Forest (RF) model attained an accuracy of 86% using 608 features and achieved 85% accuracy with 30 features. Mishra et al. [144] introduced an innovative application within the medical Internet of Things (IoMT) domain. They utilized a Recurrent convolutional neural network (Rec-CONVnet) to accurately estimate the risk of heart disease. The system design compiles various data points such as age, gender, symptoms of chest discomfort, blood sugar levels, blood pressure (BP), and other relevant clinical factors. Through comprehensive simulations and evaluations, the Rec-CONVnet demonstrated remarkable performance, achieving an impressive F1 score of 97%. Jayasudha et al. [145] designed a Social Water Cycle Driving Training Optimization (SWCDTO) ensemble classifier for heart disease detection. The classifier showed outstanding performance across specificity, accuracy, and sensitivity, reaching 95.84%, 94.80 and 95.36% in each metric. Levine et al. [146] investigated the performance of a large model (GPT-3) in diagnosing and triaging diseases like heart disease. The findings indicated that GPT-3's performance nearly approached that of professional medical practitioners.

Genetic data. Peng et al. [147] employed a Support Vector Machine (SVM), Random Forest (RF), and Logistic Regression (LR) to develop a classification model for coronary atherosclerosis heart disease (CAD). This model utilized datasets GSE12288, GSE7638, and GSE66360 from the GEO database. The ROC curve analysis revealed for SVM, RF, and LR in validation to be 75.58%, 63.57%, and 63.95%, respectively. Their respective areas under the curve were 81.3% (95% CI 0.761–0.866, $p < 0.0001$), 72.7% (95% CI 0.665–0.788,

$p < 0.0001$), and 78.3% (95% CI 0.725–0.841, $p < 0.0001$). Liu et al. [148] created a classification model for Coronary Artery Disease (CAD) using LASSO logistic regression, random forest, and SVM. They used data from the GEO dataset GSE113079, achieving an AUC of 97.1% in the training set and 98.9% in the testing set. Zhang et al. [44] introduced the Integration Machine Learning (IML) algorithm, incorporating a SVM, neural network (NN), RF, gradient boosting machine (GBM), decision trees (DT), and LASSO. This algorithm was applied to classify patients with Acute Myocardial Infarction (AMI) and stable coronary artery disease (SCAD), using GEO datasets GSE60993, GSE62646, GSE48060, and GSE59867, achieving an AUC over 90%. Hou et al. [149] utilized SVM for classifying CAD without heart failure (CAD-non HF), CAD complicated with heart failure (CAD-HF), and healthy controls, using GEO datasets GSE20681 and GSE59867. The study achieved an AUC of 0.944. Finally, Samadishadlou et al. [150] applied SVM for classifying myocardial infarction (MI), stable CAD, and healthy individuals, using datasets GSE59867, GSE56609, and GSE54475 from GEO. Their model demonstrated an AUC-ROC of 96% and an accuracy of 94%.

Electrocardiogram. The integration of Convolutional Neural Networks (CNN) into the analysis of Electrocardiogram (ECG) data has marked a significant leap forward in detecting Hypertrophic Cardiomyopathy (HCM) and other cardiovascular diseases (CVDs) [151]. Among the notable contributions, Tison et al. [152] developed an automated and highly interpretable method for analyzing patient ECG features. This method processed and analyzed 36,186 ECG datum from the University of California, San Francisco (UCSF) database. Researchers utilized Hidden Markov Models (HMM) to extract ECG vector representations containing 725 features, which were then trained using CNNs to estimate cardiac structural and functional indices and classify diseases. Compared to traditional neural network models, this vectorized processing approach better retained meaningful features in ECGs, thus enhancing the interpretability and accuracy of diagnostic results. Similarly, Dai et al. [151] used a deep CNN to classify five cardiovascular diseases (CVDs) using standard 12-lead ECG signals. The study utilized the public Physiobank (PTB) ECG database. The researchers have segmented ECG signals into different intervals—1 s, 2 s, and 3 s—without detecting individual waves, thus forming three distinct datasets. They applied ten-fold cross-validation on one-second-long ECG signals and tested on the other two datasets (two and three seconds long). The proposed CNN model achieved an accuracy, sensitivity, and specificity of 99.59%, 99.04%, and 99.87%, respectively, for one-second signals, demonstrating superior performance. For two-second signals using pre-trained models, the system achieved an overall accuracy, sensitivity, and specificity of 99.80%, 99.48%, and 99.93%. For three-second signal detection, the accuracies were 99.84%, sensitivity 99.52%, and specificity 99.95%. These results indicate that the proposed system achieved high performance while maintaining simplicity and flexibility, suggesting its potential for real-time application in medical settings.

Furthermore, Tison et al. [153] highlighted the application value of AI-enhanced ECG (AI-ECG) in assessing disease states and treatment responses for obstructive HCM. The study noted that AI-ECG could extract more physiologically and pathophysiologically relevant information related to obstructive HCM from ECGs, surpassing traditional manual interpretation methods. Moreover, the study mentioned the potential of AI-ECG for remote monitoring through smartphone electrodes to assess disease states and treatment responses. The authors also foresaw the future application of this technology in medication adjustment and enhancing treatment safety.

Another impressive study is conducted by the Mayo Clinic [154]: they used digital 12-lead ECGs from 2448 diagnosed HCM patients and 51,153 age and gender-matched non-HCM controls to train and validate a CNN. The algorithm performed impressively in adult HCM patient ECG detection, with an AUC of 0.96, sensitivity of 87%, and specificity of 90%. The algorithm's performance in a test of 300 children and over 18,000 age and gender-matched controls was equally impressive: the HCM detection model achieved an AUC of 0.98, sensitivity of 92%, specificity of 95%, Positive Predictive Value (PPV) of 22%, and Negative Predictive Value (NPV) of 99%. The study found that the algorithm generally performed better in the adolescent group than in the pediatric group.

Table 6. Summary of different medical features for heart disease diagnosis.

Literature	Feature Name	Modality	Dataset	Results
Almadani et al. [138]	Echocardiography	echocardiogram videos	Stanford EchoNet-Dynamic echocardiogram dataset	ACC: 93.13%, F1-score: 92.98%, Positive Predictive Value (PPV): 94.64%, specificity: 94.87%, AUC: 93.13%
Madani et al. [139]	echocardiography	Original echocardiograms	Private dataset	92.3% accuracy: binary left ventricular hypertrophy classification
Nasimova et al. [140]	Echocardiography	Clipped echocardiogram video frames	(1) EchoNet database; (2) Echo videos from the Internet	ACC: 98.2% (dilated cardiomyopathy vs. hyper-trophic cardiomy-opathy (HCM))
Zhang et al. [141]	Echocardiography	Original echocardiograms	Private dataset	AUC: 0.93
Ghorbani et al. [142]	Echocardiography	Cropped echocardiogram regions (inside of the scanning sector)	Private dataset	AUC: 0.75
Sundaram et al. [143]	Word Embedding, Part of Speech (POS)	Text	CMR	86% (ACC) for 608 features and 85% (ACC) for 30 features
Mishra et al. [144]	Word Embedding	Text	Real clinical records in hospital databases	97% F1 score, FPR of 64.6%, accuracy of 96.4%, and accuracy of 76.2%
Levine et al. [146]	Multivariate Features	Text	Recruited participants	Brier score = 0.18 for disease, Brier score = 0.22 for triage
Peng et al. [147]	Gene-based	RNA-seq	GEO	SVM: 81.3% (ACC); RF: 72.7% (ACC); LR: 78.3% (ACC)
Liu et al. [148]	Gene-based	RNA-seq	GEO	Training: 97.1% (AUC), test: 98.9% (AUC)
Zhang et al. [44]	Gene-based	RNA-seq	GEO	90% (AUC)
Hou et al. [149]	Gene-based	RNA-seq	GEO	94.4% (AUC)
Samadishadlou et al. [150]	Gene-based	MicroRNA	GEO	96% (AUC), 94% (ACC)
Dai et al. [151]	End-to-end Auto-learned Features	ECG	Physiobank (PTB) Public Dataset	Accuracy: 99.84%, Sensitivity: 99.52%, Specificity: 99.95%
Tison et al. [152]	725 Features Extracted using Hidden Markov Models	ECG	UCSF Database	AUR: Range 0.94 to 0.77
Tison et al. [153]	End-to-end Auto-learned Features	ECG	UCSF Database	-
Ko et al. [154]	End-to-end Auto-learned Features	ECG	Public Mayo Clinic Developed Database	AUC: 0.96, Sensitivity: 87%, Specificity: 90%

3.5. Diagnosis of Epilepsy

Epilepsy, a prevalent neurological disorder affecting approximately 60 million people worldwide [155], poses significant diagnostic challenges. A range of symptoms characterizes it, and an effective diagnosis requires a multidisciplinary approach. This article explores various diagnostic methods employed in epilepsy detection, utilizing advanced technology and medical imaging. This chapter will explore auxiliary diagnostic techniques for epilepsy based on images, medical text data, and electroencephalography (EEG). These methods play a crucial role in improving the accuracy and efficiency of epilepsy diagnosis, providing us with a new perspective to understand this complex disease and bringing better medical services to patients. This section provides a summarized overview of the model and its features, as detailed in the accompanying Table 7.

Medical video. Using video data for computer-assisted diagnosis has become essential for the timely detection of epilepsy. Karácsony et al. [156] employed clinical Motion Capture (MoCap) to quantitatively analyze seizure-related symptoms such as ictal head turning and upper limb automatisms, marking a pioneering discovery in differentiating epilepsy syndromes, providing clinical localization and lateralization information. Maia et al. [157] applied a threshold-based approach to first detect regions of interest (beds) in video data, aligning them vertically for consistency, then utilized Convolutional Neural Networks and Multilayer Perceptrons to classify epileptic seizures, achieving 65% AUC. Achilles et al. [158] recorded 52 seizures at 15 frames per second using infrared and depth imaging sensors, training distinct Deep Convolutional Neural Network architectures (CNNs) on video frames (one CNN for infrared frames, another for depth frames). Combining outputs from both networks, they achieved the prediction of ictal or interictal epilepsy phases, with their method demonstrating high sensitivity (87%) and specificity (81%) for generalized tonic-clonic seizures.

Building upon these advancements, Ahmedt-Aristizabal [159] unveiled an innovative network approach that integrates 3D facial reconstruction with deep learning. The design of this approach aims to detect and measure orofacial semiotics in a collection of 20 seizure videos, featuring recordings from patients with temporal and extra-temporal lobe epilepsy. The developed network demonstrated its capability to differentiate between two types of epileptic seizures, achieving an average classification accuracy of 89%. It marks a significant advancement in computer vision and deep learning within non-contact systems, particularly for identifying common semiotics in real-world clinical environments. Significantly, this method departs from earlier epilepsy monitoring techniques by moving beyond the reliance on single-angle image information. In contrast, Kunekar et al. [160] proposed improving accuracy by utilizing information from multiple modalities instead of relying solely on features from a single viewpoint. Ahmedt-Aristizabal et al. [161] proposed a new modular, hierarchical, multi-modal system aimed at detecting and quantifying semiotic signs recorded in 2D monitoring videos. This method combines computer vision with deep learning architectures to learn semiotic features from facial, body, and hand movements.

MRI. MRI-generated 2D or 3D images enable a better understanding of the brain's internal structure, pinpointing brain issues associated with epileptic seizures. fMRI has become indispensable tools in the detection and understanding of epileptic seizures by providing detailed images of the brain's internal structure. Garner et al. [162] applied a machine learning approach using a Random Forest classifier, trained with resting-state functional MRI (fMRI) data, to predict epilepsy outcomes. The model achieved a 69% accuracy rate in predicting epilepsy outcomes on the test set after 100 stratified cross-validation rounds, using 70% of resting-state fMRI scans for training and 30% for testing. Similarly, Sahebzamani et al. [163] employed the Gram-Schmidt orthogonalization method alongside a unified tissue segmentation approach for segmenting brain tissues in MRI images. They calculated first-order statistical and Gray Level Co-occurrence Matrix (GLCM) texture features and trained SVM classifiers using features from either the entire brain or the hippocampus to diagnose epilepsy. This comprehensive segmentation and whole-brain analysis methodology yielded a 94% accuracy rate.

In the quest for early and accurate diagnosis, researchers like Si et al. [164] have turned to diffusion MRI techniques to detect subtle brain changes in conditions such as Juvenile Myoclonic Epilepsy. They emphasized the importance of early diagnosis in Juvenile Myoclonic Epilepsy (JME), a disorder that predominantly affects adolescents and poses significant developmental challenges. They utilized two advanced diffusion MRI techniques—High Angular Resolution Diffusion Imaging (HARDI) and Neurite Orientation Dispersion and Density Imaging (NODDI)—to create connectivity matrices that capture subtle white matter changes. By adopting transfer learning, they trained sophisticated Convolutional Neural Network (CNN)-based models for JME detection. Pominova et al. [165] explored various deep 3D neural architecture building blocks for epilepsy detection, using both structural and functional MRI data. They experimented with 12 different architectural variants of 3D convolution and 3D recurrent neural networks. Santoso et al. [166] proposed a novel integrated Convolutional Neural Network approach for classifying brain abnormalities (epilepsy vs. non-epilepsy) using axial multi-sequence MR images. The model comprised base learners with distinct architectures and lower parameter counts. By aggregating the outputs and predictions of these base models (through methods like majority voting, weighted majority voting, and weighted averaging) and feeding them into a meta-learning process with a SVM, they significantly enhanced the final classification performance.

Medical text data. Hamid et al. [167] showcased the potential to differentiate epileptic patients from those with psychogenic non-epileptic seizures (PNES). They developed an NLP tool based on an annotator modular pipeline to analyze electronic medical records, identifying grammatical structures and named entities. This algorithm was proficient in detecting concepts indicative of PNES and those negating its presence. Taking a different approach, Pevy and colleagues [168] utilized written records of conversations between patients and doctors to distinguish between epileptic seizures and PNES. They employed an NLP toolkit to extract specific features of speech formulation efforts, such as hesitations, reformulations, and grammatical repairs, from these transcripts. The algorithm then trained machine learning classifiers with these features, enabling it to distinguish patients based on their verbal expression patterns. Connolly et al. [169] further affirmed the effectiveness of NLP in differentiating among various epilepsy types, including partial epilepsy, generalized epilepsy, and unclassified epilepsy. By analyzing text features extracted from electronic medical records, their algorithm successfully classified different subtypes of epilepsy with remarkable accuracy.

EEG. Researchers frequently use CNN (Convolutional Neural Network) architectures, which can extract features automatically, unlike traditional machine learning classifiers that require manual extraction of features for detecting and classifying epileptic seizures effectively. Clarke et al. [170] developed a deep Convolutional Neural Network (CNN) for detecting epileptic seizure discharges, trained using a dataset comprising over 6000 marked events from a group of 103 patients diagnosed with Idiopathic Generalized Epilepsy (IGE). This newly proposed automatic detection algorithm showcased exceptional performance in identifying epileptic seizures from clinical EEGs. The system achieved an impressive average sensitivity of 95% and kept the average false positive rate to just one per minute. These results indicate that AI-powered computer-assisted EEG analysis could significantly improve the speed and precision of EEG assessments, thereby potentially enhancing treatment outcomes for epilepsy patients. Fürbass et al. [171] employed the Fast R-CNN method for object detection, using deep regression for localization estimation of EDs (negative peaks) and the UDA training process to handle noise and artefacts in EEG. The authors used EEG data from 590,000 epochs of 289 patients for unsupervised training and tested it against 100 proprietary datasets. The experimental results indicated that the DeepSpike algorithm attained a sensitivity of 89%, a specificity of 70%, and an overall accuracy rate of 80%, showcasing its high effectiveness in identifying EEG discharges. Thara et al. [172] used a two-layer stacked bidirectional Long Short-Term Memory (LSTM) technique for detecting epileptic seizures. The researchers built a model with two LSTM layers, dropout and dense layers, and trained and optimized it using activation functions such as sigmoid and softmax, achieving good results with an accuracy of 99.89% on the training set and 99.08% on the test set. Yao et al. [173] experimented with ten different and independently improved RNN (IndRNN) architectures, achieving the best accuracy with a 31-layer Dense IndRNN with attention (DIndRNN).

Multi-modality. Torres-Velázquez et al. [174] evaluated the performance of multi-channel deep neural networks in Temporal Lobe Epilepsy (TLE) classification tasks under single and combined datasets. They trained, validated, and tested several multi-channel deep neural network models using brain structural indices from structural MRI, MRI-based region of interest correlation features, and personal demographic and cognitive data (PDC). Results indicated that PDC alone provided the most accurate TLE classification, followed by the combination of PDC with MRI-based brain structural indices. These findings affirm the potential of deep learning methods, like mDNN models, in TLE classification when combined with multiple datasets.

Table 7. Summary of different medical features for epilepsy diagnosis.

Literature	Feature Name	Modality	Dataset	Results
Karácsony et al. [156]	Medical video	2D + 3D video feature	Neuro- Kinect	-
Maia et al. [157]	Medical video	Original Infrared video feature	Private data	0.65 (AUC)
Achilles et al. [158]	Medical video	infrared and depth video frames	ADNI, AIBL	sensitivity (87%) specificity (81%)
Ahmedt-Aristizabal et al. [159]	Medical video	Regions of interest by 3D face reconstruction from the original video sequences	Private dataset	0.89 (ACC)
Ahmedt-Aristizabal [161]	Medical video	2D monitoring videos	Private dataset	83.4 % (ACC: face); 80.1% (ACC: body) body; 69.3% (ACC:hand)e
Garner et al. [162]	MRI	functional magnetic resonance imaging (fMRI) data	REDCap	0.69 (ACC)
Sahebzamani et al. [163]	MRI	first-order statistical and volumetric gray-level co-occurrence matrix (GLCM) texture features from structural MRI data	Private dataset	0.94 (ACC)
Si et al. [164]	MRI	the connectivity matrix which can describe tiny changes in white matter	Private dataset	75.2% (ACC) and the 0.839 (AUC)
Pominova et al. [165]	MRI	3D + 4D MRI data	Private dataset	0.73 (AUC)
Santos et al. [166]	MRI	axial multi-sequences of MRI	Private dataset	86.3% (ACC) 90.75% (F1-score)
Hamid et al. [167]	stemming features, POS, bag of concepts	Text	VA national clinical database	The accuracy, sensitivity, and F-score are 93%, 99%, and 96%
Pevy et al. [168]	Word embedding	Text	Recording, transcribing, and writing records of interview corpora	71% (ACC)

Table 7. *Cont.*

Literature	Feature Name	Modality	Dataset	Results
Connolly et al. [169]	N-gram	Text	DrWare- house (DrWH)	0.708 (F1) for partial epilepsy (PE), generalized epilepsy (GE), and unclassified epilepsy (UE), 0.899 (F1) for PE and GE
Clarke et al. [170]	End-to-end Auto-learned	EEG	Public Ad-hoc	Average Sensitivity: 95%
Fürbass et al. [171]	End-to-end Auto-learned	EEG	Private Dataset (Test); 590,000 Epochs from 289 Patients in Temple University's Public EEG Corpus (Training)	Sensitivity: 89%, Specificity: 70%, Overall Accuracy: 80%
Thara et al. [172]	End-to-end Auto-learned	EEG	Private Dataset	Accuracy: 99.89%
Yao et al. [173]	End-to-end Auto-learned	EEG	CHB-MIT Dataset	Average Sensitivity: 88.80%, Specificity: 88.60%, Precision: 88.69%
Torres-Velázquez et al. [174]	Multi-modality	brain structure metrics from structural MRI, MRI-based region of interest correlation features, and personal demographic and cognitive data (PDC)	Private Dataset	Acc = 69.46% ± 20.82%, AUC = 70.00% ± 26.00%

3.6. Discussion

Modality distinction. In our comprehensive review, we examine the different methods used to automatically diagnose five specific diseases: Alzheimer's disease (AD), breast cancer, depression, heart disease, and epilepsy. The medical data produced from different disease diagnosis processes has commonalities, mainly encompassing image, text, genetic, signal, and voice modalities. Distinctive preferences for specific modalities exist across different diseases. Even within the realm of single medical imaging, nuanced differences become apparent. For Alzheimer's disease diagnosis, Magnetic Resonance Imaging (MRI) and Positron Emission Tomography (PET) images emerge as the predominant modalities, supplemented by the inclusion of voice data. The widespread use of MRI and PET stems from their effectiveness in capturing the structural and functional brain changes associated with Alzheimer's disease (AD). The unique characteristics of neurodegenerative alterations make these imaging modalities particularly suitable for early detection and monitoring of disease progression.

Contrastingly, in breast cancer diagnostics, a multifaceted approach involves genetic data, X-ray imaging, ultrasound, and a notable amount of textual information. The rationale behind this approach lies in the heterogeneity of breast cancer itself, necessitating a comprehensive analysis of genetic predispositions, coupled with various imaging techniques and textual data to enhance diagnostic accuracy. Each modality contributes valuable insights into different aspects of breast cancer pathology, collectively enhancing the overall diagnostic efficacy. In the context of depression diagnosis, the emphasis shifts toward textual data and Electroencephalogram (EEG). The reliance on text data could be attributed to the subjective nature of depression symptoms, requiring a nuanced analysis of linguistic patterns and sentiment. EEG captures brain wave activity and complements textual data by providing physiological markers that indicate depression.

For heart disease diagnosis, the prevalent modalities include echocardiography, electrocardiography, and medical texts. The dominance of ultrasound-based echocardiography comes from its ability to provide real-time images of the heart's structure and function, which is essential for assessing cardiac health. Electrocardiography contributes information on the heart's electrical activity, while medical texts further contextualize the diagnostic process. For epilepsy diagnostics, a comprehensive strategy incorporates Magnetic Resonance Imaging (MRI), video data capturing patient movements, Electroencephalogram (EEG), and relevant textual information. The utilization of these diverse modalities is driven by the intricate nature of epilepsy itself, demanding a thorough examination of various aspects. MRI provides structural insights, video data offers observations of seizures and associated movements, EEG captures electrical activity in the brain, while textual information contributes contextual details.

In conclusion, the selection of modalities for automated diagnosis is intricately tied to the unique characteristics and pathological features of each disease. Understanding the rationale behind the prevalence of specific modalities facilitates a targeted and effective approach to automated disease diagnosis.

Modality fusion. Contemporary diagnostic methodologies increasingly favour the integration of multi-modal approaches. The advantages of the multi-modal paradigm lie in its ability to provide a more comprehensive and accurate understanding of complex phenomena by integrating diverse data modalities. This approach enhances robustness, improves interpretability, and allows for personalized and optimized solutions across various domains.

In diagnosing Alzheimer's Disease (AD), where subtle but significant changes in language patterns and cognitive function are markers, combining speech and text analysis is extremely valuable. This multi-modal approach adeptly captures the intricate linguistic nuances and potential confusion in communication exhibited by AD patients. Integrating genetic data and electroencephalogram (EEG) as supplementary information enriches the diagnostic process, addressing the multifaceted nature of AD symptoms and facilitating a more accurate and holistic understanding. In cancer research, there is a significant emphasis on combining imaging and genetic data. Since genetic mutations play a pivotal role in the development and progression of various types of cancer, identifying specific genetic alterations associated with different types of cancer can provide insights into their molecular mechanisms and potential therapeutic targets.

Besides, specific genetic mutations may present as unique visual patterns. For example, specific genetic alterations in breast cancer, such as those in the BRCA genes, may result in characteristic radiographic features observable in mammograms or other imaging modalities. Therefore, combining genetic data with medical imaging enhances our molecular-level understanding of cancer and supports the creation of tailored, accurate methods for its diagnosis and treatment. Depression diagnosis predominantly relies on speech modalities, with supplementary integration of text or video data. This emphasis on speech is justified by the distinct changes in vocal patterns and tone often exhibited by individuals with depression. Adding text or video data enhances the diagnostic process by providing extra information on the patient's emotional and behavioural conditions.

For diagnosing heart disease, it's common to combine ultrasound imaging with medical texts. The rationale behind this lies in the need to comprehensively assess both structural and functional aspects of the heart. Ultrasound provides real-time visualizations of cardiac anatomy, while medical texts offer additional clinical context, creating a synergistic diagnostic approach. Epilepsy diagnosis currently benefits from the mutual utilization of various imaging modalities, such as Magnetic Resonance Imaging (MRI) and Positron Emission Tomography (PET) images. This approach acknowledges the diverse epileptic manifestations and leverages the strengths of multiple imaging techniques to achieve a more comprehensive and accurate diagnosis. In essence, the choice of modalities for fusion explicitly correlates with the diverse manifestations of patients' conditions. The reasonable multi-modal fusion approach can capture the intricacies of symptoms, ensuring a more nuanced and effective diagnostic outcome tailored to the specificities of each medical condition.

Performance improvement. The evolution of research in automated disease diagnosis is accompanied by the continual improvement of performance. This progression has transitioned from machine learning dominance to primary reliance on deep learning, complemented by innovative techniques such as attention mechanisms and transfer learning. Initially, disease diagnosis methods focused on developing feature engineering within machine learning studies, where manually identifying and selecting pertinent features was vital for the model's performance. However, this process had limitations, often requiring domain expertise and not fully exploiting the richness of complex datasets. In response to these challenges, the subsequent embrace of deep learning has become a transformative force in medical diagnostics. The distinctive advantage of deep learning lies in its capability to automatically extract hierarchical and intricate features from raw data, eliminating the need for explicit feature engineering. This automated feature extraction significantly enhances the diagnostic model's performance by allowing it to discern intricate patterns and relationships within the data.

Deep learning has improved the accuracy and efficiency of disease detection. Within the domain of deep learning for medical diagnostics, scholars have proposed innovative techniques to elevate model performance. Inspired by how we humans see, attention mechanisms in deep learning models allow a focus on areas within the data for better analysis. It mimics the human ability to prioritize relevant information, improving the model's ability to capture subtle or critical features. Attention mechanisms have shown effectiveness in different medical imaging tasks, leading to diagnoses that are more precise and aware of the context. Transfer learning has also become a technique to overcome the issue of scarce medical data samples. In transfer learning, a model pre-trained on a large dataset, often from a related domain, is fine-tuned on a smaller target dataset, which is typically scarce in medical applications. This approach leverages the knowledge gained from the source domain to enhance the model's performance on the target task, even when training samples are limited. Transfer learning has proven effective in scenarios where acquiring a large, labeled medical dataset is impractical, thus facilitating the development of robust diagnostic models. The evolution from traditional machine learning, reliant on explicit feature engineering, to deep learning, with its automated feature extraction capabilities, has significantly improved disease diagnosis models. Combining attention mechanisms with transfer learning highlights scholars' dedication to enhancing model performance, improving interpretability, and tackling the problem of limited data in medical contexts. These advancements collectively contribute to the ongoing refinement and enhancement of state-of-the-art diagnostic systems.

Large model application. The emergence of large models in AI has revolutionized many industries, particularly in healthcare. These models, often trained on vast datasets, can analyze complex patterns that lead to more accurate and efficient disease diagnosis. With the increasing use of electronic health records and the integration of various data sources, medical institutions now have access to more information. This dataset comprises patient histories, symptomatology, and genetic profiles, among other details, offering a rich reservoir. Large models can analyze this data to discern patterns and correlations. Currently, most large-scale models in healthcare focus on text, analyzing medical records, discharge summaries, and other types of written data. However, there is potential for models to analyze additional forms of medical data, including images, voice recordings, genetic data, and physiological signals.

As technologies improve and datasets grow, we can expect to see more diverse applications of large models in healthcare. For example, image analysis models can process medical images such as X-rays or CT scans to detect diseases or lesions more accurately. The speech analysis model can process the patient's speech records and extract useful information from them, such as the severity of symptoms or the development trend of the condition. Genetic analysis models can predict a patient's response to specific drugs or disease risks based on their genomic data. The physiological signal analysis model can track the patient's vital signs, like heart rate and blood pressure, identify any irregularities swiftly, and take appropriate action. Notably, some challenges need to be solved. One major challenge is data privacy. Training and refining large models necessitates significant data volumes, yet it is essential to safeguard the privacy and security of medical information. Creating strong encryption and access management systems is crucial for patient data. It's imperative to address ethical considerations when integrating AI into healthcare practices. It is essential to ensure that AI algorithms do not discriminate against any particular group and that their use complies with ethical standards. Overall, the rise of large models in healthcare can contribute to improving patient outcomes and reduce the burden on the healthcare system in the future.

4. Challenges and Future Works

Despite the commendable achievements in artificial intelligence (AI) technology within the realm of disease diagnosis and analysis, it is crucial to acknowledge that notable limitations still prevail in many other facets. Exploring solutions to overcome these limitations emerges as a pivotal concern for the future trajectory of this field. Consequently, herein, we delineate the extant constraints and proffer potential resolutions to these challenges.

4.1. Medical Multimodality Data Imbalance

Typically, data imbalance encompasses two dimensions: the imbalance within classes in a single modality, and the distributional imbalances across different modalities. This aspect describes the unequal representation of various classes within a single data category. For instance, in an MRI dataset, there might be a notable discrepancy in the number of scans illustrating Alzheimer's disease compared to scans indicative of normal conditions. For the latter, there is a disproportionate representation of data from one modality compared to others: There could be a surplus of imaging data yet a scarcity of genetic or textual data about Alzheimer's diagnosis. Some strategies are needed to solve the problem of imbalanced samples:

Transfer learning: Leveraging pre-existing labelled datasets from related medical domains and applying transfer learning techniques can partially address the data scarcity. One can refine pre-trained models by fine-tuning them on smaller, specialized datasets that cater to specific diagnostic challenges.

Synthetic data generation: Employing techniques for generating synthetic data, where new data points are artificially created based on existing labelled samples, can augment the available dataset. This approach helps address limitations arising from insufficient data volume.

Ensemble methods: You can enhance the accuracy of a model by combining predictions from multiple weakly supervised models or by incorporating different sources of weak supervision. Ensemble methods help compensate for the lack of detailed annotations by aggregating diverse model outputs.

4.2. Weak Model Generalization Ability

The core technologies and algorithms of AI models designed for different diseases are typically general. For instance, a Convolutional Neural Network (CNN) has been widely applied in the diagnosis of AD [80], breast cancer [96], depression [121], heart disease [140], and epilepsy [158]. However, deploying AI models developed for specific diseases to other disease predictions often demonstrates limited generalization ability. The primary reason lies in the fact that AI diagnostic models tailored for a specific disease tend to focus exclusively on the features unique to the particular disease, overlooking broader patterns. Some state-of-the-art techniques can address this issue:

Considering multi-centre cross-institutional data collection: Encouraging healthcare institutions to collaborate on data collection is to create more diverse and representative datasets. Such collaborative efforts involve pooling data from various sources, encompassing different geographical locations, demographic profiles, and medical practices. Models trained on datasets with this heightened diversity are more likely to generalize effectively across a spectrum of patient populations and healthcare scenarios.

Adversarial training: Adversarial training involves the introduction of adversarial examples during model training. By exposing the model to perturbed or deceptive samples, it learns to become more robust and exhibits improved generalization performance when faced with unseen or unexpected data. This technique can fortify the model against variations in the input space, enhancing its adaptability to a broader range of medical scenarios.

Reinforcement learning: Reinforcement learning is a paradigm where an agent interacts with an environment to learn optimal decision-making strategies. In medical diagnosis, one can use reinforcement learning to develop policies that help the model make more generalized decisions across diverse contexts. Through trial and error, the model hones its ability to navigate complex environments and adapt its behaviour to new and varied scenarios.

4.3. Lack of Model Interpretability

AI has demonstrated tremendous potential in health and medicine, yet research on the interpretability of AI decision outcomes is limited. This review found that only 28 of the included studies directly or indirectly tackled the crucial aspect of interpretability. These studies sought interpretability through methods like logistic regression, decision trees, naive Bayes, and support vector machines, known for their inherent clarity, or by applying techniques such as incorporating prior knowledge and using attention mechanisms to improve model interpretability. However, regrettably, the majority of studies did not adequately consider this crucial factor. Future research directions urgently need to delve into the interpretability of artificial intelligence models, utilizing interpretable models to enhance trust in AI and assist clinical practitioners in making informed decisions [175,176], thereby promoting the better integration of these models into clinical practice. Some solutions may be leveraged to enhance model interpretability:

Combining inherently interpretable model architectures. Several models such as decision trees or linear models, can be integrated with machine and deep learning frameworks thus enhancing transparency. These models provide explicit rules and feature importance, making their decision-making process more understandable.

Visual heatmaps generation. Generating heatmaps is a common technique for visualizing the importance or activation of specific regions in data. For instance, gradient-based methods like guided backpropagation or gradient-weighted class activation mapping (Grad-CAM) can identify influential regions, revealing which parts of the input most significantly contribute to the output.

4.4. Data Privacy and Security

Ensuring data privacy and security has always been a critical issue awaiting resolution in medical artificial intelligence. The development of robust AI models relies on extensive training and validation datasets. Because local data is often scarce, it's usually necessary to centralize the data. However, centralized solutions come with inherent drawbacks, including concerns about data ownership, confidentiality, privacy, and security, as well as the potential for data monopolies biased towards data aggregators [177]. Means to mitigate these pitfalls include:

Anonymization and de-identification. This method is primarily achieved by removing or blurring information in the data that identifies individuals, thereby reducing the link between the data and specific persons. This method is widely employed in current research to safeguard patient privacy. However, studies indicate that even desensitized data may still be re-identifiable through sophisticated analysis methods [178].

Federated learning. Federated Learning [179] is a decentralized learning approach that pushes the model training process to local devices, forming a global model through local updates, thereby preventing sensitive data from leaving the original devices. This method of decentralized learning emerges as a progressive approach to tackle the challenges of data anonymization and de-identification, offering a proactive strategy for maintaining data privacy and security.

Swarm learning. Swarm learning [180] extends the principles of federated learning to scenarios involving multiple participants, facilitating the integration of data from various sources through collaborative learning. This approach ensures a more comprehensive and accurate learning outcome while safeguarding privacy.

4.5. Ethical and Moral Considerations

From an ethical and moral standpoint, it is vital to guarantee that developed models mitigate "bias" and "inequality" across individuals and demographic categories. It is particularly crucial to address disparities linked to gender, age, race, income, education, and geographic location to promote fairness. In most studies reviewed, the persistence of differences often stemmed from not having enough data to achieve mitigation. However, for the deployment of AI models in clinical practice, ensuring fairness and generalizability [181,182] is also essential to guarantee the ethical and effective implementation of these technologies in a clinical setting [183].

There are at least two common scenarios where ethical issues arise in medical data. The first scenario is when the data source itself cannot reflect the true epidemiological situation within a given population, such as population data bias resulting from overdiagnosis of schizophrenia in African Americans [184]. The second scenario is when the dataset used for algorithm training lacks members from specific demographic groups. For example, an algorithm primarily trained on data from elderly white males might yield poor predictions for young black females. If algorithms trained on datasets with these characteristics are adopted in healthcare, they may exacerbate health disparities [185]. Effective solutions include:

Balanced data sampling. When constructing the training dataset, employ methods such as undersampling, oversampling, adaptive sampling, etc., to ensure a relatively balanced number of samples from different groups. This helps prevent the model from overly focusing on a specific population, thereby reducing data bias.

Removal of sensitive attributes. Eliminate potentially sensitive attributes (e.g., gender, race, age, etc.) from the data to ensure that the training dataset for the model does not contain direct or indirect ethical information.

Establishment of best practices by scientific societies and regulatory bodies. Scientific societies and regulatory bodies should develop data assessment standards, allowing datasets to comprehensively and accurately represent the societal, environmental, and economic factors impacting health [186]. The aim is to identify and minimize bias in training datasets, thereby fostering the development of algorithms that mitigate bias and promote fairness. As a notable example of bias reduction, the U.S. Food and Drug Administration (FDA), within the context of its Digital Health Innovation Action Plan, initiated a pre-certification pilot program. They evaluate developing medical software based on five established excellence principles, including quality standards and other similar regulatory criteria [187]. These standards can be extended to encompass the risk of bias in training datasets, thereby addressing issues related to data "bias" and "inequality".

4.6. Future Works

Application of AI on mobile devices. Integrating AI programs on mobile devices injects a more efficient and intelligent element into the management of patient diseases, early warnings, and promotion of healthy behaviours [188,189]. Equipping various sensors and AI programs on devices such as watches and smartphones enables real-time monitoring, recording, and analysis of patients' vital signs (such as heart rate, blood pressure, oxygen levels, etc.), medication usage, dietary habits, and exercise data. This capability facilitates patients' current physical conditions and future trends, enabling timely responses to potential health risks and offering personalized treatment recommendations.

Brain-machine interfaces. Brain-machine interfaces (BMIs) [190] are poised to play a crucial role in the diagnosis of neurological disorders in the future. BMIs, through direct interaction with signals from the brain, hold the potential to identify diseases related to the nervous system, such as Parkinson's disease or stroke. BMIs are anticipated to advance brain diagnostics, particularly in the field of neuroimaging.

Collaboration of diverse teams. The application of AI in the health and medical field involves three types of parties, i.e., healthcare professionals, researchers, and AI experts. Facilitating collaboration among these three parties contributes to the advancement of AI in the health and medical domain. Healthcare professionals possess rich clinical experience and specialized medical knowledge, providing profound insights into the pathology, physiology, and other aspects of diseases. They can offer unique perspectives and high-quality annotated data for researchers and AI experts, thereby contributing to more interpretable and accurate AI models for disease diagnosis. Secondly, healthcare professionals recognize the significance and delicate nature of medical data, as well as the need to maintain its privacy and security. They can ensure the privacy protection and compliance of data, ensuring that researchers and AI experts, in the process of refining AI models, mitigate bias and promote fairness. Reciprocally, researchers and AI experts possess proficient technical development experience, enabling them to provide healthcare professionals with adaptive AI models for the ever-evolving medical environment. These models assist healthcare professionals in clinical diagnosis, achieve early disease warning and prediction, and alleviate their workload.

5. Conclusions

In this paper, we thoroughly investigate the applications of artificial intelligence in diagnosing five distinct disorders: Alzheimer's disease, breast cancer, depression, heart disease, and epilepsy. We describe commonly used datasets to illustrate the data foundation, considering numerous multimodality data sources. Subsequently, we demonstrate the data pre-processing, feature engineering process, classification model establishment, and performance evaluation metrics. These methods automatically transform original data into valuable information highly relevant to disease lesions, representing key steps for AI-based diagnosis tasks.

We report and analyze detailed efforts on different modality-driven diagnoses, highlighting diverse strategies employed to address the complexities of each disorder. For Alzheimer's disease, we scrutinize the integration of multi-modal data such as neuroimaging, genetic markers, and cognitive assessments, emphasizing the intricate interplay between various diagnostic modalities. In the field of breast cancer, we explore imaging data from mammograms and genetic information, offering a nuanced understanding of the disease at both structural and molecular levels. Regarding depression, we investigate textual and speech data, revealing the potential of linguistic and acoustic cues in enhancing diagnostic accuracy. For heart disease, we focus on physiological signals and imaging data, providing a holistic approach to cardiovascular health assessment. Additionally, in the case of epilepsy, we meticulously examine the integration of electroencephalogram (EEG) data, showcasing the significance of real-time monitoring and data-driven insights.

Finally, we acknowledge that while AI technology has made certain achievements in the medical field, significant limitations remain in disease diagnosis applications. We describe challenges such as medical multimodality data imbalance, weak model generalization ability, and lack of model interpretability, providing corresponding solutions to guide future work. Overall, this review aims to offer a valuable resource for clinicians, researchers, and stakeholders involved in the dynamic landscape of AI in healthcare by providing a comprehensive overview of advances in multi-modality-driven AI disease diagnosis.

Author Contributions: X.X. and J.L.: conceptualization, writing original draft preparation. Z.Z., L.Z., H.W., C.S. and Y.C.: investigation, resources and writing—review and editing. Q.Z.: project administration, supervision. J.Y. and Y.P.: investigation and writing-review and editing. All authors have read and agreed to the published version of this manuscript.

Funding: This study is supported by a research project from the National Natural Science Foundation of China (No. 62266041).

Institutional Review Board Statement: Not applicable.

Informed Consent Statement: Not applicable.

Data Availability Statement: Data are available in a publicly accessible repository. Please refer to Table A1.

Conflicts of Interest: The authors declare no conflicts of interest.

Abbreviations

The following abbreviations are used in this manuscript:

AI	Artificial intelligence
AD	Alzheimer's disease
HCM	Hypertrophic cardiomyopathy
ECG	Electrocardiogram
EEG	Electroencephalograms
CT	Computed tomography
MRI	Magnetic resonance imaging
PET	Positron emission tomography
SVM	Support vector machine
RNN	Recurrent neural network
CNN	Convolutional neural network
ADNI	Alzheimer's disease neuroimaging initiative
UKB	United kingdom biobank
TCGA	The cancer genome atlas
BUSI	Breast ultrasound images
GEO	Gene expression omnibus
HCM	Hypertrophic cardiomyopathy
WHO	World health organization
SCD	Sunnybrook cardiac data
ACDC	Automated cardiac diagnosis challenge
DAIC-WOZ	Distress analysis interview corpus-wizard of OZ
MODMA	Multi-modal open dataset for mental-disorder analysis
WHO	World health organization
DICOM	Digital imaging and communications in medicine
PNG	Portable network graphics
ICA	Independent component analysis
DWT	Discrete wavelet transform
RFE	Recursive feature elimination
PCA	Principal component analysis
LDA	Linear discriminant analysis
CRF	Conditional random fields
LR	Logistic Regression
NB	Naive bayes
DT	Decision tree
LSTM	Long short-term memory
LM	Large Model
GPT	Generative re-trained transformer
PaLM	Pathways Language Model
SAM	Segment Anything Model
GLM	General Language Model
TP	True Positive
FP	False Positive
TN	True Negative
FN	False Negative
ACC	Accuracy

Sen	Sensitivity
Sp	Specificity
P	Precision
AUC-ROC	Area Under the ROC Curve
TPR	True Positive Rate
FPR	False Positive Rate
H-FCN	Hierarchical Fully Convolutional Network
MIL	Multiple Instance Learning
MCI	Mild Cognitive Impairment
CN	Cognitively normal
GRU	Gated Recurrent Unit
MLP	Multilayer Perceptrons
ASR	Automatic Speech Recognition
MM-SDPN	Multi-modal Stacked Denoising Predictive Network
RVFL	Random Vector Functional Link
SNP	Single nucleotide polymorphism
MAADf	Multi-modal AD diagnostic framework
CAD	Computer-Aided Diagnosis
NN	Neural Network
BF	Benign Fibroadenom
BPT	Benign Phyllodes Tumor
BTA	Benign Tubular Adenoma
MDC	Malignant Ductal Carcinoma
MLC	Malignant Lobular Carcinoma
MMC	Malignant Mucinous Carcinoma
MPC	Malignant Papillary Carcinoma
LASSO	Least Absolute Shrinkage and Selection Operation
RFE	Recursive Feature Elimination
mRNA	Maximum Relevance Minimum Redundancy
BUSI	Breast Ultrasound Images
EDLCDS-BCD	Integrated Deep Learning Clinical Decision Support System
USI	Ultrasound image
CKHA	Chaos Krill Herd Algorithm
CSO	Cat Swarm Optimization
LLM	Large Language Model
ELRDD	Ensemble Logistic Regression Model for Depression Detection
SR	Speaker Recognition
SER	Speech Emotion Recognition
AMI	Asymmetry Matrix Image
HFD	Higuchi's fractal dimension
SampEn	sample entropy
EM	Eye movement
VLDSP	Volume Local Directional Structure Pattern
DCNN	Deep convolutional neural networks
DNN	Deep neural network
Bi-LSTM	bidirectional long short-term memory
ZCR	Zero crossing rate
MFCC	mel frequency cepstral coefficient
LPC	linear prediction coefficient
LSP	line spectrum pair
PLP	perceptual linear prediction coefficient
LPF	Low profile filtered signal
LPR	Linear Prediction Residual Signal
HFVS	Homomorphically filtered speech source signal
ZFF	Zero frequency filtered signal
PTB	Physiobank
CVD	Cardiovascular Diseases
HCM	Hypertrophic Cardiomyopathy

IoMT	Internet of Things
Rec-CONVnet	Recurrent CONVolutional neural network
SWCDTO	Social Water Cycle Driving Training Optimization
SCAD	Stable coronary artery disease patients
CAD-non HF	CAD without HF
CAD-HF	CAD complicated with HF
MI	Myocardial infarction
UCSF	University of California, San Francisco
HMM	Hidden Markov Model
PTB	public Physiobank
AI-ECG	AI-enhanced ECG
PPV	Positive Predictive Value
NPV	Negative Predictive Value
POS	Part Of Speech
MoCap	Motion Capture
fMRI	functional MRI
GLCM	Gray Level Co-occurrence Matrix
JME	Juvenile Myoclonic Epilepsy
HARDI	High Angular Resolution Diffusion Imaging
NODDI	Neurite Orientation Dispersion and Density Imaging
PNES	psychogenic non-epileptic seizures
IGE	Idiopathic Generalized Epileps
DIndRNN	Dense IndRNN with attentio
IndRNN	Independently improved RNN
TLE	Temporal Lobe Epileps
PDC	Personal demographic and cognitive data
PE	Partial epilepsy
GE	Generalized epilepsy
UE	Unclassified epilepsy
TCGA	The Cancer Genome Atlas
ADNI	Alzheimer's Disease Neuroimaging initiative
OASIS-3	Open Access Series of Imaging Studies-3
AIBL	Australian Imaging, Biomarker and Lifestyle
SCD	Sunnybrook Cardiac Data
SAFHS	San Antonio Family Heart Study
BUSI	Breast Ultrasound Images
GEO	Gene Expression Omnibus
TLGS	Tehran Lipid and Glucose Study
SCD	Sunnybrook Cardiac Data
ACDC	Automated Cardiac Diagnosis Challenge
DAIC-WOZ	Distress Analysis Interview Corpus-Wizard of OZ
MODMA	Multi-modal Open Dataset for Mental-disorder Analysis

Appendix A

Table A1. Multi-modal datasets of diagnosis task for different disease.

Dataset	Year	Disease	Modality	Link
Alzheimer's Disease Neuroimaging initiative (ADNI)	2003	AD	Image-based	https://adni.loni.usc.edu/ (accessed on 29 November 2023)
Open Access Series of Imaging Studies-3 (OASIS-3)	2019	AD	Image-based	https://www.oasis-brains.org/ (accessed on 29 November 2023)
Australian Imaging, Biomarker and Lifestyle (AIBL)	2006	AD	Image-based	https://aibl.org.au/ (accessed on 29 November 2023)
Sunnybrook Cardiac Data (SCD)	2009	AD	Image-based	https://www.cardiacatlas.org/sunnybrook-cardiac-data/ (accessed on 29 November 2023)

Table A1. *Cont.*

Dataset	Year	Disease	Modality	Link
Automated Cardiac Diagnosis Challenge (ACDC)	2018	HCM	Image-based	https://www.creatis.insa-lyon.fr/Challenge/acdc/ (accessed on 29 November 2023)
Cardiac CT Segmentation Challenge	2020	HCM	Image-based	https://www.ub.edu/mnms/ (accessed on 29 November 2023)
Congenital Heart Disease (CHD)	2013	Heart disease	Image-based	https://www.data.gov.uk/dataset/f13fbd0e-fc8a-4d42-82ef-d40f930e4b70/congenital-heart-disease-chd (accessed on 29 November 2023)
AMRG Cardiac Atlas	-	Heart disease	Image-based	https://www.cardiacatlas.org/amrg-cardiac-atlas/ (accessed on 29 November 2023)
Multi-Ethnic Study of Atherosclerosis	2002	Heart disease	Image-based	https://www.cardiacatlas.org/mesa/ (accessed on 29 November 2023)
Breast Ultrasound Images (BUSI)	2018	Breast cancer	Image-based	https://scholar.cu.edu.eg/?q=afahmy/pages/dataset (accessed on 29 November 2023)
Breast Cancer Coimbra Dataset	2013	Breast cancer	Text-based	https://archive.ics.uci.edu/ml/datasets/ (accessed on 29 November 2023)
Oncoshare Breast Cancer Database	2016	Breast cancer	Text-based	https://med.stanford.edu/oncoshare.html (accessed on 29 November 2023)
I2B2 NLP Research Database	2014	Breast cancer, Heart disease, Depression	Text-based	https://www.i2b2.org/NLP/DataSets/Main.php (accessed on 29 November 2023)
MIMIC-III Critical Care Database	2012	Heart disease, Depression	Text-based	https://github.com/MIT-LCP/mimic-code (accessed on 29 November 2023)
eDiseases Dataset	2018	Breast cancer, Heart disease, Depression, AD	Text-based	https://zenodo.org/record/1479354#.Y8P4kexBy3I (accessed on 29 November 2023)
National Alzheimer's Coordinating Center (NACC)	1999	AD	Text-based	https://naccdata.org/ (accessed on 29 November 2023)
UK Biobank database	2010	Breast cancer, Heart disease, AD, Depression	Text-based	https://www.ukbiobank.ac.uk/ (accessed on 29 November 2023)
DementiaBank	2003	AD	Text-based	https://dementia.talkbank.org/ (accessed on 29 November 2023)
SAHS	2020	Breast cancer, Heart disease	Text-based	https://www.ncbi.nlm.nih.gov/projects/gap/cgi-bin/study.cgi?study_id=phs001215.v3.p2 (accessed on 29 November 2023)
TLGS	1999	Heart disease	Text-based	https://endocrine.ac.ir/page/Tehran-Lipid-and-Glucose-Study-TLGS (accessed on 29 November 2023)
Acute Myocardial Infarction Dataset of World Health Organization (WHO)	2023	Heart disease	Text-based	http://www.who.int/ (accessed on 29 November 2023)
UCI machine learning repository	2023	Heart disease	Text-based	https://archive.ics.uci.edu/dataset/45/heart+disease (accessed on 29 November 2023)

Table A1. *Cont.*

Dataset	Year	Disease	Modality	Link
Depression text dataset	2023	Depression	Text-based	https://www.Depression-texts.com/ (accessed on 29 November 2023)
The Cancer Genome Atlas (TCGA)	2006	Breast cancer	Gene-based	https://www.cancer.gov/ccg/research/genome-sequencing/tcga (accessed on 29 November 2023)
Gene Expression Omnibus (GEO)	2000	Breast cancer	Gene-based	http://www.ncbi.nlm.nih.gov/geo (accessed on 29 November 2023)
Online Mendelian Inheritance in Man (OMIM)	1966	Breast cancer	Gene-based	https://omim.org/ (accessed on 29 November 2023)
GenBank	1982	Breast cancer	Gene-based	https://www.ncbi.nlm.nih.gov/genbank/ (accessed on 29 November 2023)
Human Gene Mutation Database (HGMD)	1996	Breast cancer	Gene-based	http://www.hgmd.org/ (accessed on 29 November 2023)
Genome Aggregation Database (genoAD)	2016	Breast cancer	Gene-based	https://gnomad.broadinstitute.org/ (accessed on 29 November 2023)
Chinese Millionome Database (CMDB)	2017	Breast cancer	Gene-based	https://db.cngb.org/cmdb (accessed on 29 November 2023)
University of California, Santa Cruz (UCSC)	2000	Breast cancer	Gene-based	http://www.genome.ucsc.edu/ (accessed on 29 November 2023)
Distress Analysis Interview Corpus-Wizard of OZ (DAIC-WOZ)	2014	Depression	Speech-based	https://dcapswoz.ict.usc.edu/ (accessed on 29 November 2023)
Multi-modal Open Dataset for Mental-disorder Analysis (MODMA)	2020	Depression	Speech-based, ECG-based	http://modma.lzu.edu.cn/data/index/ (accessed on 29 November 2023)
Depression and Anxiety Crowdsourced corpus (DEPAC)	2023	Depression	Speech-based	https://www.mturk.com (accessed on 29 November 2023)
Bipolar Disorder Corpus	2018	Depression	Speech-based, ECG-based	https://www.aconf.org/conf_153173.html (accessed on 29 November 2023)
AVEC2014	2014	Depression	Speech-based, Image-based	http://avec2014-db.sspnet.eu/ (accessed on 29 November 2023)
AVEC2013	2013	Depression	Speech-based, Image-based	http://avec2013-db.sspnet.eu/ (accessed on 29 November 2023)
ADReSS	2020	AD	Speech-based	https://luzs.gitlab.io/adress/ (accessed on 29 November 2023)
AVEC2019	2019	Depression	Speech-based	https://www.ihp-lab.org/resources/ (accessed on 29 November 2023)
ADReSS-M	2023	AD	Speech-based, Text-based	https://2023.ieeeicassp.org/ (accessed on 29 November 2023)
ADReSSo	2021	AD	Speech-based	https://luzs.gitlab.io/adresso-2021/ (accessed on 29 November 2023)
The Carolinas Conversation Collection (CCC)	2011	AD	Speech-based, Image-based	https://www.degruyter.com/how-access-works (accessed on 29 November 2023)
ERP Core	2016	AD	EEG-based	https://osf.io/thsqg/ (accessed on 29 November 2023)
EEG Epilepsy Datasets	2016	Epilepsy	EEG-based	https://www.researchgate.net/publication/308719109_EEG_Epilepsy_Datasets (accessed on 29 November 2023)

Table A1. *Cont.*

Dataset	Year	Disease	Modality	Link
CHB-MIT Scalp EEG Database	2010	Epilepsy	EEG-based	https://physionet.org/content/chbmit/1.0.0/ (accessed on 29 November 2023)
Kaggle	2018	Epilepsy	EEG-based	https://www.kaggle.com/code/harunshimanto/machine-learning-algorithms-for-epileptic-seizures (accessed on 29 November 2023)
EEG_128channels_ERP_lanzhou_2015	2015	Depression	EEG-based	http://modma.lzu.edu.cn/data/application/ (accessed on 29 November 2023)
ECG-ID Database	2014	Heart disease	EEG-based	https://physionet.org/content/ecgiddb/1.0.0/ (accessed on 29 November 2023)
Common Standards for Electrocardiography (CSE) database	1980	Heart disease	EEG-based	http://www.escardio.org/Pages/index.aspx (accessed on 29 November 2023)
European ST-T Database	2009	Heart disease	EEG-based	https://physionet.org/content/edb/1.0.0/ (accessed on 29 November 2023)
Sudden Cardiac Death Holter Database	2004	Heart disease	EEG-based	http://physionet.org/physiobank/database/sddb/ (accessed on 29 November 2023)
Bonn EEG time series database	2001	Epilepsy	EEG-based	https://www.ukbonn.de/epileptologie/ag-lehnertz-downloads/ (accessed on 29 November 2023)
Temple University EEG corpus	2000	Epilepsy	EEG-based	https://isip.piconepress.com/projects/tuh_eeg/html/downloads.shtml (accessed on 29 November 2023)

References

1. Anto, S.; Chandramathi, S. Supervised machine learning approaches for medical data set classification—A review. *Int. J. Comput. Sci. Trends Technol.* **2011**, *2*, 234–240.
2. Marcus, D.S.; Wang, T.H.; Parker, J.; Csernansky, J.G.; Morris, J.C.; Buckner, R.L. Open Access Series of Imaging Studies (OASIS): Cross-sectional MRI data in young, middle aged, nondemented, and demented older adults. *J. Cogn. Neurosci.* **2007**, *19*, 1498–1507. [CrossRef]
3. Allen, N.; Sudlow, C.; Downey, P.; Peakman, T.; Danesh, J.; Elliott, P.; Gallacher, J.; Green, J.; Matthews, P.; Pell, J.; et al. UK Biobank: Current status and what it means for epidemiology. *Health Policy Technol.* **2012**, *1*, 123–126. [CrossRef]
4. Littlejohns, T.J.; Sudlow, C.; Allen, N.E.; Collins, R. UK Biobank: Opportunities for cardiovascular research. *Eur. Heart J.* **2019**, *40*, 1158–1166. [CrossRef] [PubMed]
5. Sudlow, C.; Gallacher, J.; Allen, N.; Beral, V.; Burton, P.; Danesh, J.; Downey, P.; Elliott, P.; Green, J.; Landray, M.; et al. UK biobank: An open access resource for identifying the causes of a wide range of complex diseases of middle and old age. *PLoS Med.* **2015**, *12*, e1001779. [CrossRef] [PubMed]
6. Wilkinson, T.; Schnier, C.; Bush, K.; Rannikmäe, K.; Henshall, D.E.; Lerpiniere, C.; Allen, N.E.; Flaig, R.; Russ, T.C.; Bathgate, D.; et al. Identifying dementia outcomes in UK Biobank: A validation study of primary care, hospital admissions and mortality data. *Eur. J. Epidemiol.* **2019**, *34*, 557–565. [CrossRef] [PubMed]
7. Mueller, S.G.; Weiner, M.W.; Thal, L.J.; Petersen, R.C.; Jack, C.; Jagust, W.; Trojanowski, J.Q.; Toga, A.W.; Beckett, L. The Alzheimer's disease neuroimaging initiative. *Neuroimaging Clin.* **2005**, *15*, 869–877. [CrossRef]
8. de Vent, N.R.; Agelink van Rentergem, J.A.; Schmand, B.A.; Murre, J.M.; Consortium, A.; Huizenga, H.M. Advanced Neuropsychological Diagnostics Infrastructure (ANDI): A normative database created from control datasets. *Front. Psychol.* **2016**, *7*, 1601. [CrossRef] [PubMed]
9. Vapnik, V.N. An overview of statistical learning theory. *IEEE Trans. Neural Netw.* **1999**, *10*, 988–999. [CrossRef] [PubMed]
10. O'Shea, K.; Nash, R. An introduction to convolutional neural networks. *arXiv* **2015**, arXiv:1511.08458.
11. Vaswani, A.; Shazeer, N.; Parmar, N.; Uszkoreit, J.; Jones, L.; Gomez, A.N.; Kaiser, Ł.; Polosukhin, I. Attention is all you need. *Adv. Neural Inf. Process. Syst.* **2017**, *30*, 1–11.
12. Hochreiter, S.; Schmidhuber, J. Long short-term memory. *Neural Comput.* **1997**, *9*, 1735–1780. [CrossRef]
13. Ouyang, L.; Wu, J.; Jiang, X.; Almeida, D.; Wainwright, C.; Mishkin, P.; Zhang, C.; Agarwal, S.; Slama, K.; Ray, A.; et al. Training language models to follow instructions with human feedback. *Adv. Neural Inf. Process. Syst.* **2022**, *35*, 27730–27744.

14. Achiam, J.; Adler, S.; Agarwal, S.; Ahmad, L.; Akkaya, I.; Aleman, F.L.; Almeida, D.; Altenschmidt, J.; Altman, S.; Anadkat, S.; et al. Gpt-4 technical report. *arXiv* **2023**, arXiv:2303.08774.
15. Touvron, H.; Lavril, T.; Izacard, G.; Martinet, X.; Lachaux, M.A.; Lacroix, T.; Rozière, B.; Goyal, N.; Hambro, E.; Azhar, F.; et al. Llama: Open and efficient foundation language models. *arXiv* **2023**, arXiv:2302.13971.
16. Touvron, H.; Martin, L.; Stone, K.; Albert, P.; Almahairi, A.; Babaei, Y.; Bashlykov, N.; Batra, S.; Bhargava, P.; Bhosale, S.; et al. Llama 2: Open foundation and fine-tuned chat models. *arXiv* **2023**, arXiv:2307.09288.
17. Chowdhery, A.; Narang, S.; Devlin, J.; Bosma, M.; Mishra, G.; Roberts, A.; Barham, P.; Chung, H.W.; Sutton, C.; Gehrmann, S.; et al. Palm: Scaling language modeling with pathways. *J. Mach. Learn. Res.* **2023**, *24*, 1–113.
18. Taori, R.; Gulrajani, I.; Zhang, T.; Dubois, Y.; Li, X.; Guestrin, C.; Liang, P.; Hashimoto, T.B. Stanford Alpaca: An Instruction-Following Llama Model (2023). Available online: https://github.com/tatsu-lab/stanford_alpaca (accessed on 29 November 2023).
19. Singhal, K.; Azizi, S.; Tu, T.; Mahdavi, S.S.; Wei, J.; Chung, H.W.; Scales, N.; Tanwani, A.; Cole-Lewis, H.; Pfohl, S.; et al. Large language models encode clinical knowledge. *arXiv* **2022**, arXiv:2212.13138.
20. Singhal, K.; Tu, T.; Gottweis, J.; Sayres, R.; Wulczyn, E.; Hou, L.; Clark, K.; Pfohl, S.; Cole-Lewis, H.; Neal, D.; et al. Towards expert-level medical question answering with large language models. *arXiv* **2023**, arXiv:2305.09617.
21. Wang, H.; Liu, C.; Xi, N.; Qiang, Z.; Zhao, S.; Qin, B.; Liu, T. Huatuo: Tuning llama model with chinese medical knowledge. *arXiv* **2023**, arXiv:2304.06975.
22. Yunxiang, L.; Zihan, L.; Kai, Z.; Ruilong, D.; You, Z. Chatdoctor: A medical chat model fine-tuned on llama model using medical domain knowledge. *arXiv* **2023**, arXiv:2303.14070.
23. Xiong, H.; Wang, S.; Zhu, Y.; Zhao, Z.; Liu, Y.; Wang, Q.; Shen, D. Doctorglm: Fine-tuning your chinese doctor is not a herculean task. *arXiv* **2023**, arXiv:2304.01097.
24. Chen, Y.; Wang, Z.; Xing, X.; Xu, Z.; Fang, K.; Wang, J.; Li, S.; Wu, J.; Liu, Q.; Xu, X.; et al. BianQue: Balancing the Questioning and Suggestion Ability of Health LLMs with Multi-turn Health Conversations Polished by ChatGPT. *arXiv* **2023**, arXiv:2310.15896.
25. Luo, R.; Sun, L.; Xia, Y.; Qin, T.; Zhang, S.; Poon, H.; Liu, T.Y. BioGPT: Generative pre-trained transformer for biomedical text generation and mining. *Brief. Bioinform.* **2022**, *23*, bbac409. [CrossRef]
26. Tomczak, K.; Czerwińska, P.; Wiznerowicz, M. Review The Cancer Genome Atlas (TCGA): An immeasurable source of knowledge. *Contemp. Oncol. Onkol.* **2015**, *2015*, 68–77. [CrossRef]
27. Arar, N.; Nath, S.; Thameem, F.; Bauer, R.; Voruganti, S.; Comuzzie, A.; Cole, S.; Blangero, J.; MacCluer, J.; Abboud, H. Genome-wide scans for microalbuminuria in Mexican Americans: The San Antonio family heart study. *Genet. Med.* **2007**, *9*, 80–87. [CrossRef] [PubMed]
28. Al-Dhabyani, W.; Gomaa, M.; Khaled, H.; Fahmy, A. Dataset of breast ultrasound images. *Data Brief* **2020**, *28*, 104863. [CrossRef] [PubMed]
29. Edgar, R.; Domrachev, M.; Lash, A.E. Gene Expression Omnibus: NCBI gene expression and hybridization array data repository. *Nucleic Acids Res.* **2002**, *30*, 207–210. [CrossRef]
30. Azizi, F.; Rahmani, M.; Emami, H.; Mirmiran, P.; Hajipour, R.; Madjid, M.; Ghanbili, J.; Ghanbarian, A.; Mehrabi, J.; Saadat, N.; et al. Cardiovascular risk factors in an Iranian urban population: Tehran lipid and glucose study (phase 1). *Soz. Präventivmedizin* **2002**, *47*, 408–426. [CrossRef]
31. Radau, P.; Lu, Y.; Connelly, K.; Paul, G.; Dick, A.J.; Wright, G.A. Evaluation framework for algorithms segmenting short axis cardiac MRI. *MIDAS J.* **2009**, *47*, 408–426. [CrossRef]
32. Bernard, O.; Lalande, A.; Zotti, C.; Cervenansky, F.; Yang, X.; Heng, P.A.; Cetin, I.; Lekadir, K.; Camara, O.; Ballester, M.A.G.; et al. Deep learning techniques for automatic MRI cardiac multi-structures segmentation and diagnosis: Is the problem solved? *IEEE Trans. Med. Imaging* **2018**, *37*, 2514–2525. [CrossRef] [PubMed]
33. Gratch, J.; Artstein, R.; Lucas, G.M.; Stratou, G.; Scherer, S.; Nazarian, A.; Wood, R.; Boberg, J.; DeVault, D.; Marsella, S.; et al. The distress analysis interview corpus of human and computer interviews. In Proceedings of the Ninth International Conference on Language Resources and Evaluation (LREC'14), Reykjavik, Iceland, 26–31 May 2014; pp. 3123–3128.
34. Cai, H.; Yuan, Z.; Gao, Y.; Sun, S.; Li, N.; Tian, F.; Xiao, H.; Li, J.; Yang, Z.; Li, X.; et al. A multi-modal open dataset for mental-disorder analysis. *Sci. Data* **2022**, *9*, 178. [CrossRef] [PubMed]
35. Goldberger, A.L.; Amaral, L.A.; Glass, L.; Hausdorff, J.M.; Ivanov, P.C.; Mark, R.G.; Mietus, J.E.; Moody, G.B.; Peng, C.K.; Stanley, H.E. PhysioBank, PhysioToolkit, and PhysioNet: Components of a new research resource for complex physiologic signals. *Circulation* **2000**, *101*, 215–220. [CrossRef]
36. Andrzejak, R.G.; Lehnertz, K.; Mormann, F.; Rieke, C.; David, P.; Elger, C.E. Indications of nonlinear deterministic and finite-dimensional structures in time series of brain electrical activity: Dependence on recording region and brain state. *Phys. Rev. E* **2001**, *64*, 061907. [CrossRef] [PubMed]
37. Obeid, I.; Picone, J. The temple university hospital EEG data corpus. *Front. Neurosci.* **2016**, *10*, 196. [CrossRef]
38. Roberts, L.; Lanes, S.; Peatman, O.; Assheton, P. The importance of SNOMED CT concept specificity in healthcare analytics. *Health Inf. Manag. J.* **2023**, *1*, 1. [CrossRef]
39. Benson, T.; Grieve, G. *Principles of Health Interoperability: FHIR, HL7 and SNOMED CT*; Springer: Berlin/Heidelberg, Germany, 2021.

40. Li, X.; Morgan, P.S.; Ashburner, J.; Smith, J.C.; Rorden, C. The first step for neuroimaging data analysis: DICOM to NIfTI conversion. *J. Neurosci. Methods* **2016**, *264*, 47–56. [CrossRef]
41. Lafferty, J.D.; McCallum, A.; Pereira, F. Conditional Random Fields: Probabilistic Models for Segmenting and Labeling Sequence Data. In Proceedings of the International Conference on Machine Learning, Beijing, China, 21–26 June 2001; Volume 22, pp. 1–28.
42. Shah, Z.; Qi, S.A.; Wang, F.; Farrokh, M.; Tasnim, M.; Stroulia, E.; Greiner, R.; Plitsis, M.; Katsamanis, A. Exploring Language-Agnostic Speech Representations Using Domain Knowledge for Detecting Alzheimer's Dementia. In Proceedings of the ICASSP 2023—2023 IEEE International Conference on Acoustics, Speech and Signal Processing (ICASSP), Rhode Island, Greece, 4–6 June 2023; pp. 1–2.
43. Martinc, M.; Haider, F.; Pollak, S.; Luz, S. Temporal integration of text transcripts and acoustic features for Alzheimer's diagnosis based on spontaneous speech. *Front. Aging Neurosci.* **2021**, *13*, 642647. [CrossRef]
44. Zhang, L.; Liu, Y.; Wang, K.; Ou, X.; Zhou, J.; Zhang, H.; Huang, M.; Du, Z.; Qiang, S. Integration of machine learning to identify diagnostic genes in leukocytes for acute myocardial infarction patients. *J. Transl. Med.* **2023**, *21*, 761. [CrossRef]
45. Ho, T.K. Random decision forests. In Proceedings of the 3rd International Conference on Document Analysis and Recognition, Montreal, QC, Canada, 14–16 August 1995; Volume 1, pp. 278–282.
46. Zeng, Q.T.; Goryachev, S.; Weiss, S.; Sordo, M.; Murphy, S.N.; Lazarus, R. Extracting principal diagnosis, co-morbidity and smoking status for asthma research: Evaluation of a natural language processing system. *BMC Med. Inform. Decis. Mak.* **2006**, *6*, 30. [CrossRef] [PubMed]
47. Zhu, J.; Wang, Y.; La, R.; Zhan, J.; Niu, J.; Zeng, S.; Hu, X. Multimodal mild depression recognition based on EEG-EM synchronization acquisition network. *IEEE Access* **2019**, *7*, 28196–28210. [CrossRef]
48. Anbarasi, M.; Anupriya, E.; Iyengar, N. Enhanced prediction of heart disease with feature subset selection using genetic algorithm. *Int. J. Eng. Sci. Technol.* **2010**, *2*, 5370–5376.
49. Yang, X.; Liu, G.; Feng, G.; Bu, D.; Wang, P.; Jiang, J.; Chen, S.; Yang, Q.; Zhang, Y.; Man, Z.; et al. GeneCompass: Deciphering Universal Gene Regulatory Mechanisms with Knowledge-Informed Cross-Species Foundation Model. *bioRxiv* **2023**., 1–8. [CrossRef]
50. Sherafatian, M. Tree-based machine learning algorithms identified minimal set of miRNA biomarkers for breast cancer diagnosis and molecular subtyping. *Gene* **2018**, *677*, 111–118. [CrossRef]
51. Dong, D.; Fu, G.; Li, J.; Pei, Y.; Chen, Y. An unsupervised domain adaptation brain CT segmentation method across image modalities and diseases. *Expert Syst. Appl.* **2022**, *207*, 118016. [CrossRef]
52. Li, H.; Habes, M.; Fan, Y. Deep ordinal ranking for multi-category diagnosis of Alzheimer's disease using hippocampal MRI data. *arXiv* **2017**, arXiv:1709.01599.
53. Cheng, D.; Liu, M. Combining convolutional and recurrent neural networks for Alzheimer's disease diagnosis using PET images. In Proceedings of the 2017 IEEE International Conference on Imaging Systems and Techniques (IST), Beijing, China, 18–20 October 2017; pp. 1–5.
54. Liu, G.; Wei, Y.; Xie, Y.; Li, J.; Qiao, L.; Yang, J. A computer-aided system for ocular myasthenia gravis diagnosis. *Tsinghua Sci. Technol.* **2021**, *26*, 749–758. [CrossRef]
55. Lopez-de Ipina, K.; Martinez-de Lizarduy, U.; Calvo, P.M.; Mekyska, J.; Beitia, B.; Barroso, N.; Estanga, A.; Tainta, M.; Ecay-Torres, M. Advances on automatic speech analysis for early detection of Alzheimer disease: A non-linear multi-task approach. *Curr. Alzheimer Res.* **2018**, *15*, 139–148. [CrossRef] [PubMed]
56. Guan, Y.; Wen, P.; Li, J.; Zhang, J.; Xie, X. Deep Learning Blockchain Integration Framework for Ureteropelvic Junction Obstruction Diagnosis Using Ultrasound Images. *Tsinghua Sci. Technol.* **2023**, *29*, 1–12. [CrossRef]
57. Zhang, J.; Liu, B.; Wu, J.; Wang, Z.; Li, J. DeepCAC: A deep learning approach on DNA transcription factors classification based on multi-head self-attention and concatenate convolutional neural network. *BMC Bioinform.* **2023**, *24*, 345. [CrossRef] [PubMed]
58. Islam, M.M.; Huang, S.; Ajwad, R.; Chi, C.; Wang, Y.; Hu, P. An integrative deep learning framework for classifying molecular subtypes of breast cancer. *Comput. Struct. Biotechnol. J.* **2020**, *18*, 2185–2199. [CrossRef]
59. Zhu, W.; Sun, L.; Huang, J.; Han, L.; Zhang, D. Dual attention multi-instance deep learning for Alzheimer's disease diagnosis with structural MRI. *IEEE Trans. Med. Imaging* **2021**, *40*, 2354–2366. [CrossRef]
60. Chen, Y.; Wang, H.; Zhang, G.; Liu, X.; Huang, W.; Han, X.; Li, X.; Martin, M.; Tao, L. Contrastive Learning for Prediction of Alzheimer's Disease Using Brain 18F-FDG PET. *IEEE J. Biomed. Health Inform.* **2022**, *27*, 1735–1746. [CrossRef] [PubMed]
61. Zhu, Z.; Li, J.; Zhao, Q.; Akhtar, F. A dictionary-guided attention network for biomedical named entity recognition in Chinese electronic medical records. *Expert Syst. Appl.* **2023**, *231*, 120709. [CrossRef]
62. Bommasani, R.; Hudson, D.A.; Adeli, E.; Altman, R.; Arora, S.; von Arx, S.; Bernstein, M.S.; Bohg, J.; Bosselut, A.; Brunskill, E.; et al. On the opportunities and risks of foundation models. *arXiv* **2021**, arXiv:2108.07258.
63. Zhou, C.; Li, Q.; Li, C.; Yu, J.; Liu, Y.; Wang, G.; Zhang, K.; Ji, C.; Yan, Q.; He, L.; et al. A comprehensive survey on pretrained foundation models: A history from bert to chatgpt. *arXiv* **2023**, arXiv:2302.09419.
64. Brown, T.; Mann, B.; Ryder, N.; Subbiah, M.; Kaplan, J.D.; Dhariwal, P.; Neelakantan, A.; Shyam, P.; Sastry, G.; Askell, A.; et al. Language models are few-shot learners. *Adv. Neural Inf. Process. Syst.* **2020**, *33*, 1877–1901.
65. Chung, H.W.; Hou, L.; Longpre, S.; Zoph, B.; Tay, Y.; Fedus, W.; Li, Y.; Wang, X.; Dehghani, M.; Brahma, S.; et al. Scaling instruction-finetuned language models. *arXiv* **2022**, arXiv:2210.11416.

66. Zhang, C.; Liu, L.; Cui, Y.; Huang, G.; Lin, W.; Yang, Y.; Hu, Y. A Comprehensive Survey on Segment Anything Model for Vision and Beyond. *arXiv* **2023**, arXiv:2305.08196.
67. Garza, A.; Mergenthaler-Canseco, M. TimeGPT-1. *arXiv* **2023**, arXiv:2310.03589.
68. Thawkar, O.; Shaker, A.; Mullappilly, S.S.; Cholakkal, H.; Anwer, R.M.; Khan, S.; Laaksonen, J.; Khan, F.S. Xraygpt: Chest radiographs summarization using medical vision-language models. *arXiv* **2023**, arXiv:2306.07971.
69. Tharwat, A. Classification assessment methods. *Appl. Comput. Inform.* **2020**, *17*, 168–192. [CrossRef]
70. Lian, C.; Liu, M.; Zhang, J.; Shen, D. Hierarchical fully convolutional network for joint atrophy localization and Alzheimer's disease diagnosis using structural MRI. *IEEE Trans. Pattern Anal. Mach. Intell.* **2018**, *42*, 880–893. [CrossRef] [PubMed]
71. Baydargil, H.B.; Park, J.S.; Kang, D.Y. Anomaly analysis of Alzheimer's disease in PET images using an unsupervised adversarial deep learning model. *Appl. Sci.* **2021**, *11*, 2187. [CrossRef]
72. Ning, Z.; Xiao, Q.; Feng, Q.; Chen, W.; Zhang, Y. Relation-induced multi-modal shared representation learning for Alzheimer's disease diagnosis. *IEEE Trans. Med. Imaging* **2021**, *40*, 1632–1645. [CrossRef] [PubMed]
73. Hernández, J.B.A.; Pulido, M.L.B.; Bordón, J.M.G.; Ballester, M.Á.F.; González, C.M.T. Speech evaluation of patients with alzheimer's disease using an automatic interviewer. *Expert Syst. Appl.* **2022**, *192*, 116386. [CrossRef]
74. Yu, B.; Quatieri, T.F.; Williamson, J.R.; Mundt, J.C. Cognitive impairment prediction in the elderly based on vocal biomarkers. In Proceedings of the Sixteenth Annual Conference of the International Speech Communication Association, Dresden, Germany, 6–10 September 2015; Volume 2015, pp. 3734–3748.
75. Liu, Z.; Guo, Z.; Ling, Z.; Li, Y. Detecting Alzheimer's disease from speech using neural networks with bottleneck features and data augmentation. In Proceedings of the ICASSP 2021—2021 IEEE International Conference on Acoustics, Speech and Signal Processing (ICASSP), Toronto, ON, Canada, 6–11 June 2021; pp. 7323–7327.
76. Bertini, F.; Allevi, D.; Lutero, G.; Calzà, L.; Montesi, D. An automatic Alzheimer's disease classifier based on spontaneous spoken English. *Comput. Speech Lang.* **2022**, *72*, 101298. [CrossRef]
77. Park, D.S.; Chan, W.; Zhang, Y.; Chiu, C.C.; Zoph, B.; Cubuk, E.D.; Le, Q.V. Specaugment: A simple data augmentation method for automatic speech recognition. *arXiv* **2019**, arXiv:1904.08779.
78. Freitag, M.; Amiriparian, S.; Pugachevskiy, S.; Cummins, N. audeep: Unsupervised learning of representations from audio with deep recurrent neural networks. *J. Mach. Learn. Res.* **2018**, *18*, 1–5.
79. Shi, J.; Zheng, X.; Li, Y.; Zhang, Q.; Ying, S. Multimodal neuroimaging feature learning with multimodal stacked deep polynomial networks for diagnosis of Alzheimer's disease. *IEEE J. Biomed. Health Inform.* **2017**, *22*, 173–183. [CrossRef]
80. Sharma, R.; Goel, T.; Tanveer, M.; Suganthan, P.; Razzak, I.; Murugan, R. Conv-ERVFL: Convolutional Neural Network Based Ensemble RVFL Classifier for Alzheimer's Disease Diagnosis. *IEEE J. Biomed. Health Inform.* **2022**, *27*, 4995–5003. [CrossRef]
81. Zhou, T.; Liu, M.; Thung, K.H.; Shen, D. Latent representation learning for Alzheimer's disease diagnosis with incomplete multi-modality neuroimaging and genetic data. *IEEE Trans. Med. Imaging* **2019**, *38*, 2411–2422. [CrossRef] [PubMed]
82. Cai, H.; Huang, X.; Liu, Z.; Liao, W.; Dai, H.; Wu, Z.; Zhu, D.; Ren, H.; Li, Q.; Liu, T.; et al. Exploring Multimodal Approaches for Alzheimer's Disease Detection Using Patient Speech Transcript and Audio Data. *arXiv* **2023**, arXiv:2307.02514.
83. Mei, K.; Ding, X.; Liu, Y.; Guo, Z.; Xu, F.; Li, X.; Naren, T.; Yuan, J.; Ling, Z. The Ustc System for Adress-m Challenge. In Proceedings of the ICASSP 2023—2023 IEEE International Conference on Acoustics, Speech and Signal Processing (ICASSP), Rhode Island, Greece, 4–6 June 2023; pp. 1–2.
84. Agbavor, F.; Liang, H. Artificial Intelligence-Enabled End-To-End Detection and Assessment of Alzheimer's Disease Using Voice. *Brain Sci.* **2022**, *13*, 28. [CrossRef] [PubMed]
85. Jo, T.; Nho, K.; Bice, P.; Saykin, A.J.; Initiative, A.D.N. Deep learning-based identification of genetic variants: Application to Alzheimer's disease classification. *Brief. Bioinform.* **2022**, *23*, bbac022. [CrossRef]
86. Xu, L.; Liang, G.; Liao, C.; Chen, G.D.; Chang, C.C. An efficient classifier for Alzheimer's disease genes identification. *Molecules* **2018**, *23*, 3140. [CrossRef]
87. De Velasco Oriol, J.; Vallejo, E.E.; Estrada, K.; Taméz Peña, J.G.; Disease Neuroimaging Initiative, T.A. Benchmarking machine learning models for late-onset alzheimer's disease prediction from genomic data. *BMC Bioinform.* **2019**, *20*, 709. [CrossRef]
88. Park, C.; Ha, J.; Park, S. Prediction of Alzheimer's disease based on deep neural network by integrating gene expression and DNA methylation dataset. *Expert Syst. Appl.* **2020**, *140*, 112873. [CrossRef]
89. Golovanevsky, M.; Eickhoff, C.; Singh, R. Multimodal attention-based deep learning for Alzheimer's disease diagnosis. *J. Am. Med Inform. Assoc.* **2022**, *29*, 2014–2022. [CrossRef] [PubMed]
90. Djemili, R.; Bourouba, H.; Korba, M.A. Application of empirical mode decomposition and artificial neural network for the classification of normal and epileptic EEG signals. *Biocybern. Biomed. Eng.* **2016**, *36*, 285–291. [CrossRef]
91. Pandya, R.; Nadiadwala, S.; Shah, R.; Shah, M. Buildout of methodology for meticulous diagnosis of K-complex in EEG for aiding the detection of Alzheimer's by artificial intelligence. *Augment. Hum. Res.* **2020**, *5*, 3. [CrossRef]
92. Kim, D.; Kim, K. Detection of early stage Alzheimer's disease using EEG relative power with deep neural network. In Proceedings of the 2018 40th Annual International Conference of the IEEE Engineering in Medicine and Biology Society (EMBC), Honolulu, HI, USA, 18–21 July 2018; pp. 352–355.
93. Deepthi, L.D.; Shanthi, D.; Buvana, M. An intelligent Alzheimer's disease prediction using convolutional neural network (CNN). *Int. J. Adv. Res. Eng. Technol. (IJARET)* **2020**, *11*, 12–22.

94. Al-Antari, M.A.; Han, S.M.; Kim, T.S. Evaluation of deep learning detection and classification towards computer-aided diagnosis of breast lesions in digital X-ray mammograms. *Comput. Methods Programs Biomed.* **2020**, *196*, 105584. [CrossRef]
95. Redmon, J.; Divvala, S.K.; Girshick, R.B.; Farhadi, A. You Only Look Once: Unified, Real-Time Object Detection. In Proceedings of the 2016 IEEE Conference on Computer Vision and Pattern Recognition (CVPR), Las Vegas, NV, USA, 27–30 June 2015; pp. 779–788.
96. Brhane Hagos, Y.; Gubern Mérida, A.; Teuwen, J. Improving breast cancer detection using symmetry information with deep learning. In Proceedings of the International Workshop on Reconstruction and Analysis of Moving Body Organs, Québec City, QC, Canada, 16 August 2018; pp. 90–97.
97. Al-Tam, R.M.; Al-Hejri, A.M.; Narangale, S.M.; Samee, N.A.; Mahmoud, N.F.; Al-Masni, M.A.; Al-Antari, M.A. A hybrid workflow of residual convolutional transformer encoder for breast cancer classification using digital X-ray mammograms. *Biomedicines* **2022**, *10*, 2971. [CrossRef]
98. Abunasser, B.S.; Al-Hiealy, M.R.J.; Zaqout, I.S.; Abu-Naser, S.S. Convolution Neural Network for Breast Cancer Detection and Classification Using Deep Learning. *Asian Pac. J. Cancer Prev. APJCP* **2023**, *24*, 531–544. [CrossRef]
99. Huang, Y.; Wei, L.; Hu, Y.; Shao, N.; Lin, Y.; He, S.; Shi, H.; Zhang, X.; Lin, Y. Multi-parametric MRI-based radiomics models for predicting molecular subtype and androgen receptor expression in breast cancer. *Front. Oncol.* **2021**, *11*, 706733. [CrossRef] [PubMed]
100. Jabeen, K.; Khan, M.A.; Alhaisoni, M.; Tariq, U.; Zhang, Y.D.; Hamza, A.; Mickus, A.; Damaševičius, R. Breast cancer classification from ultrasound images using probability-based optimal deep learning feature fusion. *Sensors* **2022**, *22*, 807. [CrossRef] [PubMed]
101. Ragab, M.; Albukhari, A.; Alyami, J.; Mansour, R.F. Ensemble deep-learning-enabled clinical decision support system for breast cancer diagnosis and classification on ultrasound images. *Biology* **2022**, *11*, 439. [CrossRef]
102. Kumar, A.; Kamal, O.; Mazumdar, S. Phoenix@ SMM4H Task-8: Adversities Make Ordinary Models Do Extraordinary Things. *NAACL-HLT* **2021**, *2*, 112–114.
103. Peng, Y.; Yan, S.; Lu, Z. Transfer learning in biomedical natural language processing: An evaluation of BERT and ELMo on ten benchmarking datasets. *arXiv* **2019**, arXiv:1906.05474.
104. Chen, D.; Zhong, K.; He, J. BDCN: Semantic Embedding Self-Explanatory Breast Diagnostic Capsules Network. In Proceedings of the China National Conference on Chinese Computational Linguistics, Hohhot, China, 13–15 August 2021; pp. 419–433.
105. Zhou, S.; Wang, N.; Wang, L.; Liu, H.; Zhang, R. CancerBERT: A cancer domain-specific language model for extracting breast cancer phenotypes from electronic health records. *J. Am. Med Inform. Assoc.* **2022**, *29*, 1208–1216. [CrossRef]
106. Deng, L.; Zhang, Y.; Luo, S.; Xu, J. GPT-4 in breast cancer combat: A dazzling leap forward or merely a whim? *Int. J. Surg.* **2023**, *109*, 3732–3735. [CrossRef]
107. Hanahan, D.; Weinberg, R.A. Hallmarks of cancer: The next generation. *Cell* **2011**, *144*, 646–674. [CrossRef]
108. Sun, Y.; Yao, J.; Yang, L.; Chen, R.; Nowak, N.J.; Goodison, S. Computational approach for deriving cancer progression roadmaps from static sample data. *Nucleic Acids Res.* **2017**, *45*, e69. [CrossRef]
109. Witten, D.M.; Tibshirani, R. A framework for feature selection in clustering. *J. Am. Stat. Assoc.* **2010**, *105*, 713–726. [CrossRef]
110. Shen, R.; Olshen, A.B.; Ladanyi, M. Integrative clustering of multiple genomic data types using a joint latent variable model with application to breast and lung cancer subtype analysis. *Bioinformatics* **2009**, *25*, 2906–2912. [CrossRef] [PubMed]
111. Curtis, C.; Shah, S.P.; Chin, S.F.; Turashvili, G.; Rueda, O.M.; Dunning, M.J.; Speed, D.; Lynch, A.G.; Samarajiwa, S.; Yuan, Y.; et al. The genomic and transcriptomic architecture of 2,000 breast tumours reveals novel subgroups. *Nature* **2012**, *486*, 346–352. [CrossRef] [PubMed]
112. Xu, J.; Wu, P.; Chen, Y.; Zhang, L. Comparison of different classification methods for breast cancer subtypes prediction. In Proceedings of the 2018 International Conference on Security, Pattern Analysis, and Cybernetics (SPAC), Jinan, China, 14–17 December 2018; pp. 91–96.
113. Verma, P.; Sharma, K.; Walia, G.S. Depression detection among social media users using machine learning. In Proceedings of the International Conference on Innovative Computing and Communications: Proceedings of ICICC 2020, Delhi, India, 20–22 February 2021; Volume 1, pp. 865–874.
114. Ghosh, S.; Ekbal, A.; Bhattacharyya, P. A multitask framework to detect depression, sentiment and multi-label emotion from suicide notes. *Cogn. Comput.* **2022**, *14*, 110–129. [CrossRef]
115. Xu, X.; Yao, B.; Dong, Y.; Yu, H.; Hendler, J.; Dey, A.K.; Wang, D. Leveraging large language models for mental health prediction via online text data. *arXiv* **2023**, arXiv:2307.14385.
116. Qi, H.; Zhao, Q.; Li, J.; Song, C.; Zhai, W.; Dan, L.; Liu, S.; Yu, Y.J.; Wang, F.; Zou, H.; et al. Supervised Learning and Large Language Model Benchmarks on Mental Health Datasets: Cognitive Distortions and Suicidal Risks in Chinese Social Media. 2023. [CrossRef]
117. Liu, Z.; Yu, H.; Li, G.; Chen, Q.; Ding, Z.; Feng, L.; Yao, Z.; Hu, B. Ensemble learning with speaker embeddings in multiple speech task stimuli for depression detection. *Front. Neurosci.* **2023**, *17*, 1141621. [CrossRef] [PubMed]
118. Long, H.; Guo, Z.; Wu, X.; Hu, B.; Liu, Z.; Cai, H. Detecting depression in speech: Comparison and combination between different speech types. In Proceedings of the 2017 IEEE International Conference on Bioinformatics and Biomedicine (BIBM), Kansas City, MO, USA, 13–16 November 2017; pp. 1052–1058.
119. Jiang, H.; Hu, B.; Liu, Z.; Wang, G.; Zhang, L.; Li, X.; Kang, H. Detecting depression using an ensemble logistic regression model based on multiple speech features. *Comput. Math. Methods Med.* **2018**, *2018*, 6508319. [CrossRef] [PubMed]

120. Liu, Z.; Wang, D.; Zhang, L.; Hu, B. A novel decision tree for depression recognition in speech. *arXiv* **2020**, arXiv:2002.12759.
121. Yin, F.; Du, J.; Xu, X.; Zhao, L. Depression Detection in Speech Using Transformer and Parallel Convolutional Neural Networks. *Electronics* **2023**, *12*, 328. [CrossRef]
122. Tasnim, M.; Novikova, J. Cost-effective Models for Detecting Depression from Speech. In Proceedings of the 2022 21st IEEE International Conference on Machine Learning and Applications (ICMLA), Nassau, Bahamas, 12–14 December 2022; pp. 1687–1694.
123. He, L.; Cao, C. Automated depression analysis using convolutional neural networks from speech. *J. Biomed. Inform.* **2018**, *83*, 103–111. [CrossRef]
124. Dubagunta, S.P.; Vlasenko, B.; Doss, M.M. Learning voice source related information for depression detection. In Proceedings of the ICASSP 2019—2019 IEEE International Conference on Acoustics, Speech and Signal Processing (ICASSP), Brighton, UK, 12–17 May 2019; pp. 6525–6529.
125. Zhao, Y.; Liang, Z.; Du, J.; Zhang, L.; Liu, C.; Zhao, L. Multi-head attention-based long short-term memory for depression detection from speech. *Front. Neurorobot.* **2021**, *15*, 684037. [CrossRef]
126. Dong, Y.; Yang, X. A hierarchical depression detection model based on vocal and emotional cues. *Neurocomputing* **2021**, *441*, 279–290. [CrossRef]
127. Kang, M.; Kwon, H.; Park, J.H.; Kang, S.; Lee, Y. Deep-asymmetry: Asymmetry matrix image for deep learning method in pre-screening depression. *Sensors* **2020**, *20*, 6526. [CrossRef]
128. Čukić, M.; Stokić, M.; Simić, S.; Pokrajac, D. The successful discrimination of depression from EEG could be attributed to proper feature extraction and not to a particular classification method. *Cogn. Neurodyn.* **2020**, *14*, 443–455. [CrossRef] [PubMed]
129. Mahato, S.; Paul, S. Classification of depression patients and normal subjects based on electroencephalogram (EEG) signal using alpha power and theta asymmetry. *J. Med Syst.* **2020**, *44*, 28. [CrossRef] [PubMed]
130. Wan, Z.; Zhang, H.; Huang, J.; Zhou, H.; Yang, J.; Zhong, N. Single-channel EEG-based machine learning method for prescreening major depressive disorder. *Int. J. Inf. Technol. Decis. Mak.* **2019**, *18*, 1579–1603. [CrossRef]
131. Cai, H.; Chen, Y.; Han, J.; Zhang, X.; Hu, B. Study on feature selection methods for depression detection using three-electrode EEG data. *Interdiscip. Sci. Comput. Life Sci.* **2018**, *10*, 558–565. [CrossRef]
132. Ehghaghi, M.; Rudzicz, F.; Novikova, J. Data-driven Approach to Differentiating between Depression and Dementia from Noisy Speech and Language Data. *arXiv* **2022**, arXiv:2210.03303.
133. Diep, B.; Stanojevic, M.; Novikova, J. Multi-modal deep learning system for depression and anxiety detection. *arXiv* **2022**, arXiv:2212.14490.
134. Mao, K.; Zhang, W.; Wang, D.B.; Li, A.; Jiao, R.; Zhu, Y.; Wu, B.; Zheng, T.; Qian, L.; Lyu, W.; et al. Prediction of depression severity based on the prosodic and semantic features with bidirectional lstm and time distributed cnn. *IEEE Trans. Affect. Comput.* **2022**, *14*, 2251–2265. [CrossRef]
135. Jan, A.; Meng, H.; Gaus, Y.F.B.A.; Zhang, F. Artificial intelligent system for automatic depression level analysis through visual and vocal expressions. *IEEE Trans. Cogn. Dev. Syst.* **2017**, *10*, 668–680. [CrossRef]
136. Uddin, M.A.; Joolee, J.B.; Sohn, K.A. Deep multi-modal network based automated depression severity estimation. *IEEE Trans. Affect. Comput.* **2022**, *14*, 2153–2167. [CrossRef]
137. Yang, L.; Jiang, D.; Xia, X.; Pei, E.; Oveneke, M.C.; Sahli, H. Multimodal measurement of depression using deep learning models. In Proceedings of the 7th Annual Workshop on Audio/Visual Emotion Challenge, Mountain View, CA, USA, 23–27 October 2017; pp. 53–59.
138. Almadani, A.; Agu, E.; Sarwar, A.; Ahluwalia, M.; Kpodonu, J. HCM-Dynamic-Echo: A Framework for Detecting Hypertrophic Cardiomyopathy (HCM) in Echocardiograms. In Proceedings of the 2023 IEEE International Conference on Digital Health (ICDH), Shenzhen, China, 7–13 July 2023; pp. 217–226.
139. Madani, A.; Ong, J.R.; Tibrewal, A.; Mofrad, M.R. Deep echocardiography: Data-efficient supervised and semi-supervised deep learning towards automated diagnosis of cardiac disease. *NPJ Digit. Med.* **2018**, *1*, 59–60. [CrossRef]
140. Nasimova, N.; Muminov, B.; Nasimov, R.; Abdurashidova, K.; Abdullaev, M. Comparative analysis of the results of algorithms for dilated cardiomyopathy and hypertrophic cardiomyopathy using deep learning. In Proceedings of the 2021 International Conference on Information Science and Communications Technologies (ICISCT), Tashkent, Uzbekistan , 3–5 November 2021; pp. 1–5.
141. Zhang, J.; Gajjala, S.; Agrawal, P.; Tison, G.H.; Hallock, L.A.; Beussink-Nelson, L.; Lassen, M.H.; Fan, E.; Aras, M.A.; Jordan, C.; et al. Fully automated echocardiogram interpretation in clinical practice: Feasibility and diagnostic accuracy. *Circulation* **2018**, *138*, 1623–1635. [CrossRef]
142. Ghorbani, A.; Ouyang, D.; Abid, A.; He, B.; Chen, J.H.; Harrington, R.A.; Liang, D.H.; Ashley, E.A.; Zou, J.Y. Deep learning interpretation of echocardiograms. *NPJ Digit. Med.* **2020**, *3*, 10–20. [CrossRef]
143. Sundaram, D.S.B.; Arunachalam, S.P.; Damani, D.N.; Farahani, N.Z.; Enayati, M.; Pasupathy, K.S.; Arruda-Olson, A.M. Natural language processing based machine learning model using cardiac MRI reports to identify hypertrophic cardiomyopathy patients. In *Frontiers in Biomedical Devices*; American Society of Mechanical Engineers: New York, NY, USA, 2021; Volume 84812, pp. 1–5.
144. Mishra, J.; Tiwari, M.; Singh, S.T.; Goswami, S. Detection of heart disease employing Recurrent CONVoluted neural networks (RecCONVnet) for effectual classification process in smart medical application. In Proceedings of the 2021 4th International Conference on Recent Trends in Computer Science and Technology (ICRTCST), Jamshedpur, India, 11–12 February 2022; pp. 389–394.

145. Jayasudha, R.; Suragali, C.; Thirukrishna, J.; Santhosh Kumar, B. Hybrid optimization enabled deep learning-based ensemble classification for heart disease detection. *Signal Image Video Process.* **2023**, *17*, 4235 –4244. [CrossRef]
146. Levine, D.M.; Tuwani, R.; Kompa, B.; Varma, A.; Finlayson, S.G.; Mehrotra, A.; Beam, A. The diagnostic and triage accuracy of the GPT-3 artificial intelligence model. *medRxiv* **2023**, 1–22. [CrossRef]
147. Peng, W.; Sun, Y.; Zhang, L. Construction of genetic classification model for coronary atherosclerosis heart disease using three machine learning methods. *BMC Cardiovasc. Disord.* **2022**, *22*, 42–54. [CrossRef] [PubMed]
148. Liu, J.; Wang, X.; Lin, J.; Li, S.; Deng, G.; Wei, J. Classifiers for predicting coronary artery disease based on gene expression profiles in peripheral blood mononuclear cells. *Int. J. Gen. Med.* **2021**, *14*, 5651–5663. [CrossRef] [PubMed]
149. Hou, Q.; Sun, Z.; Zhao, L.; Liu, Y.; Zhang, J.; Huang, J.; Luo, Y.; Xiao, Y.; Hu, Z.; Shen, A. Role of serum cytokines in the prediction of heart failure in patients with coronary artery disease. *ESC Heart Fail.* **2023**, *10*, 3102–3113. [CrossRef] [PubMed]
150. Samadishadlou, M.; Rahbarghazi, R.; Piryaei, Z.; Esmaeili, M.; Avcı, Ç.B.; Bani, F.; Kavousi, K. Unlocking the potential of microRNAs: Machine learning identifies key biomarkers for myocardial infarction diagnosis. *Cardiovasc. Diabetol.* **2023**, *22*, 247. [CrossRef]
151. Dai, H.; Hwang, H.G.; Tseng, V.S. Convolutional neural network based automatic screening tool for cardiovascular diseases using different intervals of ECG signals. *Comput. Methods Programs Biomed.* **2021**, *203*, 106035. [CrossRef]
152. Tison, G.H.; Zhang, J.; Delling, F.N.; Deo, R.C. Automated and interpretable patient ECG profiles for disease detection, tracking, and discovery. *Circ. Cardiovasc. Qual. Outcomes* **2019**, *12*, e005289. [CrossRef]
153. Tison, G.H.; Siontis, K.C.; Abreau, S.; Attia, Z.; Agarwal, P.; Balasubramanyam, A.; Li, Y.; Sehnert, A.J.; Edelberg, J.M.; Friedman, P.A.; et al. Assessment of disease status and treatment response with artificial Intelligence- Enhanced electrocardiography in obstructive hypertrophic cardiomyopathy. *J. Am. Coll. Cardiol.* **2022**, *79*, 1032–1034. [CrossRef]
154. Ko, W.Y.; Siontis, K.C.; Attia, Z.I.; Carter, R.E.; Kapa, S.; Ommen, S.R.; Demuth, S.J.; Ackerman, M.J.; Gersh, B.J.; Arruda-Olson, A.M.; et al. Detection of hypertrophic cardiomyopathy using a convolutional neural network-enabled electrocardiogram. *J. Am. Coll. Cardiol.* **2020**, *75*, 722–733. [CrossRef]
155. Bhattacharyya, A.; Pachori, R.B.; Upadhyay, A.; Acharya, U.R. Tunable-Q wavelet transform based multiscale entropy measure for automated classification of epileptic EEG signals. *Appl. Sci.* **2017**, *7*, 385. [CrossRef]
156. Karácsony, T.; Loesch-Biffar, A.M.; Vollmar, C.; Noachtar, S.; Cunha, J.P.S. DeepEpil: Towards an epileptologist-friendly AI enabled seizure classification cloud system based on deep learning analysis of 3D videos. In Proceedings of the 2021 IEEE EMBS International Conference on Biomedical and Health Informatics (BHI), Athens, Greece, 27–30 July 2021; pp. 1–5.
157. Maia, P.; Hartl, E.; Vollmar, C.; Noachtar, S.; Cunha, J.P.S. Epileptic seizure classification using the NeuroMov database. In Proceedings of the 2019 IEEE 6th Portuguese Meeting on Bioengineering (ENBENG), Lisbon, Portugal, 22–23 February 2019; pp. 1–4.
158. Achilles, F.; Belagiannis, V.; Tombari, F.; Loesch, A.; Cunha, J.; Navab, N.; Noachtar, S. Deep convolutional neural networks for automatic identification of epileptic seizures in infrared and depth images. *J. Neurol. Sci.* **2015**, *357*, e436. [CrossRef]
159. Ahmedt-Aristizabal, D.; Nguyen, K.; Denman, S.; Sarfraz, M.S.; Sridharan, S.; Dionisio, S.; Fookes, C. Vision-based mouth motion analysis in epilepsy: A 3d perspective. In Proceedings of the 2019 41st Annual International Conference of the IEEE Engineering in Medicine and Biology Society (EMBC), Berlin, Germany, 23–27 July 2019; pp. 1625–1629.
160. Kunekar, P.R.; Gupta, M.; Agarwal, B. Deep learning with multi modal ensemble fusion for epilepsy diagnosis. In Proceedings of the 2020 3rd International Conference on Emerging Technologies in Computer Engineering: Machine Learning and Internet of Things (ICETCE), Jaipur, India, 7–8 February 2020; pp. 80–84.
161. Ahmedt-Aristizabal, D.; Fookes, C.; Denman, S.; Nguyen, K.; Fernando, T.; Sridharan, S.; Dionisio, S. A hierarchical multimodal system for motion analysis in patients with epilepsy. *Epilepsy Behav.* **2018**, *87*, 46–58. [CrossRef] [PubMed]
162. Garner, R.; La Rocca, M.; Barisano, G.; Toga, A.W.; Duncan, D.; Vespa, P. A machine learning model to predict seizure susceptibility from resting-state fMRI connectivity. In Proceedings of the 2019 Spring Simulation Conference (SpringSim), Tucson, AZ, USA, 29 April–2 May 2019; Volume 51, pp. 1–11.
163. Sahebzamani, G.; Saffar, M.; Soltanian-Zadeh, H. Machine learning based analysis of structural MRI for epilepsy diagnosis. In Proceedings of the 2019 4th International Conference on Pattern Recognition and Image Analysis (IPRIA), Tehran, Iran, 6–7 March 2019; pp. 58–63.
164. Si, X.; Zhang, X.; Zhou, Y.; Sun, Y.; Jin, W.; Yin, S.; Zhao, X.; Li, Q.; Ming, D. Automated detection of juvenile myoclonic epilepsy using CNN based transfer learning in diffusion MRI. In Proceedings of the 2020 42nd Annual International Conference of the IEEE Engineering in Medicine & Biology Society (EMBC), Montreal, QC, Canada, 20–24 July 2020; Volume 2020, pp. 1679–1682.
165. Pominova, M.; Artemov, A.; Sharaev, M.; Kondrateva, E.; Bernstein, A.; Burnaev, E. Voxelwise 3d convolutional and recurrent neural networks for epilepsy and depression diagnostics from structural and functional mri data. In Proceedings of the 2018 IEEE International Conference on Data Mining Workshops (ICDMW), Singapore, 17–20 November 2018; Volume 2018, pp. 299–307.
166. Santoso, I.B.; Adrianto, Y.; Sensusiati, A.D.; Wulandari, D.P.; Purnama, I.K.E. Ensemble Convolutional Neural Networks With Support Vector Machine for Epilepsy Classification Based on Multi-Sequence of Magnetic Resonance Images. *IEEE Access* **2022**, *10*, 32034–32048. [CrossRef]
167. Hamid, H.; Fodeh, S.; Lizama, A.; Czlapinski, R.; Pugh, M.; LaFrance Jr, W.; Brandt, C. Validating a natural language processing tool to exclude psychogenic nonepileptic seizures in electronic medical record-based epilepsy research. *Epilepsy Behav.* **2013**, *29*, 578–580. [CrossRef] [PubMed]

168. Pevy, N.; Christensen, H.; Walker, T.; Reuber, M. Feasibility of using an automated analysis of formulation effort in patients' spoken seizure descriptions in the differential diagnosis of epileptic and nonepileptic seizures. *Seizure* **2021**, *91*, 141–145. [CrossRef] [PubMed]
169. Connolly, B.; Matykiewicz, P.; Bretonnel Cohen, K.; Standridge, S.M.; Glauser, T.A.; Dlugos, D.J.; Koh, S.; Tham, E.; Pestian, J. Assessing the similarity of surface linguistic features related to epilepsy across pediatric hospitals. *J. Am. Med Inform. Assoc.* **2014**, *21*, 866–870. [CrossRef] [PubMed]
170. Clarke, S.; Karoly, P.J.; Nurse, E.; Seneviratne, U.; Taylor, J.; Knight-Sadler, R.; Kerr, R.; Moore, B.; Hennessy, P.; Mendis, D.; et al. Computer-assisted EEG diagnostic review for idiopathic generalized epilepsy. *Epilepsy Behav.* **2021**, *121*, 106–115. [CrossRef]
171. Fürbass, F.; Kural, M.A.; Gritsch, G.; Hartmann, M.; Kluge, T.; Beniczky, S. An artificial intelligence-based EEG algorithm for detection of epileptiform EEG discharges: Validation against the diagnostic gold standard. *Clin. Neurophysiol.* **2020**, *131*, 1174–1179. [CrossRef]
172. Thara, D.; PremaSudha, B.; Xiong, F. Epileptic seizure detection and prediction using stacked bidirectional long short term memory. *Pattern Recognit. Lett.* **2019**, *128*, 529–535.
173. Yao, X.; Cheng, Q.; Zhang, G.Q. Automated classification of seizures against nonseizures: A deep learning approach. *arXiv* **2019**, arXiv:1906.02745.
174. Torres-Velázquez, M.; Hwang, G.; Cook, C.J.; Hermann, B.P.; Prabhakaran, V.; Meyerand, M.E.; McMillan, A.B. Multi-Channel Deep Neural Network For Temporal Lobe Epilepsy Classification Using Multimodal Mri Data. In Proceedings of the 2020 IEEE 17th International Symposium on Biomedical Imaging Workshops (ISBI Workshops), Iowa City, IA, USA, 4 April 2020; pp. 1–4.
175. Asan, O.; Bayrak, A.E.; Choudhury, A. Artificial Intelligence and Human Trust in Healthcare: Focus on Clinicians. *J. Med Internet Res.* **2020**, *22*, 32–45. [CrossRef] [PubMed]
176. Rai, A. Explainable AI: From black box to glass box. *J. Acad. Mark. Sci.* **2019**, *48*, 137–141. [CrossRef]
177. Kaissis, G.; Makowski, M.R.; Rückert, D.; Braren, R.F. Secure, privacy-preserving and federated machine learning in medical imaging. *Nat. Mach. Intell.* **2020**, *2*, 305–311. [CrossRef]
178. Chia, P.H.; Desfontaines, D.; Perera, I.M.; Simmons-Marengo, D.; Li, C.; Day, W.Y.; Wang, Q.; Guevara, M. KHyperLogLog: Estimating reidentifiability and joinability of large data at scale. In Proceedings of the 2019 IEEE Symposium on Security and Privacy (SP), San Francisco, CA, USA, 19–23 May 2019; Volume 2019, pp. 350–364.
179. Kairouz, P.; McMahan, H.B.; Avent, B.; Bellet, A.; Bennis, M.; Bhagoji, A.N.; Bonawitz, K.; Charles, Z.B.; Cormode, G.; Cummings, R.; et al. Advances and Open Problems in Federated Learning. *Found. Trends Mach. Learn.* **2019**, *14*, 1–210. [CrossRef]
180. Warnat-Herresthal, S.; Schultze, H.; Shastry, K.; Manamohan, S.; Mukherjee, S.; Garg, V.; Sarveswara, R.; Händler, K.; Pickkers, P.; Aziz, N.A.; et al. Swarm Learning for decentralized and confidential clinical machine learning. *Nature* **2021**, *594*, 265–270. [CrossRef]
181. Stratigi, M.; Kondylakis, H.; Stefanidis, K. Fairness in group recommendations in the health domain. In Proceedings of the 2017 IEEE 33rd International Conference on Data Engineering (ICDE), San Diego, CA, USA, 19–22 April 2017; Volume 1, pp. 1481–1488.
182. Stratigi, M.; Kondylakis, H.; Stefanidis, K. FairGRecs: Fair Group Recommendations by Exploiting Personal Health Information. In Proceedings of the International Conference on Database and Expert Systems Applications, Regensburg, Germany, 3–6 September 2018; Volume 11030, pp. 147–155.
183. Vayena, E.; Blasimme, A.; Cohen, I.G. Machine learning in medicine: Addressing ethical challenges. *PLoS Med.* **2018**, *15*, 3–11. [CrossRef]
184. Neighbors, H.W.; Jackson, J.S.; Campbell, L.; Williams, D. The influence of racial factors on psychiatric diagnosis: A review and suggestions for research. *Community Ment. Health J.* **2004**, *25*, 301–311. [CrossRef]
185. Braveman, P. Health disparities and health equity: Concepts and measurement. *Annu. Rev. Public Health* **2006**, *27*, 167–194. [CrossRef]
186. Loftus, T.J.; Shickel, B.; Ozrazgat-Baslanti, T.; Ren, Y.; Glicksberg, B.S.; Cao, J.; Singh, K.; Chan, L.; Nadkarni, G.N.; Bihorac, A. Artificial intelligence-enabled decision support in nephrology. *Nat. Rev. Nephrol.* **2022**, *18*, 452–465. [CrossRef]
187. Bakul, P. *Developing a Software Precertification Program: A Working Model;* Food and Drug Administration: Silver Spring, MD, USA, 2018.
188. Yoo, H.J.; Shin, S. Mobile Health Intervention Contents and Their Effects on the Healthcare of Patients with Left Ventricular Assist Devices: An Integrative Review. *Comput. Inform. Nurs. CIN* **2023**, 10–1097. [CrossRef] [PubMed]
189. Girela-Serrano, B.M.; Spiers, A.D.V.; Liu, R.; Gangadia, S.; Toledano, M.B.; Simplicio, M.D. Impact of mobile phones and wireless devices use on children and adolescents' mental health: A systematic review. *Eur. Child Adolesc. Psychiatry* **2022**, 1–31. [CrossRef] [PubMed]
190. Ersaro, N.T.; Yalcin, C.; Muller, R. The future of brain–machine interfaces is optical. *Nat. Electron.* **2023**, *6*, 96–98. [CrossRef]

bioengineering

MDPI

Review

Machine Learning and Graph Signal Processing Applied to Healthcare: A Review

Maria Alice Andrade Calazans [1,*], Felipe A. B. S. Ferreira [2], Fernando A. N. Santos [3], Francisco Madeiro [4] and Juliano B. Lima [1]

1 Centro de Tecnologia e Geociências, Universidade Federal de Pernambuco, Recife 50670-901, Brazil; juliano.lima@ufpe.br
2 Unidade Acadêmica do Cabo de Santo Agostinho, Universidade Federal Rural de Pernambuco, Cabo de Santo Agostinho 54518-430, Brazil; felipe.bsferreira@ufrpe.br
3 Institute for Advanced Studies, Universiteit van Amsterdam, 1012 WP Amsterdam, The Netherlands; f.a.nobregasantos@uva.nl
4 Escola Politécnica de Pernambuco, Universidade de Pernambuco, Recife 50720-001, Brazil; madeiro@poli.br
* Correspondence: maria.calazans@ufpe.br

Abstract: Signal processing is a very useful field of study in the interpretation of signals in many everyday applications. In the case of applications with time-varying signals, one possibility is to consider them as graphs, so graph theory arises, which extends classical methods to the non-Euclidean domain. In addition, machine learning techniques have been widely used in pattern recognition activities in a wide variety of tasks, including health sciences. The objective of this work is to identify and analyze the papers in the literature that address the use of machine learning applied to graph signal processing in health sciences. A search was performed in four databases (Science Direct, IEEE Xplore, ACM, and MDPI), using search strings to identify papers that are in the scope of this review. Finally, 45 papers were included in the analysis, the first being published in 2015, which indicates an emerging area. Among the gaps found, we can mention the need for better clinical interpretability of the results obtained in the papers, that is not to restrict the results or conclusions simply to performance metrics. In addition, a possible research direction is the use of new transforms. It is also important to make new public datasets available that can be used to train the models.

Keywords: deep learning; graph signal processing; health; machine learning

Citation: Calazans, M.A.A.; Ferreira, F.A.B.S.; Santos, F.A.N.; Madeiro, F.; Lima, J.B. Machine Learning and Graph Signal Processing Applied to Healthcare: A Review. *Bioengineering* **2024**, *11*, 671. https://doi.org/10.3390/bioengineering11070671

Academic Editors: Yan Pei and Jijiang Yang

Received: 27 May 2024
Revised: 20 June 2024
Accepted: 26 June 2024
Published: 2 July 2024

1. Introduction

Graph signal processing (GSP) is an emerging research field, which focuses on generalizing the classical concepts of signal processing in order to expand them to graphs [1]. The need for GSP is related to the considerable amount of information that can be represented as a signal whose samples lie over irregular structures that can be modeled as graphs [1,2]. Among the GSP application scenarios that have attracted the attention of researchers and have been documented in recent studies, one can mention forecasting in the financial market [3], 3D point clouds [4], the Internet of Things (IoT) [5], traffic [6], and sensor, social, physical, and biological networks [7–10].

In the practical use of GSP, machine learning (ML) techniques and, in particular, deep learning (DL) techniques have been playing an important role. This is due to the fact that deep neural networks are adaptable to solving a wide range of problems, providing better or competitive results, when compared to other techniques. The extension of ML to non-Euclidean data gave rise to graph learning (GL) [11] and, consequently, to graph neural networks (GNNs). Such networks have also provided good results in several applications [12,13]. Regarding deep learning on graphs [11], specifically, we can mention the graph convolutional neural networks (GCNNs), in which deep networks with convolutional layers are proposed, such as the operations performed by the traditional convolutional neural network (CNN), but in this case, applied to problems in the non-Euclidean domain, i.e., in graphs [14–16].

Among the areas that have been highlighted in recent works involving the use of deep learning techniques and graph signal processing, one can mention the medical sciences [17]. Applications of GSP with ML for health have shown growth and been documented in a large number of works published in the literature [18]. In this scope, one identifies papers devoted to applications related to various medical specialties. There is evidence that some of these specialties, such as neurology, for example, have stood out in this context, while other areas are still little explored. The interest of researchers in using GSP and DL in neurology is due to the fact that the human brain can be modeled as a graph, so that its regions can be considered as vertices or nodes and its connectomes at functional and structural levels can be viewed as edges [19–21]. Deep networks, on the other hand, are widely used for automatic pattern recognition. In this context, the literature includes works dealing with different objectives, from early diagnosis of Alzheimer's disease [22] and autism [23] to emotion recognition [24] and imagined speech [25] and multiple sciences [26].

In general, scholars in health sciences have demonstrated interest in the development and application of techniques simultaneously based on machine learning, signal processing, and graph theory. The interpretation and analysis of complex irregular data have potential to provide a number of benefits in clinical and hospital practice as an aid in identifying the origin of diseases, the early diagnosis of medical conditions, the verification of possible treatments, and disease prevention [27]. The elements outlined above encouraged us to prepare the present paper, which corresponds to a systematic literature review focusing on machine learning-based healthcare applications, with an emphasis on deep learning applied to signal processing over graphs. The paper presents an overview of the area: the medical specialties with the most papers in GSP in recent years, the ML and GSP techniques that have been most used in healthcare, the most influential authors in the area, and challenges, gaps, and open questions that may provide opportunities for future research. To be more specific, our paper includes the following:

- A comprehensive overview of ML and GSP applied to healthcare;
- A panorama of the datasets most used in ML applied to GSP in healthcare and their corresponding description;
- The identification of gaps, open problems, and promising future research directions in ML applied to GSP in healthcare.

The remainder of the paper is organized as follows. In Section 2, the basic fundamentals of graph signal processing, machine learning, and deep learning are presented. Section 3 corresponds to the methodology adopted for the systematic review, such as the scientific databases considered, the search strings used, as well as the inclusion and exclusion criteria of the papers. Section 4 presents the main findings of the review. Section 5 brings a discussion, in which the identified gaps are addressed and future research directions in the area are presented. In Section 6, the final considerations are presented. Figure 1 summarizes the organization of the paper.

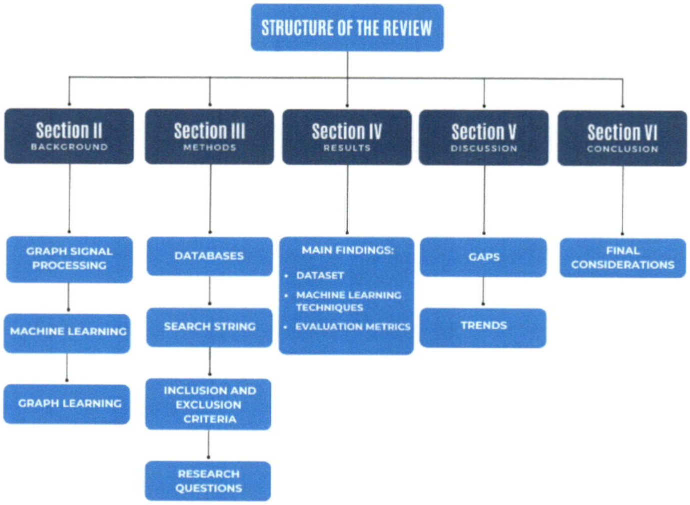

Figure 1. Organizational diagram of the paper.

2. Background

In this section, we provide a concise review of the main concepts related to graph signal processing and machine learning. In the case of GSP, the purpose is to explain what it means to consider a signal in the so-called vertex domain, as well as to indicate the main operators and approaches used in this framework. Regarding machine learning, besides listing the tasks that can be performed with its help and discussing some correlated issues, we highlight aspects of deep learning and the intersections of these tools with graph-based models.

2.1. Graph Signal Processing

Graph signal processing aims to extend classical digital signal-processing methods to signals over irregular domains represented by an arbitrary graph [27–30].

A graph is essentially a set of vertices (nodes) possibly connected by edges. Thus, each sample of a graph signal is associated with a vertex in the corresponding underlying graph; the edge weights reflect the interdependence among the signal samples [30]. In this context, the topology of a graph is inferred or determined according to the proposed application.

In terms of orientation, graphs can be directed, if the orientation of the input and output of the edge is considered, or undirected, in the opposite case. Another important characterization concerns the vertex degree. In the case of directed graphs, the vertex degree corresponds to the difference between the weight of edges that depart from it and the weight of edges that arrive at it. The degree of a vertex of an undirected graph, on the other hand, is the sum of the weights of the edges [31,32].

Additionally, a graph can be associated with an adjacency matrix, which is denoted by A and contains information about the connectivity of the corresponding graph. If there is an edge connecting vertices v_j and v_i, the entry $A_{i,j}$ in the i-th row and the j-column of the referenced matrix is filled with the value of the respective weight; otherwise, $A_{i,j} = 0$. An adjacency matrix is symmetric if and only if the associated graph is undirected. A graph can also be associated with a degree matrix, which is a diagonal matrix denoted by D and having in the entry $D_{i,i}$ the degree of the vertex v_i. Finally, the Laplacian matrix, denoted by L, is obtained by $L = D - A$ [29,32].

In the study of graph signal processing, there are two well-established approaches [2]:

- Spectral graph theory: This is based on the graph Laplacian matrix and considers signals over undirected graphs with real and non-negative weights [1];

- Algebraic signal processing theory: This considers the adjacency matrix, which assumes the role of the elementary operator. This approach is used in signal analysis of directed and undirected graphs, which may have real or complex weights [30].

2.2. Machine Learning and Deep Learning

Machine learning corresponds to a subarea of artificial intelligence (AI), which is the field of study of systems that learn problems with examples obtained by training data [33]. Thus, ML aims to propose algorithms that can learn iteratively with the available data, in order to apply such algorithms to automate the construction of models capable of performing classification, regression, and clustering. These tasks can be based, for instance, on decision trees or artificial neural networks [34,35]. The use of such techniques has shown good results for applications in the most diverse areas, including medical diagnosis in health sciences [36–38].

In ML, two main approaches can be considered: supervised learning, in which training is performed considering labeled data, and the results of the model along the training are compared with the expected (target) outputs; unsupervised learning, in which the model identifies patterns in the data, with typical applications in clustering; in the latter, the data are not labeled and, as a consequence, there is no comparison between the output of the model and the target output along training [39].

Deep learning corresponds to a subarea of ML that makes use of deep neural networks. Such networks have high computational complexity and are widely used and disseminated for automatic pattern recognition [40–45]. DL techniques have been employed as an effective solution to perform pattern recognition in images, for instance. The most used approach, in this case, employs the so-called CNNs [46]. CNNs operate similarly to the receptive fields of the visual cortex of living beings and are essentially composed of convolutional, pooling, and dense layers [47]. A characteristic of this type of network is its high connectivity, which allows it to process a large amount of input parameters, as required in image processing [48,49].

However, CNNs are designed for data with a Euclidean structure. Nevertheless, as previously mentioned, there is a latent need to extend these techniques to the non-Euclidean domain, which can be accomplished by means of their generalization to graphs [50]. This gives rise to graph learning, a field of study that encompasses graph neural networks [51]. Moreover, considering the GL scenario, one has a specific GNN approach, the graph convolutional networks. Analogous to CNNs, GCNs have high connectivity to allow the input of a high number of parameters; in this case, the inputs are graphs [15,52].

3. Methods

The review presented in this paper encompasses papers written in English and published up to 30 October 2023. No starting date was defined for the search of papers in the literature. Four databases of relevance in the field of engineering were used: Science Direct, Institute of Electrical and Electronic Engineers (IEEE Xplore), the Association for Computing Machinery (ACM), and Multidisciplinary Digital Publishing Institute (MDPI).

The strings used for the search were as follows:

1. "Graph signal processing" AND (COVID OR disease);
2. "Graph signal processing" AND (health OR medical OR medicine) AND ("Neural Network" OR "Machine Learning" OR "Deep Learning").

As a result, 396 papers were obtained. Refinements were performed to filter only the relevant papers for the purpose of this review. The first adopted strategy consisted of evaluating the title and the abstract of the papers and discarding those that did not adhere to GSP techniques applied to health. Additionally, repeated papers were also subtracted, so that 50 papers remained for analysis. Finally, 5 more papers were disregarded because they were review papers. As shown in Figure 2, a final sample of 45 papers remained for analysis.

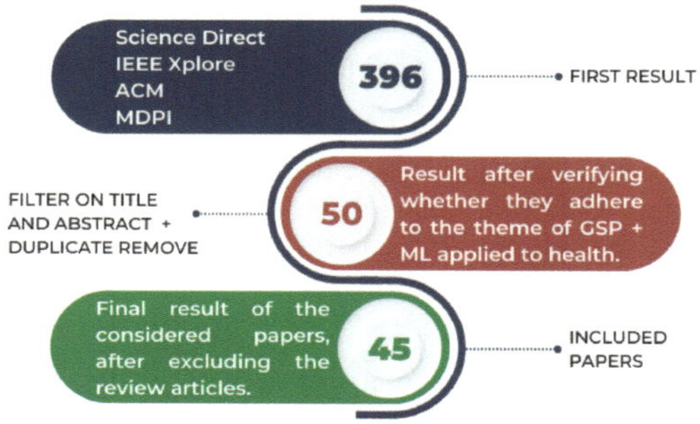

Figure 2. Flowchart of the paper selection process for the review considering exclusion criteria.

It is worth mentioning that five review papers were found, which substantially differ from our paper, both in scope and in selected works, and consequently in their findings.

In Khambhati et al. [53], for example, the selected papers concern specifically graphs on dynamic patterns of brain connectivity. In the paper of Dong, Wang, and Abbas [54], the review addresses works in the literature that use deep learning. It is not a review on graph signal processing, although there is a section dedicated to the subject. The paper by Li et al. [55] is a review on graph signal processing and neural networks in the biological data scenario. In this case, despite being a broader review, it is a study more aligned with the biological sciences, since it includes the study of molecules and proteins.

The paper by Yin and Kaiser [56] addresses neural flexibility in the human brain. To this end, they reviewed the computational approaches and suggested metrics to classify the flexibility of brain regions. In the work by Yingjie et al. [57], a specific area of the health sciences is analyzed: the work is concerned with the use of deep learning to diagnose liver diseases, and among the methods considered, one observes graph neural networks to detect liver tumors.

Unlike the aforementioned papers, our work is in the field of health in general, without a restricted medical area or specialty; we address papers on methods that use machine learning for graph signal processing in health.

After the paper selection and exclusion stages, the most relevant characteristics for carrying out the analysis of the 45 selected papers were extracted and synthesized. The information considered in the analysis are those related to the nature and metadata of the paper:

- Year of publication;
- First author's country of affiliation;
- Studied area.

Other issues considered in the analysis were the following:

- Dataset (size, type, and characteristics of sample);
- Proposed technique versus the technique used for the comparison;
- Objective of the study;
- Performance metrics.

The works were analyzed, and gaps and open challenges were identified. The results of such an analysis can serve as guidelines for future work in the area.

4. Results

Initially, lexical analyses were performed on the 45 papers included in this review. The analyses were based on the frequency of occurrence of terms in the titles and the keywords. One of these analyses is the word cloud, which consists of a simplistic visual representation to highlight the words with high recurrence in a previously defined universe [58,59]. Then, the larger the size of the word in the cloud, the more times it occurs in the text. An analysis of this type is depicted in Figure 3, which was obtained to show the co-occurrence of terms in the titles of the papers. In the presented analysis, the terms with the largest font are the most frequent ones in the area under investigation. This study was carried out using the Iramuteq software [60], which is free to use and was developed as open source using the R and Python languages. It can be inferred that terms related to GSP and DL appear very often, as expected, but there is also a considerable occurrence of terms related to neurology, such as: "fmri", "eeg", "brain", and "alzheimer".

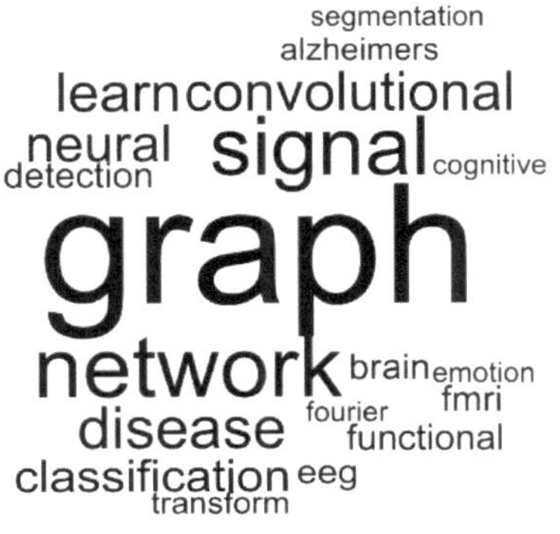

Figure 3. Word cloud obtained from the titles of the 45 papers using the Iramuteq software.

One of the encountered issues in the use of word clouds is the lack of grouping of similar terms because of grammatical variations, such as singular and plural [61]. In Iramuteq, this question is solved by the use of textual lemmatization. Thus, a certain level of variation is allowed in the terms, so that they are not considered distinct and the occurrence count is added to its most frequent equivalent term [62,63].

Another analysis that can be carried out with the Iramuteq software is the similitude analysis [64], which is based on graph theory. In this case, the most important words in the analysis are represented by vertices of a graph structure and the connections between words correspond to the edges. Thus, it is possible to identify central terms, their connections, and the grouping of words of the same theme just like a hypergraph.

Figure 4 shows a similitude analysis obtained by Iramuteq for the titles of the papers. The figure shows a central cluster with words that are frequently related; such a main cluster is connected to other clusters through its secondary terms. As a central term, one observes the word "graph", as expected. From this, branches are shown with clusters of distinct themes, but originated from the central elements.

Figure 5 allows a complementary analysis. In this case, the co-occurrence of keywords is evaluated with the VOSviewer software, which is a tool for the elaboration of bibliometric networks [65]. The most recurrent terms are "graph signal processing", "deep learning", "graph learning", "machine learning", "fmri", "connectivity", "Alzheimer's disease", "autism spectrum disorder", "brain", "mild cognitive impairment", and "graph fourier transform". The nodes were divided into four clusters, so the most frequently related terms are grouped together in the same color. According to the terms shown in the figure, once again as expected, the application focused on neurology is highlighted in the terms in evidence.

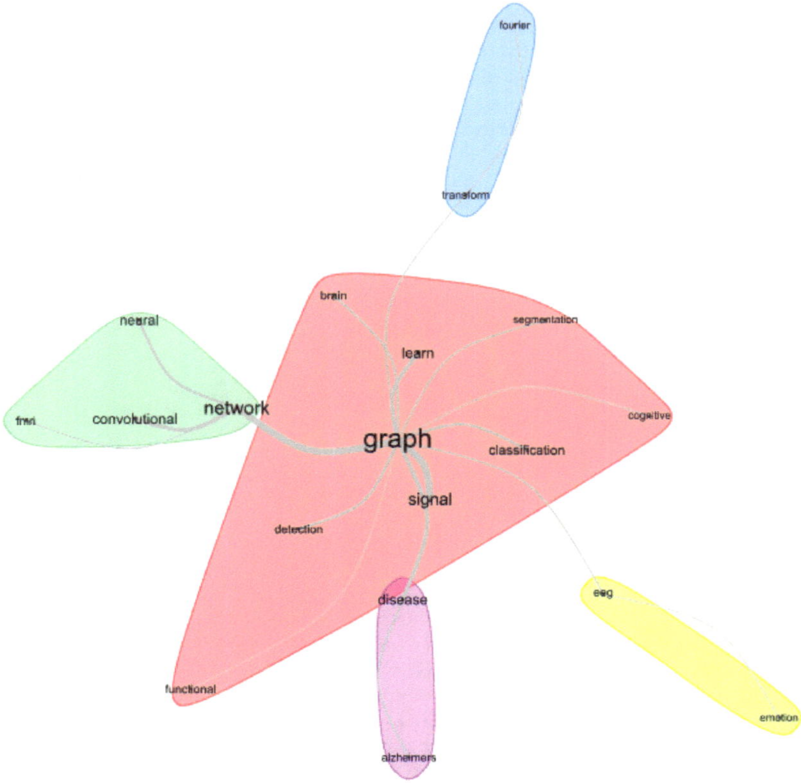

Figure 4. Similitude analysis obtained from the titles of the 45 papers using the Iramuteq software (0.7 alpha 2).

The distribution of publications by geographic location took into account the country associated with the affiliation of the first author of each paper. This made it possible to analyze the paper distribution by country and by continent, as shown in Figure 6 and Figure 7, respectively. In Figure 6, we verify that there are first authors affiliated with institutions from seventeen different countries, with emphasis on China, the United States of America (USA), Iran, and the United Kingdom; the first two countries have, respectively, ten and seven, and the last two countries have four, of the forty-five first authors.

Figure 7 presents the geographical overview from a continental point of view. It can be inferred that there is at least one first author per continent, except in Oceania. The continent with the greatest influence is Asia, corroborating the strong impact provided by China. It is followed by the European continent, which has the United Kingdom and France among the most influential countries according to the number of affiliated first authors. The next continent in this sequence is America, which, despite the strong influence of the USA, has only one other country with two affiliated first authors, Canada. Among the continents with publications, the last is Africa, with only one first author. Europe and Asia together hold 77.8% of first author affiliations.

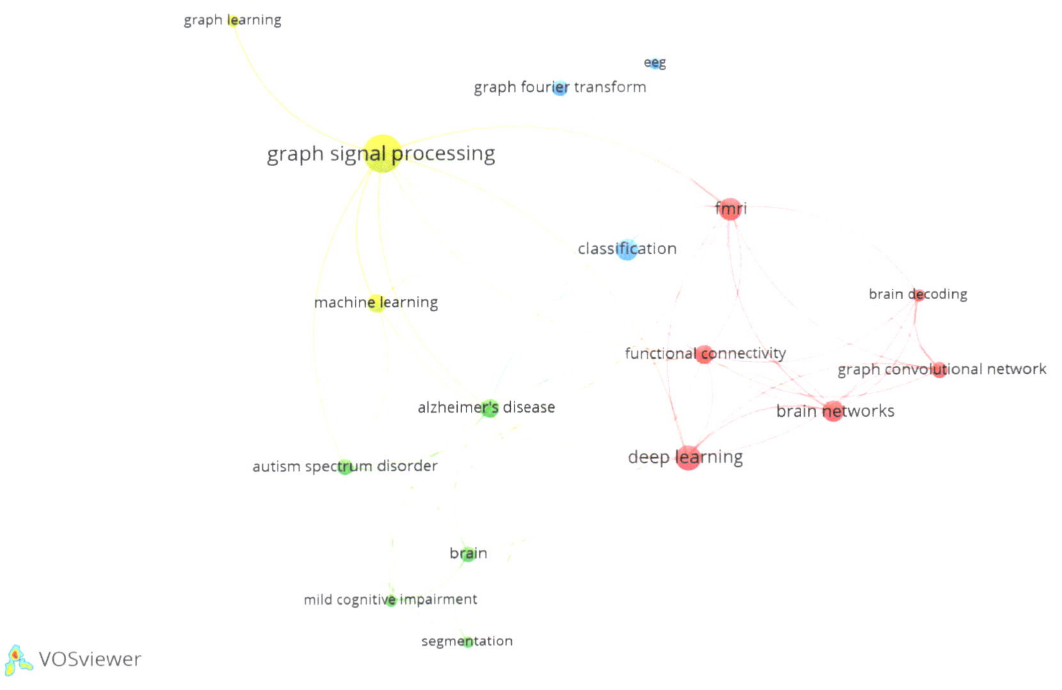

Figure 5. Graph analysis considering the occurrence of keywords.

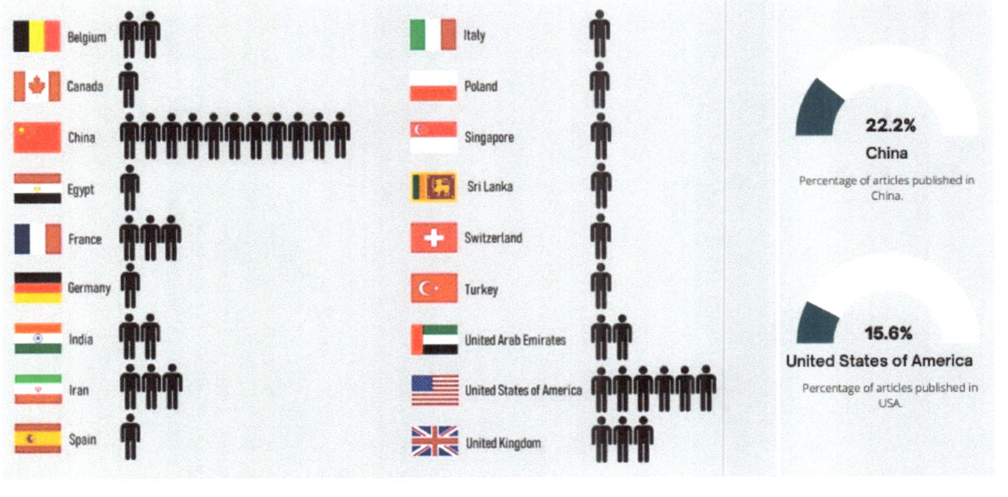

Figure 6. Country associated with the affiliation of the first author of the papers included in the review.

Figure 7. Continents associated with first author affiliation of the papers included in the review. The continents are separated by color and the numbers indicate the number of publications per continent.

The trend of publications by year was also analyzed in this paper. As illustrated in Figure 8, among the 45 considered papers, the first one was published by Toutain et al. [66], in 2015, being the only paper that year. In 2016, there was again only one publication. In 2017 and 2018, the number increased to two publications per year. In 2020, with eight papers published, the growth was 166.67% compared to the previous year. A growth in the number of publications was observed in 2021, when ten papers were published. In 2022, one observes eleven publications. It can also be inferred that the recent development of the research field that makes use of GSP and ML techniques is evident, which can be observed with the beginning of publications in 2015 and the growth in subsequent years.

Figure 8 also presents the number of papers published per year by specialty; it corroborates the emergence of papers that use GSP, ML, and DL for neurology applications, which represents 66.7% of the 45 evaluated studies. However, it is evident that the research field that makes use of GSP and DL techniques is very recent, since the first paper found in this study was published in 2015. On the other hand, it can be said that the area is under consolidation, with the remarkable growth in the number of publications in recent years: in the period from 2020 to 2022, 64.4% of papers were published.

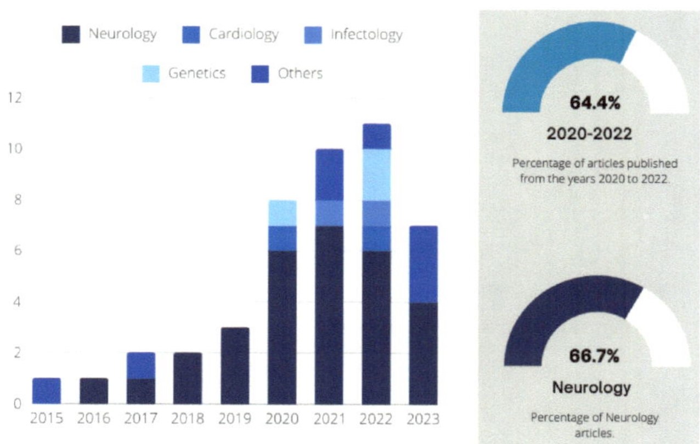

Figure 8. Distribution of publications by year and medical specialty.

Figure 9 shows the eleven areas with publications by means of the tree map, in which the sizes of the squares of the specialties are proportional to the number of publications. Thus, considering the universe of 45 papers selected for this review, neurology is the most prominent (30), followed by genetics (3), cardiology (2), infectology (2), oncology (2), gastroenterology (1), medical clinic (1), cytology (1), psychiatry (1), pneumology (1), and hepatology (1).

Figure 9. Tree map with the distribution of papers by medical specialty.

Figure 10 shows a bar chart of the number of Web of Science citations of the five most cited papers. The paper by Parisot et al. [67] is indicated as the most cited, with 242 citations. Pervaiz et al. [68] ranks second, with 95 citations. There are 38 works that use the study by Sardellitti, Barbarossa, and Lorenzo [69] as a reference, a number reasonably close to the fourth most cited, the work by Hu et al. [70], which has 29 citations. Finally, Zhang et al. [71] ranks fifth, with 22 citations. It can be inferred that there is a considerable difference in the number of citations of [67] compared to the others, which may indicate this work as recommended reading in the area.

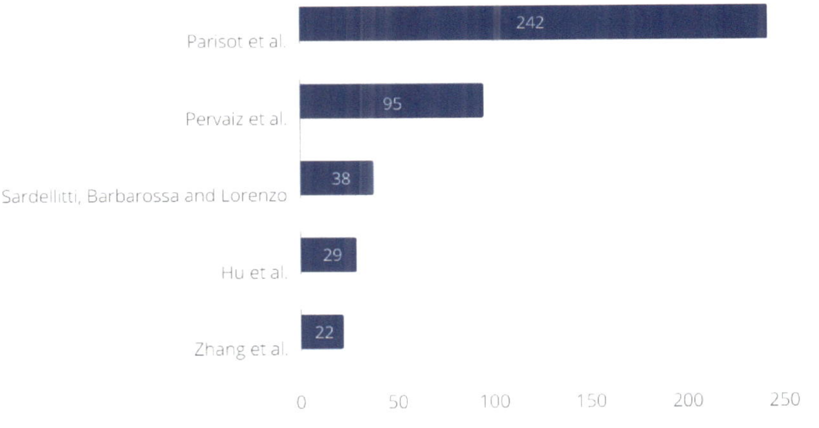

Figure 10. The five most cited papers according to the Web of Science database.

Figure 11 shows a map of citations obtained with VOSviewer [72]. The map is made up of spheres, labeled with the names of the first authors of the most cited papers and with sizes related to the number of citations received. It is also possible to see the five most cited papers, as shown in the previous figure. In general, the other nodes have similar sizes, indicating that they have received a similar number of citations, reaching a maximum of 21.

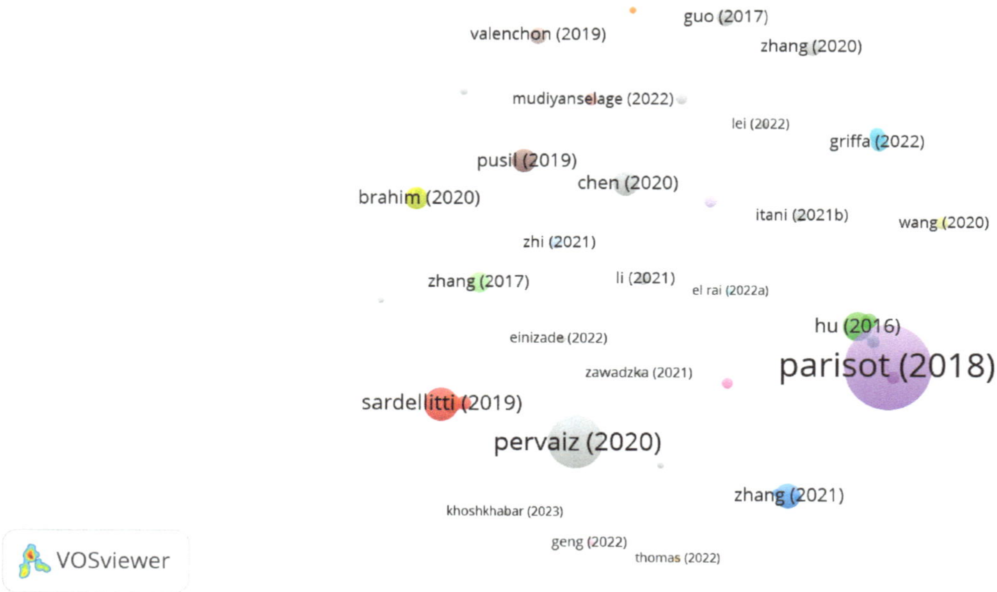

Figure 11. Most cited papers represented by the first author.

Figure 12 shows a bibliometric coupling network obtained through analysis in VOSviewer. The nodes of the graphs represent the first authors of the papers, and the size of the vertex is related to the number of citations of the paper. The edges connect the nodes that are bibliographically linked when there is another publication that is cited by the simultaneously linked papers.

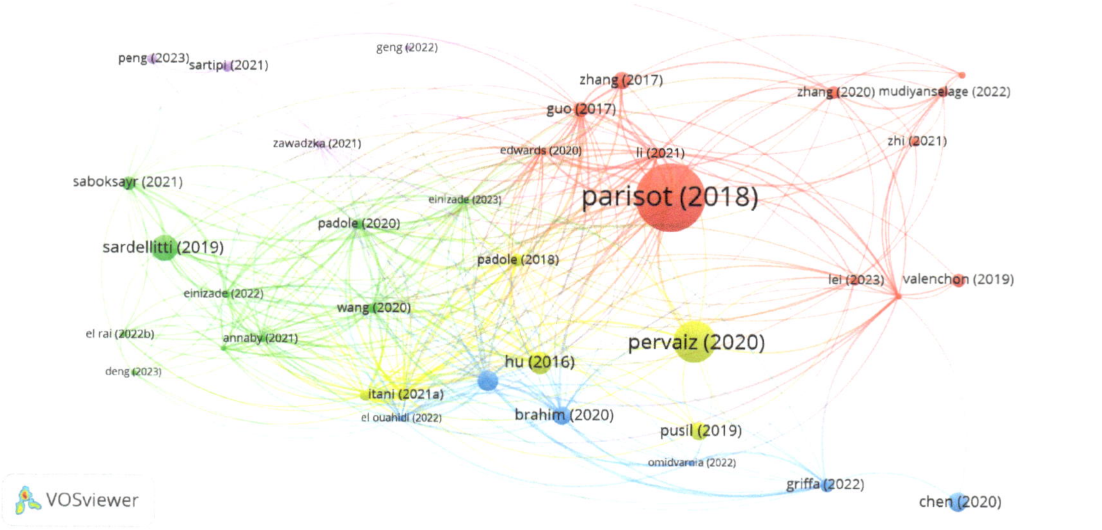

Figure 12. Graph network representation for bibliometric coupling analysis.

In order to establish a relationship between each paper and its respective application in the studied health area, Table 1 allocates the 45 evaluated papers to their respective specialty among the eight identified specialties.

Table 1. Distribution of the papers according to the medical specialty related to the proposed application.

Medical Specialty	Ref.
Neurology	[22–25,67–69,71,73–94]
Cardiology	[95,96]
Infectology	[97,98]
Genetics	[99–101]
Oncology	[102,103]
Gastroenterology	[104]
Medical Clinical	[105]
Cytology	[66]
Psychiatry	[106]
Pneumology	[107]
Hepatology	[108]

Table 2 presents the main information extracted from the studied papers. The presented descriptive data refer to the year of publication, the objective of the developed research, the technique proposed in the paper, and with whom it was compared to in order to evaluate its performance. Table 3 presents a set of information on the dataset used in the selected studies, such as the dataset used, the sample size used, and finally, the metric used to evaluate performance.

According to Table 3, it is possible to identify the five most used metrics in evaluating the performance of the proposed models, as shown in Figure 13. Accuracy occupies the first place. It is used to assess performance in 26 out of the 45 studies analyzed. In the second place, we observe the F1-score, which appears in 12 works. The AUC holds the third position. It is used in nine papers. In the fourth position, the measures precision and recall are tied. They are used in 8 out of the 45 selected papers. Finally, in the fifth position, we have sensitivity and specificity, which were used in six works.

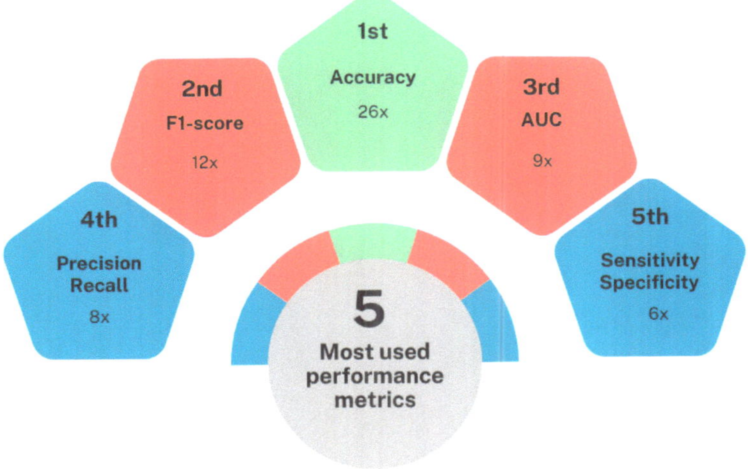

Figure 13. Diagram with the 5 most frequent performance measures in the works analyzed.

Table 2. Information from the papers included in the systematic review.

Ref.	Year	Objective of the Study	Proposed Technique	Techniques Used for Comparison
[66]	2015	Image preprocessing, segmentation, and classification of Feulgen- and Papanicolaou-stained slides.	Partial difference equations on weighted graphs.	Graph cuts, random walks, shortest-path algorithms, maximum spanning forests, and power watershed algorithm.
[79]	2016	Alzheimer's detection.	Matched signal detection (MSD) theory for signals on graphs (simple-MSD, constrained-MSD, probabilistic-MSD).	Principal component analysis (PCA), support vector machine (SVM), and linear discriminant analysis (LDA).
[82]	2017	Brain image data modeling and extraction.	Autoencoders for analysis of high-dimensional graph signals.	PCA, robust PCA (RPCA), graph-based filtering (GBF), and stacked autoencoder (SAE).
[105]	2017	Prediction of risk of comorbidities.	Heterogeneous convolutional neural network (HCNN), based on predictive learning.	Logistic regression (LR) and standard CNN.
[67]	2018	Autism and Alzheimer's classification	GCN for population analysis.	Random forest (RF) and multi-layer perceptron (MLP).
[78]	2018	Alzheimer's detection.	Graph frequency analysis for highly discriminative feature extraction and GCNN-based classifier.	MSD-G [79], RsBN-DL [109], Sparse-Cov [110], and EN-LogReg [111].
[69]	2019	Orthonormal data transformation applied to images of patients with epilepsy.	Orthonormal sparsifying transform and graph Fourier transform (GFT).	SpecTemp [112], Kalofilias [113], and Dong et al. [114]
[80]	2019	Alzheimer's detection.	Multiple feature-specific adjacency matrices for learning using GCNN.	Linear SVM, MLP, RF, Parisot et al. [115], and Vivar et al. [116].
[89]	2019	Predicting cognitive impairment in Alzheimer's disease (AD).	Multifrequency dynamic network analysis for building a connectome biomarker.	PCA.
[83]	2020	Attention deficit hyperactivity disorder (ADHD) detection.	GSP and GL to obtain structural and functional characteristics.	MLP with double input symmetrical relevance (DISR) and MLP with minimum redundancy maximum relevance (mRMR).
[77]	2020	Detection of central brain regions.	GFT based on Laplacian learning for analyzing graphs in the frequency domain.	Radial basis function (RBF) kernel and Pearson correlation methods for calculating the graph Laplacian.
[22]	2020	Alzheimer's detection.	Graph coarsening in a GCNN.	Heavy Edge [117], Kron Reduction [118], and spectral approximation [119].
[74]	2020	ADHD classification.	Dual-subspace classification algorithm using individual resting state functional connectivity.	RMf, fusion fMRI, R-Relielf, L1BioSVM, FCNet, 3D-CNN, and Deep fMRI
[73]	2020	Autism classification.	GFT and ML for analyzing the test and time series to calculate descriptive statistics for the region of interest.	[120–126].
[68]	2020	Classification of neurological function.	Graph-based modeling of the brain's functional connectivity with elastic net and independent component analysis (ICA).	RF, Dictionary Learning, and Higher Dimensional YEO parcellation.
[100]	2020	Prediction of RNA association with disease.	Graph attention adversarial network (GAAN), based on the integration of state-of-the-art GCN and the attention mechanism.	Ding's method [127], RWRMDA [128], TPGLDA [129], RLSMDA [130], GCN, GAT, and GAN.

Table 2. *Cont.*

Ref.	Year	Objective of the Study	Proposed Technique	Techniques Used for Comparison
[96]	2020	Aortic root segmentation.	Multi-resolution graph using irregularly spaced patch sampling and a graph-based CNN as a classifier.	Hand-crafted and Fully connected graph.
[102]	2021	Gene selection for cancer detection.	Algorithm for selecting significant genes with GSP techniques, using the Laplacian matrix of the graph.	Locally linear embedding (LLE) and PCA.
[81]	2021	Emotion recognition.	Spatio-temporal attention neural network with GFT signals as input.	Multi-column convolutional neural network (MCNN) [131] and bidirectional long short-term memory (BiLSTM) [132].
[23], [84]	2021	Autism classification.	Connectivity matrix with GFT values, extension of the Fukunaga–Koontz transform for feature extraction to train the decision tree (DT).	Spatial filtering method and the GFT.
[103]	2021	Lung cancer detection.	Multi-graph neural network (MGNN) with three models: GIAN, GIAT, and SGCA.	ML algorithms, RF and support vector regression (SVR).
[71]	2021	Multidomain brain decoding.	Multidomain decoding model on short time series incorporating Laplacian graph with GCN.	Classical brain decoding model, which applies multi-class linear SVM.
[97]	2021	Identification of the focus of disease spread.	GSP, GCN, and neighborhood loss calculation to optimize the average error distance.	Label propagation framework for source identification, Unbiased Betweenness.
[88]	2021	Motor imagery classification.	Graph-theoretic models of multichannel EEG signals with multivariate autoregressive models for directed graphs and extreme learning machine classifiers.	SVM, K-nearest neighbor classifiers (KNN), and Extreme Learning Machines (ELMs).
[90]	2021	Emotion classification/epileptic seizure analysis.	GFT for the extraction of discriminative features used in learning tasks and the proximal gradient method for data acquired in real time.	[113,133] and SVM.
[91]	2021	Emotion recognition and analysis.	GSP to integrate emotion recognition and analysis of signals.	–
[25]	2022	Brainwave decoding.	Fusion of GSP and GL resources for a method called graph-based imagined speech BCI decoder (GraphIS).	–
[75]	2022	Task decoding and individual fingerprinting	SVM classification and GSP functional data filtering for functional connectivity and structural connectome decomposition.	–
[85]	2022	Elimination of noise from epileptic EEG signals.	Unified objective function for GraphJADE with GL and use block coordinate descent to optimize it.	Unified objective function GraDe with GL and the blind separation methods.
[95]	2022	Left ventricular segmentation in echocardiography videos.	GraphECV with GSP for semi-supervised learning and minimization of the Sobolev norm of graph signals.	PReMVOS [134], TMANet [135], Accel [136], and OSVOS [137].
[86]	2022	Evaluation of how brain activity changes over time.	GSP, SVM, and multiscale entropy.	–
[87]	2022	Brain Activity Classification.	End-to-end GCN structure with three convolutional layers.	NetMF [138], RandNE [139], Node2Vec [138], and Walklets [140].

Table 2. *Cont.*

Ref.	Year	Objective of the Study	Proposed Technique	Techniques Used for Comparison
[99]	2022	Detection of metabolic diseases.	GCN to infer potential metabolite–disease association, named MDAGCN.	Traditional methods based on biological experiments.
[98]	2022	Study on the contagion dynamics of COVID-19.	Wavelet transform of spectral graph to process data in dynamic graph for spatio-temporal pattern detection.	–
[101]	2022	Prediction of circRNA association with diseases.	Two GCN-based prediction models: Node Classification and Link Prediction.	Other baseline models.
[104]	2022	Early diagnosis and detection of gastrointestinal polyps.	Semi-supervised segmentation called SemiSegPolyp, based on GSP. It is divided into instance segmentation, construction of graphs based on nearest neighbors, and semi-supervised semantic segmentation.	Mean-Teacher, generative adversarial networks (GANs), Cross-Consistency Training, and Wu [141].
[93]	2022	Understanding what are the most useful graph frequencies to decode fMRI signals.	Spectral ResNet, in which the frequencies of the graphs define the convolutions.	MLP pattern, where the input domain is the frequency domain of the graph.
[76]	2023	Detection of mild cognitive impairment (MCI).	Multiscale enhanced GCN.	SVM, two-layer GCNs, and multi-scale GCN with the same normalized adjacency matrix.
[92]	2023	Clinical follow-up to assist in the diagnosis of inflammatory bowel diseases.	GFT, GSP, and classical SVM are used to classify the features.	Graph theory analysis method.
[24]	2023	Emotion analysis in EEG.	Coding of relative temporal transformation and attention to the channel.	GCNN, SVM, CNN + recurrent neural networks (RNNs).
[94]	2023	Classification of sleep stages.	Adaptive GCN, named ProductGraphSleepNet, which exploits GSP and product graph learning (PGL).	SVM, RF, MLP+LSTM, DeepSleepNet, CNN, RF+ Hidden Markov Model (HMM), U-Sleep, SeqSleepNet, SleepECL, fractional Fourier transform (FRFT), catBoost, LR, second-order blind source separation (SOBI)-wavelet Transform (WT), ProductGraphSleepNet, SSL-ECG, SimCLR, TS-TCC, time–frequency features, multitaper spectral + CNN, intra-/inter-epoch BiLSTM, FRFT, NAS, Cascaded CNN+LSTM.
[106]	2023	Discover how default mode network (DMN) alignment is related to symptoms of depression and rumination.	Graph signal processing-based analyses in a transdiagnostic cohort.	-
[107]	2023	Evaluation of the quality of the photoplethysmography (PPG) signal.	Analysis of graph signals using six machine learning classifiers: RF, DT, SVM, MLP, CNN, and Naive Bayes (NB).	Comparison of the six classification techniques mentioned.
[108]	2023	Identification of liver organs and segmentation of liver tumors.	Simple Linear Iterative Clustering (SLIC) algorithm for clustering liver computed tomography (CT) images and convolutional graph networks with four Chebyshev graph convolution layers and one fully connected layer to detect liver organs and segment liver tumors.	Modified U-Net and Shortcut CNN.

Table 3. Summary of information regarding the dataset used and metrics considered in the performance evaluation of the proposed models.

Ref.	Dataset Used	Dataset Description	Evaluated Metrics
[66]	GrabCut, MNIST, OPTDIGITS, and PENDIGITS.	MNIST, OPTDIGITS, and PENDIGITS datasets are composed of handwritten digits.	Error measures and classification rates.
[79]	PIB-PET dataset and ADNI.	PIB-PET dataset is composed of PET neuroimages and consists of 30 patients with Alzheimer's disease (AD) and 40 healthy control (HC) subjects; ADNI dataset is public and consists of resting-state fMRI, containing images from 30 individuals with early MCI and 20 NC subjects.	Accuracy, sensitivity, specificity, and area under the curve (AUC).
[82]	Real MEG datasets.	MEG signals collected by 306 sensors were considered. Brain activity was captured by the participants' reaction to seeing 322 images of human faces and 197 images of objects that were shown randomly.	Accuracy.
[105]	Electronic Health Record (EHR) data.	The data consist of the medical records of 3048 patients with congestive heart failure; 18,451 with diabetes; 3948 with chronic kidney disease; 7700 patients with chronic obstructive pulmonary disease.	Precision, recall, and F1-score.
[67]	ABIDE; ADNI.	ABIDE is a public dataset of functional NMR and phenotypic data. It considered 403 individuals with spectrum disorder and 468 HC; in ADNI, 1675 samples were available, with 289 individuals (843 samples) diagnosed with AD.	AUC.
[78]	ADNI.	It considered 100 subjects with MCI and 100 HC subjects.	Accuracy.
[69]	One synthetic dataset and one real dataset [142].	The real dataset has only one epilepsy patient and 76 time series.	Correlation coefficient, percentage of recovery errors, F1-score, precision, and recall.
[80]	TADPOLE.	779 subjects, 296 MCI converters, and 483 MCI non-converters.	AUC.
[89]	Collected for the paper.	MEG recordings were obtained in 54 patients with MCI aged 65-80 years. They were divided into two groups according to their clinical outcome: (1) the "progressive" MCI group ($N = 27$) was composed of the individuals who met the criteria for probable AD; (2) the "stable" MCI group ($N = 27$) was composed of the participants who still met the criteria for a diagnosis of MCI.	Classification performance, sensitivity, and specificity.
[83]	Online dataset.	Public dataset with EEG signals from normal and ADHD children aged 7–12 years.	Accuracy.
[77]	Collected for the paper.	Task-based resting-state fMRI images. The participants were divided into two categories: young adults, aged 18–22 (119 women, 79 men); children, aged 8–12 (108 women, 83 men).	F1-score, recall, and precision.
[22]	ADNI.	Public, over 800 participants, including HC individuals with MCI and individuals with AD. The dataset included several classes of imaging: structural MRI, functional MRI, and PET scans, as well as clinical and cognitive assessments.	Operator dissimilarity index and cut index.
[74]	TDAH-200.	The resting state fMRI (rs-fMRI) data used to investigate the binary classification performance between ADHD and HC subjects.	Accuracy.
[73]	ABIDE.	fMRI images of 871 subjects were considered, 403 subjects with autism spectrum disorder (ASD) and 468 HC.	Accuracy, sensitivity, and specificity.
[68]	UKB; HCP.	fMRI data from the UK Biobank (UKB), which consists of 13,301 individuals; HCP of 1003 HC.	Accuracy/correlation.

Table 3. *Cont.*

Ref.	Dataset Used	Dataset Description	Evaluated Metrics
[100]	HMDD; LncRNADisease.	HMDD is a public dataset on miRNA diseases. A miRNA–disease network with 208 miRNAs, 250 diseases, and 3644 links was considered; LncRNADisease dataset is public and provides information on lncRNAs and diseases with over 200,000 lncRNA–disease associations across 529 diseases and 19,166 lncRNAs.	AUC and prediction results.
[96]	An example on aortic valve.	Human torso CT samples are considered for studying the aortic root.	Accuracy.
[102]	Three datasets [143].	Public genetic datasets. In the prostate cancer dataset, there are 50 normal prostate samples and 52 prostate tumor samples, each sample with 10,509 different genes. The gastric cancer dataset contains 40 samples, 20 of which are from normal patients and another 20 from gastric cancer patients, each sample with 10,519 genes. In the brain dataset, two classes are considered, both brain tumors, glioblastoma with 20 samples and oligodendroglioma with 30 samples, each sample with 10,367 genes.	Accuracy.
[81]	DEAP.	EEG of 32 subjects, each having rated 40 music videos of a one-minute duration.	Accuracy.
[23]	ABIDE I.	Dataset includes eyes open rs-fMRI. It considered 251 HC and 201 ASD, all adolescents. Adults, 67 HC and 63 ASD, were also included.	Accuracy.
[84]	ABIDE I.	Dataset in which patients with eyes open during the fMRI session were considered; less than 18 years old; resulting in 251 HC subjects and 201 subjects with ASD.	Accuracy.
[103]	STRING (version 11.0).	Ten proteins were considered to build the protein–protein interaction (PPI) network, which was generated and visualized from the STRING database.	Root-mean-squared error (RMSE).
[71]	HCP.	Task-MRI and rs-MRI acquired from 1200 HC, corresponding to the response to different cognitive tasks.	Accuracy, precision, and recall.
[97]	USC-TIMIT.	rtMRI videos of the upper airway in the mid-sagittal plane and the corresponding speech waveforms of 5 female and 5 male subjects.	Accuracy, precision, false positive, and false negative.
[88]	BCI Competition II; Dataset 1 from BCI Competition IV.	2003 BCI competition dataset EEGs were collected from 1 HC. BCI Competition IV dataset. Continuous EEGs were obtained from 6 HC.	Accuracy and AUC.
[90]	DEAP and synthetic dataset.	Public, peripheral EEG and physiological signal data from 32 participants. Participants watched 40 videos and rated them according to the levels of valence, arousal, liking/disliking, dominance, and familiarity.	Classification accuracy and similarity between the learned graph and the ground truth.
[91]	AMIGOS; ASCERTAIN; DEAP.	The AMIGOS dataset consists of data collected from 40 participants and stores EEG, ECG, and GSR signal data; the ASCERTAIN dataset contains experimentally sourced data from 58 users viewing affective videos, along with EEG, ECG, GSR, and facial activity data; the DEAP dataset has data from 32 participants, and 40 1-min clips of music videos were used as stimuli for the participants.	Accuracy and F1-score.
[25]	iBCIC2020 Competition.	EEG signals from 15 individuals (5 females). The mean age was 31 years, and all subjects were healthy and right-handed.	Accuracy.
[75]	HCP.	100 HC HCP unrelated subjects from the HCP U100 dataset, fMRI acquired with 8 different task conditions (resting state and 7 tasks: emotion, play, language, motor, relationship, social, working memory).	Accuracy.

Table 3. *Cont.*

Ref.	Dataset Used	Dataset Description	Evaluated Metrics
[85]	Epileptic EEG Data; TSP speech dataset.	For the EEG database, 50 tests of pre-ictal/epileptic ictal EEG signals were carried out. TSP speech is a public dataset, and an utterance of about 2 s duration uttered by a male and a female speaker was considered.	Interference-to-source ratio (ISR), relative graph estimation error (RGEe), AUC, F1, and MD
[95]	Econet-Dynamic; CAMUS.	EchoNet-Dynamic Dataset with 10,030 echocardiography videos; CAMUS dataset contains the medical exams of 500 patients.	Dice coefficient (DC) or F1-score.
[86]	HCP1200 release.	Consists of functional magnetic resonance imaging (fMRI) recordings from 20 HC adult participants. The dataset includes four rs-fMRI recordings, seven task-based fMRI recordings, and one diffusion fMRI recording.	Two measures of temporal complexity: the Hurst exponent and multiscale entropy.
[87]	HCP 1200 Subject Release (S1200).	fMRI data for 302 participants, consisting of 164 females and 138 males (22–35 years, mean = 28.7 ± 3.6). The fMRI data were collected while the participants performed 7 different tasks: emotion, game, working memory, language, relational, social, and motor.	Accuracy, balanced accuracy, F1-scores (macro, micro, and weighted), Matthews correlation coefficient (MCC), precision, and recall.
[99]	HMDB 4.0; CTD; DisGeNET.	The HMDB dataset has 1478 metabolites, 237 diseases, and 3460 known metabolite–disease associations, removing missing and duplicate data. For information on disease-related genes, obtained 3102 genes from the comparative toxicogenomics dataset (CTD) and DisGeNET.	AUC, area under precision–recall (AUPR), F1-score, accuracy, recall, specificity, and precision.
[98]	[144].	It includes data on COVID contamination in the population of the city of Massachusetts from 6 December 2020 to 25 September 2021, for 41 weeks in total, which is collected from the official website.	Anomaly score (a-score).
[101]	circR2Disease.	It considered 431 circRNA-disease associations, which included 365 circRNAs related to 100 diseases from circR2Disease.	Accuracy, precision, recall, F1-score, and AUC.
[104]	Kvasir-SEG; CVC-ClinicDB.	Kvasir-SEG is an open-access dataset of gastrointestinal polyp images, which contains 1000 polyp images; the public and open-access CVC-ClinicDB is composed of 612 image frames extracted from 31 different colonoscopy.	Mean intersection-over-union (mIOU).
[93]	Neurovault; HCP.	Functional MRI signals consisting of 13 subjects with many task experiments and 788 HCP subjects.	Accuracy.
[76]	ADNI.	Total number of 184 subjects in this study. 40 late MCI (LMCI) patients, 77 early MCI (EMCI) patients, and 67 HC.	Accuracy, sensitivity, specificity, F1-score, and AUC.
[92]	Collected for the paper.	It includes 30 patients with inflammatory bowel disease, 13 men and 17 women, mean age (35.3 ± 5.2) years, all right-handed. At the same time, there were 30 HC patients, including 16 males and 14 females, mean age (31.5 ± 2.9) years, all right-handed.	Accuracy, sensitivity, specificity, and F1-score.
[24]	DEAP.	Public dataset with EEG signals from 32 participants when watching 40 60-s video clips. Subjects (50% men and 50% women) were between 19 and 37 years old.	Accuracy.
[94]	Montreal Archive of Sleep Studies (MASS) SS3; SleepEDF.	Full-night polysomnographic recordings. In MASS-SS3, 62 and in SleepEDF 20 healthy individuals were considered	Accuracy, F1-score, and Kappa.
[106]	Collected for the paper.	A total of 79 participants with complete data, with 19 HCs and 60 patients, of which 31 in the cognitive behavioral therapy (CBT) group and 29 in the selective serotonin reuptake inhibitor (SSRI) group.	The statistics are Pearson's r and p values.

Table 3. *Cont.*

Ref.	Dataset Used	Dataset Description	Evaluated Metrics
[107]	MIT-BIH; Medical Information Mart for Intensive Care (MIMIC); Beth Israel Deaconess Medical Center (BIDMC) PPG and Respiration Dataset; Wrist PPG During Exercise dataset; CAPNOBASE—TBME RR benchmark dataset; Complex System Laboratory (CSL) Pulse Oximetry Artifact Labels.	Datasets with different types of normal and abnormal PPG signal patterns, as well as noisy PPG signals in real time.	Accuracy, processing time, and model size.
[108]	Liver Tumor Segmentation 2017 (LiTS17).	The dataset contains images of 130 patients with a maximum number of CT slices of 623 for each patient. For this study, CT volumes of 10 patients were considered.	Accuracy, Dice coefficient, mean intersection-over-union (IoU), sensitivity, precision, and recall.

Another analysis obtained from Table 3 concerns the most used databases. Considering the area of neurology, which corresponds to 30 of the 45 articles included in the review, Figure 14 shows that the most used databases were HCP, used 6 times, followed by ADNI with 5 uses, and then ABIDE and DEAP, which were used in 4 papers each. It is important to note that those databases are publicly available.

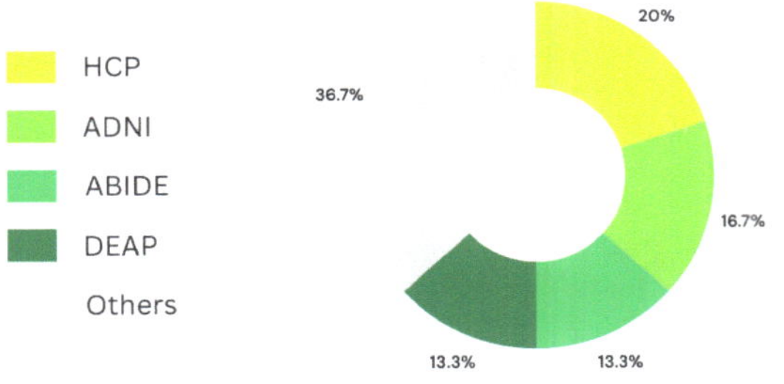

Figure 14. Graph showing the most frequently used databases in neurology.

5. Discussion

One of the challenges reported by the analyzed papers is related to the difficulty of accessing health-related datasets. The limited amount of data (whether images, signals, or medical records, among others) may lead to a lack of generalization of the proposed approaches in the detection or classification of pathologies. Another challenge is related to the reproducibility of research, since different research groups are unable to evaluate new methodological proposals for the reported problem if a common dataset is not available. There is need for more publicly available datasets.

Among other limitations addressed by the papers, we can mention data imbalance. In [67], for example, it is mentioned that, in future studies, one of the intentions is to verify the use of graph convolutions to achieve good prediction rates in problems that present data imbalance, since it is considered a factor that hinders the learning of intelligent systems.

Regarding the dataset, one possibility to achieve better performances would be to include complementary information to signals or images. This is due to the fact that, in health-related problems, it is relevant to use dataset with a combination of data, such as phenotypic information, because diagnoses may be related to morphological characteristics or conditions and clinical parameters. In [78], for example, it is stated that the work has a limitation because it uses only brain image data; better results could be obtained if the referred additional information would have been employed.

Another issue to be considered concerns the medical interpretation of the results obtained by systems using GSP and computational intelligence techniques. Although many proposals achieve good performance, considering the evaluation of objective metrics, which are quite widespread in engineering, it is of paramount importance that there is an understanding of the addressed problem, based on the understanding of what the result means, and also how it impacts the analysis in the health sciences, in order to achieve a broader and more complete analysis for diagnosis. In this context, Valenchon and Coates [80] report their intention that, in further research, rather than the outcome of the proposal indicating whether or not an individual with dementia will progress to Alzheimer's disease, it presents a mechanism that provides a value of the probability of progression, which guarantees a more complete medical analysis. Another example is [68] in which it is suggested that future work may address possibilities beyond the prediction of clinical diagnosis, one of them being to investigate and suggest possible treatments for the found medical condition.

In relation to this, an area that addresses such issues and presents possible solutions to minimize these difficulties is explainable artificial intelligence (XAI) [145]. This is a recent field that concerns the explanations and interpretability necessary in processes that use artificial intelligence techniques in predictions, so that there are justifications for and credibility of the obtained results [146]. In any case, interpretability and explainability are terms that encompass standards and criteria, which must be taken into account according to the associated context [147]. In the case of this review, the context to consider is the medical specialty of the application, and then make the proposal based on computational intelligence understandable to health professionals, knowledgeable about the nature of the problem, reducing the gap between the proposal and clinical practice. In [79], for example, a system for detecting and evaluating signals of neurological examinations for the early detection of Alzheimer's disease was developed; according to the authors, it would be interesting for health professionals to understand the approach devoted to signal detection, with the aim of enabling the use of programs that validate the proposal in a real medical context.

In this context, such issues fall under health 4.0 (H4.0), a term used to relate health advances to industrial technological revolutions. It corresponds to a field that investigates the use of technologies in favor of patient care, based on the use of technology to promote better and faster diagnostic capacity, equipment portability, and greater data management capacity [148]. Thus, the use of technology is aimed at clinical care itself, and can be supported with the use of artificial intelligence, including deep learning techniques for care aimed at early diagnosis, the prevention of the progression of health conditions, and early identification of effective treatments [149,150].

An important challenge verified in the analysis of the papers included in the review is the extraction of characteristics from the data used, since this step is not restricted to the extraction itself, but to the selection of more relevant characteristics so that the proposed system for the intended application is able to identify health changes due to the selection of more significant characteristics. In this case, the use of convolutional networks can be considered, since the convolution layers play the role of the extractor of features.

Another challenge is the selection of optimal hyperparameters for the proposed techniques, because although they present a good performance, as reported in [22], it is possible to achieve superior results with an assertive selection of hyperparameters. The choice of hyperparameters can be made using grid search, Bayesian optimization, or random search and swarm intelligence.

Many applications addressed in the selected papers are in neurology, in which the data evaluated are examinations converted into time series. In this sense, an important analysis to aid diagnosis is to check the regularity of the series and identify noise. That analysis can be carried out by using information theory metrics on graphs, such as permutation entropy and dispersion entropy [151,152]. This could be a promising research area in graph signal processing.

Regarding the works in which the specialty of neurology is considered, a frequently considered analysis employs functional connectivity. Therefore, accurate pattern recognition is essential, which can be achieved through the use of robust graph learning techniques, acting in the identification and analysis of connectivity between brain areas. Another possibility of analysis is structural connectivity. According to [75], in future work, a specific structural connectivity for each individual should be considered. This would lead to the definition of multiple spectral domains for brain signals, and would enable the analysis of inter-subject structural variability. Still in this specialty, many studies report the use of atlases to divide the brain into areas. However, there is no consensus on the use of a single atlas to carry out the referred division. It would, therefore, be interesting to verify and test the use of different atlas options for the same dataset, since this choice has a high potential impact on the final classification stage.

Due to the good results presented with different techniques that combine GSP and ML, it is possible to use graph neural networks and test the proposed methods in different medical applications of high complexity and that have data available in the literature, as suggested in [66,79]. Considering high-complexity problems, one of the future proposals reported in [95] concerns real-time processing for echocardiogram videos. This could be a major advance in early diagnosis with artificial intelligence, and would represent a significant impact for health sciences.

The proposal described in [74], which employs the modified Laplacian matrix to classify attention deficit hyperactivity disorder, presents a promising result, so its use can be considered as an alternative mathematical framework in other medical applications. Likewise, in [103], it is recommended to explore the potential of Multi-GNNs, which consist of combining the characteristics of individual GNNs.

An interesting consideration concerns the use of new transforms, because, although the Fourier transform is quite widespread and leads to good results, it is important that different transformation techniques be examined and tested. In [88], for example, the investigation of new transforms is pointed out as a future proposal, as the authors mention the fact that new transformation techniques can lead to improvements in the classification rates.

Finally, an important issue is the lack of standardization, so it would be interesting to standardize metrics and evaluation techniques for comparison purposes.

6. Conclusions

In healthcare, GSP has been used to analyze problems related to signals lying in non-Euclidean domains. In addition, ML techniques have been used for pattern recognition and early disease classification and identification. Considering the 45 papers included in the systematic review, 30 of these presented applications for neurology problems, with many of them focused on the diagnosis of cognitive impairment and Alzheimer's disease. In these cases, most of the data correspond to fMRI and EEG images. However, limitations are reported regarding the number of samples and the number of publicly available dataset.

From the presented data regarding the number of publications, it is clear that, despite GSP with ML applied to health being a recent field of study, it has shown an increase in the number of publications, which may indicate an interest of the scientific community in the area. Advances in the scope of GSP with ML in health have attracted the attention of health professionals, since the proposed methods have a high capacity to assist early diagnosis and, consequently, provide speed in decision making by specialists. In any case, there are gaps to be solved, such as a better integration between computational intelligence techniques and clinical practice.

This systematic review synthesized the information from selected papers and pointed out the trends of applications that are emerging in the area, as well as methodologies that combine artificial intelligence, graph theory, and health sciences, presenting subsidies for researchers to explore gaps in future work, as well as to reproduce existing work. A limitation of this work is the number of scientific databases considered. Although our study has considered four relevant scientific databases (IEEE Xplore, Science Direct, MDPI, and ACM), it is possible that there are other papers that fit the scope of the review and that have not been included. It is also possible that new papers have been published after the period defined for the inclusion of papers, which was October 2023.

In the future, further updates of this literature review can be carried out, including more databases and also revisiting those considered in this paper, since, with the identified trend of publications, there should soon be new research published in the area.

Author Contributions: Conceptualization, M.A.A.C. and F.M.; methodology, M.A.A.C., F.A.B.S.F. and F.M.; validation, F.M. and F.A.B.S.F.; formal analysis, M.A.A.C., F.A.B.S.F., F.A.N.S., F.M. and J.B.L.; investigation, M.A.A.C., F.A.B.S.F. and F.M.; writing—original draft preparation, M.A.A.C.; writing—review and editing, M.A.A.C., F.A.B.S.F., F.A.N.S., F.M. and J.B.L.; visualization, M.A.A.C.; supervision, F.M. and J.B.L.; project administration, F.M. and J.B.L.; funding acquisition, J.B.L. All authors have read and agreed to the published version of the manuscript.

Funding: This research was funded by Conselho Nacional de Desenvolvimento Científico e Tecnológico (CNPq) under grants 140150/2022-6, 442238/2023-1, 312935/2023-4, and 405903/2023-5, Coordenação de Aperfeiçoamento de Pessoal de Nível Superior (CAPES) under grant 88881.311848/2018-01, and Fundação de Amparo à Ciência e Tecnologia do Estado de Pernambuco (FACEPE) under grant APQ-1226-3.04/22.

Institutional Review Board Statement: Not applicable.

Informed Consent Statement: Not applicable.

Data Availability Statement: Not applicable.

Conflicts of Interest: The authors declare no conflicts of interest.

References

1. Sandryhaila, A.; Moura, J.M. Big data analysis with signal processing on graphs: Representation and processing of massive data sets with irregular structure. *IEEE Signal Process. Mag.* **2014**, *31*, 80–90. [CrossRef]
2. Chen, S.; Varma, R.; Sandryhaila, A.; Kovačević, J. Discrete Signal Processing on Graphs: Sampling Theory. *IEEE Trans. Signal Process.* **2015**, *63*, 6510–6523. [CrossRef]
3. Giudici, P.; Sarlin, P.; Spelta, A. The interconnected nature of financial systems: Direct and common exposures. *J. Bank. Financ.* **2020**, *112*, 105149. [CrossRef]
4. Shi, W.; Rajkumar, R. Point-GNN: Graph neural network for 3d object detection in a point cloud. In Proceedings of the IEEE/CVF Conference on Computer Vision and Pattern Recognition, Seattle, WA, USA, 13–19 June 2020 ; pp. 1711–1719.
5. Dong, G.; Tang, M.; Wang, Z.; Gao, J.; Guo, S.; Cai, L.; Gutierrez, R.; Campbel, B.; Barnes, L.E.; Boukhechba, M. Graph neural networks in IoT: A survey. *ACM Trans. Sens. Netw.* **2023**, *19*, 1–50. [CrossRef]
6. Jiang, W.; Luo, J. Graph neural network for traffic forecasting: A survey. *Expert Syst. Appl.* **2022**, *207*, 117921. [CrossRef]
7. Ferrer-Cid, P.; Barcelo-Ordinas, J.M.; Garcia-Vidal, J. Data reconstruction applications for IoT air pollution sensor networks using graph signal processing. *J. Netw. Comput. Appl.* **2022**, *205*, 103434. [CrossRef]
8. Majeed, A.; Rauf, I. Graph theory: A comprehensive survey about graph theory applications in computer science and social networks. *Inventions* **2020**, *5*, 10. [CrossRef]
9. Mateos, G.; Segarra, S.; Marques, A.G.; Ribeiro, A. Connecting the dots: Identifying network structure via graph signal processing. *IEEE Signal Process. Mag.* **2019**, *36*, 16–43. [CrossRef]
10. Deng, S.; Wang, S.; Rangwala, H.; Wang, L.; Ning, Y. Cola-GNN: Cross-location attention based graph neural networks for long-term ILI prediction. In Proceedings of the 29th ACM International Conference on Information & Knowledge Management, Virtual Event, Ireland, 19–23 October 2020; pp. 245–254.
11. Xia, F.; Sun, K.; Yu, S.; Aziz, A.; Wan, L.; Pan, S.; Liu, H. Graph learning: A survey. *IEEE Trans. Artif. Intell.* **2021**, *2*, 109–127. [CrossRef]
12. Wu, Z.; Pan, S.; Chen, F.; Long, G.; Zhang, C.; Philip, S.Y. A comprehensive survey on graph neural networks. *IEEE Trans. Neural Netw. Learn. Syst.* **2020**, *32*, 4–24. [CrossRef]
13. Zhang, Z.; Cui, P.; Zhu, W. Deep learning on graphs: A survey. *IEEE Trans. Knowl. Data Eng.* **2020**, *34*, 249–270. [CrossRef]
14. Zhang, S.; Tong, H.; Xu, J.; Maciejewski, R. Graph convolutional networks: A comprehensive review. *Comput. Soc. Netw.* **2019**, *6*, 1–23. [CrossRef] [PubMed]
15. Kipf, T.N.; Welling, M. Semi-supervised classification with graph convolutional networks. *arXiv* **2016**, arXiv:1609.02907.
16. Wu, F.; Souza, A.; Zhang, T.; Fifty, C.; Yu, T.; Weinberger, K. Simplifying graph convolutional networks. In Proceedings of the International Conference on Machine Learning, Long Beach, CA, USA, 10–15 June 2019; pp. 6861–6871.
17. Sun, X.; Man, Y.; Zhao, Y.; He, J.; Liu, N. Incorporating description embeddings into medical knowledge graphs representation learning. In Proceedings of the Human Centered Computing: 4th International Conference, HCC 2018, Mérida, Mexico, 5–7 December 2018; Revised Selected Papers 4; Springer: Cham, Swizerland, 2019; pp. 188–194.
18. Wu, X.; Duan, J.; Pan, Y.; Li, M. Medical Knowledge Graph: Data Sources, Construction, Reasoning, and Applications. *Big Data Min. Anal.* **2023**, *6*, 201–217. [CrossRef]
19. Bullmore, E.; Sporns, O. The economy of brain network organization. *Nat. Rev. Neurosci.* **2012**, *13*, 336–349. [CrossRef] [PubMed]
20. Van den Heuvel, M.P.; Sporns, O. Network hubs in the human brain. *Trends Cogn. Sci.* **2013**, *17*, 683–696. [CrossRef]
21. Sporns, O. Structure and function of complex brain networks. *Dialogues Clin. Neurosci.* **2022**, *15*, 247–262. [CrossRef]
22. Padole, H.; Joshi, S.D.; Gandhi, T.K. Graph Wavelet-Based Multilevel Graph Coarsening and Its Application in Graph-CNN for Alzheimer's Disease Detection. *IEEE Access* **2020**, *8*, 60906–60917. [CrossRef]
23. Itani, S.; Thanou, D. Combining anatomical and functional networks for neuropathology identification: A case study on autism spectrum disorder. *Med. Image Anal.* **2021**, *69*, 101986. [CrossRef] [PubMed]
24. Peng, G.; Zhao, K.; Zhang, H.; Xu, D.; Kong, X. Temporal relative transformer encoding cooperating with channel attention for EEG emotion analysis. *Comput. Biol. Med.* **2023**, 106537. [CrossRef]
25. Einizade, A.; Mozafari, M.; Jalilpour, S.; Bagheri, S.; Sardouie, S.H. Neural decoding of imagined speech from EEG signals using the fusion of graph signal processing and graph learning techniques. *Neurosci. Inform.* **2022**, *2*, 100091. [CrossRef]
26. Bouman, P.M.; Noteboom, S.; Nobrega Santos, F.A.; Beck, E.S.; Bliault, G.; Castellaro, M.; Calabrese, M.; Chard, D.T.; Eichinger, P.; Filippi, M.; et al. Multicenter evaluation of AI-generated DIR and PSIR for cortical and juxtacortical multiple sclerosis lesion detection. *Radiology* **2023**, *307*, e221425. [CrossRef] [PubMed]
27. Ortega, A.; Frossard, P.; Kovačević, J.; Moura, J.M.; Vandergheynst, P. Graph signal processing: Overview, challenges, and applications. *Proc. IEEE* **2018**, *106*, 808–828. [CrossRef]

28. Sandryhaila, A.; Moura, J.M. Discrete signal processing on graphs. *IEEE Trans. Signal Process.* **2013**, *61*, 1644–1656. [CrossRef]
29. Ribeiro, G.B.; Lima, J.B. Graph signal processing in a nutshell. *J. Commun. Inf. Syst.* **2018**, *33*. [CrossRef]
30. Shuman, D.I.; Narang, S.K.; Frossard, P.; Ortega, A.; Vandergheynst, P. The emerging field of signal processing on graphs: Extending high-dimensional data analysis to networks and other irregular domains. *IEEE Signal Process. Mag.* **2013**, *30*, 83–98. [CrossRef]
31. Chung, F.R. *Spectral Graph Theory*; American Mathematical Society: Washington, DC, USA, 1997; Volume 92.
32. Bondy, A.; Murty, U. *Graph Theory*; Springer: London, UK, 2008.
33. Helm, J.M.; Swiergosz, A.M.; Haeberle, H.S.; Karnuta, J.M.; Schaffer, J.L.; Krebs, V.E.; Spitzer, A.I.; Ramkumar, P.N. Machine learning and artificial intelligence: Definitions, applications, and future directions. *Curr. Rev. Musculoskelet. Med.* **2020**, *13*, 69–76. [CrossRef] [PubMed]
34. Janiesch, C.; Zschech, P.; Heinrich, K. Machine learning and deep learning. *Electron. Mark.* **2021**, *31*, 685–695. [CrossRef]
35. Topuz, B.; Çakici Alp, N. Machine learning in architecture. *Autom. Constr.* **2023**, *154*, 105012. [CrossRef]
36. Mienye, I.D.; Sun, Y. A Survey of Ensemble Learning: Concepts, Algorithms, Applications, and Prospects. *IEEE Access* **2022**, *10*, 99129–99149. [CrossRef]
37. Shailaja, K.; Seetharamulu, B.; Jabbar, M. Machine learning in healthcare: A review. In Proceedings of the 2018 Second International Conference on Electronics, Communication and Aerospace Technology (ICECA), Coimbatore, India, 29–31 March 2018; IEEE: Piscataway, NJ, USA, 2018; pp. 910–914.
38. Bhardwaj, R.; Nambiar, A.R.; Dutta, D. A study of machine learning in healthcare. In Proceedings of the 2017 IEEE 41st Annual Computer Software and Applications Conference (COMPSAC), Turin, Italy, 4–8 July 2017; IEEE: Piscataway, NJ, USA, 2017; Volume 2, pp. 236–241.
39. Alloghani, M.; Al-Jumeily, D.; Mustafina, J.; Hussain, A.; Aljaaf, A.J. A systematic review on supervised and unsupervised machine learning algorithms for data science. In *Supervised and Unsupervised Learning for Data Science*; Springer: Berlin/Heidelberg, Germany, 2020; pp. 3–21.
40. LeCun, Y.; Bottou, L.; Bengio, Y.; Haffner, P. Gradient-based learning applied to document recognition. *Proc. IEEE* **1998**, *86*, 2278–2324. [CrossRef]
41. LeCun, Y.; Bengio, Y.; Hinton, G. Deep learning. *Nature* **2015**, *521*, 436–444. [CrossRef] [PubMed]
42. Traore, B.B.; Kamsu-Foguem, B.; Tangara, F. Deep convolution neural network for image recognition. *Ecol. Inform.* **2018**, *48*, 257–268. [CrossRef]
43. Badea, M.S.; Felea, I.I.; Florea, L.M.; Vertan, C. The use of deep learning in image segmentation, classification and detection. *arXiv* **2016**, arXiv:1605.09612.
44. Deng, L.; Yu, D. Deep learning: Methods and applications. *Found. Trends® Signal Process.* **2014**, *7*, 197–387. [CrossRef]
45. Khan, M.; Jan, B.; Farman, H.; Ahmad, J.; Farman, H.; Jan, Z. Deep learning methods and applications. In *Deep Learning: Convergence to Big Data Analytics*; Springer: Singapore, 2019; pp. 31–42.
46. Cong, S.; Zhou, Y. A review of convolutional neural network architectures and their optimizations. *Artif. Intell. Rev.* **2023**, *56*, 1905–1969. [CrossRef]
47. Hubel, D.H.; Wiesel, T.N. Receptive fields, binocular interaction and functional architecture in the cat's visual cortex. *J. Physiol.* **1962**, *160*, 106. [CrossRef]
48. O'Shea, K.; Nash, R. An introduction to convolutional neural networks. *arXiv* **2015**, arXiv:1511.08458.
49. Barragán-Montero, A.; Javaid, U.; Valdés, G.; Nguyen, D.; Desbordes, P.; Macq, B.; Willems, S.; Vandewinckele, L.; Holmström, M.; Löfman, F.; et al. Artificial intelligence and machine learning for medical imaging: A technology review. *Phys. Med.* **2021**, *83*, 242–256. [CrossRef]
50. Monti, F.; Boscaini, D.; Masci, J.; Rodola, E.; Svoboda, J.; Bronstein, M.M. Geometric deep learning on graphs and manifolds using mixture model CNNs. In Proceedings of the IEEE Conference on Computer Vision and Pattern Recognition, Honolulu, HI, USA, 22–25 July 2017; pp. 5115–5124.
51. Zhou, J.; Cui, G.; Hu, S.; Zhang, Z.; Yang, C.; Liu, Z.; Wang, L.; Li, C.; Sun, M. Graph neural networks: A review of methods and applications. *AI Open* **2020**, *1*, 57–81. [CrossRef]
52. Defferrard, M.; Bresson, X.; Vandergheynst, P. Convolutional neural networks on graphs with fast localized spectral filtering. *Adv. Neural Inf. Process. Syst.* **2016**, *29*.
53. Khambhati, A.N.; Sizemore, A.E.; Betzel, R.F.; Bassett, D.S. Modeling and interpreting mesoscale network dynamics. *NeuroImage* **2018**, *180*, 337–349 [CrossRef] [PubMed]
54. Dong, S.; Wang, P.; Abbas, K. A survey on deep learning and its applications. *Comput. Sci. Rev.* **2021**, *40*, 100379. [CrossRef]
55. Li, R.; Yuan, X.; Radfar, M.; Marendy, P.; Ni, W.; O'Brien, T.J.; Casillas-Espinosa, P.M. Graph Signal Processing, Graph Neural Network and Graph Learning on Biological Data: A Systematic Review. *IEEE Rev. Biomed. Eng.* **2023**, *16*, 109–135. [CrossRef] [PubMed]
56. Yin, D.; Kaiser, M. Understanding neural flexibility from a multifaceted definition. *NeuroImage* **2021**, *235*, 118027. [CrossRef] [PubMed]
57. Tian, Y.; Liu, M.; Sun, Y.; Fu, S. When liver disease diagnosis encounters deep learning: Analysis, challenges, and prospects. *iLIVER* **2023**, *2*, 73–87. [CrossRef]

58. Heimerl, F.; Lohmann, S.; Lange, S.; Ertl, T. Word cloud explorer: Text analytics based on word clouds. In Proceedings of the 2014 47th Hawaii International Conference on System Sciences, Waikoloa, HI, USA, 6–9 January 2014; IEEE: Piscataway, NJ, USA, 2014; pp. 1833–1842.
59. Word cloud analysis of historical changes in the subject matter of public health practice in the United Kingdom. *Public Health* **2021**, *197*, 39–41. [CrossRef] [PubMed]
60. Ratinaud, P. IRaMuTeQ: Interface de R pour les Analyses Multidimensionnelles de Textes et de Questionnaires. *Téléchargeable à l'adresse.* 2009. Available online: http://www.iramuteq.org (accessed on 20 May 2024).
61. McNaught, C.; Lam, P. Using Wordle as a supplementary research tool. *Qual. Rep.* **2010**, *15*, 630–643. [CrossRef]
62. Swain, S.; Jebarajakirthy, C.; Sharma, B.K.; Maseeh, H.I.; Agrawal, A.; Shah, J.; Saha, R. Place Branding: A Systematic Literature Review and Future Research Agenda. *J. Travel Res.* **2023**, *63*, 00472875231168620. [CrossRef]
63. Mandják, T.; Lavissière, A.; Hofmann, J.; Bouchery, Y.; Lavissière, M.C.; Faury, O.; Sohier, R. Port marketing from a multidisciplinary perspective: A systematic literature review and lexicometric analysis. *Transp. Policy* **2019**, *84*, 50–72. [CrossRef]
64. Chaves, M.M.N.; dos Santos, A.P.R.; dos Santosa, N.P.; Larocca, L.M. Use of the software IRAMUTEQ in qualitative research: An experience report. In *Computer Supported Qualitative Research*; Springer: Cham, Switzerland, 2017; pp. 39–48.
65. Van Eck, N.; Waltman, L. Software survey: VOSviewer, a computer program for bibliometric mapping. *Scientometrics* **2010**, *84*, 523–38. [CrossRef] [PubMed]
66. Toutain, M.; Elmoataz, A.; Desquesnes, X.; Pruvot, J.H. A Unified Geometric Model for Virtual Slide Image Processing and Classification. *IEEE J. Sel. Top. Signal Process.* **2016**, *10*, 151–160. [CrossRef]
67. Parisot, S.; Ktena, S.I.; Ferrante, E.; Lee, M.; Guerrero, R.; Glocker, B.; Rueckert, D. Disease prediction using graph convolutional networks: Application to Autism Spectrum Disorder and Alzheimer's disease. *Med. Image Anal.* **2018**, *48*, 117–130. [CrossRef] [PubMed]
68. Pervaiz, U.; Vidaurre, D.; Woolrich, M.W.; Smith, S.M. Optimising network modelling methods for fMRI. *NeuroImage* **2020**, *211*, 116604. [CrossRef] [PubMed]
69. Sardellitti, S.; Barbarossa, S.; Lorenzo, P.D. Graph Topology Inference Based on Sparsifying Transform Learning. *IEEE Trans. Signal Process.* **2019**, *67*, 1712–1727. [CrossRef]
70. Hu, M.; Zhai, G.; Li, D.; Fan, Y.; Duan, H.; Zhu, W.; Yang, X. Combination of near-infrared and thermal imaging techniques for the remote and simultaneous measurements of breathing and heart rates under sleep situation. *PLoS ONE* **2018**, *13*, e0190466. [CrossRef]
71. Zhang, Y.; Tetrel, L.; Thirion, B.; Bellec, P. Functional annotation of human cognitive states using deep graph convolution. *NeuroImage* **2021**, *231*, 117847. [CrossRef]
72. Van Eck, N.J.; Waltman, L. Citation-based clustering of publications using CitNetExplorer and VOSviewer. *Scientometrics* **2017**, *111*, 1053–1070. [CrossRef] [PubMed]
73. Brahim, A.; Farrugia, N. Graph Fourier transform of fMRI temporal signals based on an averaged structural connectome for the classification of neuroimaging. *Artif. Intell. Med.* **2020**, *106*, 101870. [CrossRef]
74. Chen, Y.; Tang, Y.; Wang, C.; Liu, X.; Zhao, L.; Wang, Z. ADHD classification by dual subspace learning using resting-state functional connectivity. *Artif. Intell. Med.* **2020**, *103*, 101786. [CrossRef]
75. Griffa, A.; Amico, E.; Liégeois, R.; Van De Ville, D.; Preti, M.G. Brain structure-function coupling provides signatures for task decoding and individual fingerprinting. *NeuroImage* **2022**, *250*, 118970. [CrossRef] [PubMed]
76. Lei, B.; Zhu, Y.; Yu, S.; Hu, H.; Xu, Y.; Yue, G.; Wang, T.; Zhao, C.; Chen, S.; Yang, P.; et al. Multi-scale enhanced graph convolutional network for mild cognitive impairment detection. *Pattern Recognit.* **2023**, *134*, 109106. [CrossRef]
77. Wang, J.; Xiao, L.; Wilson, T.W.; Stephen, J.M.; Calhoun, V.D.; Wang, Y.P. Examining brain maturation during adolescence using graph Laplacian learning based Fourier transform. *J. Neurosci. Methods* **2020**, *338*, 108649. [CrossRef]
78. Padole, H.; Joshi, S.; Gandhi, T.K. Early Detection of Alzheimer's Disease using Graph Signal Processing on Neuroimaging Data. In Proceedings of the 2018 2nd European Conference on Electrical Engineering and Computer Science (EECS), Bern, Switzerland, 20–22 December 2018; pp. 302–306.
79. Hu, C.; Sepulcre, J.; Johnson, K.A.; Fakhri, G.E.; Lu, Y.M.; Li, Q. Matched signal detection on graphs: Theory and application to brain imaging data classification. *NeuroImage* **2016**, *125*, 587–600. [CrossRef]
80. Valenchon, J.; Coates, M. Multiple-graph Recurrent Graph Convolutional Neural Network Architectures for Predicting Disease Outcomes. In Proceedings of the ICASSP 2019—2019 IEEE International Conference on Acoustics, Speech and Signal Processing, Brighton, UK, 12–17 May 2019; pp. 3157–3161.
81. Sartipi, S.; Torkamani-Azar, M.; Cetin, M. EEG Emotion Recognition via Graph-based Spatio-Temporal Attention Neural Networks. In Proceedings of the 2021 43rd Annual International Conference of the IEEE Engineering in Medicine and Biology Society (EMBC), Virtual, 1–5 November 2021; pp. 571–574.
82. Guo, Y.; Nejati, H.; Cheung, N.M. Deep neural networks on graph signals for brain imaging analysis. In Proceedings of the 2017 IEEE International Conference on Image Processings (ICIP), Beijing, China, 17–20 September 2017; pp. 3295–3299.
83. Einizade, A.; Mozafari, M.; Rezaei-Dastjerdehei, M.; Aghdaei, E.; Mijani, A.M.; Hajipour Sardouie, S. Detecting ADHD children based on EEG signals using Graph Signal Processing techniques. In Proceedings of the 2020 27th National and 5th International Iranian Conference on Biomedical Engineering (ICBME), Tehran, Iran, 26–27 November 2020; pp. 264–270. [CrossRef]

84. Itani, S.; Thanou, D. A Graph Signal Processing Framework for the Classification of Temporal Brain Data. In Proceedings of the 2020 28th European Signal Processing Conference (EUSIPCO), Amsterdam, The Netherlands, 18–21 January 2021; pp. 1180–1184.
85. Einizade, A.; Sardouie, S.H. A Unified Approach for Simultaneous Graph Learning and Blind Separation of Graph Signal Sources. *IEEE Trans. Signal Inf. Process. Netw.* **2022**, *8*, 543–555. [CrossRef]
86. Omidvarnia, A.; Liégeois, R.; Amico, E.; Preti, M.G.; Zalesky, A.; Van De Ville, D. On the Spatial Distribution of Temporal Complexity in Resting State and Task Functional MRI. *Entropy* **2022**, *24*, 1148. [CrossRef]
87. Saeidi, M.; Karwowski, W.; Farahani, F.V.; Fiok, K.; Hancock, P.A.; Sawyer, B.D.; Christov-Moore, L.; Douglas, P.K. Decoding Task-Based fMRI Data with Graph Neural Networks, Considering Individual Differences. *Brain Sci.* **2022**, *12*, 1094. [CrossRef]
88. Annaby, M.; Said, M.; Eldeib, A.; Rushdi, M. EEG-based motor imagery classification using digraph Fourier transforms and extreme learning machines. *Biomed. Signal Process. Control* **2021**, *69*, 102831. [CrossRef]
89. Pusil, S.; Dimitriadis, S.I.; López, M.E.; Pereda, E.; Maestú, F. Aberrant MEG multi-frequency phase temporal synchronization predicts conversion from mild cognitive impairment-to-Alzheimer's disease. *Neuroimage Clin.* **2019**, *24*, 101972. [CrossRef]
90. Saboksayr, S.S.; Mateos, G.; Cetin, M. Online discriminative graph learning from multi-class smooth signals. *Signal Process.* **2021**, *186*, 108101. [CrossRef]
91. Zawadzka, T.; Wierciński, T.; Meller, G.; Rock, M.; Zwierzycki, R.; Wróbel, M.R. Graph Representation Integrating Signals for Emotion Recognition and Analysis. *Sensors* **2021**, *21*, 4035. [CrossRef] [PubMed]
92. Deng, J.; Sun, J.; Lu, S.; Yue, K.; Liu, W.; Shi, H.; Zou, L. Exploring neural activity in inflammatory bowel diseases using functional connectivity and DKI-fMRI fusion. *Behav. Brain Res.* **2023**, *443*, 114325. [CrossRef] [PubMed]
93. Ouahidi, Y.E.; Tessier, H.; Lioi, G.; Farrugia, N.; Pasdeloup, B.; Gripon, V. Pruning Graph Convolutional Networks to Select Meaningful Graph Frequencies for FMRI Decoding. In Proceedings of the 2022 30th European Signal Processing Conference (EUSIPCO), Belgrade, Serbia, 29 August–2 September 2022; pp. 937–941.
94. Einizade, A.; Nasiri, S.; Sardouie, S.H.; Clifford, G.D. ProductGraphSleepNet: Sleep staging using product spatio-temporal graph learning with attentive temporal aggregation. *Neural Netw.* **2023**, *164*, 667–680. [CrossRef] [PubMed]
95. El rai, M.C.; Darweesh, M.; Al-Saad, M. Semi-Supervised Segmentation of Echocardiography Videos Using Graph Signal Processing. *Electronics* **2022**, *11*, 3462. [CrossRef]
96. Edwards, M.; Xie, X.; Palmer, R.I.; Tam, G.K.; Alcock, R.; Roobottom, C. Graph convolutional neural network for multi-scale feature learning. *Comput. Vis. Image Underst.* **2020**, *194*, 102881. [CrossRef]
97. Li, L.; Zhou, J.; Jiang, Y.; Huang, B. Propagation source identification of infectious diseases with graph convolutional networks. *J. Biomed. Inform.* **2021**, *116*, 103720. [CrossRef]
98. Geng, R.; Gao, Y.; Zhang, H.; Zu, J. Analysis of the Spatio-Temporal Dynamics of COVID-19 in Massachusetts via Spectral Graph Wavelet Theory. *IEEE Trans. Signal Inf. Process. Netw.* **2022**, *8*, 670–683. [CrossRef]
99. Lei, X.; Tie, J.; Pan, Y. Inferring Metabolite-Disease Association Using Graph Convolutional Networks. *IEEE/ACM Trans. Comput. Biol. Bioinform.* **2022**, *19*, 688–698. [CrossRef]
100. Zhang, J.; Jiang, Z.; Hu, X.; Song, B. A novel graph attention adversarial network for predicting disease-related associations. *Methods* **2020**, *179*, 81–88. [CrossRef]
101. Bamunu Mudiyanselage, T.; Lei, X.; Senanayake, N.; Zhang, Y.; Pan, Y. Predicting CircRNA disease associations using novel node classification and link prediction models on Graph Convolutional Networks. *Methods* **2022**, *198*, 32–44. [CrossRef]
102. Wang, W.; Zhou, F.; Tay, D.B.; Jiang, J. Gene selection for cancer detection using graph signal processing. *Inform. Med. Unlocked* **2021**, *25*, 100662. [CrossRef]
103. Zhi, H.Y.; Zhao, L.; Lee, C.C.; Chen, C.Y.C. A Novel Graph Neural Network Methodology to Investigate Dihydroorotate Dehydrogenase Inhibitors in Small Cell Lung Cancer. *Biomolecules* **2021**, *11*, 477. [CrossRef] [PubMed]
104. El Rai, M.C.; Darweesh, M.; Al-Saad, M.; Mansoor, W.; Al-Ahmad, H. SemiSegPolyp: Semi-Supervised Polyp Segmentation using Graph Signals. In Proceedings of the 2022 29th IEEE International Conference on Electronics, Circuits and Systems (ICECS), Glasgow, UK, 24–26 October 2022; pp. 1–4.
105. Zhang, J.; Gong, J.; Barnes, L. HCNN: Heterogeneous Convolutional Neural Networks for Comorbid Risk Prediction with Electronic Health Records. In Proceedings of the 2017 IEEE/ACM International Conference on Connected Health: Applications, Systems and Engineering Technologies (CHASE), Philadelphia, PA, USA, 17–19 July 2017; pp. 214–221.
106. Thomas, P.J.; Leow, A.; Klumpp, H.; Phan, K.L.; Ajilore, O. Default Mode Network Hypoalignment of Function to Structure Correlates With Depression and Rumination. *Biol. Psychiatry Cogn. Neurosci. Neuroimaging* **2023**, *9*, 101–111. [PubMed]
107. Surapaneni, S.P.; Manikandan, M.S. Graph Signal Processing Based Classification of Noisy and Clean PPG Signals Using Machine Learning Classifiers for Intelligent Health Monitor. In Proceedings of the 2023 15th International Conference on Electronics, Computers and Artificial Intelligence (ECAI), Bucharest, Romania, 29–30 June 2023; pp. 1–6.
108. Khoshkhabar, M.; Meshgini, S.; Afrouzian, R.; Danishvar, S. Automatic Liver Tumor Segmentation from CT Images Using Graph Convolutional Network. *Sensors* **2023**, *23*, 7561. [CrossRef] [PubMed]
109. Ju, R.; Hu, C.; zhou, p.; Li, Q. Early diagnosis of Alzheimer's disease based on resting-state brain networks and deep learning. *IEEE/ACM Trans. Comput. Biol. Bioinform.* **2017**, *16*, 244–257. [CrossRef] [PubMed]
110. Huang, J.; Li, Y. Sparse inverse covariance network-based modeling for mild cognitive impairment classification. In Proceedings of the 2017 IEEE International Conference on Imaging Systems and Techniques (IST), Beijing, China, 18–20 October 2017; IEEE: Piscataway, NJ, USA, 2017, pp. 1–5.

111. de Vos, F.; Koini, M.; Schouten, T.M.; Seiler, S.; van der Grond, J.; Lechner, A.; Schmidt, R.; de Rooij, M.; Rombouts, S.A. A comprehensive analysis of resting state fMRI measures to classify individual patients with Alzheimer's disease. *NeuroImage* **2018**, *167*, 62–72. [CrossRef] [PubMed]
112. Segarra, S.; Marques, A.G.; Mateos, G.; Ribeiro, A. Network topology inference from spectral templates. *IEEE Trans. Signal Inf. Process. Netw.* **2017**, *3*, 467–483. [CrossRef]
113. Kalofolias, V. How to learn a graph from smooth signals. In Proceedings of the 19th International Conference on Artificial Intelligence and Statistics, Cadiz, Spain, 9–11 May 2016; pp. 920–929.
114. Dong, X.; Thanou, D.; Frossard, P.; Vandergheynst, P. Learning Laplacian matrix in smooth graph signal representations. *IEEE Trans. Signal Process.* **2016**, *64*, 6160–6173. [CrossRef]
115. Parisot, S.; Ktena, S.I.; Ferrante, E.; Lee, M.; Moreno, R.G.; Glocker, B.; Rueckert, D. Spectral graph convolutions for population-based disease prediction. In Proceedings of the Medical Image Computing and Computer Assisted Intervention—MICCAI 2017, Quebec City, QC, Canada, 11–13 September 2017; Springer: Berlin/Heidelberg, Germany, 2017; pp. 177–185.
116. Vivar, G.; Zwergal, A.; Navab, N.; Ahmadi, S.A. Multi-modal disease classification in incomplete datasets using geometric matrix completion. In Proceedings of the Graphs in Biomedical Image Analysis and Integrating Medical Imaging and Non-Imaging Modalities: Second International Workshop, GRAIL 2018, Granada, Spain, 20 September 2018; Springer: Berlin/Heidelberg, Germany, 2018; pp. 24–31.
117. Dhillon, I.S.; Guan, Y.; Kulis, B. Weighted graph cuts without eigenvectors a multilevel approach. *IEEE Trans. Pattern Anal. Mach. Intell.* **2007**, *29*, 1944–1957. [CrossRef]
118. Shuman, D.I.; Faraji, M.J.; Vandergheynst, P. A multiscale pyramid transform for graph signals. *IEEE Trans. Signal Process.* **2015**, *64*, 2119–2134. [CrossRef]
119. Loukas, A. Graph Reduction with Spectral and Cut Guarantees. *Mach. Learn. Res.* **2019**, *20*, 1–42.
120. Nielsen, J.A.; Zielinski, B.A.; Fletcher, P.T.; Alexander, A.L.; Lange, N.; Bigler, E.D.; Lainhart, J.E.; Anderson, J.S. Multisite functional connectivity MRI classification of autism: ABIDE results. *Front. Hum. Neurosci.* **2013**, *7*, 599. [CrossRef] [PubMed]
121. Dodero, L.; Minh, H.Q.; San Biagio, M.; Murino, V.; Sona, D. Kernel-based classification for brain connectivity graphs on the Riemannian manifold of positive definite matrices. In Proceedings of the 2015 IEEE 12th International Symposium on Biomedical Imaging (ISBI), Brooklyn, NY, USA, 16–19 April 2015; IEEE: Piscataway, NJ, USA, 2015; pp. 42–45.
122. Wee, C.Y.; Yap, P.T.; Shen, D. Diagnosis of autism spectrum disorders using temporally distinct resting-state functional connectivity networks. *CNS Neurosci. Ther.* **2016**, *22*, 212–219. [CrossRef] [PubMed]
123. Heinsfeld, A.S.; Franco, A.R.; Craddock, R.C.; Buchweitz, A.; Meneguzzi, F. Identification of autism spectrum disorder using deep learning and the ABIDE dataset. *Neuroimage Clin.* **2018**, *17*, 16–23. [CrossRef] [PubMed]
124. Ktena, S.I.; Parisot, S.; Ferrante, E.; Rajchl, M.; Lee, M.; Glocker, B.; Rueckert, D. Distance metric learning using graph convolutional networks: Application to functional brain networks. In Proceedings of the Medical Image Computing and Computer Assisted Intervention—MICCAI 2017, Quebec City, QC, Canada, 11–13 September 2017; Springer: Berlin/Heidelberg, Germany, 2017; pp. 469–477.
125. Abraham, A.; Milham, M.P.; Di Martino, A.; Craddock, R.C.; Samaras, D.; Thirion, B.; Varoquaux, G. Deriving reproducible biomarkers from multi-site resting-state data: An Autism-based example. *NeuroImage* **2017**, *147*, 736–745. [CrossRef] [PubMed]
126. Iidaka, T. Resting state functional magnetic resonance imaging and neural network classified autism and control. *Cortex* **2015**, *63*, 55–67. [CrossRef] [PubMed]
127. Ding, L.; Wang, M.; Sun, D.; Li, A. A novel method for identifying potential disease-related miRNAs via a disease–miRNA–target heterogeneous network. *Mol. BioSystems* **2017**, *13*, 2328–2337. [CrossRef] [PubMed]
128. Chen, X.; Liu, M.X.; Yan, G.Y. RWRMDA: Predicting novel human microRNA–disease associations. *Mol. Biosyst.* **2012**, *8*, 2792–2798. [CrossRef] [PubMed]
129. Ding, L.; Wang, M.; Sun, D.; Li, A. TPGLDA: Novel prediction of associations between lncRNAs and diseases via lncRNA-disease-gene tripartite graph. *Sci. Rep.* **2018**, *8*, 1065. [CrossRef]
130. Chen, X.; Yan, G.Y. Novel human lncRNA–disease association inference based on lncRNA expression profiles. *Bioinformatics* **2013**, *29*, 2617–2624. [CrossRef]
131. Zhang, Y.; Zhou, D.; Chen, S.; Gao, S.; Ma, Y. Single-image crowd counting via multi-column convolutional neural network. In Proceedings of the IEEE Conference on Computer Vision and Pattern Recognition, Las Vegas, NV, USA, 26 June–1 July 2016; pp. 589–597.
132. Lipton, Z.C.; Berkowitz, J.; Elkan, C. A critical review of recurrent neural networks for sequence learning. *arXiv* **2015**, arXiv:1506.00019.
133. Kao, J.Y.; Tian, D.; Mansour, H.; Ortega, A.; Vetro, A. Disc-GLasso: Discriminative graph learning with sparsity regularization. In Proceedings of the 2017 IEEE International Conference on Acoustics, Speech and Signal Processing (ICASSP), New Orleans, LA, USA, 5–9 March 2017; IEEE: Piscataway, NJ, USA, 2017; pp. 2956–2960.
134. Luiten, J.; Voigtlaender, P.; Leibe, B. PReMVOS: Proposal-generation, refinement and merging for video object segmentation. In Proceedings of the Asian Conference on Computer Vision. Springer, Perth, Australia, 2–6 December 2018; pp. 565–580.
135. Wang, H.; Wang, W.; Liu, J. Temporal memory attention for video semantic segmentation. In Proceedings of the 2021 IEEE International Conference on Image Processing (ICIP), Anchorage, AK, USA, 19–22 September 2021; IEEE: Piscataway, NJ, USA, 2021; pp. 2254–2258.

136. Jain, S.; Wang, X.; Gonzalez, J.E. Accel: A corrective fusion network for efficient semantic segmentation on video. In Proceedings of the IEEE/CVF Conference on Computer Vision and Pattern Recognition, Long Beach, CA, USA, 15–20 June 2019; pp. 8866–8875.
137. Caelles, S.; Maninis, K.K.; Pont-Tuset, J.; Leal-Taixé, L.; Cremers, D.; Van Gool, L. One-shot video object segmentation. In Proceedings of the IEEE Conference on Computer Vision and Pattern Recognition, Honolulu, HI, USA, 21–26 July 2017; pp. 221–230.
138. Qiu, J.; Dong, Y.; Ma, H.; Li, J.; Wang, K.; Tang, J. Network embedding as matrix factorization: Unifying deepwalk, line, pte, and node2vec. In Proceedings of the Eleventh ACM International Conference on Web Search and Data Mining, Marina Del Rey, CA, USA, 5–9 February 2018; pp. 459–467.
139. Zhang, Z.; Cui, P.; Li, H.; Wang, X.; Zhu, W. Billion-scale network embedding with iterative random projection. In Proceedings of the 2018 IEEE International Conference on Data Mining (ICDM), Singapore, 17–20 November 2018; IEEE: Piscataway, NJ, USA, 2018, pp. 787–796.
140. Perozzi, B.; Kulkarni, V.; Chen, H.; Skiena, S. Don't walk, skip! online learning of multi-scale network embeddings. In Proceedings of the 2017 IEEE/ACM International Conference on Advances in Social Networks Analysis and Mining 2017, Sydney, Australia, 31 July–3 August 2017; pp. 258–265.
141. Wu, H.; Chen, G.; Wen, Z.; Qin, J. Collaborative and adversarial learning of focused and dispersive representations for semi-supervised polyp segmentation. In Proceedings of the IEEE/CVF International Conference on Computer Vision, Montreal, BC, Canada, 11–17 October 2021; pp. 3489–3498.
142. Kramer, M.A.; Kolaczyk, E.D.; Kirsch, H.E. Emergent network topology at seizure onset in humans. *Epilepsy Res.* **2008**, *79*, 173–186. [CrossRef]
143. Statnikov, A.; Tsamardinos, I.; Dosbayev, Y.; Aliferis, C.F. GEMS: A system for automated cancer diagnosis and biomarker discovery from microarray gene expression data. *Int. J. Med Inform.* **2005**, *74*, 491–503. [CrossRef] [PubMed]
144. Health, M.D.P. Archive of COVID-19 Cases in Massachusetts. 2020. Available online: https://www.mass.gov/coronavirus-disease-2019-covid-19 (accessed on 20 May 2024).
145. Došilović, F.K.; Brčić, M.; Hlupić, N. Explainable artificial intelligence: A survey. In Proceedings of the 2018 41st International Convention on Information and Communication Technology, Electronics and Microelectronics (MIPRO), Opatija, Croatia, 21–25 May 2018; pp. 0210–0215.
146. Gunning, D.; Stefik, M.; Choi, J.; Miller, T.; Stumpf, S.; Yang, G.Z. XAI—Explainable artificial intelligence. *Sci. Robot.* **2019**, *4*, 44–58. [CrossRef] [PubMed]
147. Tjoa, E.; Guan, C. A Survey on Explainable Artificial Intelligence (XAI): Toward Medical XAI. *IEEE Trans. Neural Netw. Learn. Syst.* **2021**, *32*, 4793–4813. [CrossRef]
148. Tortorella, G.L.; Fogliatto, F.S.; Mac Cawley Vergara, A.; Vassolo, R.; Sawhney, R. Healthcare 4.0: Trends, challenges and research directions. *Prod. Plan. Control* **2020**, *31*, 1245–1260. [CrossRef]
149. Yang, G.; Pang, Z.; Jamal Deen, M.; Dong, M.; Zhang, Y.T.; Lovell, N.; Rahmani, A.M. Homecare Robotic Systems for Healthcare 4.0: Visions and Enabling Technologies. *IEEE J. Biomed. Health Inform.* **2020**, *24*, 2535–2549. [CrossRef]
150. Wehde, M. Healthcare 4.0. *IEEE Eng. Manag. Rev.* **2019**, *47*, 24–28. [CrossRef]
151. Fabila-Carrasco, J.S.; Tan, C.; Escudero, J. Permutation Entropy for Graph Signals. *IEEE Trans. Signal Inf. Process. Netw.* **2022**, *8*, 288–300. [CrossRef]
152. Fabila-Carrasco, J.S.; Tan, C.; Escudero, J. Dispersion entropy for graph signals. *Chaos Solitons Fractals* **2023**, *175*, 113977. [CrossRef]

MDPI AG
Grosspeteranlage 5
4052 Basel
Switzerland
Tel.: +41 61 683 77 34

Bioengineering Editorial Office
E-mail: bioengineering@mdpi.com
www.mdpi.com/journal/bioengineering

www.ingramcontent.com/pod-product-compliance
Lightning Source LLC
Chambersburg PA
CBHW041421210926
51850CB00013B/3326

* 9 7 8 3 7 2 5 8 3 5 6 9 0 *